OPERATORS AND PROMOTERS

OPERATORS AND PROMOTERS

The Story of Molecular Biology and Its Creators

170101

HARRISON ECHOLS

Edited by Carol A. Gross

UNIVERSITY OF CALIFORNIA PRESS   BERKELEY LOS ANGELES LONDON

University of California Press
Berkeley and Los Angeles, California

University of California Press, Ltd.
London, England

Library of Congress Cataloging-in-Publication Data

Echols, Harrison.
  Operators and promoters : the story of molecular
biology and its creators / Harrison Echols ; edited by
Carol A. Gross.
      p.   cm.
  Includes bibliographical references and index.
  ISBN 0-520-21331-9 (cloth : alk. paper)
  1. Molecular biology—History. I. Echols, Harrison.
II. Gross, Carol. III. Title.
QH506.E246   2001
572.8′09—dc21   00-061523

Manufactured in Hong Kong
10   09   08   07   06   05   04   03   02   01

10   9   8   7   6   5   4   3   2   1

The paper used in this publication meets the
minimum requirements of ANSI/NISO
Z39.48-1992(R 1997) (Permanence of Paper). ∞

Jacket photographs courtesy of Cold Spring Harbor
Laboratory and private collections.

To our families and our community

CONTENTS

FOREWORD

Molecular biology, like a rainbow, is seen differently when viewed from different angles: the wide spectrum of the origins and components of molecular biology from chemistry to the biology of life processes is perceived in a singular way by each observer. This wonderful, readable book by the late Hatch Echols gives his original views of the regulatory mechanisms of biologic form and function and how each of these mysteries was solved.

This book is unique among the many texts and monographs on molecular biology. In addition to giving clear descriptions of the most salient facts and patterns of the regulatory aspects of gene expression, Echols presents vignettes of the circuitous routes that led to the discoveries, leavened with revealing personal features of the people who pursued them.

Each of the ten chapters can be read independently of the others because each is presented as a whodunit mystery with surprising starts and strange twists and turns. Characters come and go, some with helpful clues, others with confusing misdirections. As aspects of a mystery are resolved, patterns emerge that lead to revelations that explain enigmatic events in seemingly remote and unrelated aspects of biology.

Beyond the emphasis on the regulation of gene expression, excellent though briefer accounts are given of other DNA transactions: replication, reverse transcription, recombination, and repair. In a book of this scope, these and other subjects are necessarily given lesser attention. Otherwise the book would have become unwieldy and lacked the intimacy of personal involvement. Echols often interposes insightful and sometimes humorous reflections on the pursuit of science and its role in society.

How tragic that Hatch Echols, who worked so hard and long on this book, could not live to see it bound and appreciated. Those of us who had the good fortune of enjoying his friendship and passion for science are grateful for this enduring memento. To all others, this book will be a precious view of the origins and development of molecular biology that will be found in no other source.

Arthur Kornberg

PREFACE

During his last six years, Harrison Echols, known to the world's biologists as "Hatch," was motivated by a single plan: to write this book—the story of the biological revolution that created contemporary molecular biology. Following an encounter with colon cancer, Hatch realized the uncertainty of his own life and, in typical fashion, clearly and rationally prioritized his goals. The idea that emerged was to write a personal account of the development of the biological paradigms that we now take for granted. Not surprisingly, this book also represents a synthesis of Hatch's own life. For he was a person so consumed with science that he stopped thinking about it only when playing tennis.

This book is based upon Hatch's own experiences, his vast knowledge of the scientific literature, and his interviews with many of the major protagonists that he taped to settle, in his own mind, how things really had happened. It was crafted during the one-month stays he made in Taos, New Mexico, in each of his last six summers, augmented with two sabbatical leaves from his professorial duties at the University of California, Berkeley. Hatch sat at his desk in Taos, synthesized each segment of the story in his mind, and then wrote it out in longhand, sentence by sentence. He wrote with the eye of a physicist, the knowledge of a biologist, and the soul of an artist. The sheltering vista of Taos mountain, visible from his desk, and the vibrancy of the Taos artistic community contributed to the serene yet energetic style of this work.

Hatch had four distinct agendas in telling this story. He wanted to show the rest of the world how people actually did science; he wanted to tell the intellectual history of the most influential discoveries in molecular biology; he wanted to show the importance of technological breakthroughs; and he wanted to share his feelings for science itself. Each of these perspectives contributes to the story's highly personal flavor. He shows the human side of science—how the personalities of scientists and their competitive and collaborative relationships affect the generation of ideas and discoveries. He recounts the intellectual development of each field, showing that eventual solutions often come from experiments performed for other reasons, and that logic and order often arise only in hindsight from the chaos of discovery. He demonstrates how some of the most important discoveries were sparked by new combinations of scientific disciplines previously thought to be separate and by the development of completely new technologies. Finally, he communicates his admiration, even awe, for the purity and simplicity with which life systems are organized.

These intertwining agendas combine to make this book a rich resource for seemingly disparate groups of people. On one level, it is a primer of the scientists and ideas that created and fueled the biological revolution; this book introduces interested nonscientists to concepts affecting our everyday lives. On another level, it is a readable, comprehensive account of the interplay between the major experiments, the technological developments, and the personalities, which together created our current conception of molecular biology. This lively account can be used to supplement traditional textbooks for students beginning their study of molecular biology, whether or not they plan to pursue a career in science. Finally, the synthesis of ideas that Hatch offers is of particular interest for those who created this field, for historians of science, and for those who aspire to be biologists.

The experiments and thinking that showed how cells use the information in genes occurred over a period of about 35 years, roughly from 1955 to 1990. Hatch tells the story of these discoveries in ten chapters. The introductory chapter presents the two central molecular protagonists of this story: DNA, which stores the genetic information, and proteins, which are created from this information to carry out the work of the cell. The nine subsequent chapters are case studies that tell the story of the molecular biology revolution. Chapters 2 and 3 present the experiments that established how information is transferred from DNA to proteins and how that process is regulated. Chapters 4–7 recount the discoveries that established how DNA is duplicated, recombined, and moved, how expression of DNA can be controlled, and how RNA, the sister molecule to DNA, performs such diverse functions in the cell. Finally, Chapters 8–10 recount important examples of how our expanding molecular knowledge is used for further discovery: they tell how the cell makes complex decisions, how a rogue virus reverses the usual flow of information in the cell, and how genetic engineering was created.

Each case study begins with a description of the intellectual milieu before the discovery and the questions that motivated the scientists who were destined to make the key breakthroughs. The story is then built around these people, allowing Hatch to show how science is actually created. Using his extensive taped interviews with many of the major protagonists, Hatch brings these people alive with the vivid sketches, anecdotes, and pictures that are sprinkled throughout the book. In the course of constructing his story, Hatch was able to do something very rare: he moved beyond the development of scientific ideas to the realities of the people who created them. He shows how scientists' subjective beliefs and emotional interactions, rather than external objective reality, sometimes influenced the thrust of their work. By making this shift of focus, he shows how the lives and beliefs of the scientists have changed the science they discovered.

Hatch tells each story with simplicity and elegance. Although he does not hesitate to use technical terms, he typically finds a way to tell the story in plain English, conveying only what is absolutely necessary to intuitively comprehend the science being described. To round out the explanation of a difficult scientific principle or experimental model, Hatch often uses a salient metaphor. References to commonplace and approachable items (a tape recorder, a train, or a stop sign, for example) illuminate the text. The most important concepts and experiments are also conveyed in simple figures, which provide an additional way for the reader to follow the story line of each chapter.

I was with Hatch during most of the time that he was writing this book, first as his partner and then as his wife. I was privileged to share those times in Taos with him, privileged to watch the joy, energy, and love that went into the creation of his book. When he knew that he was dying, he asked something of me for the first time in our entire relationship: to promise that I would finish his book. When I suggested that others, more expert in particular areas, might be better suited to the task, he demurred. He said that only I, who understood him, his motivations, and his philosophy so well would be true to his original ideas. I now believe he was right. During the past several years as I worked ceaselessly to finish his book, I have often felt him there at my side, urging me on, helping me make decisions, and facilitating a seamless transition between his own writing and mine.

Hatch left me with a nearly complete draft of the text and figures. I have done only the things necessary to turn the manuscript into a book. I edited the text for clarity and added connections, summary statements, and figure legends. Not surprisingly, these were especially needed for those portions of the book close to Hatch's own research interests, where he took for granted a basic knowledge on the part of the reader. Chapters have also been updated to reflect current ideas, but, amazingly, very little of this was necessary, attesting to Hatch's extraordinary grasp of the direction molecular biology was heading. Several small sections, central to his story line, had not been written when he died. Friends and colleagues of ours stepped up to fill this void. Tania Baker wrote about Mu transposition, Julian Davies discussed medical and industrial applications of genetic engineering, Mike Botchan contributed a brief tour of eukaryotic molecular biology, and Ann Marie Skalka wrote the section on HIV virus and AIDS. Jon Tupy, a current graduate student in my lab, served as art editor, revising Hatch's preliminary figures and drafting figure legends. These figures were then made into elegant line drawings by Laura Southworth. Deborah Cowing, a former student from my lab, created both the glossary, which is filled with wonderful analogies that make use of commonplace objects to explain scientific principles, and the timeline. Another former student, Debby Siegele, now a faculty member at Texas A&M University, did the entire research for the bibliography and timeline. Bruce Stillman gave us permission to use the Cold Spring Harbor photographic collection. Cold Spring Harbor Laboratory Archives, Clare Bunce, curator, provided photographic material that formed the basis for most of the portrait sketches scattered throughout the book. Marc Nadel, a longtime friend of my family, was the portrait artist responsible for bringing the protagonists to life. Mimi Koehl made substantial contributions to the final rendering.

Hatch believed passionately in the community of science and the collaborative approach to solving scientific problems. His own efforts to promote this modus operandi are recounted throughout this book and encapsulated in the wonderful story he tells of his almost successful efforts to convince a group of scientists to publish their work collaboratively (see Chapter 5). How appropriate, then, that his book should have been finished by the very community of people that he believed in so passionately, supplemented by efforts from both of our families. Without their help and guidance, the book that you see today would not have been published. A few people were primarily responsible for getting me started on this task. Sandy Johnson read through the text and looked at the figures and then told me what needed to be done and how to do it. (Only later, when I saw the consequences of neglecting some advice, did I realize how very wise and knowledgeable he was.) Our editor at U.C. Press, Howard Boyer, guided the entire process from start to finish. My daughter, Miriam, and Hatch's son, Bob, read the early chapters and explained what was necessary to make it comprehensible to an intelligent but biologically naive audience. Hatch's daughter Cathy, a professor of psychology at the University of Texas, Austin, took on the important task of reading the entire final version for understandability. Takashi Yura and Rick Morimoto made suggestions on the early chapters, and Sankar Adhya, Liz Blackburn, Ellen Daniell, Bob Lehman, Lucia Rothman-Denes, and Howard Temin read and commented on the entire book.

Experts in each field have read the book for accuracy and readability. Sandy Johnson spent many hours making the introductory material in Chapters 1–3 comprehensible, and John Abelson made important contributions to Chapter 2. Bob Lehman, Sue Wickner, and Mike Cox examined the DNA replication portion (Chapter 4); Sandy Johnson worked extensively on gene regulation (Chapter 5); Joan Steitz, Christine Guthrie, John Abelson, Elizabeth Blackburn, Harry Noller, and Tom Cech helped with the RNA world (Chapter 6); Bob Lehman, Mike Cox, Art Landy, Howard Nash, and Tania Baker contributed to the discussion of how DNA moves and recombines (Chapter 7); Sandy Johnson, Evelyn Witkin, Betty Craig, Graham Walker, and Mike Cox contributed to the discussion of complex regulation (Chapter 8); Howard Temin, Mike Bishop, Bill Sugden, and Ann Marie Skalka each helped, in their own way, to complete the discussion of retroviruses (Chapter 9); and Julian Davies, Mike Botchan, Dale Kaiser, Peter Lobban, Dan Nathans,

Kathy Danna, and Janet Mertz helped with aspects of the genetic engineering discussion (Chapter 10). Brian Matthews and Sydney Kustu contributed their original photos for the book.

Portions of most chapters were based on taped interviews. Francis Crick, Paul Berg, Phil Leder, Gobind Khorana, Bob Holley, and Marshall Nirenberg were interviewed for Chapter 2; Mel Cohn and Francois Jacob for Chapter 3; Frank Stahl, Matt Meselson, Bob Lehman, Arthur Kornberg, John Cairns, and Randy Scheckman for Chapter 4; Gunther Stent, Boris Magasanik, Jerry Hurwitz, Allan Campbell, Charlie Yanofsky, and Alex Rich for Chapter 5; Joan Steitz, Phil Sharp, Bob Holley, Peter Moore, Christine Guthrie, Chuck Kurland, Mike Botchan, Francis Crick, Masayasu Nomura, and Norm Pace for Chapter 6; Ethan Signer, Frank Stahl, and Allan Campbell for Chapter 7; Evelyn Witkin for Chapter 8; Howard Temin, Harry Rubin, Harold Varmus, and Mike Bishop for Chapter 9; and Dale Kaiser, Dan Nathans, Paul Berg, Stan Cohen, Ham Smith, and Peter Lobban for Chapter 10.

How wise of Hatch to have chosen this book for his final presentation to the world. As I worked on it, I was able to relive Hatch's scientific life and to hear, for the last time, his view of the connections and implications of a world that he helped create. Hatch started school as an English major and finished with a Ph.D. in physics. It was only then that he entered the nascent world of molecular biology. Undoubtedly these early experiences gave him a perspective that allowed him to become a central figure in molecular biology yet stand apart from and observe this community that he loved so much. It is somehow fitting that Hatch fashioned the first draft of this work but allowed his community to provide its final form.

Carol A. Gross

1 BEGINNINGS: SIMPLICITY AND ELEGANCE, DNA AND PROTEIN

This book describes the creation of contemporary molecular biology. In one sense, the history of molecular biology can be characterized as the repeated use of principles of physics and chemistry to understand the central problem of biology—the replication, expression, and evolution of genes. From this point of view, our current understanding of molecular biology can be presented in an impressively logical way. But molecular biology has been developed by people and not by principles, and concepts have developed in a chaotic, disorganized fashion in which the logic of nature is generally evident only in hindsight. This story of molecular biology is also about the people who have participated in its development. In each subfield, the pace, style, and major approaches have reflected the personalities of the investigators. Each of the following chapters will follow a major thread of molecular biology—the biological problem, the scientific approaches, and the people involved. My description of molecular biology is of course a personal view and is necessarily colored by my own way of viewing the scientific universe and the individuals who inhabit it.

Physics has provided the principles from which all of quantitative science is derived. Although an experimental and quantitative science, physics depends on a belief that is ultimately aesthetic and nonderivable—that natural phenomena can be described in terms that are simple, elegant, and universal. The application of this view to understanding the physical and chemical universe has been extraordinarily successful. In turn, the much younger science of molecular biology has achieved phenomenal progress in a very short time by a derived point of view—that proper application of the principles of physics and chemistry will inform all of biology. The concept of simplicity and universality also appears in a critical biological tenet of molecular biology: to start with the simplest organisms—bacteria—and the viruses that infect them.

There are two central chemical protagonists in molecular biology: DNA and proteins. These chemical entities correspond to the biological protagonists of classical genetics: genotype and phenotype. The genotype of a creature is its array of units of heredity, or genes. The phenotype is the expression of the genes in the observable traits of the organism, such as green eyes or curly tail. In the period from the 1920's to early 1950's, a small group of geneticists, chemists, and microbiologists established that DNA is the heredity material: genes are DNA. A second crucial principle was also defined during the same period: genes control phenotypes by defining proteins with a precise structure that meet a particular biological need.

I will not describe this era of discovery, which constitutes "prehistory" for purposes of this book. Accounts may be found in *The Double Helix* and *The Eighth Day of Creation*, among others. However, the fundamental properties of DNA and protein are the basis for understanding all of molecular biology. In this chapter on beginnings, I will summarize some of these properties and the origins of our knowledge about them. Although these sections may be a review to some readers and seem technical to others, they are a necessary foundation for the story to be told in the subsequent chapters.

PROTEINS

The understanding of proteins has been the central achievement of classical biochemistry. A protein is composed of subunits called amino acids, which are joined together through peptide (carbon-nitrogen) bonds to form a long chain or polymer. Each of the 20 different amino acids used in proteins has chemically distinct properties. The order of the amino acid subunits in these chains is known as the "primary structure" of the protein (Figure 1–1). Because the 100 to 1,000 amino acids of a typical protein are arranged in a different order and folded in a different way from every

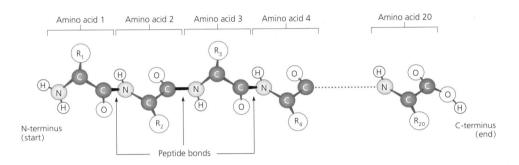

Proteins are long chains of amino acids joined through peptide bonds

Amino acid 1 — Amino acid 2 → Dipeptide

N-terminus (start) — Peptide bond — C-terminus (end)

H_2O

Amino acid 1 — Amino acid 2 — Amino acid 3 — Amino acid 4 — Amino acid 20

N-terminus (start) — Peptide bonds — C-terminus (end)

Primary structure of a polypeptide chain of 20 amino acids

other protein, the possibilities for different chemical interactions are enormous. The capacity for unique recognition and molecular association, or specificity, is the key property of proteins. This specificity is most clearly exemplified by the antibodies. There are more than a hundred million distinct antibodies. Despite being extremely similar structurally, each antibody recognizes and binds to a different molecule. Proteins recognize not only small molecules, such as sugars, but also large molecules, such as DNA or other proteins.

Proteins often use their specific capacity to recognize and bind to other molecules to accelerate (catalyze) chemical reactions. The recognition of this "enzyme activity" of proteins to catalyze reactions was the first great triumph of the principles of chemistry in explaining biology. Chemical reactions occur in fractions of seconds in the cells of our bodies. These same reactions would not occur within our lifetimes if the molecules were placed free in a test tube. In 1897, Eduard Buchner showed that this activity did not require a living cell, providing the first indication that a chemical reaction

FIGURE 1–1. Each of the 20 amino acids starts with an amino (NH_2) group, ends with a carboxyl (COOH) group and is distinguished by having a different chemical group (R). The carboxyl group of one amino acid joins to the amino group of another amino acid to form a dipeptide. This carbon-nitrogen (C-N) linkage is called a peptide bond. When more amino acids are added, the dipeptide grows into a polypeptide chain. Because the amino acids are joined end to end in a linear array, the polypeptide chain has a direction: it starts with an amino group at the N-terminus and ends with a carboxyl group at the C-terminus. In the illustration of the polypeptide chain shown, the C-N peptide bonds are darkened, the carbon atoms are indicated by a C and the nitrogen atoms by an N. The hydrogen atoms attached to these nitrogens are used to make ordered structures in proteins.

rather than a "vital force" was responsible for catalysis. We now know that the ability of proteins to accelerate chemical transitions of their target molecules, termed substrates, is responsible for the rapid reactions within the cell. The way in which proteins carry out catalysis is through their capacity for specific recognition. The specific enzyme binds best to a rare form of the substrate molecule that is already close to the chemical transition and pushes the substrate through this transition, thus speeding up the process enormously. In this way, the speed of the reaction can be accelerated from many years to a fraction of a second.

Once the key cellular role and fascinating catalytic property of proteins had been established, an understanding of protein structure became a major research topic. Although the chain of amino acids does not break on exposure to treatments such as heat and mild acid or alkali, protein chemists observed that the folded three-dimensional ("globular") structure of proteins was generally very sensitive to these treatments. They therefore supposed that the folded structure of proteins depends on certain structural features involving weak chemical bonds, in contrast to the strong bonds connecting the amino acids in their long chain.

What features of proteins give them their structure? In 1951, theoretical chemist Linus Pauling proposed that the structure of proteins derives from the simple and elegant extension of the structural features of the amino acid subunits. The Pauling Principles are three: (i) the subunits are each used in the same way (subunit equivalence); (ii) the geometric properties of the subunits are preserved in the larger structure; (iii) the critical weak bond is the hydrogen bond, a hydrogen atom that is shared by two amino acids in the chain. Since the geometries of the amino acid subunits were known from studies with the technique of X-ray crystallography, only very few possible structures provided subunit equivalence and multiple hydrogen bonds. Pauling and his collaborator Al Corey picked two structures that would allow amino acids to associate with each other in an ordered way as the most likely—the β-sheet and the α-helix. This prediction has turned out to be completely correct. The α-helix and the β-sheet are the major elements of "secondary structure" in proteins, defining the contour of the amino acid chain (Figure 1–2). Each of these folding patterns uses hydrogen atoms to connect peptide bonds. In the β-sheet, a shared hydrogen bond connects peptide bonds in adjacent polypeptide chains together, rigidly positioning them relative to each other. In the α-helix, a shared hydrogen bond

Linus Pauling

FIGURE 1–2.   In a β-sheet (*left*), polypeptide chains are stretched out, and adjacent chains are connected. The adjacent chains either run in the same direction (a parallel β-sheet) or in the opposite direction (an antiparallel β-sheet). The example shown is an antiparallel β-sheet, formed when the polypeptide chain folds back upon itself. Hydrogen bonds (dashed lines) between peptide bonds in adjacent strands are used to connect both parallel and antiparallel β-sheets. In an α-helix (*right*), the amino acid chain adopts a coiled conformation with 3.6 amino acids per turn. Hydrogen bonding within the structure between every fourth peptide bond (dashed lines) stabilizes the helix. Representations of both structures are schematic. The carbon atoms are darkly shaded and the nitrogen atoms are lightly shaded. The peptide bonds between amino acids are darkened. The hydrogen atoms attached to the nitrogens form the hydrogen bonds that stabilize both β-sheets and α-helices. The R groups of the individual amino acids are omitted for clarity.

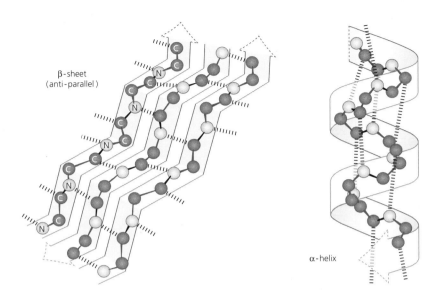

β-sheet (anti-parallel)

α-helix

connects every fourth peptide bond in the same chain, resulting in a helix with regular dimensions. In turn, these secondary structures are folded into the compact "tertiary structure" of the typical protein. Although indirect experiments argued for the existence of α-helices in proteins, the direct proof came in the early 1960's from the first high-resolution determinations of the three-dimensional structure of proteins by X-ray crystallography, achieved by John Kendrew for myoglobin and by Max Perutz for hemoglobin, two oxygen-carrying proteins.

In the meantime, based on the tendency of certain amino acids to cluster together to avoid the aqueous environment, much like a fat droplet in water, protein chemists had become convinced that an additional kind of weak interaction was important in protein structure. The high resolution structures provided convincing evidence for this "hydrophobic" interaction; water-hating "hydrophobic" amino acids clustered in the center of the protein, whereas water-loving "hydrophilic" amino acids appeared on the surface of the protein, in contact with the cellular milieu, composed mostly of water. The hydrophobic interaction probably determines the major features of tertiary structure in most proteins. The complex, precisely folded three-dimensional structure of the protein that results

Max Perutz

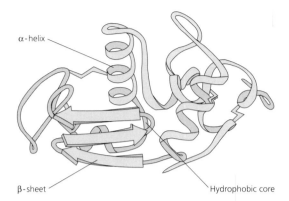

α-helix

β-sheet

Hydrophobic core

FIGURE 1–3. This is a schematic representation of the tertiary structure of a protein. α-helices and β-sheets are joined together by unstructured regions. A protein is organized so that oily, hydrophobic ("water-hating") amino acids lie in the center of the protein.

James Watson

Francis Crick

from these various forces then provides appropriately shaped and suitably reactive sites for the biological functions of the protein (Figure 1–3).

DNA

The structure of DNA was proposed in 1953 by James Watson and Francis Crick. This discovery is described in detail in Watson's book, *The Double Helix*. Watson and Crick's approach was to apply the chemical and aesthetic Pauling Principles to the problem of DNA. The results were revolutionary. It is somewhat ironic that, although Linus Pauling never worked on DNA, one of his most important biological achievements was to figure out how to approach its structure.

The fundamental problem in applying Pauling Principles to DNA was to decipher the nature of the equivalent subunits. Just as a protein is a long chain of amino acids, so DNA is a long chain or polymer of subunits, called nucleotides. Each nucleotide subunit consists of three chemical parts: a base, a sugar, and a phosphate. Nucleotides have one, two, or three phosphates. The nucleotide subunits are joined to the chain by linking the sugar and phosphate into a polynucleotide chain. The sugar (deoxyribose) and the phosphate groups are the same for all nucleotides, but the presence of four different bases provides structural diversity. The nucleotides are notably different in structure; the purine bases adenine (A) and guanine (G) are bulky, whereas the pyrimidine bases thymine (T) and cytosine

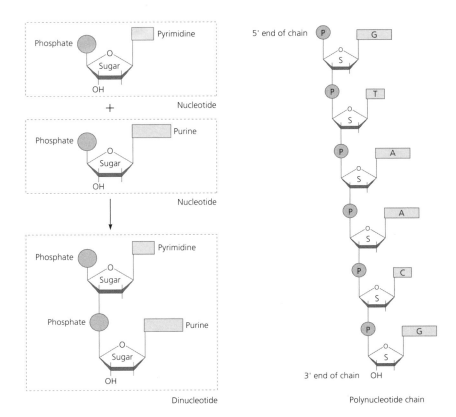

(C) are considerably more svelte (Figure 1–4). The X-ray work on DNA by Rosalind Franklin and Maurice Wilkins indicated a helical structure with at least two polynucleotide chains. Somehow the dissimilar nucleotides had to be aligned in an orderly way to generate two uniform helices with repeating subunits.

The critical insight depended on the biochemist Erwin Chargaff's observation that the purine base A and the pyrimidine base T were always present equally in DNA from various organisms, as were the purine base G and the pyrimidine base C. As he shuffled cardboard models of the bases, Watson realized the answer. "Suddenly I became aware that an adenine-thymine pair held together by two hydrogen bonds was identical in shape to a guanine-cytosine pair held together by at least two hydrogen bonds." The equivalent subunits are not the individual nucleotides, but the base pairs A-T and G-C (Figure 1–5).

FIGURE 1–4.   Each of the four nucleotides is distinguished by having a different base. Two of the nucleotides have bulky purine bases, and two have smaller pyrimidine bases. When two nucleotides are joined together by a phosphate-sugar linkage, a dinucleotide is formed. When more nucleotides are added, the dinucleotide grows into a polynucleotide chain. Because the nucleotides are joined end to end in a linear array, the polynucleotide chain has a direction: it starts with a phosphate at the 5' end and ends with a hydroxyl (OH) at the 3' end. In the schematic illustration of a polynucleotide chain shown, A (adenine) and G (guanine) are the purine bases, T (thymine) and C (cytosine) are the pyrimidine bases, P is the phosphate and S is the sugar.

Adenine
(purine)

Thymine
(pyrimidine)

Guanine
(purine)

Cytosine
(pyrimidine)

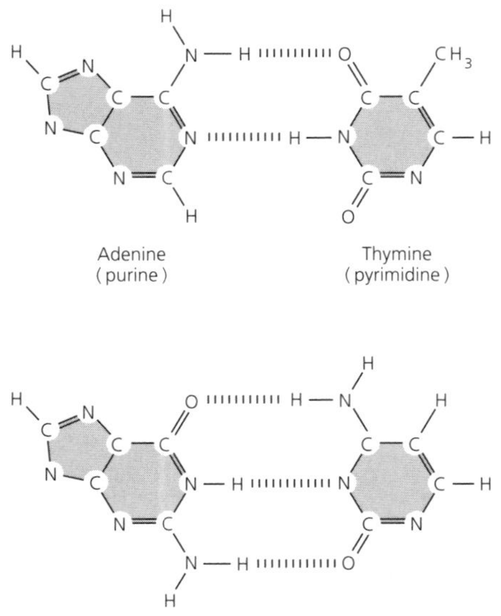

FIGURE 1–5.   When the purine adenine (A) is paired with the pyrimidine thymine (T), the two are almost identical in shape to the purine guanine (G) paired with the pyrimidine cytosine (C). These base-pairs constitute the equivalent subunits from which the duplex structure of DNA can be formed.

The double helical, base-paired structure of DNA is a scientist's dream—simple, elegant, and universal for all organisms (Figure 1–6). "The structure was too pretty not to be true," in the words of Watson. Most wonderful of all, the structure immediately suggested a means by which DNA might guide the production of more DNA or of its companion nucleic acid, RNA, which, as we will see later, is the informational copy of DNA that is used to make protein. Because the bases in one strand of DNA are paired with the partner or "complementary" bases in the other chain, each of the two DNA chains has all of the genetic information. Therefore, one DNA chain can guide the transfer of genetic information to new DNA or to RNA by ordering the complementary bases into an information-bearing partner. This information transfer from old DNA to new DNA and from DNA to RNA is a complex process dependent on the specific recognition and catalytic capacity of proteins. Much of the rest of this book is about these processes, which underlie life as we know it.

From the structure and properties of DNA and proteins, we derive two critical principles that allow us to understand the replication, expression, and evolution of genes. DNA gives us the concept of base-pair complementarity by which nucleic acids recognize each other. Proteins give us the principle of the active site, a site at which specific binding of the substrate catalyzes its chemical transformation. Based on these two principles, the temporally ordered selection of specific chemical reactions provides for the development and reproduction of each individual organism.

SIMPLE ORGANISMS

The principle of studying simple organisms to understand molecular biology developed from several directions, but the idea was enunciated most explicitly by Max Delbrück, who championed the cause of the bacterial virus (also called bacteriophage or phage for short). Delbrück, Salvador Luria, and Alfred Hershey developed the biology of phage and formed the nucleus of the "phage group," a collection of talented scientists who made many contributions to molecular biology. Phage were ideal for genetic and molecular analysis because of their small genomes and their simple structure composed only of DNA (or occasionally RNA) and a protein coat. These properties meant that phage had few genes and that the DNA was easy to prepare free of protein. In addition, to everyone's sur-

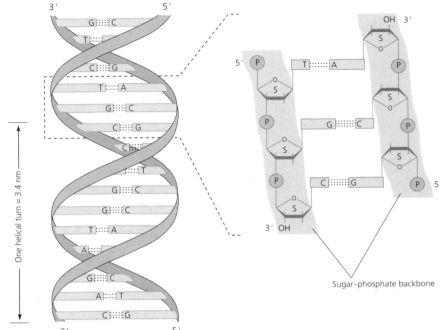

FIGURE 1–6. DNA has two polynucleotide chains whose sugar-phosphate backbones (shown as ribbons) wind around a single helical axis (*left*). The chains are held together by hydrogen bonding between the AT and GC base-pairs. Note that the two polynucleotide chains run in opposite directions: one chain goes 5′ to 3′ (the forward direction) whereas the other chain goes 3′ to 5′ (the backward direction). This is best shown by a schematic representation of the sugar-phosphate backbone of DNA (*right*). The two chains in DNA are said to be "antiparallel."

Sugar-phosphate backbone

One helical turn = 3.4 nm

prise, phage turned out to be remarkably diverse in their lifestyles, and so provided simple model systems for a number of molecular studies. In the early days, Delbrück attempted to unify the field by defining the seven phage (T1 through T7) that should be studied. Fortunately, this admonition was ignored.

Like all viruses, phage reproduce only in their host cells, the bacteria. So the study of phage is really a window through which the scientist peers at a selected realm of the bacterial cell. The bacteria of choice have been *Escherichia coli*, a major inhabitant of the intestinal tract of humans (and other mammals). The variety of *E. coli* bacteria used in the laboratory is harmless to scientists and easy to grow in large quantities, so various proteins can be purified and studied biochemically. In addition, because *E. coli* is capable of sex (of sorts), it is a convenient subject for genetic study. Thus bacteria and phage provided a means to combine genetics and biochemistry in the easiest way.

It is impossible to overestimate the importance of genetics for molecular biology. The science of genetics defined the questions

Max Delbrück

that molecular biologists have tried to answer at a chemical level and has guided this pursuit by allowing the molecular biologist to correlate the action of proteins in the test tube with biological events in cells.

This combined ease of genetic and biochemical study made phage and bacteria the major objects of study in molecular biology for many years. Underlying this intense pursuit of the simplest creatures has been the belief that the fundamental genetic mechanisms are likely to be universal—that larger creatures with more complex and differentiated cells will function according to the same rules. This seeming article of faith follows directly from a belief in evolution; not surprisingly, it has turned out to be correct. The complexity of the higher organism appears to be explained by a more involved use of the same principles used by the simplest organisms.

FURTHER READING

Cairns, J., G. S. Stent, and J. D. Watson, eds. (1992 [1966]) *Phage and the Origins of Molecular Biology,* expanded ed. Cold Spring Harbor, N.Y.: Cold Spring Harbor Laboratory Press.

Crick, F. (1974) The double helix: A personal view. Nature 248, 766–69.

Judson, H. F. (1979) *The Eighth Day of Creation: Makers of the Revolution in Biology.* New York: Simon & Schuster.

Schrödinger, E. (1945) *What Is Life?* Cambridge, Engl.: Cambridge University Press.

Watson, J. (1968) *The Double Helix.* New York: Atheneum.

2 THE CODE FOR LIFE: DNA TO PROTEIN

Life depends on the orderly transfer of genetic information from the genes to the proteins. The mechanism for storing information in DNA, the "genetic code," first came under intense study in 1953, when the Watson–Crick model for DNA structure defined the gene in chemical terms. Within a dozen years, the code was known. This short time span is astounding even in retrospect, because nothing was known at the beginning, nor were there any clear approaches to the problem. The spectacular achievement of solving the code depended on several currents. The excitement of the problem attracted an exceptionally talented group of scientists; the fusion of different experimental philosophies yielded new approaches; and the belief that a rapid solution would reward the correct insight led to a liberated quest for fresh ideas.

A gene is the segment of DNA that defines the structure of a protein. In turn, proteins determine cellular activity. For example, the *E. coli* gene called *lacZ* specifies (or codes for) a protein named β-galactosidase, which splits the sugar lactose into two simpler sugars, glucose and galactose. Other proteins specified by other genes act to convert glucose and galactose into smaller molecules. This process yields energy, which is stored as adenosine triphosphate (ATP) in cells, and building materials that can be used to make new proteins, nucleic acids, and other cellular constituents. The overall process is called the metabolism of the cell. Proteins that catalyze chemical transformations are termed enzymes (originally meaning "in yeast" because the first examples were studied in wine produced by yeast).

Understanding the relationship between genes and proteins began with the "one gene–one enzyme" hypothesis that resulted from a fruitful collaboration between the geneticist George Beadle and the biochemist Edward Tatum. The geneticist learns about genes mainly from mutations that inactivate them. During the 1940's, Beadle and Tatum isolated many mutations in both the bread mold *Neurospora crassa* and the bacterium *E. coli* that produced a new nutritional requirement for the organism. For example, *E. coli* normally makes its own amino acids and so does not require an external source of amino acids. However, mutations could be isolated that made the growth of *E. coli* require an amino acid such as tryptophan, because it was no longer produced within the cell. Each sequential chemical step needed to make tryptophan from simpler compounds is carried out by a specific enzyme. Therefore, a mutation interfering with the synthesis of tryptophan was likely to inactivate one of the specific enzymes for this metabolic pathway. In a number of cases, Beadle, Tatum, and their colleagues showed that a single mutation led to a failure of a single enzyme-catalyzed reaction. Based on these observations, they proposed in 1941 that each gene determined the structural specificity of a corresponding enzyme. Beadle and Tatum really rediscovered (with better evidence) an idea proposed in 1909 by Archibald Garrod, an English physician studying a human genetic defect leading to dark-colored urine, called alkaptonuria.

The one gene–one enzyme concept provided a simple and revolutionary link between genetics and biochemistry. However, the idea was slow to catch on, even in the 1940's. Most geneticists of the time were not much interested in biochemistry, and biochemists were not interested in genes. Moreover, how could the wondrous complexity of biology emanate from such a simple-minded notion? Beadle commented that, in 1951, "I have the impression that the number whose faith in one gene–one enzyme remains steadfast could be counted on the fingers of one hand—with a couple of fingers left over."

The gene began to acquire a molecular reality when Oswald Avery and Alfred Hershey each showed in different ways that DNA carried genetic information. Avery's work, published in 1944, showed that bacterial genes could be transferred between bacterial cells as free DNA, a process termed "transformation." Hershey demonstrated in 1952 that only the DNA portion of phage T4 entered the host bacterial cell; the protein that composed the rest of the virus remained outside. One example might be a fluke, but two distinct demonstrations that DNA was the hereditary material were convincing. Shortly after this, the Watson-Crick model for DNA suddenly proclaimed that genes are understandable in chemical terms and that a molecular pathway for information transfer from gene to protein might be at hand. A small number of physicists, geneticists, and biochemists took up the quest to understand the information transfer pathway, in a search that came to be known as "the coding problem."

The precise definition of the coding problem and its eventual solution came from two very different approaches, one focused on information transfer and one focused on the biochemistry of protein synthesis. The information transfer group studied mostly the two end points of the coding pathway, the gene and the protein. The nucleic acid and protein biochemists sought to follow the biochemical pathway of nucleotides into nucleic acids and amino acids into proteins. The two groups differed not only in how they approached the coding problem but also in their way of thinking about science, and they sometimes said unkind things about each other. The information people, mostly physicists and geneticists, formulated hypotheses or "models" with a number of assumptions about how it all worked, and they hoped that they could figure out the code without doing too much biochemistry, which many of them regarded as mindless drudgery. The protein synthesizers largely followed the biochemical fashion of the time, attempting to figure out a pathway with the fewest assumptions. They viewed the information crowd as naive and overbearing, with an approach unadmittedly derivative of biochemistry. In fact, the fusion of these two approaches is the basis of contemporary molecular biology: precise models based on genetic and physical insights, combined with a rigorous biochemical analysis.

Francis Crick was the center of the information group. During his long stay at the Medical Research Council (MRC) Lab in Cambridge, England, Crick was the most flamboyant of molecular biologists. This brilliant, enthusiastic, witty, and sometimes overbearing man generated many of the ideas and much of the excitement that fueled the early days of the coding studies. Watson begins *The Double Helix* with "I have never seen Francis Crick in a modest mood." Watson then goes on to say, "Anything important would attract him, and he frequently visited other labs to see which new experiments had been done. Though he was generally polite and considerate of colleagues who did not realize the real meaning of their latest experiments, he would never hide this fact from them." Although his role as a roving theorist was irritating to some, Crick was exceedingly insightful and charismatic, with a phenomenal capacity to integrate disparate information into a coherent picture and to generate excitement about a topic that he found interesting. To Crick, science was a joyous intellectual game played on the stage of a lecture hall or around the MRC cafeteria table. In a series of talks at meetings during the 1950's, some of which were later published, Crick summarized the coding problem and the available data about the flow of information from gene to protein. As experimental approaches became more defined, life as a theorist became more tenuous because the people doing the experiments were eminently capable of interpreting the results themselves. In his introductory remarks for the 1966 Cold Spring Harbor Symposium on The Genetic Code, Crick (in a somewhat modest mood) says, "One of the reasons that I enumerated, in this introduction, something of the early history of the code was to show how little theory was able to contribute."

Paul Zamecnik pioneered the biochemical understanding of protein synthesis and thereby opened the eventual road to the code. Zamecnik was not flamboyant, but he was a spectacularly successful experimental innovator. In a deliberate, determined, and precise manner, he defined the route of amino acids into protein with a methodology that became standard for biochemical studies of protein and nucleic acid synthesis. Trained as a physician, Zamecnik began to work on the problem of protein synthesis at Massachusetts General Hospital in Boston in 1948. A useful radioactive isotope of carbon (^{14}C), discovered by Martin Kamen, became available after World War II, and Zamecnik and colleagues set out to use ^{14}C-labeled amino acids as a rapid and convenient way to track the path of amino acids into protein. After ^{14}C-labeled amino acids were added to cells (or later to extracts from broken cells), the appearance of ^{14}C-protein could be measured by a simple "assay"—the insolubility of the protein in acid. As described below, Zamecnik and his coworkers unveiled the biochemical steps from amino acid to protein with an inspirational precision and clarity. In a typically understated summary years later, Zamecnik wrote of "some of the doors

Paul Zamecnik

FIGURE 2–1. In a triplet DNA code, a linear sequence of three bases specifies a single amino acid. In this figure, S-P represents the sugar-phosphate backbone and A (adenine), T (thymine), G (guanine), and C (cytosine) are the four bases.

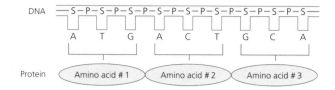

which were consciously or inadvertently opened for others to pass through." These doors were the correct outlines of the most complex biochemical event in the cell—the pathway of protein synthesis.

DEFINING THE CODING PROBLEM: THE INFORMATION PEOPLE

The first explicit formulation of the coding problem seems to be mainly due to the physicist George Gamow, who in 1953 expressed in cryptographic terms some current ideas about gene and protein. The structure of DNA has only limited possibilities for storing genetic information. The sugar-phosphate "backbone" is monotonous, like the hum of a telephone wire, and therefore the variable sequence of the bases must be the voice that carries the genetic message. How does this base code define the specific structure of proteins? The simplest notion was the "sequence hypothesis": the linear sequence of bases in DNA defines a corresponding linear sequence of amino acids in a protein. Remarkably, the sequence hypothesis preceded the Watson-Crick structure of DNA, which gave the first clear physical picture of the gene. A glimmer of the idea is contained in a paper by Cyril Hinshelwood in 1950, and the sequence-matching concept was presented explicitly by Alexander Dounce somewhat later.

George Gamow

Given the sequence hypothesis, Gamow noted that gene and protein were connected informationally by a simple code. Since there were known to be about 20 different amino acids in protein (we now know the number is exactly 20), and only four bases in DNA, the coding problem was defined. How do the four bases in the DNA of a gene, taken in sequence, specify the 20 amino acids of a protein, taken in sequence? Each of the four bases taken singly could specify only four amino acids, and the four bases taken two at a time could define only sixteen (4 x 4) of the 20 amino acids. Thus, a three-base sequence (or "triplet code") was the minimal coding unit that made sense (Figure 2–1). Triplet sequences, however, clearly provided more potential information than was needed be-cause 64 amino acids could be specified (4 x 4 x 4). Thus a triplet code would presumably have more than one coding unit (codon) for an amino acid—the code was overspecified (or "degenerate," using a term borrowed from physics).

The sequence hypothesis and the precise statement of the coding problem was a giant conceptual leap—almost as spectacular in intel-lectual terms as the Watson-Crick model because there was so little evidence for this coding concept. Defining the coding problem was of course vastly easier than solving it, or even than knowing that the problem was correctly stated. In 1953, the gene could be presented as a linear array of bases because of the Watson-Crick model. A for-mulation of the coding problem in terms of the sequence hypothe-sis was therefore plausible for the DNA. But accepting the sequence hypothesis required a leap of faith for the protein side. At that time, not everyone was even convinced that each protein had a specific amino acid sequence, to say nothing of accepting the next giant step that the sequence of the amino acids of a protein suffices to deter-mine its complex three-dimensional structure. In 1951, Fred Sanger figured out the amino acid sequence of insulin, the smallest protein known at the time, and, before long, enough work was done on se-quences of other proteins to give confidence that the amino acid se-quence was a unique property of each protein. Eventually, in 1959, Chris Anfinsen found that a native protein could be unfolded and then direct its own refolding to the original precise three-dimen-sional structure. Thus the information for correct folding of a pro-tein indeed resided in the linear sequence of its amino acids.

How might the sequence hypothesis be verified and the coding problem solved? Learning the *sequence* of the bases in the DNA of a gene was clearly not possible at that time. However, learning the *ar-rangement* of information in a gene was possible using genetics. Geneticists study mutations, which change a base in the DNA to produce altered proteins. With several different mutations in the same gene, genetics could establish their relationship to each other. The idea that the code is carried in a linear sequence of bases im-plies that disruptions in that code, mutations, should also occur in a linear sequence.

The order of mutations is determined by the procedure of ge-netic mapping. In this technique, the order and relative separation of mutations is inferred from the frequency with which two mutant DNAs exchange segments by genetic recombination to restore a

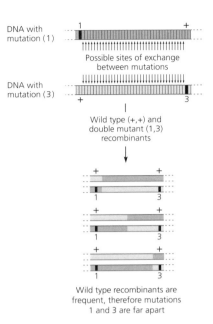

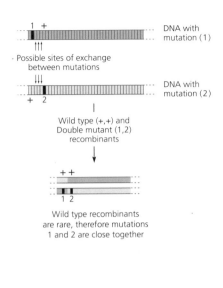

FIGURE 2–2. When a single recombination event between two DNA molecules occurs in the interval between two mutations, a normal "wild-type" recombinant and a double mutant recombinant are produced. Methods are available to enumerate the wild-type recombinants, allowing their frequency to be determined. When two mutations are far apart on the DNA, many different recombination events can occur between them, so wild-type recombinants are frequent. When two mutations are closely spaced, very few recombination events can occur between them, so wild-type recombinants are rare. Note that Benzer's experiments to define the rII gene of phage T4 were done in the absence of any explicit understanding of gene structure.

normal gene (a "wild-type" gene in the parlance of geneticists). As diagrammed in Figure 2–2, two mutations far away from each other (1, 3) will give more wild-type recombinants than two mutations closer together (1, 2) because the likelihood of breaking and rejoining between the two sites is greater for larger separations.

Seymour Benzer used genetic mapping of the rII gene of phage T4 to verify the sequence hypothesis for DNA. Classical genetic mapping had focused on relatively frequent recombination between well-separated genes. Because mutations within a gene are very close together, recombinational mapping required a way to enumerate a few rare normal recombinants in a large background of mutant phage. In 1953, Benzer developed a method to select the normal "wild-type" recombinants from the rII mutants—only the normal phage could grow in a particular bacterial strain. By mapping hundreds of mutations, Benzer established that mutations in rII did indeed form a linear array. The hypothesis that a linear sequence of bases in DNA carried the code was verified.

Crick and his colleagues then realized Benzer's rII mapping system could be used to test Gamow's hypothesis that the code was triplet in nature. In a single remarkable paper published in 1961,

Seymour Benzer

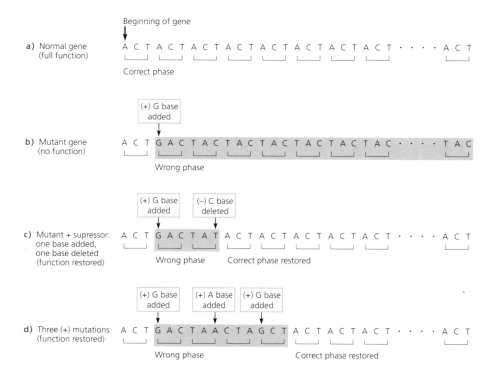

a) Normal gene
(full function)

Beginning of gene

A C T A C T A C T A C T A C T A C T A C T A C T · · · A C T

Correct phase

b) Mutant gene
(no function)

(+) G base added

A C T G A C T A C T A C T A C T A C T A C T A C · · · T A C

Wrong phase

c) Mutant + supressor:
one base added,
one base deleted
(function restored)

(+) G base added   (–) C base deleted

A C T G A C T A T A C T A C T A C T A C T A C T · · · A C T

Wrong phase   Correct phase restored

d) Three (+) mutations
(function restored)

(+) G base added   (+) A base added   (+) G base added

A C T G A C T A A C T A G C T A C T A C T A C T · · · A C T

Wrong phase   Correct phase restored

FIGURE 2–3. The experiment that led Crick to propose a triplet code is diagrammed here. (a) The sequence of a hypothetical wild-type gene is shown. For simplicity, the gene consists of a triplet of bases repeated over and over again. The gene is shown to be read from a fixed point (arrow) in triplets of three bases (bracketed). A dashed line indicates that the same sequence continues to the end of the gene. (b) Inserting a single base changes the phase of all subsequent triplets. Following insertion of a G (indicated by +), the phase of the code changes from ACT to TAC. Deleting a single base would also change the phase, in this case to CTA. Single base additions or deletions make a nonfunctional protein. (c) The change in phase caused by adding a base can be restored to normal by second mutation deleting a single base. When addition (+) is followed by deletion (–) of a base, only a small portion of the gene is read in the wrong phase. When this small region is not essential, the protein is able to function. In genetic jargon, the (–) mutation is a suppressor of the (+) mutation. If the starting mutation were a deletion, and the suppressor mutation were an insertion of a single base, then the correct phase of the code would likewise be restored. (d) Putting three (+) mutations together restores protein function. If the coding unit of DNA consists of three bases, then three mutations that each add a single base to the same DNA molecule should restore function. This proved to be true.

Crick argued that the DNA code is read three bases at a time from a fixed starting point. His evidence was the behavior of certain exceptional mutations in the *r*II gene. Most mutations still allow partial function of the mutant genes, but almost all mutations caused by the chemical acridine yellow completely eliminate the function of the gene. Acridine yellow was believed to cause mutations by either adding or deleting a base. If the code is read from a fixed point (Figure 2–3a), it is easy to see why adding or deleting a base would eliminate gene function: all triplets after the mutation would be read out of phase (Figure 2–3b). Crick found that some mutations caused by acridine yellow could be "suppressed" by a second mutation, lo–

Likewise three mutations that each delete a single base in the same DNA molecule should restore function. That prediction (not shown here) was also true. Although these results are formally compatible with a codon that is a multiple of three bases (if acridine yellow causes insertion or deletion of more than one base), this was considered very unlikely, and future experiments focused on determining the sequence of each triplet codon.

cated nearby. He hypothesized that these suppressor mutations restored the correct phase. If the initial mutation resulted from adding a base, then the suppressor mutation would delete a base and vice versa. The suppressor mutation would restore function because only a small (and nonessential) portion of the gene would be read in the wrong phase (Figure 2–3c). Crick arbitrarily called his original mutation "+," so each suppressor was a "−" mutant. When separated from the original mutation, each "−" mutant results in a nonfunctional *r*II gene. Crick isolated "suppressors of the suppressors," which would be "+" mutants. In this way, Crick obtained a large number of mutations that were either of the "+" or of the "−" class.

Crick realized that this collection of mutations could be used to determine whether the coding unit was a three base sequence. In a triplet code, putting three mutations of the "+" type together or three mutations of the "−" type together should restore the reading frame and allow *r*II function (Figure 2–3d). Crick tested this hypothesis and found that it was true. This very clever genetic experiment established the triplet nature of the code.

The relationship between the triplet DNA code of each gene and the protein it specifies was spelled out by Sydney Brenner and Francis Crick (Figure 2–4). If the base sequence of DNA specified the amino acid sequence of a protein, then the linear order of mutations in a gene should correspond to the order of amino acid changes in altered proteins produced by these mutations. This "colinearity hypothesis" focused interest on the amino acid chains in proteins.

At this point, information theory collided with a very practical biochemical problem: there were no amino acid sequences for proteins easily studied by genetics. Fortunately, there was one mutant protein that could be studied—the altered human hemoglobin produced by the mutation giving rise to sickle cell disease. Linus Pauling, with characteristic intuitive brilliance, had guessed in 1945 that the aberrant "sickle" shape of red blood cells from individuals with this genetic disease was caused by a mutation producing a damaged hemoglobin. Then, in 1949, Pauling and Harvey Itano showed that sickle cell hemoglobin did have an altered electric charge compared to the normal protein. The altered charge of sickle-cell hemoglobin encouraged Vernon Ingram to look for an altered amino acid. Ingram broke the normal and the altered hemoglobin into pieces with digestive enzymes (proteases) (a technique developed by Fred Sanger when he determined the amino acid sequence of insulin). Ingram then separated the protein

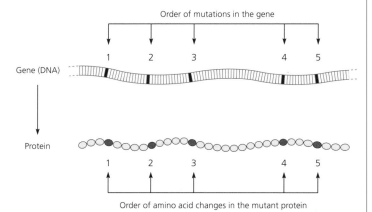

The colinearity hypothesis proposes mutations
and amino acid changes have the same order

Order of mutations in the gene

1   2   3   4   5

Gene (DNA)

Protein

1   2   3   4   5

Order of amino acid changes in the mutant protein

FIGURE 2–4. The sequence hypothesis stated that the order of DNA base-pairs in a gene should match the sequence of amino acids in the protein encoded by that gene. It followed that the order of mutations in genes should correspond to the order of amino acid changes in mutant proteins. This idea is known as the colinearity hypothesis.

segments from each other using two successive procedures, one based on differences in electrical charge and the other based on variations in solubility. The resultant pattern was called a "fingerprint" because the distribution of protein pieces was characteristic of each protein. In 1956, Ingram showed that one of the segments of the mutant protein migrated differently from that of the normal protein. This resulted from a single amino acid change between normal and mutant hemoglobin.

The demonstration that an amino acid was altered in sickle-cell hemoglobin was consistent with the sequence hypothesis but did not, by itself, verify the idea. However, Ingram's success started a world-wide quest for the right combination of gene and protein to verify the colinearity hypothesis. Some researchers dreamed that they might solve the coding problem by producing mutations in DNA with chemicals that changed only certain of the bases and then determining the resulting amino acid changes in the protein. However, the dreams of information theory quickly ran up against the harsh reality of biochemistry. The genetics of the *r*II gene of phage T4 was second to none; however, the *r*II protein could not be found. In reaction, Brenner decided to study the most abundant protein that surrounded phage T4 DNA (the "head protein"), but, unfortunately, the protein was too large for an easy study of mutant pieces. Cyrus Levinthal undertook the *E. coli* protein alkaline phosphatase (my own introduction to molecular biology); however, alkaline phosphatase was also too large. Apparently, protein biochemistry was not as easy as it appeared from Ingram's initial efforts. The most successful gene and protein approach was that of Charles Yanofsky, who combined the genetics and biochemistry of the *E. coli* protein tryptophan synthetase, the enzyme responsible for making the amino acid tryptophan. In 1964, Yanofsky did eventually demonstrate colinearity of gene and protein for tryptophan synthetase. However, by that time a direct biochemical approach to the genetic code had been discovered, and the solution of the code was not far off.

DNA TO RNA TO PROTEIN: THE BIOCHEMICAL PATHWAY PEOPLE

The molecular understanding of biological information transfer and the eventual solution to the coding problem came from studies of nucleic acid and protein synthesis, illuminated by genetic insights about the regulation of genes. Beadle and Tatum defined the end points in the 1940's—genes make proteins. Avery and Hershey showed that genes are DNA. But what is the route from DNA to protein? The critical intermediate is the other cellular nucleic acid polymer, RNA. The information group hoped to learn the genetic code by studying only genes and proteins. In actuality, the answers came from a biochemical approach, based on an understanding of the role of RNA in protein synthesis.

RNA, like DNA, is a chain of four nucleotide subunits joined together by linking sugar and phosphate. RNA differs from DNA in only two respects. The sugar in RNA (ribose) differs slightly from that in DNA (deoxyribose). The base uracil (U) is present in RNA instead of the thymine (T) of DNA, but the base-pairing properties of U are the same as T, so that an RNA copy of DNA has the same genetic information as the original DNA template.

By the 1940's, RNA was strongly implicated in protein synthesis. The early work, especially of Torbjörn Caspersson and Jean Brachet, established three major points implicating RNA as a genetic intermediate. RNA was most abundant in cells undergoing rapid growth and therefore highly active in protein synthesis; in animal cells, RNA was mostly in the cytoplasm, the site of protein synthesis, whereas DNA was confined to the cell nucleus (in chromosomes); cytoplasmic RNA was mostly in organized particles containing both RNA and protein (now called ribosomes as an abbreviation for ribonucleoprotein particles). Brachet wrote in 1946 that

FIGURE 2–5. Using cell extracts, Zamecnik showed that an amino acid is first "activated" by association with AMP, then joined to a small RNA (sRNA), and finally bound to ribosomes prior to assembly into protein. The amino acids were labeled with radioactivity (aa*) to permit their detection.

"ribonucleoprotein particles might well be the agents of protein synthesis in the cell."

The direct demonstration of Brachet's idea came from experiments in whole cells, tracking the path of radioactively labeled amino acids into protein. The cellular components with which the radioactive amino acids associated before becoming completed proteins presumably revealed intermediates in the pathway of protein synthesis. Paul Zamecnik started tracking the path of amino acids in 1954 in animal cells; Richard Roberts and Roy Britten later developed the same approach with *E. coli* cells. The two concurred in a critical conclusion—the radioactive amino acids stopped at the ribosomes on their way to becoming proteins.

Zamecnik then turned his efforts toward the monumental task of trying to reproduce protein synthesis with separated components in a test tube. Only a biochemical approach would allow components to be added or subtracted so that the molecular route from amino acid to protein could be identified. By 1953, Zamecnik and coworkers were able to achieve some protein synthesis in a "cell-free" extract. The cells were broken by grinding, and spun at low speed in the centrifuge to remove the large pieces of cell debris. Radioactive amino acids added to the remaining intracellular "soup" were incorporated into proteins. By then spinning at high speed in the centrifuge, the "soup" could be fractionated into a "soluble" fraction (containing enzymes and small molecules) and a pellet (ribosomal) fraction. Zamecnik then showed that the radioactive amino acids ended up in the pellet. This result confirmed the conclusion from whole cell experiments that ribosomes were the sites of protein synthesis (Stop III in Figure 2–5).

But did the amino acids have other critical stops to make before reaching the ribosome? To pursue this question, Zamecnik and Mahlon Hoagland examined the route of the amino acid in the "soluble" fraction. Protein synthesis required energy in the form of ATP; where did ATP and amino acids come together? The answer to this question identified another stopping point for their radioactive amino acid. The amino acid appeared in an "activated" form that included AMP, a product of ATP breakdown (Stop I in Figure 2–5). Finally, in 1955 and 1956, Hoagland and Zamecnik defined a third intermediate in protein synthesis—the activated amino acid became joined to a relatively small RNA (Stop II in Figure 2–5). They termed this type of RNA "soluble RNA" or sRNA to distinguish it from the large ribosomal RNA (rRNA), which was collected by high speed centrifugation in the pellet fraction. Bob Holley and Paul Berg also obtained evidence for this intermediate.

The discovery of sRNA provided biochemical reality to a guess about genetic information transfer made originally by Crick—that an "adapter" would be necessary to convert the base language of nucleic acids into the amino acid language that would order the amino acids for protein synthesis. Crick noted that the bases of the information-bearing RNA were unlikely to recognize the amino acids directly because no sensible chemical mechanism for such an interaction was known. However, bases were very good at hydrogen-bonding and therefore might recognize an amino acid–bearing "adapter" having appropriate hydrogen-bonding potential. Soluble RNA was clearly the biochemical manifestation of the adapter. A clear mechanism for recognition using hydrogen bonds existed: sRNA could base-pair with the informational nucleic acid in the same way that the two DNA strands base-pair with each other. As the sRNAs, each carrying their amino acid, attached to the informational nucleic acid in an ordered sequence, the amino acids would be added in the specified order to the protein synthesis machinery.

The adapter sRNA is now called transfer RNA, or tRNA, because its role in protein synthesis is to transfer amino acids to the ribosome for assembly into proteins. With the identification of tRNA, the biochemical pathway from amino acids to proteins began to take shape. The route by which the amino acid came to tRNA was highly accessible for further biochemical study because only amino acids, ATP, enzymes, and the small RNAs were needed. Experiments by Holley, Berg, and others soon clarified the essential features. Each individual amino acid is recognized by an amino acid "activating" enzyme specific for that amino acid (an aminoacyl tRNA synthetase). The activating enzyme combines the amino acid with AMP (aa•AMP), and then transfers it to the end of a specific tRNA. In fact, the activated amino acid (aa•AMP) never leaves the surface of the activating enzyme before the joining reaction to

FIGURE 2–6.   Radioactive amino acids (aa*) are "activated" by activating enzymes, and transferred first to tRNA and then to ribosomes, prior to their assembly into protein.

Free aa* ⟶ aa*AMP·enzyme ⟶ aa* tRNA ⟶ aa* tRNA·ribosome ⟶ aa* in protein

tRNA. The tRNA brings the amino acid to the surface of the ribosome, where the amino acids are joined to each other to form a protein. The amino acid stopping points in Figure 2.5 could now be assembled into a pathway for protein synthesis (Figure 2–6).

The definition of tRNA as an intermediate in protein synthesis completed a general outline of biological information transfer from gene to protein. However, a critical mystery remained. What is the information-bearing template on which the amino acids are correctly assembled for synthesis of a specific protein? In the earliest consideration of coding, some thought that the template might be the DNA itself. But DNA was in the nucleus—too far removed from the action. Given that protein synthesis occurred on the ribosome (and in the cytoplasm of animal cells, outside the nucleus), a more logical possibility was that ribosomal RNA (rRNA) was the template. Since rRNA is a long, single-strand polymer, the genetic information might be transcribed from DNA to rRNA, using the specificity of base-pair recognition. A given rRNA would therefore be an information-bearing copy of the gene for a protein. In turn, base-pairing interactions between the rRNA and tRNAs could select specific tRNAs, thereby positioning the amino acids they carried in the correct sequence for polymerization into protein (Figure 2–7).

There was one serious problem with the idea that rRNA was the information-bearing template. If rRNA is a copy of the gene, the rRNA sequence should match that of the DNA. However, that was not the case. Although sequences were not known, data were available on average base compositions of RNA and DNA in different organisms. It turned out that the rRNA of all organisms studied had roughly equal proportions of A, U, G, and C. In contrast, the base composition of DNA varied markedly between these organisms. Since the bases in rRNA clearly had a distribution different from those in DNA, rRNA was unlikely to represent copies of all of the genes. Biological information transfer had come up against a paradox. The tRNA (80 bases) was too small to carry the genetic information, the rRNA was too uniform in size and had the wrong base composition, and the DNA was too far away from the action. There

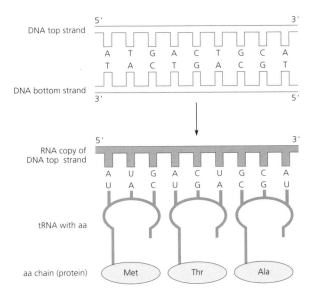

FIGURE 2–7. An RNA copy of one of the two DNA strands could base-pair with tRNA to align the amino acids in the correct order for synthesis of a protein.

were no other stable RNA species known. Francis Crick agonized about the problem: "The RNA you see is not the one you want, and the RNA you want has never been seen."

The answer to the dilemma came from the studies of François Jacob and Jacques Monod on how the gene for the *E. coli* enzyme β-galactosidase was regulated. Jacob and Monod noted that the route from gene to protein was very rapid. When the gene for β-galactosidase was switched from an inactive to an active state, cells rapidly gained the ability to make β-galactosidase; conversely, this ability was rapidly lost when the gene was switched off. These and other considerations led Jacob in 1959 to postulate the existence of yet another RNA, an unstable RNA termed messenger RNA (mRNA), which carried the genetic information from DNA to ribosome and served as the template for the assembly of amino acids. Experiments described in Chapter 3 indicate that Jacob's hypothesis was correct. One of the two DNA strands of the gene is copied into mRNA. The mRNA is then aligned on the ribosome surface by the rRNA and associated protein, and the tRNAs position the amino acids by base-pairing with the mRNA. Messenger RNA had not been identified previously in *E. coli* mainly because of its instability—mRNA is rapidly broken down to mononucleotides after its use in protein synthesis. Even though generally stable in higher organisms, mRNA was missed here as well because it is heterogeneous and relatively rare. With the discovery of mRNA, our understanding of the pathway of information transfer from DNA to protein was complete.

The idea of messenger RNA was obvious in hindsight. The same is true of nearly all major new concepts in molecular biology. Unlike physics, the fundamental theoretical framework is simple. Most experiments are easy to do, and a well-designed experiment generally gives a clear answer. The accessibility of molecular biology provides much of the joy in studying it—almost anyone can grasp the big idea. Still, it is incredibly hard to peer past the immediate conceptual framework. This deceiving simplicity is also probably responsible for many of the frequent quarrels in molecular biology about who should be credited with a discovery. The problem is usually clear, and there is generally substantial evidence that points toward the answer—the difficulty is coming up with the correct insight (which is often prodded by a fortunate experiment).

CRACKING THE CODE: SYNTHETIC mRNA

By 1961, the role of mRNA in protein synthesis had been established, and the overall information pathway from gene to protein had been determined. However, no one had found a clear route to solve the code. Marshall Nirenberg provided the answer: use a synthetic mRNA with a defined sequence. Again, this approach is obvious when you know it works. Shortly before Nirenberg announced his positive results, my friend Julius Adler asked me if I thought such an approach would succeed. Still imbued with the precise views of my previous scientific life as a physicist, I told him that the idea was unlikely to work because protein synthesis from mRNA clearly required a special start sequence and a way to bind mRNA to the ribosome. Both of these special features are true, but my answer was colossally wrong. The ribosome is not all that fussy under the conditions that Nirenberg happened to use in his biochemical experiments. When Nirenberg's work was published, Adler told me that he was entering an area of science about which I knew nothing and would therefore be unable to give him advice. (Adler inaugurated a new field—the molecular biology of bacterial chemical sensing.)

Nirenberg began to work on protein synthesis at the National Institutes of Health in 1959. He started his first independent job

Marshall Nirenberg

working alone in a research field that was suddenly enormously popular and pursued by a number of large, established research groups—a "suicidal" choice, his friends told him. Nirenberg instead followed some advice partially distilled from the organic chemist Feodor Lynen: "Pick a really exciting field, work hard, and you'll find something interesting. More important, you'll have the fun of knowing you're doing something really exciting."

Nirenberg's success story is a microcosm of the American dream. In the scientific world, he came from nowhere, had no initial assets, and yet solved one of the major scientific problems of the twentieth century. Science is a cliquish profession. A few people in a few institutions acquire preeminence in a research field and often decide that progress in the field happens only at their labs and those of their scientific friends and progeny. At Cal Tech, for example, Delbrück developed and nurtured with loving care the phage group (the "phage church," as André Lwoff termed it), but Delbrück was not kind to outsiders. He ordered one scientist who he deemed undesirable to leave a Cold Spring Harbor Phage Meeting. By 1959, the phage band had expanded to include what I have been calling the information theory group, with strongholds at Cambridge, Harvard, and MIT as well as Cal Tech. This group was the most visible force in molecular biology. Nirenberg did not represent this heritage. Nor did he really represent the mainstream of biochemical work on nucleic acid and protein synthesis. In racetrack-style betting on solving the coding problem, he would have had no takers at 100-to-1.

Nirenberg's approach to protein synthesis used cell-free extracts from *E. coli*, which by then had become the organism of choice for biochemical as well as genetic studies. A way to make the appropriate extracts had been developed by Paul Zamecnik and Alfred Tissières in 1960. Nirenberg planned to program his extract with informational DNA or RNA and to look for the synthesis of specific proteins. He began by looking for a complete protein but then switched to the simpler and less demanding Zamecnik assay that involved the insertion of radioactive amino acids into protein-like material (insoluble in acid). After two years, nothing very dramatic had happened. However, one promising lead was an observation that the best RNA by far for protein synthesis was the RNA from a plant virus, tobacco mosaic virus. (Some viruses use RNA instead of DNA as the genetic material.)

Encouraged by this evidence for a specific mRNA, Nirenberg flew to Berkeley to initiate a collaboration with a tobacco mosaic

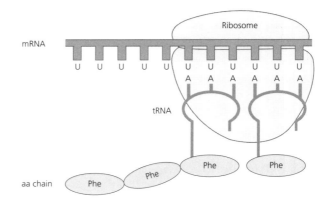

FIGURE 2–8. When an *E. coli* extract was programmed with an mRNA consisting only of U's, a protein containing only phenylalanine was produced. This established UUU as a codon for the amino acid phenylalanine.

virus worker, Heinz Fraenkel-Conrat. They wanted to test whether the viral coat protein was produced as the product of protein synthesis. Meanwhile, Nirenberg's other collaborator, Heinrich Matthaei, continued with a plan to add a variety of different RNAs to the extract to see if they would code for proteins. In his RNA collection, Matthaei included a "fake" RNA in which all of the bases were U (called polyU). Astoundingly, polyU directed efficient incorporation of radioactivity from his mixture of amino acids into a polypeptide. Matthaei sorted through his amino acid mixture and found that polyU coded for a "protein" consisting of many residues of a single amino acid, phenylalanine. The conclusion was clear: polyU codes for polyphenylalanine. Knowing that the coding unit was three bases (see Figure 2–3), Nirenberg could say that the codon in mRNA for phenylalanine was UUU (Figure 2–8). Nirenberg's first published paper on protein synthesis was the 1961 classic that described this work and opened up the coding problem. In a less supportive environment than that provided by his lab chief, Gordon Tompkins, Nirenberg might have been looking for another job after two years without a publication.

The polyU experiment gave a clearly defined route to solving the code—program the ribosome with fake messenger RNA. Nirenberg, who had just married, gave the big news in August 1961 at an international biochemistry meeting in Moscow, took a week off for a honeymoon in Denmark, and spent the next week proving that phenylalanine tRNA really bound to polyU. He then plunged into what became a race for the code. The challenge was that no methods were then available to make RNAs with a defined sequence of bases. PolyU was produced by an enzyme called polynucleotide phosphorylase (PNPase for short). This enzyme had been discovered in 1955 by Severo Ochoa and Marianne Grunberg-Manago and was briefly considered to be the enzyme responsible for RNA synthesis in the cell; in fact, its biological role is probably the reverse reaction of RNA degradation. PNPase synthesizes random polymers from a mix of nucleoside diphosphate substrates (ADP, GDP, CDP, UDP); if given one (UDP), it would make the homopolymer (polyU). The homopolymers of each of the four single nucleotides would give four codons unambiguously. But a mixed random sequence polymer (e.g., polyUC) would contain a mixture of codons (e.g., UUU, UUC, UCU, CUU, UCC, CUC, CCU, and CCC). Only categories of codons could be determined by this approach; precise codon assignments could not be deduced.

Making specific codon assignments required RNA polymers with a specific sequence. The chemical synthesis route pursued by Gobind Khorana in the 1950's and 1960's appeared initially to be the method of choice. Although it did not ultimately provide a unique answer, this work introduced chemical approaches to nucleic acid synthesis and sequence analysis that have been crucial for contemporary molecular biology. Khorana had come to Liverpool from his native India to pursue a Ph.D. in organic chemistry. Unable to find a job thereafter in India, he moved first to Zurich and then to Cambridge for postdoctoral work before settling in Vancouver. In 1953, Khorana began to work on the chemical synthesis of nucleotides such as ATP, and his lab soon became a center for scientists interested in the use of such compounds. By 1955, Khorana had begun to develop chemical methods for polynucleotide synthesis, first for homopolymers and then for mixed polymers of defined sequence. However, Khorana also faced formidable problems. His methods worked only for DNA because the ribose sugar of RNA had an extra reactive group that got in the way.

In 1960, Khorana moved to the University of Wisconsin and began to work more with enzymes involved in nucleic acid synthesis and degradation. The Nirenberg discovery of synthetic messenger RNA provided an immediate goal for Khorana's work—RNAs with a defined sequence. Since the chemical synthesis of specific DNA sequences was beginning to work, Khorana decided to first make specific DNAs and then use RNA polymerase to synthesize specific RNAs from the DNA. (The story of RNA polymerase, the

Gobind Khorana

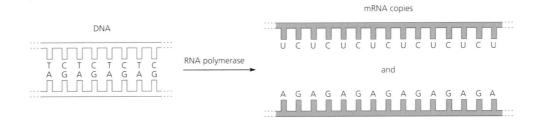

FIGURE 2–9. When given a short piece of DNA of known sequence, RNA polymerase copies each of the strands into mRNA of complementary sequence. When the RNA consists of a repeating dinucleotide, it will code for a repeating dipeptide. In the example shown, the top strand of mRNA has the repeating codons UCU and CUC, which will code for a peptide with alternating serine (UCU) and leucine (CUC). The bottom strand of mRNA will code for a peptide with alternating arginine (AGA) and glutamine (GAG).

Phil Leder

enzyme that makes mRNA, is told in Chapter 5.) Khorana and Julius Adler found that RNA polymerase worked much better than expected at converting DNAs to RNAs. For a repetitive short DNA template, RNA polymerase would dutifully copy the DNA into RNA and then recopy the DNA without terminating the RNA, eventually producing a long RNA with a defined sequence. A repeating DNA dinucleotide could be converted into two RNA copies (one for each DNA strand) (Figure 2–9). Each of these RNAs should code for a repeating sequence of two amino acids in a triplet coding scheme in which three bases specify an amino acid. This approach could answer many important questions; for example, it could provide a biochemical proof for Crick and Brenner's strong genetic argument for a triplet code (see Figure 2–3). However, the techniques were not easy, and at best only combinations of codes for amino acids could be revealed.

The experimental key to the revelation of the code came from another spectacular discovery in the Nirenberg group. Phil Leder came to the National Institutes of Health (NIH) through a program allowing recent medical graduates to carry out their military service obligation as appointees in the Public Health Service. He chose to work with Nirenberg, convinced by his new mentor's boundless enthusiasm and passion for understanding the gene. Leder quickly revealed the scientific insight, determination, and intellectual breadth that has characterized his career as a long-term, major contributor to molecular biology.

Largely because he happened to know the author, Leder read a paper by Akihiro Kaji which demonstrated that, in the presence of polyU, tRNA carrying phenylalanine was associated with the ribosome. This presumably represented the step that aligned amino acids in the order specified by the mRNA code. If short "mRNAs" suf-

Filters

Ribosomes stick

A tRNA-trinucleotide-ribosome complex sticks

Aminoacyl-tRNAs not bound to ribosomes flow through

FIGURE 2–10. The genetic code was cracked by the observation that a trinucleotide (UUU is shown here) suffices for formation of a complex between a tRNA bearing an amino acid and the ribosome. The tRNA-trinucleotide-ribosome complex binds to filters via the ribosome, whereas a free tRNA flows through the filter. This assay allowed rapid determination of which tRNA bound to each of the 64 trinucleotides.

ficed for tRNA binding to the ribosome, a direct route to determine codon identity was at hand. To test this idea, Leder made a series of short polyU chains; he also developed a rapid binding assay in which the ribosome-mRNA-tRNA complex but not the free tRNA was retained on a special type of filter (analogous to a miniature coffee filter). The first experiment showed the route to the code. A synthetic "mRNA" of only three bases was sufficient to bind phenylalanine tRNA to the ribosome—a single triplet codon would serve to reveal its identity (Figure 2–10). Leder had produced one of the great experimental breakthroughs of molecular biology with his first scientific discovery.

The trinucleotide binding system largely reduced the solution of the genetic code to the relatively straightforward problem of making the 64 possible trinucleotides: Leder, Nirenberg, and colleagues produced their magic 64 by limited random synthesis with PNPase, augmented by degradation to triplet pieces by RNA-degrading enzymes (RNases). Khorana's research group made the 64 by chemical synthesis. The results of the binding experiments were corroborated in some cases by protein synthesis experiments with Khorana's defined RNAs. By the end of 1965, the coding problem was essentially solved.

FIGURE 2–11.

The genetic code specifies each of the 20 amino acids

1st position (5' end)	2nd position U	2nd position C	2nd position A	2nd position G	3rd position (3' end)
U	Phe	Ser	Tyr	Cys	U
	Phe	Ser	Tyr	Cys	C
	Leu	Ser	STOP	STOP	A
	Leu	Ser	STOP	Trp	G
C	Leu	Pro	His	Arg	U
	Leu	Pro	His	Arg	C
	Leu	Pro	Gln	Arg	A
	Leu	Pro	Gln	Arg	G
A	Ile	Thr	Asn	Ser	U
	Ile	Thr	Asn	Ser	C
	Ile	Thr	Lys	Arg	A
	Met	Thr	Lys	Arg	G
G	Val	Ala	Asp	Gly	U
	Val	Ala	Asp	Gly	C
	Val	Ala	Glu	Gly	A
	Val	Ala	Glu	Gly	G

THE GENETIC CODE: GOD HAS AN ORDERLY MIND

Tabulated in all its glory, the genetic code is a delightful toast to simplicity and elegance, the most direct way to accomplish a difficult task with the least chance of error. All of the 64 trinucleotides carry genetic information for ordering amino acids in protein; 61 are used to specify amino acids and 3 to define the ends of the amino acid chain ("stop codons") (Figure 2–11). Two of the three "nonsense" codons were also inferred by Sydney Brenner and Alan Garen in 1965 from their ability to cause premature termination of protein synthesis when introduced by mutation into phage T4 or *E. coli* genes. Because 61 codons are used for 20 amino acids, the code is overspecified (or "degenerate") . This feature leaves no meaningless codons that somehow must be avoided and is useful in evolution because nearly all base change mutations will still produce a complete (though often altered) protein rather than no protein at all. The RNA codons of the messenger RNA are recognized on the surface

a) An RNA sequence has three possible "reading frames"

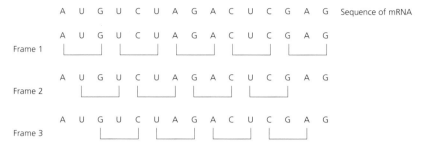

FIGURE 2–12. Because mRNA molecules are read in successive blocks of three bases, each mRNA can be read in one of three frames (a). The reading frame used is set because translation starts at the initiating AUG codon (b). How the ribosome identifies this codon is discussed in Chapter 6 (Figures 6–17 and 6–18).

b) The reading frame of a protein begins with the "initiator codon," AUG

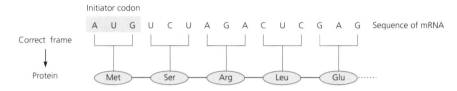

of the ribosome by transfer RNAs with the correct amino acids, and the amino acids are joined together sequentially to produce the finished protein.

One critical feature not obtained from the synthetic mRNA work was where to start reading the genetic code. Some signal is needed to tell RNA where to begin synthesis of the protein. The AUG initiator codon of the mRNA that is positioned on the ribosome and recognized by a special transfer RNA carrying the amino acid methionine provides this signal. Under conditions like those in cells, the initiator AUG is required for efficient protein synthesis and thus defines the set of mRNA codons that are used for a given protein; correct initiation is said to define the "reading frame" of the code. So, even though three reading frames are possible from the messenger RNA for a gene (Figure 2–12a), the AUG initiator ensures that only one reading frame will be used and one protein starting with methionine will be produced (Figure 2–12b). The terminal methionine is often removed from the finished protein.

A nearly miraculous aspect of solving the genetic code was how little knowledge was required. The synthetic mRNA worked like an audiotape converted into an amino acid tune by the mystical translation machinery of protein synthesis. The molecular biologist had only to insert the tape and listen to the melody—how the tape player worked was unimportant. Understanding the code ended the interest of many molecular biologists in the process of protein synthesis. The flow of information from DNA to RNA to protein, termed the "Central Dogma" by Crick, had been defined. Other molecular biologists, more biochemically inclined, turned to the workings of the tape player—the machinery for protein synthesis required for the reading of the code. There will be more about this in Chapter 6. Knowledge of the nucleic acid code for amino acids provided a spectacular achievement, but said nothing about how translation was achieved.

FURTHER READING

Crick, F. H. C. (1966) The genetic code: III. Sci. Am. 215, 55–62.

Khorana, H. G. (1968) Polynucleotide synthesis and the genetic code. Harvey Lectures Series 1966–67, vol. 62. New York: Academic Press.

Nirenberg, M. W. (1963) The genetic code: II. Sci. Am. 208, 80–94.

3 TURNING GENES ON AND OFF: GENES THAT CONTROL OTHER GENES

The genome of an organism carries vastly more genetic information than is ever used by a single cell at a given time. The genes of a cell are turned on and off in an orderly way to respond to environmental signals and to specify the temporal control needed for development of multicellular creatures. In 1959, François Jacob and Jacques Monod provided the crucial insight into how genes are regulated. Their operon model, along with the double-helical structure of DNA, stand as the two most revolutionary and far-reaching concepts in molecular biology.

The essence of the operon model is the beautifully simple notion that certain genes, called regulatory genes, exist solely to control other genes. The product of a regulatory gene is the key that turns a genetic lock to control the activity of genes involved in a common cellular activity—for example, production of the amino acid tryptophan. Before the operon concept, the world was happy with Beadle and Tatum's idea that genes coded exclusively for metabolic enzymes (which many genes do). The notion that some genes might control other genes was revolutionary. But, once stated, the idea was so simple and compelling that Jacob's wife, Lise, who was not a scientist, responded to his first excited rendition of the model by saying, "of course, it is obvious" (to the considerable irritation of her husband).

The operon concept resulted from a fortunate intellectual collaboration. Two extraordinary scientists, Jacques Monod and François

François Jacob

Jacques Monod

Jacob, working in neighboring laboratories in the Institut Pasteur in Paris, found, to their surprise, that they were studying similar problems in gene regulation. Monod had spent a number of years trying to understand the observation that certain metabolic enzymes were produced ("induced") in bacterial cells only when bacteria needed these enzymes to utilize a particular sugar for growth. He had focused his work on the enzyme β-galactosidase, induced by the presence of the sugar lactose in the medium. Down the hall, Jacob was trying to understand how phage λ (lambda) could be induced to go from the quiescent "prophage" state, during which the λ genome replicates along with that of its bacterial host cell, to the lytic state, during which λ multiplies rapidly and lyses the host cell. As Jacob recalls the event, the birth of the operon concept was engendered by an attempted escape from work. Jacob had tried to get away from preparing a talk by going to a movie, but then was trying to escape a boring film by thinking of other things. "Suddenly I realized that prophage induction and β-galactosidase induction are just the same thing." Why that insight was so crucial requires some explanation.

BACTERIAL GROWTH AND THE INDUCED SYNTHESIS OF β-GALACTOSIDASE: THE MONOD GROUP

The propagation of a bacterial species such as *E. coli* depends on efficient growth. Bacteria must be able to grow rapidly when nutrients are abundant; they must be able to grow more slowly and survive starvation when supplies are leaner. The necessity for efficient growth imposes two regulatory requirements: many different metabolic pathways must be available to capitalize on whatever nutrients might be available; conversely, unneeded pathways must be shut off to conserve energy and materials. For example, when lactose is used as the energy source for bacterial growth, the cell needs the enzyme β-galactosidase, which degrades lactose. So, this enzyme is present (or "induced") only when lactose is available. The enzymes that produce tryptophan are not needed when tryptophan is present in the growth medium; therefore, these enzymes are not present (or "repressed") when tryptophan is available. This ability to produce enzymes in response to an environmental need was originally called enzymatic adaptation.

Monod was introduced to *E. coli* and enzymatic adaptation by André Lwoff, his Ph.D. adviser. Monod came with a background in classical biology and a determination to bring rigorous, quantitative analysis to a rather anecdotal field. Upon completing his Ph.D. dissertation in 1941, Monod largely devoted himself to working with the French underground during World War II. Following the war, he obtained a position at the Institut Pasteur with the help of Lwoff. Monod remained at the Pasteur until his death in 1976.

Jacques Monod was a remarkable man, complex and contradictory in his scientific and personal characteristics. He was an imaginative scientific revolutionary and a destroyer of new ideas, a warm, supportive colleague and a domineering, condescending demagogue. All of these qualities are presented in an extraordinary book of personal reminiscences by Monod's colleagues, *Origins of Molecular Biology: A Tribute to Jacques Monod*. Perhaps because Monod's death was still recent and because no one could fail to respond strongly to this powerful personality, the accounts are far more emotional than I have ever seen in any other collection of similar essays. Jacob's balanced commentary may give the clearest summary: "The first of these individuals—let us call him Jacques—was a very warm and generous man of great charm; a man interested in people as well as in ideas, constantly available to his friends, ready to discuss their problems and find a solution; a man of great rigor and insight, always to the point, asking cogent questions, and sharply self-critical. The second individual—let us call him Monod—was incredibly dogmatic, self-confident, and domineering; a person unceasingly in quest of admiration and publicity, demanding to be the focus of attention; a person making definitive black-and-white judgments of everything and everybody, fond of teaching fellow scientists the *real* meaning of their own work but sweeping away as nonsense any objection they might timidly offer."

The successful quest for an understanding of the adaptive capacity of bacteria involved many scientists, both inside and outside of the Institut Pasteur. Largely because of his clear, logical presentations and dedication to quantitative analysis, Monod attracted a group of exceptionally talented visitors, many of them young Americans in search of postdoctoral training. These visitors were key intellectual contributors and greatly accelerated the experimental pace in the lab.

Mel Cohn, the longest visitor with an eight-year tenure, turned the Monod lab in a crucial direction toward the combination of genetics and biochemistry that eventually solved the adaptation problem. Cohn arrived in 1948 with an early background in physics and doctoral research in immunology and physical chemistry. Monod

Mel Cohn

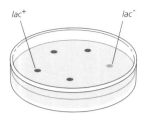

lac⁺   *lac*⁻

FIGURE 3–1.   When plated on nutrient agar containing lactose and a pH indicator, *lac*⁺ colonies are red and *lac*⁻ colonies are pink.

suggested a number of possible inducible enzymes that might be amenable to biochemistry. However, Cohn found that only β-galactosidase was stable enough outside of cells to be studied in the attic lab of the Institut Pasteur in the summer; the Pasteur did not have the traditional biochemical amenities of cold rooms and refrigerated centrifuges used to preserve delicate enzymes during their isolation as pure proteins. Partly for this reason, β-galactosidase was selected as the primary topic for study in the lab.

The choice of β-galactosidase was also influenced by the availability of a genetic approach. In the late 1940's, Joshua Lederberg had learned to distinguish normal bacteria able to use lactose, *lac*⁺ bacteria, from mutants unable to use lactose, *lac*⁻ bacteria (which were presumably defective in β-galactosidase). When the bacteria grew as individual colonies on nutrient agar dishes that contained lactose, *lac*⁺ colonies utilized lactose for energy whereas *lac*⁻ colonies grew on other nutrients in the plates. These plates were spiked with a dye whose color changed with pH. Since colonies growing on lactose produced more acid than those growing on other nutrients, *lac*⁺ colonies were dark red, whereas *lac*⁻ colonies were pink (Figure 3–1).

Lederberg had also developed a delightfully simple way to measure quantitatively the enzymatic activity of β-galactosidase by using a color change. Nitrophenyl galactoside (NPG), a compound related to lactose by having a similar chemical linkage (a β-galactoside), produced a yellow color when the β-galactoside linkage was cleaved by the enzyme. The more enzyme present, the more rapidly the yellow color developed from cleavage of the synthetic substrate. The amount of yellow dye could be measured quantitatively by a device termed a spectrophotometer (Figure 3–2).

All *lac*⁻ mutants of *E. coli* were expected to be defective in β-galactosidase. Mel Cohn set out to obtain a number of mutations in *lacZ*, the gene for β-galactosidase, and indeed, most *lac*⁻ mutants were defective in β-galactosidase. He found one "funny" mutant that gave a peculiar color on the dye-containing agar dishes. This mutant failed to grow on dishes with lactose as the sole carbon and energy source, suggesting that it was a *lac*⁻ mutant. But Cohn and Annamaria Torriani found that this mutant could produce an active β-galactosidase with an inducing compound other than lactose, so it was not defective in β-galactosidase. They proposed that the mutant might be defective in a "permeability" function required to transport lactose from the growth medium into the cell. Monod rejected the idea. "If we have to think about silly things like permeability, this field is hopeless—let's forget this mutant." This decision delayed for many years the formal discovery that lactose needed a permease enzyme to transport it into the cell. The funny mutant was in the *lacY* gene that codes for lactose permease.

In the early days of lactose regulation, Monod was not thinking that enzymatic adaptation could be explained by control over gene expression. Instead, he supposed that induction occurred when inactive subunits of enzymes were converted to active ones. In this "instructive" model, β-galactosidase was produced all the time as an inactive precursor subunit and "induction" resulted from lactose binding to subunits, which converted inactive subunits to active β-galactosidase. If this instructive model were true, any substrate of β-galactosidase should be an inducer, activating the inactive β-galactosidase precursor subunits. Cohn synthesized a number of chemical analogs of lactose, the substrate of β-galactosidase, and promptly destroyed the theory. Some excellent substrates for β-galactosidase were not inducers (such as NPG); some excellent inducers were not substrates (β-galactosides with exotic names but simple initials such as TMG and IPTG). Because these latter compounds were not substrates of β-galactosidase, they were termed "gratuitous inducers." The use of gratuitous inducers in experiments avoided complications arising because the cell uses up lactose, the natural inducer, for energy.

At the time, the major rival idea for induced enzyme production was the 1946 "plasmagene" hypothesis of Monod's competitor Sol Spiegelman, who thought that the gene might produce a cytoplasmic self-replicating unit upon induction. In this model, more enzyme was a consequence of more copies of the gene. This idea explained why addition of lactose induced β-galactosidase only after a long time lag. The lag would represent the time needed to accumulate copies of the gene. But, experiments from the Monod group argued against this idea as well. The "gratuitous inducers" TMG and IPTG induced β-galactosidase within a minute or so after they were added to the bacteria, thus eliminating Spiegelman's hypothesis, which required a long time lag before induction.

The only viable concept that seemed to remain was that addition of the inducer resulted in immediate synthesis of new β-galactosidase protein. David Hogness, another American visitor, tested this idea directly in 1955. When radioactive sulfur is added to the medium, it is incorporated into proteins as a component of the

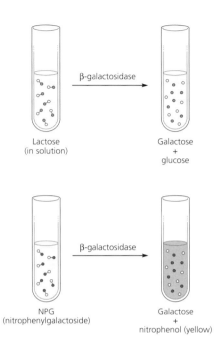

Some substrates of β-galactosidase turn yellow after cleavage by the enzyme

FIGURE 3–2.   Lactose is the natural substrate for the enzyme β-galactosidase, but the synthetic substrate NPG (nitrophenylgalactoside) is also cleaved, yielding nitrophenyl and galactose. Nitrophenyl is yellow, allowing the activity of β-galactosidase to be measured by the rate of appearance of this color.

sulfur-containing amino acids cysteine and methionine. Hogness added radioactive sulfur either before addition of inducer or only after inducer had been added and determined whether the β-galactosidase protein became radioactive. Only radioactive sulfur added after inducer addition appeared in β-galactosidase. The conclusion was clear—the enzyme was newly synthesized from amino acids in response to the environmental induction signal. Moreover, the synthesis of new enzyme occurred with no discernible lapse in time. The problem of induced enzyme production now became a problem of how the gene was regulated. The inducer somehow turned on expression of the *lacZ* gene that coded for β-galactosidase.

To celebrate this enormous clarification, Monod renamed enzymatic adaptation "induced enzyme synthesis." Monod was anxious for the name switch for political as well as scientific reasons. In the 1950's, many Marxist French biologists, including a number of Monod's former colleagues at the Sorbonne, supported Trofim Lysenko's ideas that the environment could change heredity. Monod's early instructive ideas about β-galactosidase were very popular with this group, who imagined that the supposed capacity of lactose to modify an enzyme could then somehow become a genetic change. Moreover, adaptation was the evolutionary term for genetic change, which further confused the issue, at least for people who preferred not to think too clearly about it in the first place. Induced enzyme synthesis defined the true nature of the regulatory process and laid to rest the ghost of discredited ideas.

THE PROBLEM OF LYSOGENY AND PHAGE λ: THE JACOB GROUP

Like any creature, the phage genome has evolved to reproduce itself successfully. A virus has a particularly chancy existence because its propagation depends on an available host. Viruses make more viruses by taking over the biochemical machinery of the host cell to replicate their own nucleic acid and to produce their new protein coats. The reproductive process typically results in the demise of the host cell. In fact, lytic bacterial viruses generally burst (or lyse) the infected bacterium to release a hundred or so new phage particles. But life is not so simple for the lytic virus. If the phage is too voracious, only those mutant host cells to which the virus can no longer attach will remain (evolution goes rapidly in a bacterial population that doubles in number every half hour).

Rather than following this lethal, lytic pattern slavishly, many phage have adopted a more varied lifestyle. In addition to the lytic response, these "temperate" phage can also achieve a harmonious, symbiotic state in which the viral genome replicates along with that of the host. The bacterial cell carrying the virus is said to be "lysogenic" because such cells can be induced to give rise to the lytic pathway of viral growth. The latent viral genome is termed a prophage. The existence of the temperate phage defines two fascinating regulatory problems. What determines the choice of the lytic or lysogenic pathway? How is the lysogenic state established and maintained? Phage λ (lambda) has been the primary source of information about these questions from 1953 to the present, and it is phage λ that brought François Jacob to the operon.

François Jacob came to lysogeny much as Monod came to enzyme adaptation: he sought out André Lwoff as a mentor. However, Lwoff was not initially interested in such an obviously untrained and evidently unfocused person. Jacob had spent two years in medical school intending to be a surgeon but left France in the dark days of World War II to join the Free French Army. After four years in Africa, he was severely wounded in the Normandy invasion, leaving him with shoulder damage that ended hopes of a career in surgery. Jacob then tried a succession of careers—newspaper reporter, actor, city administrator—before approaching Lwoff for training in biology. Lwoff refused; Jacob persisted. Finally, in June 1950 with a new phenomenon to study, Lwoff relented. "We have found prophage induction—are you interested in that?" "I am dying for it," replied Jacob with absolutely no idea what a prophage was.

Jacob had found his niche. Although never trained in genetics, he had the ideal combination of visual, analytical, and intuitive abilities to revolutionize microbial genetics and regulatory biology. A lover of new ideas and new experiments, he was always eager for the next round of fresh data. Monod liked the role of research director, analyzing experiments done by others of his group to test his ideas. Jacob preferred to participate directly in the experiments, but his ability to listen made him an ideal collaborator, and his success depended on the biological systems introduced by his collaborators. Prophage induction and phage λ were introduced to him by Lwoff and Elie Wollman, induced enzyme synthesis by Monod. Jacob was the right person in the right place at the right time.

The very existence of lysogeny was controversial for many years. A population of lysogenic bacteria always produced a small amount of virus. How could it be established that, in the lysogenic state, the

André Lwoff

Elie Wollman

virus was truly quiescent in most cells with only an occasional cell producing phage? Maybe virus was just growing slowly in all cells. To distinguish these possibilities, in the period 1949–51, Lwoff actually observed single cells with a microscope. He found that only rare cells produce virus and do so by lysis. Moreover, Lwoff found that irradiation of his lysogenic bacteria with ultraviolet light caused the entire population of cells to lyse and release viruses. He termed this event prophage induction. So not only is the prophage normally a quiescent genome, but there must also be some control mechanism that allows the alternative states of lytic or lysogenic growth. Later work showed that all agents that interfere with bacterial DNA replication are inducing agents—the virus maintains an option of divorce from a potentially doomed host cell. The early experiments are described in depth in a delightful essay by Lwoff in *Phage and the Origins of Molecular Biology*.

Lwoff's early work was done with a bacterium named *Bacillus megaterium* in which no genetic system was available. The problem of quiescence and the hiding place of the prophage could clearly profit from the genetic analysis of mutants defective in these processes. *E. coli* had a genetic system and, most fortunately, a temperate *E. coli* phage called lambda (λ) had been discovered by Esther Lederberg in 1951. λ was a prophage in the original *E. coli* strain used by Joshua Lederberg for the discovery that bacteria have sex, transferring genetic markers from one strain to another. A temperate phage such as λ is not easily noticed because a cell with a λ prophage is immune to infection by another λ; no one knew λ was there until a variant strain of *E. coli* was accidentally "cured" of the prophage by the ultraviolet (UV) irradiation procedure used to introduce mutations. The lysogenic strain was written as *E. coli* (λ) to show that it carried the prophage. λ soon emerged as the favorite temperate phage for study (Figure 3–3).

The inspiration for the focus on phage λ at the Institut Pasteur came from Wollman, after he returned in 1950 from a postdoctoral visit with Max Delbrück at Cal Tech. Wollman was impressed by the rigorous, quantitative genetic approach of the Cal Tech group and wanted to apply similar techniques to a lysogenic phage. He was further inspired by Delbrück's pronouncement that lysogeny did not exist, as Wollman's parents had contributed to the study of lysogeny many years before. In a bit of irony, many years later Ethan Signer and I switched the subject matter of the phage course (originated by Delbrück) from Delbrück's favorite lytic phage T4 to

The temperate phage λ (lambda) has varied life styles

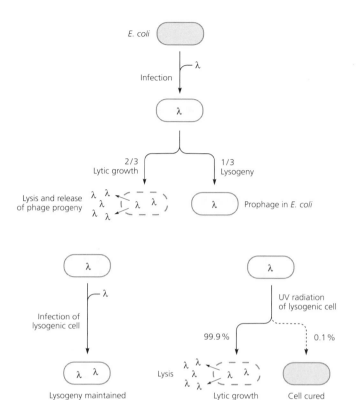

FIGURE 3–3.   Upon infecting E. coli, λ usually follows the lytic pathway, producing more phage and then lysing the host cell to release its progeny (*top, left*). Sometimes, the phage follows the lysogenic pathway, remaining in the host cell in a dormant form called a prophage (*top, right*). Infection of lysogenic cells by λ does not induce the lytic pathway (*bottom, left*). However, irradiation of lysogenic cells by UV does induce the lytic pathway and, very rarely, cures the lysogenic cell of its prophage (*bottom, right*).

the lysogenic phage λ because so many more people were interested in it.

Somewhat to everyone's surprise, a most important flurry of work with phage λ also began at Cal Tech. Jean Weigle, a physicist from Switzerland, had come to visit Cal Tech for a few days in 1953 and stayed until his death in 1970. Because prophage induction was really the definitive proof of lysogeny, Delbrück wanted Weigle to show that prophage λ was not inducible. Weigle demonstrated the opposite. In 1953, he also found that some of the λ phage in the induced population were mutants. Even more mutants of λ occurred if phage were first irradiated outside of the bacteria and then used to infect UV-irradiated E. *coli*. Not only was the prophage inducible by UV light, but mutagenesis itself was also inducible (a most important observation discussed in Chapter 8).

Jean Weigle

FIGURE 3–4. When a virulent phage (e.g., T4) infects a cell in a bacterial lawn, progeny released from the lysed cell infect neighboring bacteria. As this process continues, the death of host cells leaves a hole or clear "plaque" in the surrounding bacterial lawn. Infection by a temperate phage (e.g., λ) allows survival of some cells as lysogens. Because lysogenic cells continue to grow in the presence of released phage, they are found within the plaque. This results in a translucent, or "turbid" plaque.

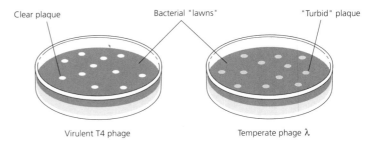

Clear plaque   Bacterial "lawns"   "Turbid" plaque

Virulent T4 phage   Temperate phage λ

Genetics is a visual profession. Just as bacterial genetics depended heavily on mutants that altered the morphology or color of bacterial colonies (see Figure 3–1), so phage genetics depended heavily on mutants that alter the classical plaque assay by which phage are recognized. If many bacteria are spread on an agar dish with a few phage, each virus infects a bacterium and produces more viruses, which in turn infect many neighboring bacteria, lysing them to release more phage. At the end of a growth period (typically overnight), the original phage has produced a small circular hole, called a plaque, in an otherwise uniform "lawn" of bacteria. Virulent phage, which can execute only lytic growth, produce clear plaques because there are no survivors of infection (except for extremely rare phage-resistant bacterial mutants); temperate phage, which carry out a frequent lysogenic response, produce "turbid" plaques because surviving bacteria grow within the ring of lysed cells (Figure 3–4).

Among Weigle's collection of phage λ mutants were some that made small plaques, some that made minute plaques, and most spectacularly some that made clear plaques like the virulent phage. The clear-plaque mutants (*c* for short) were obviously defective in the ability to execute the lysogenic response. Dale Kaiser, a graduate student with Delbrück, began a systematic study of λ*c* mutations. Kaiser brought to the problem a good eye, a precise, methodical mind, and a superlative sense of how to apply genetics to a complex biological problem. Kaiser came to Paris in 1954 with his collection of mutants, and picked out the critical gene, *cI*, required to maintain the quiescent, lysogenic state. This gene was the key for the "genetic lock" that shut off λ.

I will return to this story after a brief interlude describing the de-

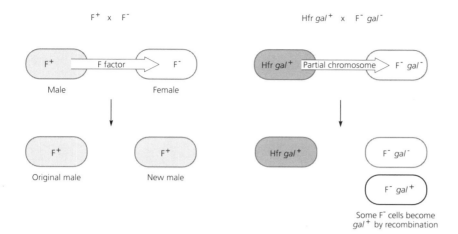

FIGURE 3–5. During mating, F+ male bacteria transfer the F factor to the recipient females, transforming them into F+ males. Males also retain a copy of their F factor for themselves (*left*). When Hfr (or high frequency recombination) males mate, they transfer a portion of their bacterial chromosome to the F− cell. The transferred chromosome can be incorporated into the F− chromosome by recombination, causing incorporation of genetic markers (e.g., *gal*+) from the male into the female.

velopment of techniques for gene transfer in *E. coli*. These techniques provided the experimental means to carry out the genetic experiments that established the Jacob-Monod model of gene regulation.

BACTERIAL SEX AND EROTIC INDUCTION

In contrast to higher eukaryotes, bacteria carry only one copy of their chromosome—they are "haploid" organisms. As will become apparent, to distinguish between various regulatory models, the cell must contain two copies of the regulatory genes (that is, be in a "diploid" state). Bacterial sex, discovered by Joshua Lederberg and Edward Tatum, permitted the establishment of this state, but only transiently. With additional manipulations, experimenters were able to create strains that were permanently diploid for small regions of the chromosome. The rather peculiar properties of bacterial sex (peculiar at least by human standards) also contributed to our understanding of how genes are regulated.

Mainly through the efforts of William Hayes in London, the sexual diversity of male and female bacteria was discovered, and special male strains were developed that were efficient in genetic exchange. There are three sexual types (Figure 3–5) of *E. coli*: F+ "male" bacteria, Hfr "male" bacteria, and F− "female" bacteria. F+ bacteria transfer a nonchromosomal piece of DNA (the "F factor") to F− bacteria, converting them to F$^+$ males. But F$^+$ bacteria do not transfer the bacterial chromosome. Hfr (for high frequency of recombination) males transfer their bacterial chromosome to F$^-$ bacteria; parts of the male chromosome are then incorporated into the F$^-$ chromosome by recombination. For example, galactose-defective (*gal$^-$*) F$^-$ can be converted to *gal$^+$* by mating with an Hfr that is *gal$^+$*.

Wollman and Jacob entered the emerging field of *E. coli* gene transfer in 1953 to learn where the λ genome went in the lysogenic state. They wound up using lysogeny to help develop modern bacterial genetics. Earlier, Esther and Joshua Lederberg had shown that the λ genome was often transmitted to F$^-$ (female) bacteria together with the genes for galactose utilization—in genetic terms, λ was linked to *gal*. But genetic exchange was extremely rare and hard to analyze with the Lederberg strains.

When Jacob and Wollman used the Hayes Hfr strains to check the linkage of *gal* and λ, they found the surprising phenomenon of "erotic induction." If the Hfr carried λ and the F$^-$ did not, then the F$^-$ population lysed after the mating, producing λ phage. Apparently after transfer to F$^-$ cells, the prophage was induced to lytic growth. However, erotic induction occurred only for this one combination of bacteria. When λ was carried by both strains or only by the F$^-$ strain, the prophage was not induced to lytic growth. In this latter case, recombination with the nonlysogenic Hfr could remove the λ in the F$^-$ cells from the chromosome. This shows that λ behaved as a typical chromosomal gene under these conditions (Figure 3–6). Apparently the prophage was kept quiescent by a cytoplasmic "immunity" substance, which was not transferred to the female along with the Hfr chromosome. Safe sex required that the female have the immunity substance (later called repressor protein) preexisting in its cytoplasm. For publication, the terminology was cleaned up to "zygotic induction" (zygote is the genetic term for the product of a mating).

The erotic induction experiment showed that every F$^-$ conjugated with an Hfr. Therefore, the large differences in genetic recombination observed for different chromosomal genes probably reflected differences in the frequency with which each gene is transferred. Wollman verified this point by the rather cruel experiment of separating the happy mating pairs in a blender, revealing that genes are transferred in a precise time sequence. However, the sequence of gene transfer was different for each Hfr strain. Transfer appeared to start from an "origin" that differed from one Hfr to another.

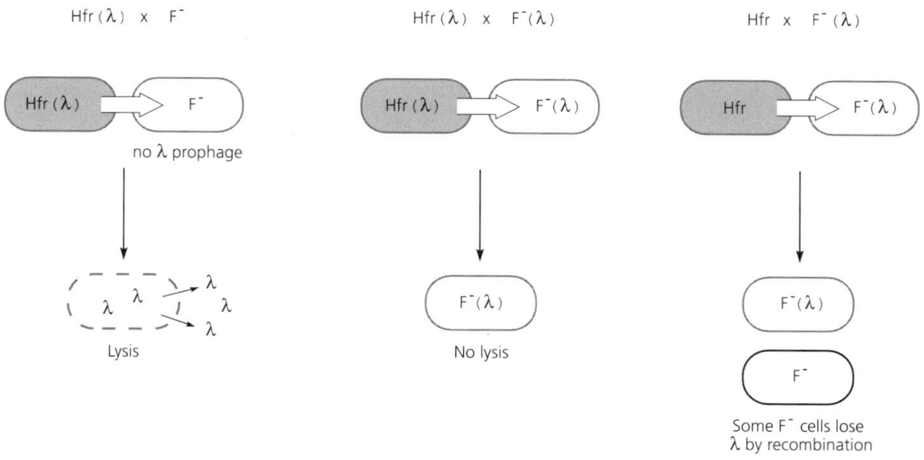

Hfr (λ) x F⁻   Hfr (λ) x F⁻(λ)   Hfr x F⁻ (λ)

FIGURE 3–6. When an Hfr strain transfers λ to F⁻ cells, the recipient cells lyse (*left*). However, when the F⁻ cells have a prophage, lysis does not occur (*middle*). When a nonlysogenic male mates with a lysogenic female, λ behaves like a standard chromosomal gene and can be lost from the female by recombination with the transferred nonlysogenic Hfr chromosome (*right*). In this figure, only the fate of female recipients is shown.

From experiments with different Hfrs, the genes could be ordered along the *E. coli* chromosome in a scale of "minutes of transfer." Remarkably, the array of genes appeared to form a circle. Jacob and Wollman supposed that the genome of *E. coli* was circular and that Hfr strains arose from the insertion of the mating factor F into this circular chromosome. The F factor would bring its origin of transfer (*oriT*) along with itself, allowing transfer of nearby chromosomal genes. Different Hfrs would have the F factor inserted at different places in the chromosome.

The general picture drawn by Jacob and Wollman was completely correct. Later molecular experiments and the genetic insights of Allan Campbell (discussed in Chapter 5) have revealed the mechanisms in more detail. The F factor found in F⁺ cells is a small circular DNA, which encodes the proteins necessary to transfer itself to the F⁻ cell. Transfer begins at the *oriT* site and transmits the entire F⁺ DNA to F⁻ cells upon mating (see Figure 3–7, top). In Hfr strains, F DNA is integrated in the chromosome. Now, DNA transfer initiating at *oriT* carries along chromosomal genes; those near *oriT* are transferred efficiently (Figure 3–7, middle). The chromosomal DNA acquired by an F⁻ in a mating produces only a transient diploid (two copy) state. The transferred partial chromosome is unable to replicate along with the F⁻ chromosome. Only the Hfr genes that are integrated into the F⁻ genome by genetic recombination are inherited by the F⁻ cell and can be passed on to progeny cells.

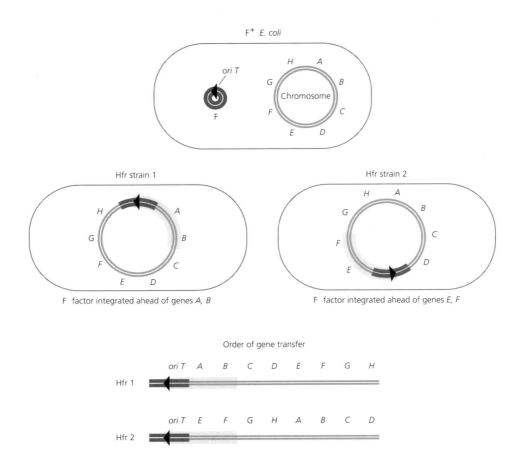

FIGURE 3–7. F+ cells have a small circular piece of DNA, called the F factor, in addition to chromosomal DNA (*top*). Hfr strains arise when F inserts into the bacterial chromosome, carrying along its own origin of transfer (*oriT*) (*middle*). During mating, the genes closest to *oriT* will be transferred to recipient cells first (*bottom*). Since F can integrate at several different places in the chromosome, each Hfr strain transfers different *E. coli* genes first. By observing the results of matings using different Hfrs, it was concluded that the genes in *E. coli* are arranged in a circle.

Interestingly, because *oriT* is located in the middle of the inserted F DNA, the entire chromosome would need to be replicated before all F DNA is transferred (Figure 3–7, middle). Most of the time, the mating pairs separate before the entire chromosome enters the F⁻ (complete transfer requires 100 minutes of uninterrupted mating, not easy even for bacteria). Because F⁻ bacteria rarely acquire the entire F factor after mating, they rarely become Hfrs.

With the tools in hand to create genetically marked strains and cells that transiently became diploid for some chromosomal genes, a full-scale genetic assault on the nature of regulatory genes was now possible. In the next section, I sketch the exciting story of the parallel genetic experiments on repression in bacteriophage λ and regulation of β-galactosidase synthesis in *E. coli* that led to the Jacob-Monod model of gene regulation.

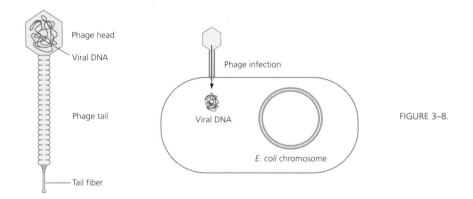

Phage DNA is injected into the bacterial cell through its tail

Phage head

Viral DNA

Phage tail

Tail fiber

Phage infection

Viral DNA

E. coli chromosome

FIGURE 3–8.

TURNING OFF RELATED GENES:
THE ROAD TO THE REPRESSOR AND THE OPERON

The varied lifestyle of a temperate phage like λ has obvious advantages for its survival but clearly requires some sophisticated regulation. Producing new phages is not an easy task, nor is knowing how and when not to produce them. The first view of a phage in the electron microscope by Tom Anderson in 1942 revealed a rather complicated structure with a head and a tail (Figure 3–8, left). The experiments of Alfred Hershey in 1953 demonstrated that this protein shell is left bound to the outside of the bacterial host, whereas the viral DNA is injected. Later work showed that the free virus has the DNA packaged into a protein head connected to a hollow tail made of different proteins. After the phage binds to a bacterium by means of a tail fiber of another distinct protein, the viral DNA is injected through the tail into the host cell (Figure 3–8, right). For lytic growth, there must be extensive DNA replication, production of new viral heads and tails, packaging of viral DNA into the complicated virus particle, and, finally, lysis of the host cell to release some 100 new phage. This process is termed lytic "development" because "temporal" (timed) control of gene expression is a general feature of developmental processes.

For λ to execute the lysogenic response to infection and become a prophage, the many genes needed for lytic growth must be shut off. Using the clear-plaque mutants of λ that he brought to Paris from his graduate studies with Delbrück, Kaiser collaborated with Jacob to show in 1957 that this regulatory process was surprisingly

a)

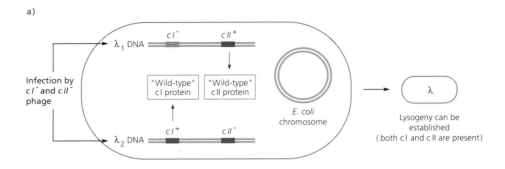

b)

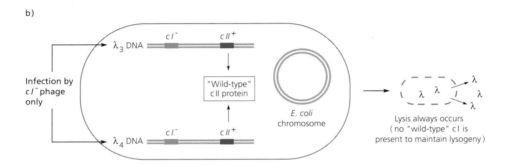

FIGURE 3–9. Joint infection by phage with clear plaque mutations in different genes (e.g., cI and cII) allows the establishment of lysogeny in the host cell because each phage supplies the gene product missing from the other (a). When joint infection is by phage with clear plaque mutations in the same gene (e.g., cI), lysis will occur because neither phage can produce the missing product (b).

simple. Kaiser found that all of his clear-plaque mutants could be classified into three genes, which he called *cI*, *cII*, and *cIII*. This determination was based on a genetic test for separation of function, called a "complementation test" introduced into phage genetics by Seymour Benzer. In this test, two mutant phage infect the same host cell. If the two phage have mutations in different genes, each mutant phage can supply the function that the other mutant lacks. Thus, these two phage can help ("complement") each other and normal ("wild-type") growth is restored (Figure 3–9a). However, if the two infecting phage have mutations in the same gene, then neither phage can supply the necessary function and no complementation (or wild-type growth) ensues (Figure 3–9b). Complementation tests among various clear-plaque mutants indicated that only three genetic functions, encoded by *cI*, *cII*, and *cIII* were required for executing the lysogenic response. In each complementation test, the normal, turbid-plaque phenotype prevailed, indicating that all three clear-plaque mutant phenotypes were "recessive" to the wild-type phenotype.

Kaiser found that *cI*, *cII*, and *cIII* gene functions could be further classified into those required for establishing lysogeny and those required for maintaining lysogeny. The *cII* and *cIII* gene products are "establishment" functions: *cII⁻* and *cIII⁻* mutants lysogenize rarely, but, once attained, the lysogenic state is as stable for the mutants as for normal λ. In contrast, the *cI* gene product is required for maintaining lysogeny; λ defective for this gene product cannot lysogenize at all. Even when *cI⁻* mutants formed lysogens with the help of *cII⁻* mutants (Figure 3–9a), no lysogens containing exclusively *cI⁻* λ could be found. Moreover, *cI⁻* lysogens always had a *cII⁻* phage as well, to provide cI. Thus, the *cI* product is the single phage-coded agent required to maintain lysogeny. Kaiser and Jacob found that the *cI* gene product is also the essential determinant of prophage "immunity"—the ability of the λ prophage to prevent lytic growth by an infecting λ (see Figure 3–3). Finally, cI is required in the recipient cell to prevent "erotic" induction after Hfr transfer of λ into the recipient F⁻ cell (see Figure 3–6). Taken together, this work led to a spectacular conclusion: the cytoplasmic product coded by the *cI* gene is able to turn off all the genes of lytic development—a "repressor," in later regulatory terminology.

Also in the mid-1950's, Mike Levine arrived in Paris with news of his similar experiments with a temperate phage of *Salmonella* bacteria called P22, which defined three genes with properties similar to those of the λ clear mutants. (Years later, P22 turned out to be a relative of λ, with many closely similar genes.) Other temperate phages related to λ isolated by Jacob and Wollman also looked similar. By 1956, some mechanistic insights into phage development seemed at hand. Unfortunately, at that time, the prevalent ideas about lysogeny were based on the incorrect notion that hooking the phage DNA to the chromosome was sufficient to turn off lytic development. Thus the *cI*, *cII*, and *cIII* genes were thought to be involved in this docking function, rather than in directly turning off expression of other genes. The correct picture did not emerge until Jacob went to his bad movie in 1958.

Mutations similar to λ*cI⁻* had also been isolated for the lactose system of *E. coli*. These mutant bacteria produced abundant β-galactosidase all the time and were termed "constitutive." Because the mutants no longer produced β-galactosidase only when induced by the presence of lactose in the growth medium, these mutations

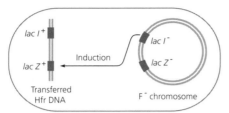

a) *lac I⁻* dominance model: *lac I⁻* in the F⁻ cell induces the transferred *lac Z⁺* gene (β-galactosidase produced)

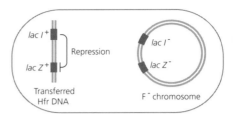

b) *lac I⁺* dominance model: the transferred *lac I⁺* gene represses the transferred *lac Z⁺* gene (no β-galactosidase produced)

FIGURE 3–10. By assaying expression of β-galactosidase immediately after an Hfr transferred its *lacI⁺lacZ⁺* chromosome to a *lacI⁻lacZ⁻* F⁻ cell, Pardee was able to discover whether the constitutive *lacI⁻* state or the inducible *lacI⁺* state of the regulator was dominant. Contrary to the predictions of an "internal inducer" hypothesis that stated that *lacI⁻* was dominant (*a*), expression of β-galactosidase was turned off by the newly synthesized I⁺ gene product, establishing that *lacI⁺* was dominant to *lacI⁻* (*b*).

were given the designation *lacI⁻* for induction-defective. In 1955, Georges Cohen and Howard Rickenberg had finally convinced Monod to accept the concept of a lactose permease, an enzyme required to bring lactose into the cell. The activity of this enzyme was defined by its ability to concentrate radioactive TMG, a lactose analog that functioned as an inducer. Using this assay, Cohen found that lactose permease (coded by the *lacY* gene) was also expressed constitutively in the *lacI⁻* mutant. So, the *lacY* gene achieved a place of honor next to *lacZ* (the gene previously shown to code for β-galactosidase), and the notion was born that a two-gene system was subject to some sort of unified regulation by lactose.

What might be the molecular basis for the coordinate regulation of the lactose enzymes? The consistently high amounts of β-galactosidase and permease in *lacI⁻* mutant cells provided a possible clue. Monod thought that *lacI⁻* mutants probably produced an "internal inducer," thus bypassing the need for lactose, the external inducer. This idea provided a simple but not revolutionary explanation for the *lacI⁻* phenotype—production of a new small molecule by the cell to replace the normal, external inducer. To test this notion, Monod and Jacob needed to study bacteria carrying both the *lacI⁺* and the *lacI⁻* genes. The internal inducer idea gave a clear prediction: cells with both *lacI⁺* and *lacI⁻* genes should produce β-galactosidase at the constitutive rate because the *lacI⁻* gene would make the internal inducer. In classic genetic terminology, the *lacI⁻* allele is dominant over the *lacI⁺* allele (Figure 3–10a).

It was not trivial to carry out this experiment. Bacteria are normally haploid (as they have only one chromosome per cell). At the time, there was no way to produce stable diploids, but bacterial sex permitted the establishment of cells transiently containing both the *lacI⁺* and the *lacI⁻* genes. Arthur Pardee developed the functional analysis of such partially diploid bacteria. Pardee came to Paris from Berkeley on sabbatical leave in 1957 and promptly initiated a crucial series of experiments that Monod christened the PaJaMa experiment (for Pardee, Jacob, and Monod). The PaJaMa experiment examined expression of the *lacZ* gene immediately after partial transfer of the *lacI⁺lacZ⁺* chromosome from an Hfr male cell to a *lacI⁻lacZ⁻* F⁻ female cell. In these experiments, there was no β-galactosidase expression prior to mating: the Hfr had not been induced and the F⁻ did not have a functional *lacZ* gene to produce β-galactosidase. After mating, an experimental trick was used to selectively block protein synthesis in the Hfr cells, so that all β-galactosidase production originated in the F⁻ cells (see Figure 3–10).

Arthur Pardee

Using this system, Pardee could ask the key regulatory question: Does the constitutive *lacI⁻* or inducible *lacI⁺* phenotype prevail in bacteria with both genes? Initially, the expectations of the internal inducer idea seemed to be fulfilled. β-galactosidase synthesis began promptly after transfer of the *lacI⁺lacZ⁺* genes into a *lacI⁻lacZ⁻* recipient, as if an internal inducer in the *lacI⁻* recipient cells induced β-galactosidase production from the newly transferred *lacZ* gene (Figure 3–10a). However, during the course of the experiment, joy turned to consternation. Production of β-galactosidase stopped after about 60 minutes when sufficient *lacI⁺* product accumulated to do its job. So in the presumed stable regulatory state, the *lacZ⁺* gene was turned off by the presence of the *lacI⁺* product. Contrary to the predictions of the internal inducer idea, *lacI⁺* inducible gene expression was dominant to *lacI⁻* constitutive gene expression (Figure 3–10b). Clearly, a new regulatory paradigm was needed.

The next view of regulation was the brainchild of Leo Szilard, a physicist turned biologist who happened to be visiting the Pasteur at the time. Szilard argued for a "repressor" model in which the product of the *lacI⁺* gene normally turned off the *lacZ* gene. In this model, lactose, the external inducer, would somehow eliminate or antagonize the action of the repressor (Figure 3–10b). How the putative repressor might work was very vague. The repressor might work on the DNA itself, on a previously unobserved unstable RNA, or on a special rapidly assembled class of ribosomes. By the middle of 1958, the fundamental problems of regulation had been identified, but not sharply focused.

In the burst of insight described in the beginning of the chapter, François Jacob provided the focus. Jacob realized that λ was regulated just like *lac*. In both systems, regulation required turning off a cluster of genes (a very large number in the case of λ). Both systems were regulated by a single gene product. Most important, genetic tests indicated that, in both systems, the inducible state (*lacI*$^+$, λ*cI*$^+$) was dominant to the constitutive state (*lacI*$^-$, λ*cI*$^-$), indicating that both regulators were repressors (see Figures 3–9 and 3–10). What was the repressor produced by the *lacI* and *cI* genes and how did it work? Jacob observed: "It seemed impossible that the phage would make 30 types of ribosomes—there must be some key to a lock acting at the DNA level on a group of genes." This idea was the essence of what became the operon model of regulation. A group of adjacent genes serving a related function are controlled together at a single site by a cytoplasmic repressor.

Initially, Monod was not impressed with the operon model. Among other objections, he pointed out that β-galactosidase and permease should always be induced to the same extent ("coordinate induction"), and there was one inducer that worked for permease but not for β-galactosidase. But by the second day of discussion, Monod became a believer and decided to ignore the funny inducer, which became "a skeleton in the closet." Eventually the permease activity induced by this compound turned out to be a different permease altogether, which was regulated by a different system. This was not the only time that data had to be ignored to formulate a model. Sometimes, as in this case, the eventual explanation for the discrepancy did not necessitate revising the model. As we will soon see, the alternative is also true.

Jacob describes the next stage of his collaboration with Monod as a "ping-pong game of ideas and experiments," and progress was rapid. Monod pointed out that, if a repressor really was the key that fit into a genetic lock, then the lock itself should be able to be mutated so that the key would no longer fit. Most important, although both the *lacI*$^-$ mutants and the lock mutants would result in constitutive expression of β-galactosidase, their genetic properties should differ. Whereas the *lacI*$^-$ constitutive mutations were recessive, the lock constitutive mutations should be dominant. Because *lacI*$^+$ would not be able to turn off the mutated genetic lock, expression of β-galactosidase should continue even when the *lacI*$^+$ repressor is present.

Jacob realized that he and Wollman had already found lock mutations for λ in 1953 but had not realized their significance. When wild-type λ infects a lysogenic host, the *cI* repressor present in the

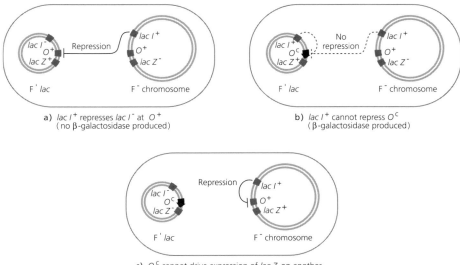

a) $lac\ I^+$ represses $lac\ I^-$ at O^+
(no β-galactosidase produced)

b) $lac\ I^+$ cannot repress O^c
(β-galactosidase produced)

c) O^c cannot drive expression of $lac\ Z$ on another
chromosome (no β-galactosidase produced)

cell shuts off gene expression from the incoming phage, preventing lytic growth. In contrast, when "λ virulent" mutants infect a lysogenic host, they undergo lytic growth and kill the cell. These phage behaved as if the genetic lock were mutated, so that *cI* repressor could no longer work. The lock for the repressor was termed the "operator" or *O* site.

Could similar mutations be isolated for *lac*? To answer this question, Jacob needed an easy test for whether the constitutive phenotype was dominant. Luckily, he and Ed Adelberg had recently learned how to make variants of F^+ containing *lac*, allowing him to construct *E. coli* strains that were diploid for the *lac* region. To carry out dominance tests, these "F'*lac* factors" could simply be transferred to F^- bacteria with different genotypes. These diploid bacteria could then easily be tested for β-galactosidase on the agar dish on which bacterial colonies were growing. Using a perfume atomizer, the dish was sprayed with the color-producing substrate ONPG; colonies producing β-galactosidase turned yellow. Jacob was now in his ideal visual element. The world of regulation could be seen with yellow colonies and phage plaques. In a few weeks, the *lac* operator-constitutive (O^c) mutations were isolated. The O^c mutations had the predicted property of lock mutations: they were dominant (Figure

FIGURE 3–11.   Cells diploid for the *lac* genes were used to test whether constitutive mutants were dominant. *lacI⁻* constitutive mutations are recessive because they can be repressed by a *lacI⁺* allele at O^+ (a). O^c constitutive mutations are dominant because they cannot repressed by *lacI⁺* (b). O^c mutations are said to be "*cis*-dominant" (*cis* is from the Latin for *on this side of*) because they can relieve repression only for *lac* genes that are next to the O^c mutations and on the same chromosome (c).

3–11b), expressing β-galactosidase even in the presence of $lacI^+$ repressor. In addition, O^c mutations possessed the special characteristic of relieving repression only for the adjacent *lac* genes on the same chromosome; in genetic terminology they are "cis-dominant," cis meaning "next to" (Figure 3–11c). Cis-dominance is an expected characteristic of mutations that alter the site of action of a regulator. Only those genes on the same DNA duplex as the altered regulatory site will be affected.

The operon model was now nearly complete. The repressor produced by the *I* gene acts at the operator *O* site to coordinately turn off expression of the *lacZ* and *lacY* genes. A third gene, called the transacetylase or *lacA* gene, immediately downstream of the *lacZ* and *lacY* genes, was also subject to coordinate control by *lacI* and *O* (Figure 3–12). Jacob and Monod then realized that there must also be some site where the proteins required for expression of the operon could bind once the repressor was removed. They termed this the promoter or *p* site. If this site were mutated, it should be more difficult to initiate expression of the operon. Thus, the *p* site should be identified by mutations that coordinately reduced enzyme synthesis from the *lacZ*, *lacY*, and *lacA* genes (in contrast to O^c mutations, which coordinately increased expression of these genes). The cis-acting p^- mutations were identified in 1966 by Jon Beckwith and John Scaife, after Beckwith cleared up a considerable amount of confusion caused by a special class of mutations within the *lacZ* gene that, for other reasons, also diminished enzyme synthesis from *lacY* and *lacA*.

THE DISCOVERY OF MESSENGER RNA

From the beginnings of the operon idea, Jacob believed that regulation was exerted on the DNA. However, the beautiful series of genetic experiments that developed the operon model really said nothing about molecular mechanisms. Genetics is extremely powerful for general conclusions but by itself says nothing about biochemistry. The operator might have been the DNA, the RNA product, or conceivably even the protein produced from the RNA, provided that synthesis of all proteins of the operon required the presence of the first protein. If the point of regulation was the DNA, the most likely mechanism was inhibition of DNA transcription into a multi-gene RNA, but this idea made sense only if the regulated RNA is unstable. Unfortunately, there was no evidence for this type of RNA.

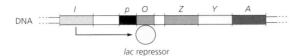

DNA

lac repressor

FIGURE 3–12. The *lac* repressor (made by the *lacI* gene) inhibits expression of the three genes in the operon by interacting with the operator (*O*) site. Expression of operon genes begins at the promoter site (*p*). In this figure, *O* and *p* are shown at a very expanded scale.

In the fall of 1959, Jacob spoke at a symposium on bacterial genetics attended by most of the leading molecular biologists, and he proposed the idea of an unstable RNA as an intermediate in protein synthesis. No one apparently took notice. In the spring of 1960, Jacob brought up the idea again in a small discussion that included Francis Crick and Sydney Brenner (who had also been at the earlier meeting). Suddenly, Brenner realized that unstable informational RNA had already been observed but not identified as such. Elliot Volkin and Larry Astrachan worked with the "virulent" phage T2, which exhibits a lytic lifestyle. In 1956, Volkin and Astrachan had reported finding an unstable RNA fraction in cells infected with T2. Remarkably, the base composition of "T2 RNA" mimicked that of T2 DNA and not *E. coli* ribosomal RNA.

Brenner, who was working with the closely related phage T4, saw that virulent phage infection provided the perfect experiment to determine whether the informational intermediate was the RNA component of a newly synthesized ribosome or an unstable "messenger" RNA (mRNA for short) unrelated to ribosomal RNA. If an RNA component of a ribosome was informational RNA, phage RNA would be found only in ribosomes synthesized after infection ("new" ribosomes). If informational RNA was mRNA, it would not be restricted to ribosomes synthesized after infection and would be found with both "old" and new ribosomes. New and old ribosomes could be distinguished with a density-labeling procedure worked out a few years earlier by Matt Meselson and Frank Stahl at Cal Tech to study DNA replication (see Figure 4–2). By a fortunate chance, both Brenner and Jacob had been invited independently to Cal Tech; they decided to spend 30 days with Meselson performing the experiments.

Before phage infection, cells were grown in nutrients with "heavy" atoms of nitrogen and carbon (^{15}N and ^{13}C), so that the old ribosomes were "heavy." After infection, the bacteria were grown in "light" medium with the normal atoms of nitrogen and carbon (^{14}N and ^{12}C), so that the new ribosomes were "light." The "light"

Sydney Brenner

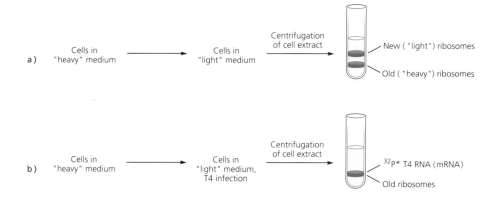

FIGURE 3–13. By growing cells in medium with heavy isotopes of nitrogen and carbon, and then shifting them to a medium with lighter atoms, newly synthesized ribosomes are lighter than old ribosomes and the two can be separated from each other by density gradient centrifugation (a). This allowed Brenner to ask whether newly synthesized informational RNA is the RNA component of a new ribosome or messenger RNA (mRNA) that could associate with "old" ribosomes. To test this, T4 phage was added to cells at the time of the shift to light medium. The radiolabeled informational phage RNA was found to be associated with old ribosomes (b), showing that the informational RNA is mRNA, not ribosomal RNA. We now know that no ribosomes are synthesized after T4 infection, explaining why no T4 mRNA is found where newly synthesized, light ribosomes would be located.

medium also contained radioactive phosphorus atoms (^{32}P) to label phage RNA. Heavy and light ribosomes could be separated by centrifugation in a dense solution of a salt compound, cesium chloride. Because of the centrifugal force, the cesium chloride solution is more dense at the bottom of the centrifuge tube than at the top. Heavy and light ribosomes migrate to a position at which their density matches that of the surrounding salt solution (the procedure is analogous to floating on very salty water). The experiments yielded the hoped-for result. The newly synthesized radioactive viral RNA was associated with old "heavy" ribosomes that were present before infection (Figure 3–13). Messenger RNA moved from an idea to a defined informational molecule.

The actual experiments were more complicated than just described. It turns out that ribosomes do not like to hold together in dense cesium chloride. For 24 of the planned 30 days nothing worked; no intact ribosomes could be found in the gradient. Then Brenner realized that, in the high concentrations of cesium salts found in these gradients, more magnesium than usual might be needed to hold the ribosomes together and to bind mRNA to the ribosomes. The experiment worked—another demonstration that most successful molecular biology depends on knowing how to do good experiments as well as on having good ideas.

Based on the T4 experiments, mRNA might still have been a special product of phage infection. However, Jim Watson's group at Harvard found that uninfected *E. coli* possessed an unstable RNA with the properties expected of mRNA. Charles Kurland had separated two large, stable rRNAs by velocity centrifugation, in which larger, heavier molecules move faster in a centrifugal field (a more standard separation technique than the density gradient). Kurland and François Gros, a visitor from the Institut Pasteur, found that an unstable, heterogeneous class of RNA could be detected by labeling the RNA with radioactive phosphorus (^{32}P). This unstable RNA associated with ribosomes and revealed the distribution of sizes expected for informational copies of many operons. The demonstration of mRNA did not prove that regulation occurred at the DNA to RNA level, but this idea now became the accepted view.

WHAT IS THE REPRESSOR?

The nature of the cytoplasmic repressor now became the major molecular question posed by the operon model. Jacob and Monod initially supposed that the repressor was an RNA molecule—a sort of regulatory "adapter" RNA that recognized the operator DNA by base-pairing. Some complicated experiments with inhibitors of protein synthesis were rather emphatically interpreted in favor of this notion (or at least against the idea that repressor was a protein). However, a regulatory protein seemed to be necessary somewhere to recognize the external inducer. After a period of confusion, "nonsense-type" mutations were isolated in the λcI and $lacI$ genes. Since nonsense mutations were known to act by terminating protein synthesis (see Figure 6–5), the repressor was almost surely a protein. The isolation and biochemical activity of regulatory proteins are described in Chapter 5.

SOME COMMENTS ON THE ROAD TO THE OPERON

The pathway to the operon model has been presented in considerable detail for two reasons. First, the model is universal: the major control mechanisms for all organisms depend on regulatory proteins that act at specific DNA sites to control the genes of a metabolic or developmental pathway. Second, the lengthy development of the operon concept is an excellent example of how molecular biology is done. The final picture can be presented in a wonderfully logical way (and it is, unfortunately, in most text books on molecular biology). But the real pathway to this general view of the biological universe was exceedingly tortuous. Major insights often came from experiments designed with assurance to test an incorrect idea. Correct ideas obvious in retrospect were sometimes discarded or overlooked. Experimental results were sometimes ignored, helping or hindering the rate of progress. Eventually, the quest for simplicity and elegance prevailed as the experiments became more and more sharply focused through the combination of genetics and biochemistry.

Many years later, Jacques Monod said to Mel Cohn, "Mel, we were never wrong." Mentally reviewing the history, Cohn replied "We were always wrong—we just got it right in the end." And that is great science—going for the big news and getting it right in the end.

POSITIVE AND NEGATIVE REGULATION: TWO WAYS TO RUN AN OPERON

For lactose and λ, regulation is termed negative—the LacI and λcI regulatory proteins act to turn off the genes of the operon. Jacob and Georges Cohen found that the concept of negative regulation also applied to the *trp* operon genes that controlled production of the amino acid tryptophan. This work extended negative regulation to "repressible" systems. Repressible systems had a cute twist. Here, the environmental signal ("effector") prevents enzyme production by acting as a "co-repressor." The co-repressor converts the repressor to its active state and turns off gene expression. As we have seen, for inducible systems, the environmental signal performs the opposite function. It inactivates the repressor, turning on expression of the genes that are regulated. Gerard Buttin, a student with Monod, found that the operon for galactose utilization also obeyed the rules of negative regulation. Monod decided that all operon regulation was negative. His fervent adherence to that position delayed the acceptance of the possibility of positive regulation for many years.

A large international meeting on "Cellular Regulatory Mechanisms" was held in 1961 at the Cold Spring Harbor Laboratory. Jacob gave a triumphant presentation of the evidence for the operon model, and Monod concluded the meeting with a general discussion of regulation, including possible applications of the operon mechanism to developmental biology. Somewhere in between, I gave my first scientific talk to a large audience, a brief presentation about regulation of the enzyme alkaline phosphatase, which I had been studying with Allan Garen and Annamaria Torriani. I commented that control of alkaline phosphatase might be positive—the regulatory product might be an "inducer" required

FIGURE 3–14. *Left panel:* In a positive regulatory system, the regulator turns on expression of operon genes (*a*), inactivating the regulator by mutation results in loss of ability to express the operon (*b*). Providing a copy of the wild-type regulator to the mutant operon restores the capacity to express the genes of the operon (*c*). *Right panel:* In a negative regulatory system, the situation is exactly the opposite. The regulator turns off expression of operon genes (*d*), inactivating the regulator by mutation results in high constitutive expression of the operon (*e*). Providing a copy of the wild-type regulator to the mutant operon turns off constitutive expression of the operon genes (*f*).

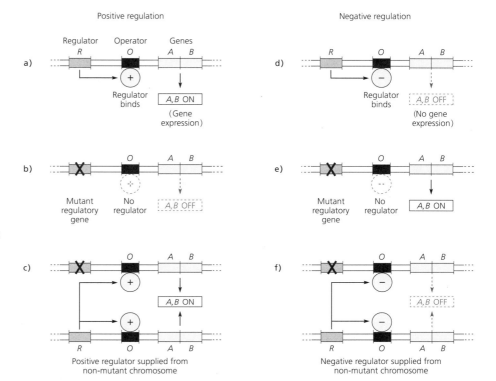

to turn on the phosphatase gene. I expected some mild interest in this generalization of the operon model from my heroes Jacob and Monod. Instead, Monod ran to the front of the room, proclaiming, "No, no, we know that all regulation is negative—moreover, we have mutants like yours and know how to interpret them." I slunk from the stage, and Garen and I retired to the bar to heal our wounds. Subsequent experiments demonstrated that phosphatase, as well as many other biological regulatory systems, exhibit positive regulation. Ironically, even the lactose operon and λ use dual control, positive and negative.

The essence of positive regulation is to use a regulatory protein to turn ON the genes of the operon, when the appropriate inducer is supplied (Figure 3–14, left panel). As a consequence, the genetic predictions for positive regulation are the opposite of those for negative regulation. Mutations inactivating a positive regulatory protein turn off expression of all of the genes in an operon, a "pleiotropic negative" phenotype (Figure 3–14, left panel), whereas mutations inactivating a repressor turn on expression of all operon genes, a "constitutive" phenotype (Figure 3–14, right panel). Supplying the normal regulatory gene (R^+) to cells lacking the regulator (R^-) restores enzyme production in positively regulated systems (Figure 3–14, left panel) but, as the PaJaMa experiment demonstrated, turns off constitutive expression in negatively regulated systems (Figure 3–14, right panel).

Although Garen and I believed (correctly) that phosphatase exhibited positive regulation, this system was complicated and had only one enzyme to study. Fortunately, Ellis Englesberg came up with an excellent experimental demonstration of positive regulation in the operon controlling use of the sugar arabinose. Many mutants of the regulatory *araC* gene were defective in production of all three enzymes specific to the *ara* operon. When the normal gene was supplied in a partial diploid, production of all three enzymes was restored. Thus, the arabinose operon provided a textbook case of positive regulation, but Monod insisted that Englesberg had misinterpreted his data. So strong was Monod's influence that Englesberg's beautiful experiments between 1962 and 1966 did not achieve much recognition. In fact, he had difficulty even publishing his experiments. Positive regulation was not generally accepted until Bill Dove, René Thomas, and I showed that phage λ had two positive regulatory genes. The younger generation of "lambdologists" (as λ workers were termed) were numerous and vocal, and objections to positive regulation disappeared. Positive regulation even appeared at the Institut Pasteur through Maxime Schwartz's work on the maltose operon. This interesting period is recounted in an historical essay by Jon Beckwith in the general book called *Escherichia coli and Salmonella: Cellular and Molecular Biology*.

MULTI-OPERON REGULATION: THE GLUCOSE EFFECT

Of all the sugars that *E. coli* can use for food, glucose is the favorite, giving the most rapid growth because it is used by the most direct metabolic pathway. When fed a mixture of sugars including glucose, the bacteria use the glucose first and do not waste energy and materials making the enzymes required for using the other sugars. So long as glucose is present, operons such as lactose, arabinose, and maltose are not induced efficiently by their respective sugars. This phenomenon was termed the "glucose effect" by Monod, who discovered it in 1941. Somewhat ironically, in 1970 the major basis of glucose control was shown to be the activity of a positive regulatory protein, the cyclic AMP receptor protein, which is required for activity of the lactose, arabinose, and maltose operons (and many other operons as well). Thus, even the *lac* operon is subject to both positive and negative regulation. This story is discussed in more detail in Chapter 5 (see Figure 5–15). As discussed in Chapter 8, multi-operon regulation controls adaptive responses to various environmental conditions as well as temporal regulation of developmental programs.

REGULATORY DIVERSITY: REVIEW AND PREVIEW

The general operon model developed so far includes positive and negative regulatory proteins combined with positive and negative environmental effectors that control the activity of the regulatory proteins themselves. I have implied that regulation of operons is exerted at the DNA level by determining whether or not a multi-gene RNA for the operon is produced from DNA. This feature is true for the examples considered so far.

Regulatory biology, however, is much broader than an on-off switch for messenger RNA. As we shall see, regulatory mechanisms can control the termination of RNA synthesis as well as its initiation; the initiation of protein synthesis and even its completion; and the activity and stability of proteins. Control over protein activity represents a most important regulatory response. Operon regulation depends on this mechanism because the external environmental signals that control the operon do so by altering the biochemical activity of the regulatory protein.

Regulation of protein activity also controls metabolic pathways by reducing or enhancing the flow of precursors into products such as amino acids or nucleotides. Control of the catalytic activity of an enzyme provides a more rapid means of directing small molecule substrates down the appropriate pathway than increasing or decreasing the amount of an enzyme. A most important example of this type of regulation is end-product (or feedback) inhibition, in which the final product of a metabolic pathway inhibits the activity of the first enzyme specific to the pathway. As soon as tryptophan is supplied to a growth medium, the bacteria will stop making this amino acid and all of its precursor compounds because tryptophan inhibits the catalytic activity of the first enzyme of the pathway. The enzyme

FIGURE 3–15. The end-product of tryptophan synthesis inhibits the first enzymatic step in its synthesis. This is an example of feedback inhibition.

subject to feedback inhibition is called anthranilate synthetase. This enzyme catalyzes the synthesis of the first specific compound of tryptophan production, anthranilate (Figure 3–15).

The study of feedback inhibition with purified enzymes and substrates has yielded an important regulatory principle, initially revealed in the early 1960's by John Gerhardt and Arthur Pardee at Berkeley and by Jean-Pierre Changeux and Jacques Monod. Enzymes subject to end-product inhibition and many other proteins have highly specific regulatory sites as well as the precisely defined catalytic sites. The interaction of effector molecules at these regulatory sites controls the activity of the enzymes. Monod termed this type of control mechanism "allosteric" regulation. Although this book focuses on events involving the genes, it is important to keep in mind that the precisely controlled activity of proteins is what makes everything in cells work.

FURTHER READING

Beckwith, J. (1996) The operon: An historical account, in F. C. Neidhardt, ed., *Escherichia coli and Salmonella: Cellular and Molecular Biology*, 2nd ed. Washington, D.C.: ASM Press.

Cohn, M. (1978) In memoriam, in J. H. Miller and W. S. Reznikoff, eds., *The Operon*. Cold Spring Harbor, N.Y.: Cold Spring Harbor Laboratory Press.

Jacob, F. (1966) Genetics of the bacterial cell. Science 152, 1470–78.

Lwoff, A. (1966) The prophage and I, in J. Cairns, G. S. Stent, and J. D. Watson, eds., *Phage and the Origins of Molecular Biology*. Cold Spring Harbor, N.Y.: Cold Spring Harbor Laboratory Press.

Monod, J. (1966) From enzymatic adaptation to allosteric transitions. Science 154, 575–83.

Ullmann, A., and A. Lwoff, eds. (1979) *The Origins of Molecular Biology: A Tribute to Jacques Monod*. New York: Academic Press.

4 REPLICATING THE GENOME

DNA POLYMERASE REPLICATES DNA

The revolutionary news from the structure of DNA proposed by Watson and Crick was a possible mechanism for transfer of genetic information—not only from gene to protein, as we have just documented, but also from old genes to new genes. "It has not escaped our notice that the specific pairing that we have postulated immediately suggests a possible copying mechanism for the genetic material," wrote Watson and Crick in their paper announcing their discovery. In theory, each strand of the DNA helix could serve as the template for synthesis of a new strand having a complementary base at each position; an A in the template strand would pair with a T, a G with a C, and so on. With this idea, the previously mysterious duplication of genes suddenly acquired a plausible chemical mechanism. The chemical idea is correct, but 35 years were needed to achieve understanding of the complex array of enzymes needed to execute this chemical scheme. The quest for how genes duplicate involved many people and a variety of approaches. However, the central figure for nearly all of those 35 years was Arthur Kornberg.

Arthur Kornberg

There are actually two Arthur Kornbergs, and knowing about one helps in understanding the other. I learned about Arthur Kornberg II from a dinner conversation in which Arthur asked Jack Griffith, his former postdoctoral associate, "How is Arthur Kornberg?" Since I thought I was sitting with Arthur Kornberg, I was mystified until I learned that Arthur II is a Bengal tiger, owned by Griffith and named after Arthur I. Then I remembered the collection of pictures on the wall of Arthur's office, featuring a tiger cub in various poses, playing with the *DNA Replication* book of Arthur I, or being petted by the master. The animal analogy evidently seemed appropriate to Jack and to Arthur, and I rather like it myself. Kornberg generally presents a rather quiet, polite, and pleasant demeanor. However, he can be ferocious toward an adversary and is fiercely demanding of his associates in the relentless pursuit of his quarry—the mechanism of DNA replication. Like the tiger, Kornberg very rarely has to use his strength; his mere presence is sufficient to produce feverish activity in his lab and a clear path from potential scientific competitors.

Kornberg's approach to DNA replication has been the traditional one of quantitative biochemistry: develop a way to measure the reaction in a cell-free system (assay); separate and purify each active component (fractionation); put the components back together to understand the detailed mechanism of the reaction (reconstitution). However, the way in which Kornberg has applied this traditional approach to the enormously complex problem of DNA replication has been both spectacularly successful and enormously influential for other biochemical studies in molecular biology. Kornberg took to heart Efraim Racker's famous aphorism, "don't waste clean thoughts on dirty enzymes." His insistence on precise assays, rigorous control experiments, and pure components has provided the guidelines for a generation of younger scientists devoted to fathoming the mysteries of genetic information transfer at a molecular level. Kornberg combines rigorous and imaginative biochemistry with an intense and demanding devotion to success. The period of a high batting average for most successful scientists is about 10–15 years, the same as that of a good baseball player. Kornberg's career has been somewhat analogous to the baseball hitter who bats at least .400 every year for 35 years.

The biochemist's goal is to study a process by mixing pure proteins and the molecules with which they interact (substrates). To undertake this adventure with a process as complex as replicating the DNA content of the cell (the genome) requires a belief that the essence of the reaction is simple. And, of course, the chemical idea of base-pairing invokes an exquisitely simple replication mechanism. A single enzyme, DNA polymerase could order and link the new complementary bases of the DNA. In contrast, many biologists and physical chemists were fascinated by the obvious overall complexity of genome duplication. There must be starting and stopping points, and the DNA strands must be unwound to be copied without producing a hopeless knot ahead of the replication fork (try pulling apart the strands of a rope while holding the other end with your feet). The entire process must be regulated in concert with cell division. Both points of view are correct. Although the process of DNA synthesis is done by DNA polymerase, the polymerase itself has several separate polypeptide chains (subunits), and some 20 other proteins are required to do it all right. The continuing interaction between the two views of genome duplication, biochemical and cellular, has given us our current rather good understanding of the mechanism and regulation of DNA replication.

Kornberg began the biochemical study of DNA replication in 1955 in a pragmatic way. After several years of work on the cellular route of nucleotide synthesis, he began a major effort on RNA synthesis. However, word came that Severo Ochoa had found the enzyme that makes RNA (later this turned out to be untrue: Ochoa's enzyme was polynucleotide phosphorylase, which puts nucleotides together randomly rather than making biologically significant RNA). Since the RNA problem was thought to be in hand, Kornberg decided to switch from RNA to DNA. He walked down the hall to speak with his postdoctoral fellow, Robert Lehman, and said, "Bob, how would you like to switch your project to DNA synthesis?" So began a most productive career with DNA for both. Kornberg and Lehman continued to study DNA replication as next-door professors at Stanford.

One special chemical quality of DNA that distinguishes it from RNA is the presence of thymine (T) in DNA versus uracil (U) in RNA. Thymidine (the base thymine plus deoxyribose sugar) was known to work as a precursor for DNA in cells. Therefore, to look for DNA synthesis in cell extracts, the Kornberg group looked for the incorporation of thymidine into DNA. Incorporation could be detected because DNA, a long chain nucleic acid, is insoluble in acid, whereas thymidine itself is acid soluble. The thymidine was radioactive so that incorporation into DNA could be followed with sensitivity and precision. In the "soup" freed from broken bacterial cells, a tiny amount of radioactive thymidine appeared in an acid insoluble product. This product was judged to be DNA by its sensitivity to the enzyme DNase, which degrades DNA. This small start initiated many years of fruitful research—the infield single that led to a very big inning.

The synthesis of DNA by the cell extract was much more efficient if a "high energy" form of thymidine, one with three phosphate groups attached (thymidine triphosphate or dTTP) was used. With this more efficient reaction, the simple assay of acid insolubility allowed Lehman and others in the Kornberg group to divide the extract into different fractions, each of which was required for the synthesis of DNA. Many subfractions were identified, and the process appeared horrendously complex.

Up to this point, the Kornberg group followed the traditional biochemical approach of tracking precursor to product with the fewest assumptions. However, the frightening complexity of the biochemical replication reaction was abruptly clarified by considering the Watson-Crick scheme of DNA replication. Perhaps many of the fractions might be needed only to make the triphosphate form of the three other bases (dCTP, dGTP, and dATP) besides thymine found in DNA. When Lehman tested whether addition of these purified substrates would bypass the need for most of the complex mixture, the radioactivity counter used to determine the amount of DNA product "went off scale." Moreover, the process was not only more efficient but also much simpler. The test-tube synthesis of DNA required all four bases (as triphosphates), a single protein fraction, and DNA—the fundamental process of gene copying could be done outside of cells. Molecular biology is a strange profession in which the cheers for a homerun are often like this spasm from a radioactivity counter on a quiet Saturday afternoon.

By 1958, the critical protein, DNA polymerase, had been separated from other proteins. The biochemical reaction in the test tube had exactly the properties expected for DNA replication according to the Watson-Crick scheme. Not only were a DNA template and nucleotides corresponding to each of the four bases required, but the DNA product also mimicked the base composition of the DNA template (a template with a high A:T content yielded high A:T DNA). Kornberg could write an enzyme-catalyzed chemical reaction for DNA replication. The DNA polymerase carried out DNA replication by adding the incoming base (together with its sugar and a single phosphate) to the end of one DNA chain, called the primer, with the choice of incoming base determined by the complementary template strand. Energy for the polymerization reaction into DNA derives from splitting off the first two phosphates from the triphosphate precursor during the reaction (Figure 4–1).

The initial Kornberg studies showed that DNA can be dupli-

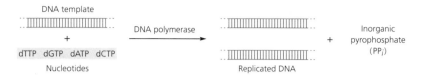

DNA polymerase replicates DNA

DNA template

+

dTTP dGTP dATP dCTP

Nucleotides

DNA polymerase →

Replicated DNA

+

Inorganic
pyrophosphate
(PP$_i$)

FIGURE 4–1.   Starting with a DNA template and deoxynucleotides with three phosphates ("triphosphates") as substrates, DNA polymerase makes new DNA by joining together the deoxynucleotides. Deoxynucleotides with one phosphate ("monophosphates") are incorporated, and the other two phosphates are released as inorganic pyrophosphate (PP$_i$) in this process.

cated in the test tube but did not specify the detailed biochemical mechanism because the role of the DNA template was unclear. How do DNA chains start? Are both strands replicated at the same time? How is an entire chromosome replicated? Work on DNA replication in cells by the biologists was required to define the process of genome duplication, which in turn tossed the problem back to the biochemists with a clear challenge.

THE PROBLEMS OF REPLICATING A GENOME

The study of DNA replication in living cells has involved many people over many years. However, the initial view of the overall process and its complexity was mainly the work of John Cairns, who sought to understand genome duplication by developing ways to visualize the ongoing process of DNA replication within the cell. He came to this effort with a background of research on the growth of viruses in his native Australia and an exposure to the new world of molecular biology from an encounter with the animal virus group at Cal Tech in 1957, followed by a sabbatical leave with Alfred Hershey at Cold Spring Harbor in 1960. Intuitive and clever, Cairns possessed a vision of simplicity and elegance and a lust for the bold, revolutionary experiment. He loved the quest for new biological insights and shared the traditional suspicion of the phage group for biochemical approaches. Cairns and Kornberg were at opposite poles of molecular biology.

When Cairns started, only one thing was really known about DNA replication—that the Watson-Crick scheme was likely to be correct. This conclusion followed from Kornberg's biochemical work and from a clever physical experiment by Matt Meselson and Frank Stahl, which demonstrated that DNA replication in bacterial cells does indeed follow the proposed mechanism of Watson and Crick—each strand serves as the template for a new strand. This

John Cairns

FIGURE 4–2.   To determine whether replicated DNA has one or two newly synthesized strands, cells were first grown in ^{15}N medium for many doublings so that both strands of DNA were heavy ("H") and then transferred to ^{14}N medium so that newly synthesized DNA strands were light ("L"). After one doubling, all of the DNA is half-heavy, containing one H and one L strand. Since only one of the two parental strands is conserved in the replicated product, DNA replication is said to be "semi-conservative."

Matt Meselson

Frank Stahl

mode of replication is called "semi-conservative," meaning that the newly replicated molecule is composed of one parent and one daughter strand; only one of the two parental DNA strands is conserved in the product of replication. To perform the experiment, Meselson and Stahl labeled the old (parental) and new (daughter) strands with different kinds of nitrogen atoms: old DNA was synthesized with the heavy isotope ^{15}N, whereas new DNA was synthesized with the normal (light) isotope ^{14}N. They invented the technique of density-gradient centrifugation (described in the previous chapter; see Figure 3–13) to determine the density of newly synthesized DNA. Recall that, during centrifugation, salt density increases toward the bottom of the centrifuge tube. When a molecule hits its density, it ends up "floating" at that position in the tube. By this criterion, the newly replicated DNA was "half-heavy"—intermediate in density between DNA with both strands heavy and DNA with both strands light. Thus the newly replicated DNA carried one parental and one daughter DNA chain (Figure 4–2).

Cairns wanted to understand how an entire genome duplicated itself. His approach was to catch DNA in the act of replication by methods that would allow a direct view of the process. To do this, he used radioactive thymidine as the precursor for DNA replication in *E. coli* cells. As replication proceeded, the radioactive thymidine was incorporated into DNA, marking the path of the newly replicated strand. The radioactive DNA was then isolated from cells by a very gentle procedure, and exposed to a highly sensitive photographic film. The decays of the radioactive atoms incorporated into DNA produced a series of dots in the developed film, allowing the path of the replicating DNA to be inferred by mentally "connecting the dots."

In one type of experiment, Cairns exposed cells to radioactive thymidine for a very short time (a "pulse" label). Since the labeling period was much less than the time needed to replicate all of the bacterial DNA, this experiment followed events at the point of

FIGURE 4–3. Cairns added radioactive thymidine, a DNA precursor, to growing cells for a short time to label newly synthesized DNA, broke the cells open gently, fixed their contents in place, and exposed it all to sensitive photographic film. The radioactive DNA emitted energy, which was visualized as dots when the film was developed. The experimental data (*left side*) demonstrated that DNA is synthesized using a symmetric ("fork") mechanism in which both strands elongate in the same direction at about the same time. DNA is synthesized continuously from the same fork, as the longer labeling period (*b*) gives a longer fork than the shorter labeling period (*a*). The right side of the figure redraws the data as double-stranded DNA, showing the nonradioactive parental strand as a continuous line. An arrow indicates the overall direction of the progress of the replication fork.

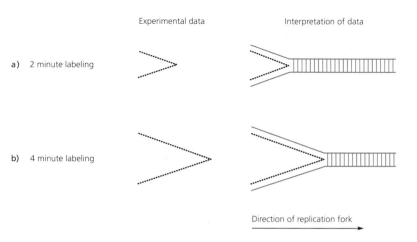

Replication occurs by a symmetric mechanism

Experimental data   Interpretation of data

a)   2 minute labeling

b)   4 minute labeling

Direction of replication fork

DNA synthesis: conversion of nonradioactive parental DNA to radioactive newly replicated DNA (which has one radioactive and one nonradioactive strand). The dots on the film (represented schematically by the dots in Figure 4–3, left) show the path of the newly replicated radioactive strand. Although the nonradioactive template DNA strand is not seen on the film, its position can be inferred from the properties of Watson-Crick base pairing (Figure 4–3, right). Two major conclusions follow. First, DNA replication is a sequential, continuous process because a four-minute labeling time gives a replicated stretch twice as long as the two-minute labeling time. Second, replication occurs by a "fork" or symmetric mechanism in which each of the two parental strands is replicated in the same direction at about the same time. (Note that, in the Cairns experiment, each dot on the film is some ten thousand bases in DNA, so the conclusions do not pertain to the "fine-structure" biochemistry of DNA replication.)

In a second type of experiment, Cairns visualized the structure of the entire replicating chromosome. He grew the cells in radioactive thymidine for a long enough time to completely label both of the DNA strands and then shifted to growth in nonradioactive thymidine so that the newly replicated DNA would have one radioactive and one nonradioactive strand. In this experiment, the dot density along the newly replicated DNA (one radioactive strand) is half that

Experimental data   Interpretation of data

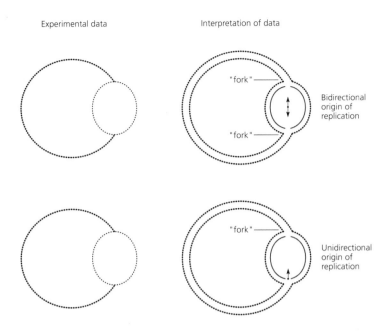

FIGURE 4–4.   To observe the number of junctions between newly replicated and old DNA, cells growing for a long time in radioactive thymidine were shifted to nonradioactive medium for a short time. Newly replicated DNA (one radioactive strand) can be distinguished from old DNA (two radioactive strands) because newly replicated DNA produces only half of the dot-density of old DNA upon exposure to photographic film. The experimental data (*left side*) demonstrates that there are only two junctions between old and newly replicating DNA. The right side of the figure redraws the data as double-strand DNA, showing the nonradioactive newly replicated strand as a continuous line. The two DNA junctions (forks) could have arisen from a single bidirectional origin between the forks (double arrow), or from a unidirectional origin (single arrow). Only one of the two possible undirectional origins is shown.

of the parental nonreplicated DNA (two radioactive strands). This allowed Cairns to "see" each juncture between the two DNA types and count the number of replication forks in the entire bacterial chromosome, provided that there were any fortunate cases in which the DNA was not broken during preparation or hopelessly tangled. Anyone other than Cairns would have considered it a technical impossibility to visualize the entire *E. coli* chromosome. It had been extraordinarily difficult to prepare unbroken DNA from phage T4, which has a genome that is much smaller than that of *E. coli*. What would be the chance of obtaining the unbroken *E. coli* genome? Even if some unbroken *E. coli* chromosomes could be extracted from gently broken cells, the chance seemed remote that they would display themselves in untangled form. While waiting two months for radioactive decays to give him his answer, Cairns tried to guess the probability of success by looking for untangled nets washed up on the beach—he never found any.

Against all odds, the experiment worked. Some rare complete genomes could be seen, and the number of replicating points could be determined. This experiment allowed the remarkable conclusion that there are very few replicating points on the entire chromosome

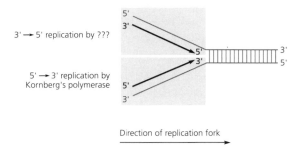

3' → 5' replication by ???

5' → 3' replication by
Kornberg's polymerase

Direction of replication fork

FIGURE 4–5. The Cairns experiment indicated that both DNA strands are replicated symmetrically—in the same direction at about the same time. Kornberg's polymerase can replicate DNA only in the 5' to 3' direction. How is replication accomplished in the opposite direction (3' to 5')? Newly replicated DNA is shown as a bold arrow.

of *E. coli*. (Figure 4–4, left). As can be seen, two possibilities cannot be distinguished: an origin between the two forks and replication in both directions (bidirectional replication), or a single origin of replication at one of the forks and replication in one direction (unidirectional replication) (Figure 4–4, right).

Cairns guessed that replication was unidirectional, and a rush of experiments with other techniques seemed to back up his guess. However, Ross Inman and Maria Schnös used a clever experiment with electron microscopy to show that the bacterial virus λ showed bidirectional replication. In turn, this report triggered a second rush of experiments on the *E. coli* chromosome, which resulted in the conclusion that replication is bidirectional for bacteria as well. Nature tends to be rather unkind to scientists who do experiments with a conclusion too firmly in mind.

Cairns' experiments, published in 1963, defined the problems for the next 20 years of biochemical study. Symmetric replication demanded that the two DNA chains be copied in the same direction at nearly the same time. This conclusion posed a serious biochemical problem. Recall that the two DNA strands in duplex DNA run in opposite directions (which are designated by the chemical structures associated with each end): one strand has a 5' to 3' direction and the other has a 3' to 5' direction (see Figure 1–6). Kornberg's DNA polymerase copied DNA in only one direction (5' to 3') by adding a nucleotide to the 3' end of the growing DNA chain. Elongation of the other strand in the same direction requires adding a nucleotide to the 5' end of the growing DNA chain. Did cells use a different polymerase to replicate this strand, or was there a yet undiscovered DNA polymerase that could carry out both types of reactions (Figure 4–5)?

A second critical challenge for the biochemist was the nature of the origin, the site where chromosomal DNA replication initiates. How is a site identified that starts replication once and only once each cell division? What molecular events at this site told the polymerases to start their multimillion base journey around the bacterial genome? How was this immense stretch of DNA unwound and the replicated molecules segregated? Clearly, DNA replication at the genome level was not an easy problem.

THE GENETICS OF REPLICATION

As for all of molecular biology, the science of genetics provided the framework for the eventual solution of the mysteries of DNA replication. Mutations can identify the genes required for a particular process and can also provide some insight into the normal function of the protein product. By identifying a particular protein with a specific biological function, mutations correlate biochemistry with biology.

Genetic analysis begins with the isolation of mutations that block the process under study by inactivating a protein essential for that process. But mutations blocking bacterial DNA replication posed a particular problem—they would kill the cell. The problem of studying lethal mutations was solved by isolating "conditional-lethal" DNA synthesis mutations in which replication proceeded normally at lower temperature but stopped at elevated temperature. Cells could then be grown at low temperature, and the consequences of the mutation analyzed at high temperature. These *dna*(ts) mutations (*ts* for temperature-sensitive) had the added advantage that they might be expected to produce a protein that was temperature-sensitive when used for replication studies in the test-tube. Temperature-sensitive replication in the test-tube reaction provides direct proof that the protein in question is the product of a gene directing DNA replication.

The isolation of *dna*(ts) mutations was not easy because mutations in many other activities of the cell besides DNA replication are lethal. However, the most enjoyable part of genetics is devising clever selections to pick out the desired class of mutations. One approach for *dna*(ts) mutations, devised by Walt Fangman, was a "suicide" approach that killed normal cells able to replicate DNA at high temperature. When Fangman raised the temperature of a growing culture of *E. coli* bacteria, he added extremely radioactive thymidine to the medium. Normally replicating bacteria incorporate this radioactive thymidine into DNA. When the bacteria were removed from growth medium and stored to allow the radioactive thymidine to "decay," the strands of the DNA duplex of normally replicating bacteria were broken and the bacteria were killed. Because the *dnats* mutant bacteria stopped replication at high temperature, they incorporated very little radioactive thymidine and were spared from killing. In this way, the *dnats* bacteria were selected as survivors of this "radioactive suicide" experiment.

The immediate impact of the genetic work was confusion—there seemed to be a depressingly large number of genes! The genes for DNA replication were given alphabetical names: *dnaA*, *dnaB*, *dnaC*, etc. As the letters poured in, some scientists started from the other end with *dnaZ*, trying to avoid the problem of two people choosing the same letter for different genes. After a few years of sorting things out, some mutations were found to affect other things, and the picture brightened. An important simplification was the division of mutations into those that stop replicating their DNA rapidly after temperature upshift (the "quick stop" mutants) and those that ceased replication only after prolonged incubation at high temperature (the "slow stop" mutants). If a gene had only quick stop mutations, the protein product was likely to be required for elongation of DNA chains—as soon as the temperature was raised, the *ts* protein would be inactivated and replication would stop. If all mutations in a gene were slow stop, the protein product was likely to be required only during initiation. Inactivation of such proteins would affect replication only after each chromosome finished its round of replication and prepared to start a new round of replication from the origin region. In this way, *dnaA* and *dnaC* were defined as initiation genes, and *dnaB*, *dnaE*, *dnaG*, *dnaN*, *dnaX*, and *dnaZ* as elongation genes (Figure 4–6).

None of the *dna*(ts) mutations affected the activity of the Kornberg DNA polymerase. Cairns was disturbed by this finding as well as by the fact that the Kornberg polymerase copied in only one of the two directions (5′ to 3′) required for overall genome duplication. Simplicity and elegance seemed to demand a more sophisticated enzyme that would duplicate both DNA strands in the same direction at the same time. Cairns therefore questioned whether the Kornberg polymerase was the central replicative enzyme.

Cairns tried a bold experiment—a "brute-force" screen for a bacterial cell with a mutation in the gene for Kornberg's

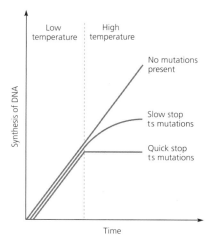

Temperature sensitive mutants defective in DNA synthesis can be divided into two classes

FIGURE 4–6.   *dnats* mutants growing in radioactive thymidine at low temperature were shifted to high temperature. By measuring the amount of radioactivity incorporated into DNA after this temperature shift, it was obvious that some mutants stopped DNA synthesis immediately ("quick stop" mutants), whereas others continued DNA synthesis for a while ("slow stop" mutants). Nonmutant cells continue DNA synthesis at high temperature.

polymerase, so that its contribution to replication could be determined. This experiment was technically feasible because of the availability of potent mutagenic agents that would mutate every *E. coli* cell at least once. Because *E. coli* has about 4,000 genes, roughly 1 in 4,000 cells should have a mutation in any given gene after mutagenesis. Paula DeLucia in Cairns' lab laboriously checked each of several thousand survivors of such a mutagenic treatment for the enzyme activity of the Kornberg DNA polymerase; she found one mutant defective in this activity. The gene identified by the mutation was termed *polA* (for polymerase and Paula), and Kornberg's polymerase was renamed Pol I. The *polA⁻* mutant grew normally but was more sensitive than normal cells to the lethal effects of UV irradiation, a DNA-damaging agent. The initial guess from this remarkable experiment has turned out to be largely correct. The Kornberg polymerase is a vital cellular enzyme but is not used for elongation of chromosomal DNA. Instead, it is used for replicating short stretches of DNA, such as would occur when damaged DNA is repaired.

By 1970, the field of DNA replication seemed to be going backward. Of the many genes required for DNA replication in the cell, none had a defined biochemical activity (such as polymerase) in a test-tube reaction. The only defined biochemical activity (Pol I) seemed not to be essential for chromosomal DNA replication. The light at the end of the tunnel seen earlier was apparently only an oncoming train. However, a period of remarkable progress was about to begin.

Major advances in science are sometimes portrayed as colossal intellectual leaps surmounting hopeless paradoxes and sometimes as especially useful insights derived from continuous accumulation of information by careful testing of logical hypotheses. Our understanding of molecular biology (and I suspect science in general) derives more often from a series of semi-logical jumps. A set of experiments is carried out within a given conceptual framework (model), and the results of these experiments pose some problems for the existing viewpoint. The model can be changed slightly, or a new framework introduced. There is rarely a clear paradox that demands something totally new. Generally, most scientists in a field prefer to continue within an old but troubled framework, partly because the basic conservatism of scientific culture provokes a scornful reaction to a new idea that turns out to be wrong. This is unfortunate for both science and the scientists. The greatest joy of science is in its creativity, and the most rapid progress comes when the clash of ideas defines the issues and generates new experiments.

The isolation of the *polA⁻* mutant introduced such a time in the study of DNA replication. The search for the mutation was motivated by the notion that the "real" replication enzyme should be vastly more sophisticated than the Kornberg polymerase—it should know how to start DNA chains and add nucleotides to both the 3′ and 5′ ends of the DNA at the same time. In fact, the real replication enzyme (DNA polymerase III, or Pol III) also replicates solely in the 5′ to 3′ direction and does not start chains. In the end, Kornberg was wrong in thinking that Pol I was the enzyme that replicated the genome, and Cairns was incorrect in his assumptions about how replication ought to occur (simplicity and elegance failed in this case). But, of course, the important points for molecular biology are that Pol I defined how an enzyme that copies genetic information works, and the problems posed by Cairns and his mutant produced an exciting era of new experiments.

THE FINE STRUCTURE OF DNA REPLICATION

The experiments on genome duplication by Cairns showed that DNA replication is symmetric—each parental DNA strand is copied in the same direction at nearly the same time. This posed a clear biochemical problem. If the Kornberg polymerase (or a similar enzyme) elongated in the 5′ to 3′ direction by adding to the 3′ end, then what enzyme elongates in the 3′ to 5′ direction by adding to the 5′ end? There were three possible solutions: another polymerase for the 5′ end; the same polymerase for both ends with some fancy biochemistry; or elongation only in the 5′ to 3′ direction. In this last scenario, one strand of the DNA would have to be synthesized in small pieces in a direction opposite to the direction of overall chain growth and then joined together ("discontinuous replication") (Figure 4–7).

To distinguish these hypotheses, a higher resolution picture of DNA replication was required than the one provided by the Cairns experiments. In 1968, Reiji Okazaki carried out experiments that provided the necessary resolution. His experiments said that the discontinuous mechanism was probably correct. Okazaki used radioactive thymidine to label DNA, as had Cairns, but he used several experimental strategies that allowed him to see the intermediate stages in the very rapid replication process. He used a very short

Discontinuous replication allows a polymerase with only 5′ to 3′ activity to work symmetrically

Discontinuous replication

Direction of replication fork

FIGURE 4–7. In the discontinuous replication model, one DNA strand is elongated continuously while the other strand is elongated for short distances (discontinuously). This allows a single polymerase with only 5′ to 3′ activity to replicate both strands. Although the small pieces are synthesized in a direction opposite to that of the replication fork, joining the fragments together results in overall chain growth in the same direction for both strands.

Reiji Okazaki

FIGURE 4–8. Cells were "pulse" labeled with radioactive thymidine for a very short time and either harvested immediately or "chased" by growing in a nonradioactive medium for a short time before harvest. Strands of the DNA duplex were separated and the mixture sedimented to separate the strands by size. When replication is analyzed immediately after the pulse, both small and large pieces of radioactive DNA have been synthesized, as would be expected if one DNA strand were synthesized continuously and the other discontinuously (a, left). When replication is analyzed after the chase, all of the previously synthesized radioactive DNA is long, as would be expected if the small pieces of DNA had been joined together. Note that, in this experiment, the DNA synthesized during the chase is not observed because it is nonradioactive (b, left). An interpretation of the data is shown on the right side of each figure.

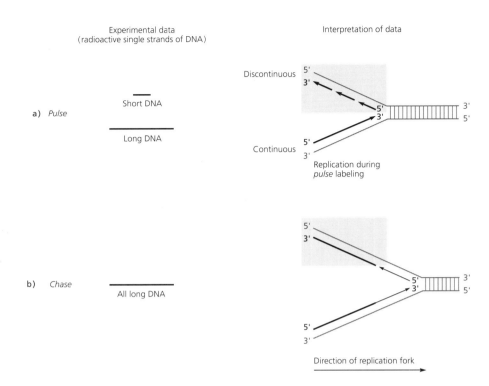

labeling time. He slowed replication by incubating cells at low temperature (at normal temperatures, 1,000 nucleotides per second are polymerized into DNA). Finally, he used a sedimentation technique that is a sensitive way of detecting short single-strand DNA chains. After a very short labeling period, Okazaki found that the radioactive DNA was detected as short DNA chains. When the short labeling period was followed by a time in nonradioactive thymidine, the radioactive DNA was now in long pieces. This observation indicated that the short pieces of radioactive DNA had been joined into a longer DNA chain, as expected if the short DNA was an intermediate stage in the production of the continuous DNA chain that shows overall growth in the 5′ direction (Figure 4–8). At about the same time, a DNA-joining enzyme called DNA ligase was discovered independently by Marty Gellert, Bob Lehman, Jerry Hurwitz, and Charles Richardson. This enzyme was a candidate for stitching together the DNA segments into a continuous chain. Because it can join preformed pieces of DNA, DNA ligase later became a key enzyme in genetic engineering.

Initially, the interpretation of the Okazaki experiments was controversial. The short DNA might have come from DNA undergoing repair (some does) or even have been generated by the experimental procedure itself if the newly replicated DNA was very fragile. Unfortunately, Okazaki did not live to see the eventual biochemical proof of his picture of replication; he died from leukemia in 1975 at the age of 42, most likely a late victim of the atom bomb attack on his native Hiroshima during World War II.

Okazaki's experiments answered the puzzle of symmetric growth of DNA chains and also brought into sharp focus another problem of DNA polymerase—how does it start DNA chains? The Kornberg polymerase could only add onto the end of a preexisting DNA chain (or primer); it did not start synthesis "do novo." Starting a chain was clearly a problem at the initiation of chromosomal replication. Had this been an event occurring only at the beginning of replication, then some special complicated mechanism could be invoked. However, if one of the DNA strands replicated by a discontinuous mechanism, then initiation of DNA synthesis must occur throughout elongation. A mechanism for starting new DNA chains must be part of the ongoing replication process.

The thinking of the time was constrained by the notion that DNA polymerase must do everything, and a number of complicated models were proposed. The answer turned out to be very simple (in hindsight)—a different enzyme makes the primer, and DNA polymerase elongates from the primer. The primer-synthesizing enzyme is an RNA polymerase, which provides a clear separation between the biochemistry of the priming reaction and that of the elongation reaction. Kornberg and his student Doug Brutlag provided the critical insight and first experiments in 1971.

Kornberg, struggling with the vexing biochemical problem of how to start DNA chains, became intrigued with the fact that RNA polymerase of *E. coli* was able to start RNA chains. What if RNA polymerase started the DNA chains as well? Kornberg and Brutlag decided to test the idea with a small, single-strand DNA phage named M13. The earliest stage of M13 replication was the conversion of the parental single-strand DNA (SS) into a double-strand form (DS). In a few weeks of phenomenal success, the SS to DS conversion was demonstrated first in infected cells, then in a crude extract made from infected cells, and finally in a soluble enzyme fraction to which M13 DNA was added. For the first time, a system for studying replication in a soluble extract had been developed. The

FIGURE 4–9.   An *E. coli* extract replicates M13 DNA by converting the single-stranded (SS) phage DNA to a double stranded (DS) DNA duplex. When RNA polymerase function was inhibited with the drug rifampicin, no synthesis of DS DNA was observed, demonstrating that an RNA chain made by RNA polymerase is required for replication.

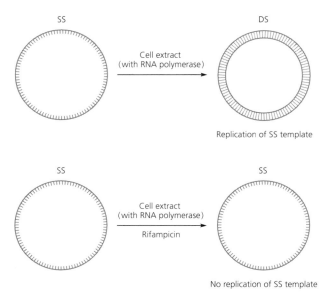

critical experiment was to ask whether blocking the action of RNA polymerase with the antibiotic drug rifampicin would prevent the SS to DS replicative conversion (Figure 4–9). The answer was yes! DNA synthesis started from an RNA primer.

The conceptual birth of RNA priming required good fortune as well as exceptional insight. Although RNA priming is a general route for starting DNA chains, not all replicons use *E. coli* RNA polymerase to make the primer. A special enzyme for primer formation, DNA primase, is used to initiate DNA replication for other single-strand and double-strand DNA phages and for *E. coli*.

The concepts of discontinuous replication and RNA priming defined the probable fine structure of ongoing DNA replication. DNA polymerase adds nucleotides only to the 3′ end of a growing chain. The chain growing in the 5′ to 3′ direction, called the leading strand, is elongated continuously from a single RNA primer. The chain growing in the 3′ to 5′ direction, or "lagging strand," is elongated discontinuously. For the lagging strand, RNA primers are elongated for a short distance from their 3′ end by DNA polymerase. The primers are then removed by a degradative enzyme acting from the 5′ end (5′ exonuclease). To complete the lagging strand synthesis, gaps in the DNA strands are filled in and the DNA segments joined together by DNA ligase. Although the detailed mechanisms differ for each strand, both strands grow symmetrically in the same direction overall as defined by the Cairns experiments (Figure 4–10).

THE "REAL" REPLICATION ENZYME IS JUST LIKE POL I

Cairn's isolation of the mutant defective in the Kornberg polymerase (Pol I) opened a clearly defined biochemical route to isolation of the polymerase responsible for genome duplication. With the most active polymerase gone from mutant extracts, it now seemed possible to isolate the replicative polymerase. Not one but two new DNA polymerases were found (Pol II and Pol III). But which one was responsible for genome duplication? At this point, genetics and biochemistry finally came together for DNA replication. Malcolm Gefter and Tom Kornberg measured the activity of Pol II and Pol III in extracts from each of the *dna*(ts) mutants. Two *dna*(ts) mutants, both having mutations in the *dnaE* gene, were defective in Pol III activity at high temperature. None of the *dna*(ts) mutants were defective in Pol II. The conclusion was clear: Pol III, specified by *dnaE*, is required for genome duplication. This experiment provided the crucial connection between biology and biochemistry.

The association of another Kornberg with another DNA polymerase was the sort of historical quirk that would be considered too unlikely for a fictional description. Arthur's son Tom was at the Julliard School of Music, seemingly headed toward a career as a cellist, when an injury to his hand nudged him toward biological science. Tom Kornberg began to work with Gefter, a beginning professor at Columbia, on a challenge. Gefter was lecturing on the importance of genetics in relating biochemistry to biology. He cited Pol I as an example in which the wrong enzyme had been associated with chromosomal DNA replication because of the absence of a genetic analysis. Coming to his father's defense, Tom Kornberg told Gefter that he was wrong; Gefter responded by offering Tom the chance to work in his lab and find out the answer directly.

After establishing the biological importance of Pol III, Kornberg and Gefter finished characterizing the enzyme. Their somewhat surprising conclusion was that Pol III is remarkably similar to Pol I. Like Pol I (and Pol II), Pol III requires a template DNA strand and a primer and uses the high energy deoxynucleotides containing three phosphates (dATP, dGTP, dCTP, and dTTP) as substrates. It makes

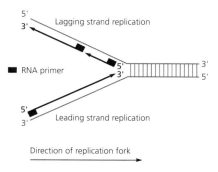

DNA replication utilizes different mechanisms for leading and lagging strand synthesis

FIGURE 4–10.   Whereas the entire leading strand is replicated continuously from a single RNA primer, the lagging strand is replicated discontinuously for a short distance from many RNA primers. To complete lagging strand synthesis, RNA primers are removed, the gaps in DNA are filled in, and DNA segments are ligated together.

FIGURE 4–11.   Like Pol I, the DNA polymerase Pol III adds nucleotides only to the 3' end of a growing DNA chain, allowing replication only in the 5' to 3' direction. Since replication of the *E. coli* chromosome is accomplished by Pol III, this process must make allowances for discontinuous replication.

new DNA by incorporating a deoxynucleotide with one phosphate exclusively at the 3' end and releasing the other two phosphates (Figure 4–11). Also like Pol I, Pol III has the ability to remove nucleotides from the 3' end of DNA (3' exonuclease activity). At first sight, this 3' exonuclease activity seems silly because it reverses the effects of the polymerase. However, the exonuclease preferentially removes incorrectly inserted bases and therefore corrects mistakes in polymerization (the polymerase gets another chance to do its job correctly). This "editing function" is important for lowering the frequency of mutations caused by incorporation of incorrect bases.

The only fundamental enzymatic difference between Pol I and Pol III is that only Pol I has a 5' exonuclease activity. An interesting historical footnote is that, though the Pol I defective mutant isolated by Cairns lacked polymerase activity, the 5' exonuclease activity of Pol I was intact. Moreover, Pol I mutants that lack 5' exonuclease activity are dead. The 5' exonuclease activity of Pol I probably plays an essential role in genome replication by removing RNA primers from the discontinuous DNA chains.

The properties of Pol III left the biochemical mysteries of chromosomal DNA replication unchanged from the Pol I era. However, the association of Pol III with the *dnaE* gene for DNA replication made two critical points: discontinuous replication to accommodate Pol III was probably the way of the biological world; and an understanding of chromosomal replication would require determining the role of many proteins in this complex process—the products of the lengthy list of *dna* genes identified by temperature-sensitive mutants.

A SIMPLE TASK THAT NEEDS MANY PROTEINS:
PRIMING AND ELONGATING

To reconstruct DNA replication, there was a clear need to fractionate the cell contents. If replication required a complex structure associated with the cell membrane, as many thought, then the

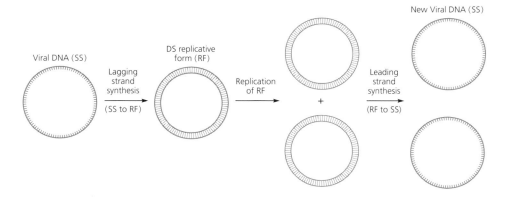

FIGURE 4–12.   The general replication cycle of single-strand phages is illustrated. By a happy coincidence, the conversion of SS viral DNA to the DS replicative form (RF) uses the proteins involved in lagging strand synthesis, and the conversion of RF to SS uses the proteins involved in leading strand synthesis. By studying these two simple processes, the functions of many proteins involved in DNA replication were identified.

starting point should be broken cells with everything still present. Yet if the soluble "soup" of free enzymes would do the job, then the best starting point would be to program such an extract with free DNA, presumably viral DNA that could be prepared in intact form. The intermediate approach, which kept all options open, was to infect cells with virus, make an extract of the infected cells, and assay for DNA replication. Initially, the proper approach was not obvious. Brutlag and Kornberg's success in programming a soluble "soup" with single strand phage DNA made that the system of choice (see Figure 4–9). Kornberg would no longer have to "waste clean thoughts and clean enzymes on dirty DNA."

The first single-strand DNA phage to be identified was φX174. Much to his surprise, Robert Sinsheimer found that φX174 was not only single-strand but also circular; this was unexpected because other viral DNAs were linear. He then went on to show that this virus replicated by an interesting sequential mechanism. The parental single-strand (SS) DNA is copied to a double-strand (DS) form, which then replicates to make more double-strand circles, termed RF, for replicative form. The RF circles serve as templates for a different mode of replication (asymmetric replication), giving rise to new viral single strands that are packaged into phage (Figure 4–12).

The single-strand phages had generally been considered interesting oddities, decidedly not part of mainstream molecular biology. However, the world of single-strand DNA phage turned out to have just the right amount of biological diversity to help

Jerry Hurwitz

biochemists resolve the complex mixture of replication proteins. Three single-strand phages—M13, φX174, and G4—each played a unique role in understanding of DNA replication. Each stage of their life cycle provided a model for a different aspect of replication. The SS to RF reaction provided a model for lagging strand DNA synthesis (discontinuous chains), and the RF to SS reaction provided a model for leading strand synthesis (continuous chains).

The door was now open for a frontal attack on DNA replication. The Kornberg group set out to purify the proteins required for the SS to RF conversion by M13, φX174, and G4. The quest was joined by Jerry Hurwitz, a former faculty associate of Kornberg's, who had become interested in DNA replication. The first experiments on φX174 gave spectacular results. Not only did cell extracts carry out SS to RF replication of φX174, but this reaction failed at a high temperature in an extract from various *dna*(ts) strains. In experiments published within two months of each other, Randy Schekman, Bill Wickner, and Kornberg showed that φX174 replication depended on the *dnaB* gene product; Reed Wickner, Sue Wickner, and Hurwitz showed that it depended on the *dnaC, dnaE,* and *dnaG* gene products. Suddenly, there was hope of identifying and understanding the "alphabet soup" of DNA replication enzymes identified by the various *dna*(ts) mutations.

The involvement of three different Wickners in this story has confused many people over the years. Reed and Sue in the Hurwitz lab were the brother and sister-in-law of Bill in the Kornberg lab. Although the two groups were competing, the three talked all the time, sharing ideas, procedures, and results, undoubtedly to the mutual benefit of both groups. Reed Wickner soon moved on to his own position at the NIH, and Sue Wickner undertood the task of protein purification in a tense race with the Kornberg group. As with all fierce competitions in science, the immediate problem of replication of single-strand phage was largely solved in five years, and the people involved went in separate directions to attack new unresolved problems.

The alphabet soup programmed with single-strand DNA was a biochemist's heaven—a zoo of fascinating enzymes, with an assay to allow their purification free from other proteins. The fractionation of the individual proteins was not easy, however, because many replication proteins are required, and there are not many copies of each protein in a bacterial cell.

Fractionation identified the proteins required for DNA replication

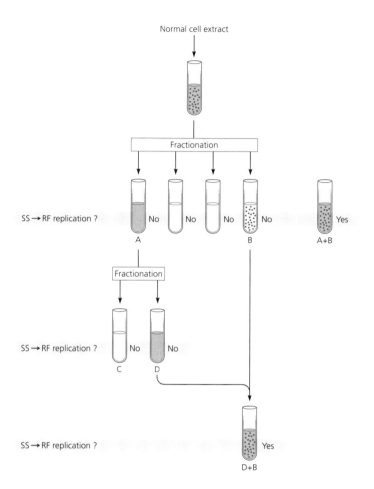

FIGURE 4–13. A cell extract capable of the SS to RF conversion was fractionated to identify the activities required for this process. Although no single fraction could support replication, when two fractions were combined, the reaction proceeded. The fractionation process was repeated until the activity could no longer be subdivided and a single protein or protein complex was present in the tube. This strategy allowed purification of all the components required for DNA replication.

Two fractionation approaches were used for these and other complex biochemical systems. The classical biochemical method involved separation of the extract into different fractions, each of which was required for DNA replication (Figure 4–13).

A second approach depended on using an extract from a *dna*(ts) mutant to provide all of the enzymes except the one defective in the mutant. The missing enzyme was supplied by a fraction taken from a normal cell extract (Figure 4–14). This is called a "complementation assay" by analogy to the standard genetic procedure in which a mutant function is supplied in cells by a normal gene (compare Figure 4–14 and Figure 3–9). The complementation assay is easier for the

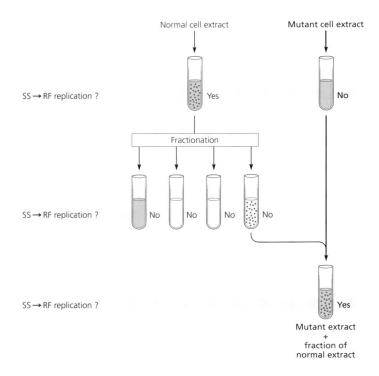

FIGURE 4–14. A mutant cell extract made from a *dna*(ts) strain is defective in converting SS to RF phage. This extract was "complemented" by fractions from a wild-type cell extract in order to identify the protein missing or defective in the mutant extract.

purification of an individual protein but suffers from the problem that determining the special role of that protein in DNA replication may be difficult (because the whole complex extract is required). In addition, the complementation assay can be used only for proteins for which a mutant gene is available. In practice, both approaches were used to provide the eventual "clean" biochemical systems with pure proteins.

As fractionation of the alphabet soup proceeded in the two laboratories, the biochemical separation between the process of priming DNA synthesis and that of DNA chain elongation became defined. As we discussed earlier, the first experiments on conversion of M13 SS to DS led to the immediate discovery that its RNA primer is made by RNA polymerase (see Figure 4–9). Because M13 bypassed the normal priming event, it became the system of choice to identify the proteins required for elongation from the alphabet soup. In contrast, study of G4 and ϕX174 resulted in an understanding of the priming event that starts DNA synthesis.

FIGURE 4–15.   When SSB binds to SS M13 DNA, it leaves a single hairpin exposed that serves as a binding site for RNA polymerase. RNA polymerase synthesizes an RNA primer, which is then elongated by Pol III "holoenzyme." The Pol III holoenzyme contains Pol III and many associated proteins. All of the holoenzyme components were identified by studying this reaction.

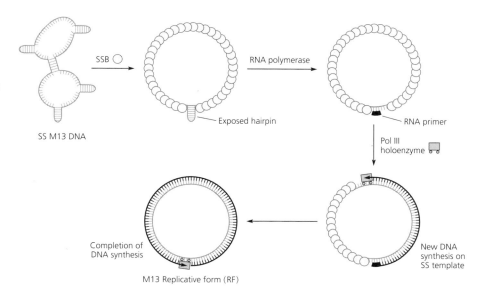

SS M13 DNA

SSB ◯

Exposed hairpin

RNA polymerase

RNA primer

Pol III holoenzyme

New DNA synthesis on SS template

Completion of DNA synthesis

M13 Replicative form (RF)

For the SS to RF conversion, M13 DNA requires only RNA polymerase, the DNA elongation proteins purified from the alphabet soup and one additional protein—a single-strand binding protein (SSB for short) (Figure 4–15). SSB helps replication by binding to single-strand DNA, thereby preventing complementary regions from pairing with each other. When SSB coats M13 DNA, it leaves one stretch of the M13 DNA as a double-strand "hairpin." This hairpin is recognized as a binding site by RNA polymerase. RNA polymerase makes the RNA chain that is used as a primer for subsequent DNA replication. The DNA elongation proteins, part of a large multi-protein ensemble called Pol III holoenzyme (much larger than the original minimal replication enzyme Pol III), carry out elongation. The additional proteins in holoenzyme cause the replication enzyme to recognize primers efficiently and to copy very long stretches of DNA without leaving the template, a property called processivity. The protein components (subunits) of Pol III holoenzyme have been given Greek letter names. This tradition for such ensembles may have been meant to show scholarship but mostly confuses typists and typesetters. So Pol III holoenzyme is an ensemble of at least α, β, γ, δ, δ', ε, θ, and τ (not alphabetical, but only theoretical physicists can write the letters that directly follow

FIGURE 4–16.   When SSB binds to SS G4 DNA, it leaves a single hairpin that serves as a binding site for DnaG primase. DnaG primase synthesizes an RNA primer, which is then elongated by Pol III "holoenzyme." This reaction identified DnaG as a cellular "primase," an enzyme that makes short RNA chains used to initiate replication.

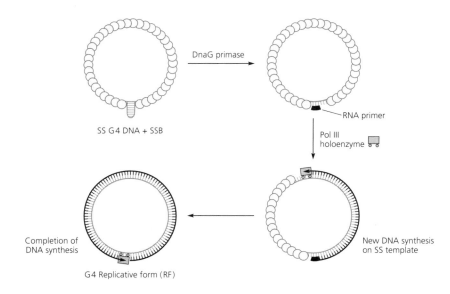

ε). The alphabet soup genes, *dnaE*, *dnaN*, *dnaQ*, *dnaX*, and *dnaZ* code for five of these subunits.

When coated with SSB, G4 DNA requires only the DnaG protein and the chain elongation machinery provided by Pol III holoenzyme to carry out the SS to RF conversion. This identifies DnaG as a priming protein. DnaG, the DNA primase, binds to a special site on G4 DNA left by SSB and synthesizes short RNA primers when given ribonucleoside triphosphates. These primers are then elongated by Pol III holoenzyme (Figure 4–16). DnaG is also the primase used in replication of the *E. coli* chromosome.

In addition to the DnaG primase, φX174 DNA needs the DnaB and DnaC proteins and four additional proteins not represented by alphabet genes to carry out the priming reaction. We now know that these cellular proteins are used to restart replication when the replication fork is stalled by damage to DNA. This collection of proteins builds the φX174 priming structure to which DnaG primase binds. As for the other cases, the RNA primer synthesized by DnaG is elongated by Pol III holoenzyme (Figure 4–17).

In sum, the study of SS to RF replication by single-strand phage provided a rather pleasing conceptual picture. The DNA prepared by SSB becomes a target for a priming enzyme, and the primer is

FIGURE 4–17. φX174 uses holoenzyme to synthesize RF from SS but requires a number of proteins for RNA priming. In addition to DnaG primase, φX174 requires DnaB, DnaC, and four other proteins in order to synthesize the RNA primer used to start DNA synthesis.

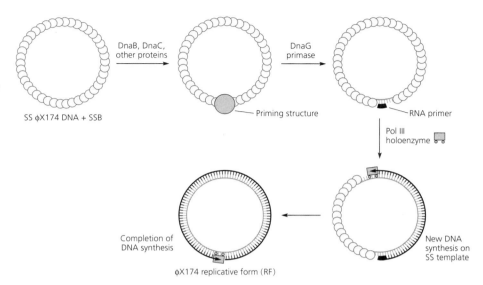

SS φX174 DNA + SSB

DnaB, DnaC, other proteins

Priming structure

DnaG primase

RNA primer

Pol III holoenzyme

New DNA synthesis on SS template

Completion of DNA synthesis

φX174 replicative form (RF)

elongated rapidly and continuously by a DNA polymerase with a special class of associated proteins ("processivity factors") that keep the polymerase from wandering off the replicating DNA.

SPECIAL PROTEINS NEEDED FOR GENOME REPLICATION: STRETCHING, UNWINDING, UNTWISTING

Besides the proteins required for priming and elongation, several other proteins are needed to prepare DNA for efficient replication, without themselves participating in the synthesis of DNA.

The first of these proteins to be identified was the single-strand binding protein, SSB. The discovery of the protein that binds to single-stands of DNA and its early understanding came from studies of phage T4 by Bruce Alberts, an exceptionally innovative molecular biologist both conceptually and technically. Alberts devised a simple way to identify proteins that operate on DNA: he passed a cell extract through a solid matrix to which DNA had been attached and isolated proteins that bind tightly to the DNA. He found a protein produced in T4-infected cells that stuck like glue to single-strand DNA. This SSB is the product of gene 32 of phage T4. It binds so tightly to single strands of DNA that it actually converts

Bruce Alberts

FIGURE 4–18.   In the vicinity of the replicating fork, DNA is predominantly single stranded but also has short DNA duplexes arising from fortuitous base-pairing. The single-strand binding protein, SSB, binds both to single-strand DNA and to short DNA duplexes and hairpins, converting the entire region to a stretched out single strand of DNA that is easy for Pol III to replicate.

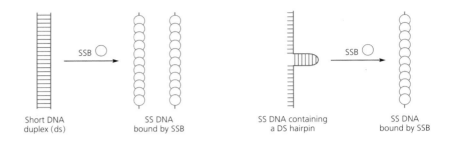

Short DNA duplex (ds)   SS DNA bound by SSB   SS DNA containing a DS hairpin   SS DNA bound by SSB

double-strand DNA to single strands under certain conditions in the test tube. The reason for this surprising reaction is that SSB binds better to single-strand DNA than the two DNA strands bind to each other. Part of the success of SSB in separating the DNA strands comes from its preference to bind next to its neighbor to form a long array of such proteins, a process termed cooperative binding (Figure 4–18, left).

In the cellular environment, double-strand DNA is more stable and SSB binding is not so cooperative. So, under cellular conditions, SSB does not separate the long duplex strands of the genome and convert them to single-strand DNA. However, SSB does bind to single-strand regions of DNA and to short DNA duplexes (which are less stable than long duplexes), both of which are found near the replication fork. This prevents complementary regions of the SS DNA from pairing and binding to each other and converts the DNA to a stretched, single-strand form (see Figure 4–18). This "DNA-stretching" property is presumably the contribution of SSB to DNA replication. DNA polymerases are very good at copying single strands (even those coated by SSB) but work poorly if they encounter even short stretches of double-strand regions. After the importance of SSB was established for T4 replication, the *E. coli* version of SSB was discovered. As we have seen, the bacterial SSB turned up as one of the protein fractions required for efficient replication of single-strand M13 DNA (see Figure 4–15).

The problem of unwinding the DNA duplex became apparent as soon as Watson and Crick suggested that replication might proceed by each of the two strands serving as the template for the other one. There are really two problems. First, the long DNA duplex is very stable because of the multitude of base pairs, and so a great deal of

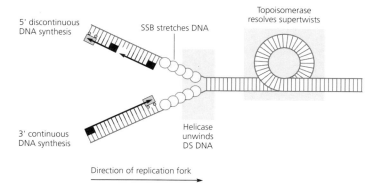

5' discontinuous
DNA synthesis

SSB stretches DNA

Topoisomerase
resolves supertwists

3' continuous
DNA synthesis

Helicase
unwinds
DS DNA

Direction of replication fork

FIGURE 4–19.   In addition to SSB, two proteins facilitate the smooth and rapid passage of DNA polymerase. "Helicases" unwind the duplex DNA, creating the single-stranded regions to which SSB binds. "Topoisomerases" untwist DNA so that polymerase can continue to move forward.

energy is required to separate long stretches of double helix. Second, as the strands are pulled apart, the DNA ahead of the unwinding point will twist up ("supertwist"). The reason for the supertwists is that the two DNA strands are interwound, and the number of windings cannot change when the ends of the two strands are not free to rotate. This is the case for bacteria and many phage whose ends are joined in a circle. This point is hard to visualize (or draw) but is fun to demonstrate to yourself with two interwound strands of string with the ends tied into a circle. The more the DNA twists up ahead of the unwinding point, the more energy is required to unwind the duplex.

The biochemical solution to the unwinding problem depends on two types of proteins called helicases and topoisomerases. The helicase separates the double helix into single strands (which are then captured by SSB). The topoisomerase removes the supertwists by first nicking and then closing the DNA strand, leaving the DNA in a more relaxed state (Figure 4–19). The first topoisomerase was discovered by Jim Wang, who was looking for the enzyme that removes the supertwists that should form in DNA during its replication. As a result of helicases and topoisomerases, the only task of the DNA polymerase is running along the template DNA inserting nucleotides opposite the complementary bases.

Helicases appear to work by binding repeatedly at the junction between single- and double-strand DNA. The proteins accomplish this task by switching between a DNA-grabbing and a relaxed form in which the switch (and the energy for unwinding) comes from

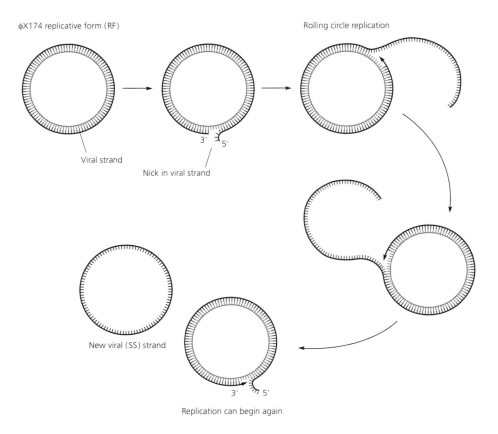

FIGURE 4–20. In RF to SS replication of phage φX174, one of the two DNA strands of the circular duplex RF molecule is first nicked and then replicated continuously from the 3' end. Simultaneously, the 5' end is unwound from the circular molecule. When slightly more than the whole chromosome has been replicated by this "rolling circle" mechanism, a genome-sized piece of the displaced SS DNA is released and circularized. Continuing replication generates many SS circles.

φX174 replicative form (RF)

Rolling circle replication

Viral strand

Nick in viral strand

New viral (SS) strand

Replication can begin again

hydrolysis of the energy-rich molecule ATP to ADP. Helicases were discovered by Ham Smith and Hans Hoffmann-Berling, who wondered why some proteins that associate with DNA use up ATP. As described below, by studying the replication of phage φX174, Arthur Kornberg and his colleagues demonstrated that helicases are used in DNA replication.

REPLICATING DUPLEX DNA

The M13, G4, and φX174 single-strand phages seemed to have been designed in heaven to lead biochemists stepwise through the mysteries of genome replication. The SS to RF replication event just described was the simplest possible system for studying DNA replication. As a single-strand phage, the DNA was already un-

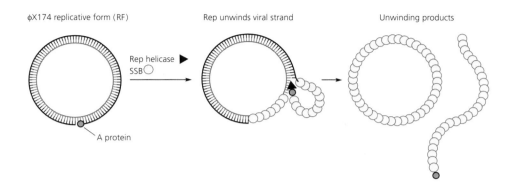

φX174 replicative form (RF)   Rep unwinds viral strand   Unwinding products

Rep helicase ▶
SSB ○

A protein

wound—neither a helicase nor a topoisomerase was required for replication. By studying the SS to RF conversion, the reaction for starting DNA chains was elucidated and the elongation proteins were identified.

The next level of complexity was the RF to SS replication pathway in which duplex DNA was copied into the single-strand DNA that was packaged in the virus particle. Although RF to SS replication is clearly atypical because the process is asymmetric—only one of the two DNA strands is copied—it allows the study of the 3′ (continuous) elongation characteristic of the leading strand of duplex DNA in the absence of the 5′ (discontinuous) lagging strand reaction. This asymmetric mode of replication is accomplished by a "rolling-circle" mechanism. One of the two strands is nicked and then elongated continuously from its 3′ end, which displaces the other end of the strand (the 5′ end) from the circle. This is followed by closure of genome-sized circles (Figure 4–20). RF to SS replication requires a helicase to unwind strands but does not need a topoisomerase. Because the phage DNA is linear, it has a free end, allowing the interwound DNA to unwind ahead of the helicase. In contrast, the circular bacterial chromosome has no free end and requires a topoisomerase to assist in its replication.

Studies of replication in infected cells indicated that there were two special proteins required for RF to SS replication in φX174: the A protein of φX174 and the *E. coli* Rep protein (Figure 4–21). As expected from its previous good deeds, when the *E. coli* extracts were supplemented with φX174 A protein, they carried out replication of φX174 RF to SS DNA. Happily, purification of the proteins

FIGURE 4–21.   Two new proteins were required for RF to SS replication of φX174: φX174 A protein, and Rep protein of *E. coli*. When only these proteins were incubated with φX174 DNA, the A protein nicked one strand of the DNA duplex, and Rep protein unwound the two strands, functioning as a helicase. SSB stabilized the SS DNA products. The A protein also binds to Rep, keeping the nicked end in close proximity to the helicase.

FIGURE 4–22.   In RF to SS replication, the φX174A protein makes the nick from which Rep helicase unwinds the two strands. Pol III holoenzyme initiates replication from the 3′ end of the nick and travels continuously around the circle. The A protein, which binds both the 5′ nick and the Rep helicase, also passes around the circle. When a whole φX174 genome has been replicated, A protein cuts out a genome-size piece and ligates it together to makes SS φX174 circles.

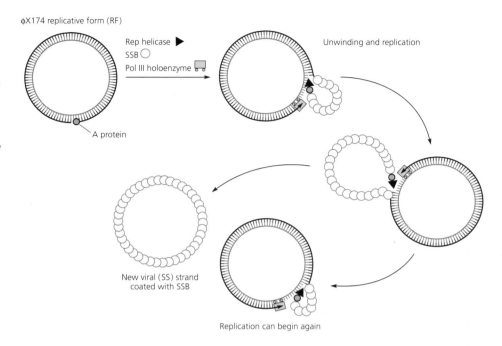

φX174 replicative form (RF)

Rep helicase
SSB
Pol III holoenzyme

A protein

Unwinding and replication

New viral (SS) strand coated with SSB

Replication can begin again

needed for this process indicated that replication required only φX174 A protein, Pol III holoenzyme, SSB, and Rep protein. The φX174 A protein turned out to make a single-strand cleavage (or "nick") at a special site in double-strand RF DNA. Remarkably, in the presence of only the φX174 A, Rep, and SSB proteins (omitting DNA polymerase from the reaction), the nicked RF unwound to give a circular and linear single-strand molecule. So, the Rep protein is the helicase that unwinds the duplex DNA; the resultant single-strands are stabilized by SSB (Figure 4–21). When the Rep protein reaction was visualized in the electron microscope by Jack Griffith, an even more remarkable picture emerged. After nicking the DNA, the φX174 A protein holds onto both the 5′ end of the nicked DNA and the Rep protein in the unwinding reaction, so that the 5′ end is in close proximity to the helicase.

In the presence of Pol III holoenzyme, DNA replication starts at the 3′ end generated by the nick and follows the helicase around and around the circle. Because circular SS viral strands are produced, the φX174 A protein must not only nick the viral strand to start replication but also rejoin it once the polymerase has produced a bit more than a complete circle (Figure 4–22).

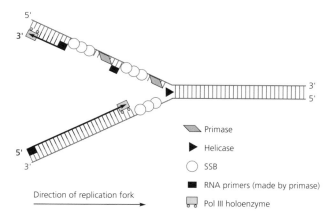

Primase

Helicase

SSB

RNA primers (made by primase)

Pol III holoenzyme

Direction of replication fork

FIGURE 4–23.   In chromosomal replication of *E. coli,* the DnaB helicase unwinds strands and the DnaG primase, which is closely associated with the helicase, makes the RNA primer. SSB stabilizes SS DNA. The RNA primer is elongated continuously on the leading strand and discontinuously on the lagging strand. Three stages of discontinuous replication are shown: a primase about to make the RNA primer; a primase just completing the RNA primer; and a completed Okazaki fragment.

Except for the gymnastics of the ϕX174 A protein, RF to SS replication has the general properties of leading strand replication. A helicase unwinds the DNA duplex, and Pol III holoenzyme runs along the freed single strand, copying the DNA. Now, the problem became one of understanding the coupled reaction in which the leading strand and the discontinuous lagging strand replicate at the same time. Although this process is now defined in an overall way, at the present time it is not understood in detail.

Our current understanding of duplex DNA replication has depended especially on work with the large DNA phages, T4, T7, and λ, as well as with *E. coli.* Phages T4 and T7 make their own replication proteins in the infected cell, and so they have provided useful comparative views of the replication process through parallel studies by Bruce Alberts with T4 and Charles Richardson with T7. For these two phages, the helicase and primase activities are closely associated (in fact, in T7, they are both in one protein). So, as the helicase unwinds the duplex, the primase is positioned to lay down RNA primers. For *E. coli* and phage λ, as described below, the critical helicase is the DnaB protein. The DnaB helicase also probably helps to bring the DnaG primase to the DNA for RNA-priming of discontinuous lagging strand replication, just as it does in the ϕX174 priming reaction (see Figure 4–17). In sum, the helicase unwinds the duplex for leading strand replication and also guides the priming process for lagging strand replication (Figure 4–23). In detail, the process may be more complicated than in the diagram because the leading and lagging strand polymerases are probably joined to each

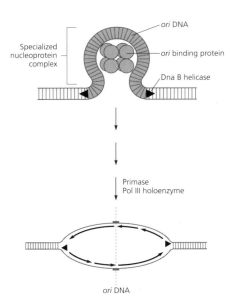

The origin of replication (*ori*) is a complex nucleoprotein structure

FIGURE 4–24. A specialized assemblage of proteins binds to *ori* DNA, creating a structure that permits initiation of DNA synthesis. The process of initiation is dynamic, with many sequential steps culminating in the binding of primase and polymerase. The origin pictured initiates DNA replication bidirectionally.

Supression analysis can identify proteins that act together

Bacterial strain	λ P⁺ growth?	λ P* growth?
dnaB(ts)	No	Yes
dnaB⁺	Yes	Some yes, some no

FIGURE 4–25. Phage λ is unable to grow in a *dnaB*(ts) mutant strain at high temperature. However, some mutations in the λP gene (P* mutations) permit growth. The P* mutations "suppress" *dnaB*(ts) mutations. Some P* mutants can no longer work with normal *dnaB*.

other, coordinating copying of the two strands (a mechanism that requires a looping of the lagging strand).

STARTING GENOME REPLICATION: A SPECIAL STRUCTURE

Studies on DNA replication in bacterial cells pointed out two fascinating properties of the initiation site (or origin) of genome replication. First, the site is unique on the chromosome, so somehow it must be selected from the five million base pairs of the genome. Second, the origin of replication is the only point at which the frequency of genome replication is regulated. Once started, DNA replication moves around the genome at about 1,000 base pairs per second. However, bacteria can grow with generation times of anywhere from 30 minutes to many hours, and so there must be a way to control precisely how often the origin fires to start a round of DNA replication. Unraveling the mysteries of the replication origin depended on genetics to identify the proteins, biochemistry to determine their function, and finally on electron microscopy to follow the dynamics of this complex molecular pathway. Comparison of *E. coli* and viral replication has aided understanding of origin events, just as it aided understanding of duplex DNA elongation. For the study of initiation of replication, a particularly useful virus has been phage λ, which has its own unique origin of replication but uses the elongation machinery of its *E. coli* host.

As might be expected from its complicated role, the origin of replication (termed *ori*) involves some special molecular properties. Many copies of an origin-binding protein bind to the origin and *ori* DNA is then wound around this protein aggregate. The resultant DNA–protein complex, termed a nucleoprotein structure, functions in a fashion somewhat analogous to a train station. Proteins move in and out in a controlled sequential pathway to provide for regulated initiation of DNA replication. For phage λ, six proteins must before the priming and elongation reactions can begin. Once these proteins act, the eventual biochemical reaction for *E. coli* and λ is quite simple; the DnaB helicase begins unwinding the duplex DNA at the origin, and DNA primase and Pol III holoenzyme can then act to begin DNA replication (Figure 4–24).

Genetics identified the protein players at the origin. For *E. coli* replication, the previously described "slow stop" mutations identified *dnaA* and *dnaC* as genes required for initiation. Likewise, a genetic approach also identified the λ *O* and *P* genes as initiation

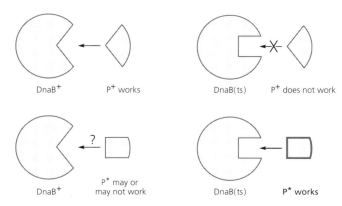

DnaB+ P+ works   DnaB(ts) P+ does not work

DnaB+   P* may or may not work   DnaB(ts)   P* works

FIGURE 4–26.   One interpretation of the suppression of a *dnaB*(ts) mutation by P* mutants is that DnaB and P normally interact, but that this interaction site is altered in the DnaB(ts) mutant. Suppression by P* may result from a P* protein that is altered to fit the *ts* form of DnaB. P* may not work with DnaB+ protein.

factors for λ replication. This approach also led to the development of a powerful general genetic method for analyzing how proteins interact with each other. Costa Georgopoulos found that phage λ was unable to grow in *dnaB*(ts) mutants at high temperature, indicating that λ replication depends on DnaB. However, he found mutations in the λ *P* gene (*P★* mutations) that were able to grow at high temperature in the *dnaB*(ts) mutant strain. In genetic terminology, the *P★* mutation "suppressed" the *dnaB*(ts) mutation (Figure 4–25).

One interpretation of the *P★* mutation is based on the assumption that λ P protein normally binds to DnaB. If so, the altered DnaB ts protein may no longer interact with λ P, and the compensatory *P★* mutant protein may now bind successfully to the wounded DnaB (Figure 4–26). For λ P and *E. coli* DnaB and for many other cases, this interpretation has proven correct. Developed initially for λ by Costa Georgopoulos, Ira Herskowitz, and Jun-Ichi Tomizawa, this type of suppression analysis is now used as a general genetic approach to identify proteins that interact with each other. Suppression analysis provides a valuable tool for figuring out a complex biochemical pathway.

The biochemical breakthrough needed for the functional study of initiation at the *E. coli* replication origin was provided by the Kornberg research group in 1981. By this time, experimental life was somewhat easier because the origin DNA sequence (*oriC*) had been placed on a smaller plasmid DNA of about 5,000 base pairs, using molecular cloning techniques (described in Chapter 10). As a result, an easily prepared "good" DNA substrate was available. Even so, the initial experiment successfully demonstrating *oriC*-dependent DNA replication in a cell extract came only after three years of unsuccessful struggles by Bob Fuller and others in the Kornberg lab.

Major new insights in molecular biology have a defined time and place, but the overall timescale for notable progress is measured in years and not days. This slow acquisition of new knowledge provides an odd contrast to the frantic pace with which most scientists pursue their quest. Yet those who are not in a hurry rarely find out anything, for the pace of pursuit reflects the passion of the investigator, and Mother Nature loves lovers.

Replication of *oriC* was the first model system available to study how duplex DNA replication was initiated from a genome origin. One essential trick turned out to be an ingredient left in for no planned reason from another protocol, the industrial polymer polyethylene glycol. This molecule, which takes up a lot of space in solution, concentrated the dilute replication proteins so that they could work. The availability of a biochemical approach to origin events allowed the isolation and functional characterization of the *E. coli* initiation proteins that had been identified genetically. As the same cell extract initiated replication from the phage λ origin when the λ O and λ P proteins were added, it could be used to study λ initiation as well. By 1985, the Kornberg group was studying *E. coli* origin events with a set of purified proteins, and Roger McMacken and his associates were doing the same for λ.

My student Mark Dodson and I got into the λ origin business with McMacken because we wanted to see if a special structure was involved in marking the origin, by analogy to another precisely located and controlled reaction that I had studied, site-specific recombination. Our approach was to use electron microscopy, which has turned out to be an extremely powerful technique for figuring out reactions involving nucleoprotein structures. The use of the electron microscope to see nucleoprotein structures was pioneered by Robley Williams, who began visualizing viruses with electron micrsopy in 1946. In theory, the electron microscope is capable of visualizing even small proteins. However, biological materials are not ideal subjects both because they are easily destroyed by the electron beam and because they are composed of light atoms (the heavier an atom, the more it deflects electrons and therefore the more it contrasts with the background sample-holder). Williams and others learned to coat nucleic acid and protein with heavy uranium and tungsten atoms so that these biological materials could be "seen" by the electron microscope. The special value of this approach for complicated DNA-protein associations is that the size of the structure can be estimated and the path of the DNA followed. As my grandmother, an enthusiastic bridge player, said, "Sometimes one peek is worth a thousand finesses." The observations from electron microscopy coupled with information derived from biochemical studies provided an overall structural picture and a reaction pathway for replication.

Based on genes required for λ replication, an alarming number of proteins emerged as candidates for initiating DNA replication by the virus. In addition to the O and P proteins encoded by λ, the host DnaB, DnaJ, and DnaK proteins were implicated even before primase makes the RNA primer to start replication. However, suppression studies and biochemical tests for protein-protein and DNA-protein interactions allowed us to guess the pathway with remarkable accuracy. This sequential pathway has been determined by work from several research groups—Roger McMacken's, Costa Georgopoulos's, Sue Wickner's, and my own (figure 4–27, right). The O protein binds to the origin region of DNA. There are four nearly identical DNA sequences in the origin, each of which binds two O molecules. P then binds to O, and DnaB binds to P. P also interacts with DnaJ and DnaK.

Observing this pathway in the electron microscope provided a visual picture of the initiation process (Figure 4–27, left). O does not simply bind linearly to its four binding sites at the origin. As judged by the electron microscope, the O proteins associate with each other as well as with the DNA to produce the O–some, a rather symmetric specialized nucleoprotein structure in which the DNA is wound. The P protein and DnaB add to this structure to generate a larger and more asymmetric structure. In the presence of the DnaJ, DnaK, and SSB proteins, DnaB begins to act as a helicase, unwinding the DNA starting at the origin. DnaJ and DnaK work as disassembly proteins to take apart the multi-protein aggregate that was initially needed to localize DnaB at the origin sequence. Once these proteins are removed, DnaB is free to carry out its helicase activity. Yet another protein, GrpE, functions with DnaJ and DnaK to make disassembly more efficient, and it is probably the GrpE assisted reaction that occurs in the cell. SSB binds to the single-strand DNA generated by the helicase.

For the replication origin of *E. coli*, a closely similar initiation

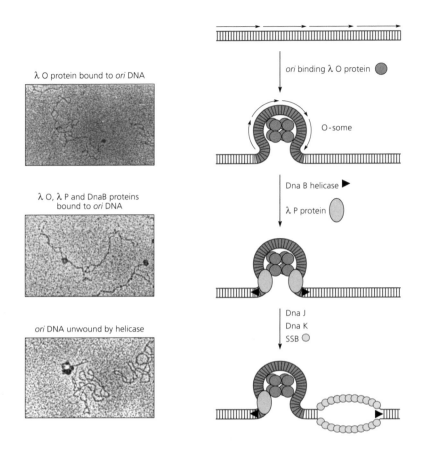

The initiation of DNA replication at λ *ori* can be seen with an electron microscope

λ O protein bound to *ori* DNA

λ O, λ P and DnaB proteins bound to *ori* DNA

ori DNA unwound by helicase

ori binding λ O protein

O-some

Dna B helicase

λ P protein

Dna J
Dna K
SSB

FIGURE 4–27.   λ *ori* DNA was incubated sequentially with various proteins and prepared for observation in the electron microscope. The left side of the figure shows the electron micrographs; the right side shows an interpretation of the data. First, λ O protein dimers bind to four sites at λ *ori*, forming a circular structure called an "O-some." Then λP and the DnaB helicase bind, forming a more asymmetric structure. Host proteins DnaK and DnaJ remove the λP protein, allowing DnaB helicase to unwind DNA. SSB protein coats the single strands stabilizing the unwound structure. The interpretive cartoons are based on many biochemical and genetic experiments as well as on the electron microscopic images. (Reprinted with permission from *Science* 233: 1053. Copyright 1986, American Association for the Advancement of Science.)

pathway was determined in parallel studies by the Kornberg group. DnaA binds to four nearly identical DNA sequences in the origin region, which then associate with each other and more copies of DnaA to produce a DNA-wound nucleoprotein structure with 20–30 molecules of DnaA. DnaC and DnaB add to produce a larger structure, and DnaB begins to carry out its helicase activity to yield single-strand DNA that is coated by SSB. There is not so far a clear need for DnaJ and DnaK to disassemble this initiation structure.

Why is the procedure so complicated? The answer probably lies in the special properties of the replication origin noted earlier. First, building a structure involving both DNA and protein confers a great deal of precision to a process that must select a site from five million base pairs. There is very little chance of error in a reaction dependent on such a highly specified localization mechanism.

Second, the complicated initiation pathway allows many potential regulatory signals to be assimilated into the decision of whether or not to fire a replication origin. These features of precise localization and diverse regulation benefit from the initial formation of a stable nucleoprotein complex at the origin. However, such a stable complex requires a disassembly pathway to fire the helicase and start DNA replication. The disassembly reaction for λ depends on DnaK, DnaJ, and GrpE, which are ubiquitous cellular regulators of protein structure and assembly (described further in Chapter 8).

This mechanism of bacterial chromosome replication has been beautifully conserved in evolution. Using the 5,000 base pair Simian virus 40 (SV40) minichromosome, which produces tumors in monkey (and human cells) as a model replication system, Tom Kelly, Bruce Stillman, and Jerry Hurwitz have shown its replication requirements to be very similar to those of *E. coli*. Like *E. coli*, SV40 replication requires initiator proteins, polymerases, primases, SSB, helicases, and topoisomerase.

FURTHER READING

Cairns, J. (1966) Autoradiography of DNA, in J. Cairns, G. S. Stent, and J. D. Watson, eds., *Phage and the Origins of Molecular Biology.* Cold Spring Harbor, N.Y.: Cold Spring Harbor Laboratory Press.

Cairns, J. (1966) The bacterial chromosome. Sci. Am. 214, 36–51.

Kornberg, A. (1960) Biologic synthesis of deoxyribonucleic acid. Science 131, 1503–8.

MAKING RNA FROM DNA

As we have seen, DNA carries the information for life. DNA not only makes more of itself with exquisite accuracy to produce multi-cellular organisms and to perpetuate the species; it also specifies the formation of proteins in a precisely regulated fashion. By 1961, the general pathway for transfer of biological information from DNA to protein had been defined; this process is termed gene expression. DNA is copied ("transcribed") into the informational intermediate mRNA; the mRNA binds to the ribosome; and the triplet code in the mRNA is translated into an amino acid sequence on the surface of the ribosome. The specific interaction between the mRNA triplets and the tRNAs with their attached amino acids selects the appropriate amino acids to synthesize the protein encoded by the mRNA. By 1961, the fundamental properties of biological regulation had also been recognized. The products of regulatory genes act to control the flow of information in order to respond to environmental signals or to carry out the timed expression of genes.

At this point, the key questions for molecular biologists became those of molecular understanding. What protein activities do the job of regulating gene expression? How does a cell recognize a particular gene? How does it change expression of a gene? How do the complicated recognition mechanisms work? What regulates the regulators? The pursuit of these questions engendered surprising biological insights and the new industry of biotechnology. In describing

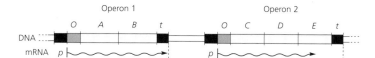

RNA is transcribed from DNA (circa 1960)

FIGURE 5–1. DNA transcription into messenger RNA (mRNA) was thought to begin at a promoter ("*p*"), end at a terminator ("*t*"), and be regulated by an operator ("*O*"). Genes in an operon were thought to be transcribed from the same promoter into a common mRNA. In this figure, and all figures in this chapter, *O, p,* and *t* are presented on an expanded scale.

the next 30 years of the quest for understanding gene expression, I will follow two major threads. This chapter considers the mechanism and regulation of transcription of DNA into RNA. The following chapter describes the subsequent pathway of RNA "maturation" and protein synthesis. The search for a detailed understanding of transcriptional control mechanisms has yielded both an expanded view of the regulatory universe and an impressive molecular picture of how proteins, RNA, and DNA work in concert to govern gene expression.

THE TRANSCRIPTION ENZYME: RNA POLYMERASE

The problem of copying ("transcribing") DNA into RNA came under intense study with the realization that mRNA was the informational intermediate, carrying information from the gene to the ribosome. Presumably, mRNA was produced by an enzyme that copied the base sequence of the DNA, using the ribonucleotides of RNA instead of the deoxyribonucleotides of DNA. Unlike DNA replication, RNA synthesis had to be fully conservative, freeing the newly produced RNA and restoring the original DNA duplex. Moreover, RNA synthesis should have multiple starting and stopping points, corresponding to the units of gene expression—the operons (Figure 5–1). Experiments proceeded in two directions: identifying the enzymatic machinery needed to make RNA and defining the properties of mRNA in cells.

The enzymatic mechanism for RNA synthesis had been under study for a number of years using Severo Ochoa's enzyme, polynucleotide phosphorylase, but eventually it became clear that this enzyme did not make mRNA. Rather, under physiological conditions, this enzyme had the reverse function—that of RNA degradation. At this point, the realization that mRNA was the informational intermediate in gene expression and the search for the enzyme that made mRNA evolved in tandem. In 1957, Elliot Volkin found that, immediately after phage T2 infected *E. coli* cells, an RNA with a base composition matching that of T2 DNA (and distinct from that of *E. coli* DNA) suddenly appeared. Lacking the conceptual framework of mRNA, Volkin postulated that T2 RNA was the precursor of DNA. Perhaps T2 DNA synthesis occurred by first joining ribonucleotides into a long RNA chain and then converting the ribose sugars in the chain into deoxyribose sugars. Jerry Hurwitz looked into this possibility as a new faculty member in Arthur Kornberg's department at Washington University. Hurwitz indeed found conditions where ribonucleotides entered DNA, but the mechanism turned out to involve an aberrant reaction of Pol I. The ubiquitous Mel Cohn, then a member of Kornberg's department, grasped the right idea from the wrong experiment. When first told of Hurwitz's results, he proclaimed, "You've got the enzyme that makes RNA from DNA."

In 1960, the time was right for the discovery of RNA polymerase, the enzyme that transcribes DNA into RNA. In 1959, Sam Weiss had described for the first time an activity in rat liver nuclei that incorporated all four ribonucleotides. Then, in 1960 and 1961, several groups reported increasing evidence for RNA polymerase in many organisms. Resituated in New York University and refocused by the concept of mRNA, Hurwitz renewed his work on RNA synthesis. He found an enzyme in extracts of *E. coli* that produced an RNA polymer from ribonucleotides. The reaction depended on DNA, and the RNA product had the base composition of the DNA template. Similar experiments were done at the same time by Audrey Stevens, also working with *E. coli*. Weiss reported further characterization of the enzyme both in animal cells and in the bacterium *M. leuteus*. Finally, Jose Bonner reported the activity in peas.

The chemical reaction of RNA polymerase is highly similar to that of DNA polymerases. The high energy ribonucleotides containing three phosphates (ATP, GTP, CTP, and UTP) are used as substrates; ribonucleotides containing one phosphate (AMP, GMP, CMP, and UMP) are incorporated into the RNA polymer; and the two extra phosphates are released as pyrophosphate (Figure 5–2). As for DNA replication, RNA synthesis proceeds by adding nucleotides to a growing 3' end. However, in contrast to DNA polymerases, RNA polymerase can start RNA chains. The identification of RNA polymerase opened the door for subsequent biochemical studies of gene regulation.

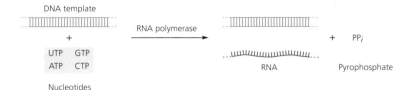

DNA template

+

UTP	GTP
ATP	CTP

Nucleotides

RNA polymerase →

RNA   + PP$_i$

Pyrophosphate

FIGURE 5–2.   Using DNA as a template and ribonucleotides with three phosphates as substrates, RNA polymerase joins the ribonucleotides together to make RNA. This reaction releases the diphosphate compound pyrophosphate (PP$_i$). The DNA template remains unchanged.

THE PROPERTIES OF mRNA

As a complement to the search for the enzyme that makes RNA from DNA, some scientists began to study features of mRNA itself. Much of the technology to study the mRNA produced in cells was developed by Sol Spiegelman and his associates. Even before Brenner and Jacob linked "Volkin T2 RNA" (the unstable RNA produced after infection of the host cells by the virulent T2 phage) to informational RNA (see Figure 3–13), Spiegelman had realized the importance of this unstable RNA and had begun to study its properties. With Masayasu Nomura and Ben Hall, he found that RNA of the related T4 phage was associated with intact ribosomes but could be released under appropriate conditions. These experiments provided strong evidence that ribosomal RNA itself was not the informational RNA but only served as a stopping point for mRNA, the true informational RNA. "The ribosome is a garbage pail," proclaimed Spiegelman, dismissing the idea that rRNA was the informational RNA with characteristic reticence.

If this unstable T4 RNA were truly mRNA, then it must be complementary in base sequence to T4 DNA. To test this idea, Hall and Spiegelman developed the technique of "DNA-RNA hybridization." Julius Marmur and Paul Doty had shown that duplex DNA could be separated into single strands by heat, and that the two complementary strands could reassociate during slow cooling. The slow cooling process allowed the DNA strands to base-pair with each other without the hindrance of imperfect base-pairing within the separated individual strands. By allowing the separated DNA strands of phage T4 to reassociate in the presence of T4 RNA, Hall and Spiegelman showed that DNA-RNA as well as DNA-DNA duplexes formed during the slow cooling process. The DNA-RNA "hybrids" could be distinguished from DNA-DNA duplexes

Sol Spiegelman

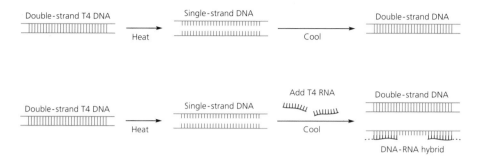

FIGURE 5–3. The two strands of double-strand DNA separate upon heating and reassociate upon cooling. When phage T4 RNA is added to separated strands of T4 DNA during cooling, DNA-RNA as well as DNA-DNA duplexes are formed. DNA-RNA hybrids proved that RNA complementary to DNA existed and was one proof of the "mRNA" concept.

by their difference in density after centrifugation in a cesium chloride density gradient. By labeling DNA with radioactive phosphorus (^{32}P) and RNA with radioactive hydrogen (^{3}H)—a radioactive "double label" experiment—Hall and Spiegelman were able to confirm that both DNA and RNA were present in this new band found after centrifugation. Base-pairing of T4 RNA with T4 DNA demonstrated that the unstable "mRNA" fraction was a complementary copy of the DNA (Figure 5–3).

The cesium chloride procedure was slow, and the results were messy and difficult to quantify. However, Hall discovered a simpler way to detect DNA-RNA hybrids that allowed rapid and quantitative measurements of messenger RNA. For reasons that are still obscure, single-strand DNA adheres to nitrocellulose filters, but RNA alone does not (probably because the DNA was much longer, but no one bothered to figure out why this useful miracle happened). A DNA-RNA hybrid also binds, providing a simple assay for the complementarity of RNA to DNA: "complementary" RNA sticks to the DNA on the filter, and noncomplementary RNA flows through. This process could easily be monitored by using radioactive RNA. This was the first of the "miracle" filter assays that converted very difficult measurements into easy ones, and it is similar in concept to the one that eventually solved the coding problem (see Figure 2–10).

The technique of DNA-RNA hybridization opened the way to study RNA from a specific bacterial operon provided that the corresponding pieces of bacterial DNA could be obtained. Fortunately, phage variants existed that carried specific bacterial genes: λ*gal* provided a source of *gal* DNA, and ϕ80*lac* phage provided a source of *lac*

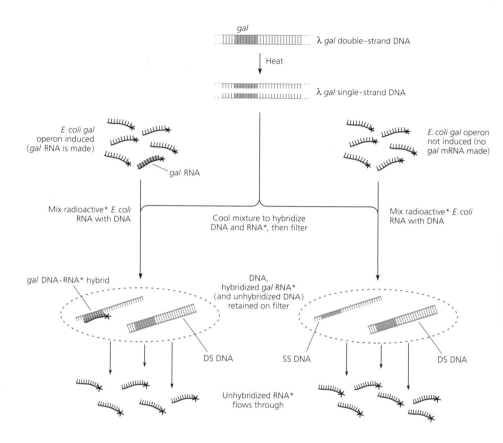

FIGURE 5–4.   To prove that *gal* mRNA was produced only after induction of the *gal* operon, Gros mixed single-strand λ DNA containing the *gal* genes (λ*gal* DNA) with radioactive *E. coli* RNA (RNA*) made from cells that either were or were not induced by galactose. RNA from the *gal* operon was retained on a nitrocellulose filter because it bound to λ*gal* DNA; other cellular RNAs flowed through the filter because they could not hybridize to the DNA bound to the filter. Measuring the amount of radioactive RNA retained on the filter revealed that induction by galactose was required for radioactive RNA to be bound to the filter. Cultures that had not been induced for galactose showed very little RNA binding to the filter.

DNA for the early experiments on these operons. These naturally occurring phage were the prototype for the new technology of "molecular cloning," developed years later, in which other bacterial or eukaryotic genes were inserted into the "vector" phage allowing the reproduction of specific ("cloned") DNA segments. By DNA-RNA hybridization, François Gros, Waclaw Szybalski, and their associates demonstrated that induction of the *gal* and *lac* operons resulted in increased amounts of *gal* and *lac* RNA. These experiments verified the predicted operon-specific RNA and indicated that the expression of these operons was regulated by controlling the amount of RNA (Figure 5–4).

With the development of quantitative hybridization procedures, the concept of discrete groups of genes transcribed together in a particular mRNA (the "operon" of Jacob and Monod) rapidly became a biochemical reality. Spiegelman and associates further showed that rRNA and tRNA are also produced as transcription products from DNA, and that their abundance in the cell results not only from their efficient transcription but also from their stability. In contrast, mRNA is unstable, with a lifetime of only a few minutes before it is broken down into mononucleotides. This instability is important for regulation: it allows the pool of "informational RNA" to change rapidly with changing growth conditions. Because mRNA is unstable, it does not accumulate, so only about 5% of the RNA in cells is mRNA. The remainder (>95%) is stable RNA, mostly composed of the rRNA in ribosomes. Messenger RNA is most readily detected by a short labeling period with a radioactive precursor, tagging the RNA as it is synthesized.

The availability of quantitative hybridization assays also opened the way for a rigorous investigation of phage development, both in infected cells and in cell-free systems. Because these were the only developmental systems accessible to molecular study at this time, phage became a focus of intensive investigation. We will take a brief digression from the major story line of this chapter to recount these findings, which were interesting in their own right and also had profound implications for our understanding of how transcription is regulated.

DEVELOPMENTAL PATHWAYS OF PHAGE

Phage growth provides a simple biological system for the study of development. Like development in multicellular organisms, phage development results both from a time-dependent pattern of gene expression and a mechanism for choosing between alternative pathways. The study of phage development constitutes a classical problem in gene-protein relationships, which can be approached by isolating mutants and then determining their phenotypes. There were clearly two special problems. First, a mutation blocking lytic development would be lethal—how could a dead phage be preserved? Second, based on genome size, some 40–100 gene products were likely to be involved in the development of large DNA phages, a considerably more complex problem than a single operon.

The lethality problem was solved here, as in other systems, by isolating "conditional-lethal" mutations, which allowed the mutant phage to grow in one cellular environment but not another. The successful conditional-lethal solution was accidentally developed by

Allan Campbell for phage λ and by Richard Epstein for phage T4. Campbell found certain mutants of λ that would grow on one bacterial host but not another; he named such mutants host-dependent, or *hd*. Epstein, Charles Steinberg, and Harris Bernstein isolated T4 mutants with the same properties; they named the mutations *amber* as a promised award to Bernstein for helping with the experiment (in German, bernstein translates to amber). Both the *hd* and the *amber* mutants were able to grow on the same host strains.

An understanding of the *hd/amber* mutations came shortly thereafter, from the seemingly unrelated experiments of Allan Garen and Seymour Benzer, who figured out that certain *E. coli* strains had "suppressor genes" that reversed the effects of "nonsense" mutations. Recall that a nonsense mutation causes translation to terminate during synthesis of a protein. Suppressor strains have an altered tRNA that "misreads" the terminating nonsense codon as a sense codon. The altered tRNA inserts an amino acid at the position of the nonsense codon so that a full-length protein can be translated. The hosts that allowed the *hd/amber* phage mutants to grow were those that carried suppressor tRNAs. Therefore, the *hd* or *amber* mutations were nonsense mutations in various phage genes.

Bob Edgar introduced the use of temperature-sensitive mutations as another class of conditional-lethals; the mutant gene produced an inactive protein at elevated temperature, but an active (or partially active) protein at lower temperature. Interestingly, temperature-sensitive lethals had been known to fruit-fly geneticists for years, but isolation of the phage *ts* lethal mutations marked the first time this technique was seriously used by molecular biologists.

Epstein and Edgar used the conditional-lethal mutations for the systematic study of viral development in phage T4. They enlisted the efforts of Edward Kellenberger, an electron microscopist, to look for phage components in cells infected with different mutant phage. They identified many genes that affected development, allowing them to construct a rudimentary developmental pathway. The genes could be grouped into those required for viral DNA replication, formation of phage heads, and formation of phage tails. DNA replication occurred before the formation of phage components and was necessary for the appearance of phage heads and tails, and for cell lysis. In addition to DNA replication, at least one additional gene was needed for any of the "late" functions: head and tail formation and lysis. The process of lytic development by phage T4 thus involved an early period devoted to viral DNA replication and a late

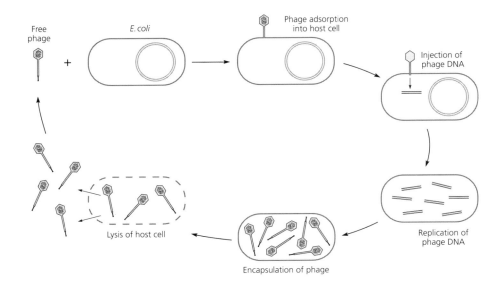

FIGURE 5–5. The lytic developmental cycle of many DNA phages (such as T4) is controlled by timing gene expression. The early phase is devoted to replication. In the late phase, head and tail proteins are made, phage are assembled, and cells are lysed. When replication of phage T4 is prevented, late processes are blocked. Although not shown, the *E. coli* chromosome is fragmented during growth of T4.

phase devoted to DNA encapsulation and cell lysis. In addition, some early events were required before the late ones could occur (Figure 5–5).

Phage λ and other large DNA phages exhibited temporal pathways of lytic growth similar to T4; λ turned out to be the example studied in the greatest detail. It had many fewer genes than T4, some technical advantages for biochemical work, and the fascinating developmental choice between lytic growth and the lysogenic state, where it exists as an inactive prophage integrated into the host chromosome.

LAMBDOLOGY: THE DEVELOPMENTAL BIOLOGY OF A FIELD

Starting in 1963, there was a sudden flow of young scientists into research on phage λ. An annual meeting on phages was a tradition at the Cold Spring Harbor Laboratory. In 1963, the talks were nearly all on phage T4 and other virulent phages; by 1968, the meeting was much larger, and the topic was almost entirely λ and other temperate phages. In 1970, the exploding information about λ was incorporated into a book, *The Bacteriophage* λ. Another book about λ, *Lambda II*, was produced in 1983. The number of researchers in lambdology, as the field was termed, remained roughly constant in size from 1970 to 1980, before dwindling in the 1980's as many went on to newly opening fields. So the humble phage λ became the topic of two large books (an old and new testament), a large number of review articles, and well over a thousand research papers.

The flowering of lambdology was a new phenomenon in molecular biology. Whereas the old phage school grew up around Delbrück and the biochemistry of DNA replication centered around Kornberg, lambdology did not grow up around one or two individuals. Rather, there seemed to be a sudden worldwide sense that λ had arrived and that the disciples should gather. Although it is difficult to assess clearly an area in which I was so deeply involved, I believe that the early years of lambdology were also truly extraordinary in the degree of communication, mutual enthusiasm, and trust. Both ideas and new phage mutants were exchanged as soon as they were created, and debates over whose idea was right were passionate but good-natured. Not everyone went along with these egalitarian ideas, but enough did to define the mood of the field. Early lambdology was science at its most joyous and productive.

The immediate progenitors for lambdology were François Jacob, Dale Kaiser, and Allan Campbell. Jacob contributed the operon model, and Kaiser established the framework for thinking about the switch between lysis and lysogeny. Campbell not only contributed the λ nonsense mutations but also developed the correct concept about how prophage λ associated with the host DNA. Jacob and Elie Wollman had proposed that the phage DNA somehow "hooked" itself as free viral DNA to the bacterial chromosome. Later, when the DNA in phage particles was shown to be a linear molecule, the hooking notion appeared even more likely to be correct. In one set of genetic experiments that almost no one took seriously, Enrico Calef reported that the order of genes in the prophage was different from that in the virus. In 1962, Campbell realized that, if λ DNA formed a circle inside the cell and inserted into the chromosome with a single genetic recombination event, then the gene order in the prophage would be permuted, as reported by Calef. This idea turns out to be true. Upon entering the cell, the linear viral DNA forms a circle and a breaking and rejoining event inserts the phage DNA into the host DNA between the *gal* and *bio* genes. Starting with the work of Naomi Franklin, it gradually became clear that the inserted prophage has a gene order *N R A J* that is permuted from the gene order *A J N R* found in phage DNA (Figure 5–6).

The mechanism proposed by Campbell represents one of the great concepts of molecular biology. The idea explains prophage localization, how λ could pick up bacterial genes (a rare "cut" in the

Allan Campbell

FIGURE 5–6. Campbell hypothesized that following injection of a linear λ DNA molecule into host cells, the λ genome circularizes. The λ and host chromosomes then pair and recombine at the break-join sites on each chromosome. This inserts the phage genome into the host and changes (permutes) the order of markers in the integrated phage DNA (or "prophage") from that of linear λ. Reversal of this recombination event leads to normal excision; rare excision events at different places incorporate nearby bacterial markers, as shown for formation of λgal.

In the Campbell model, a circularized λ genome is inserted into the host chromosome

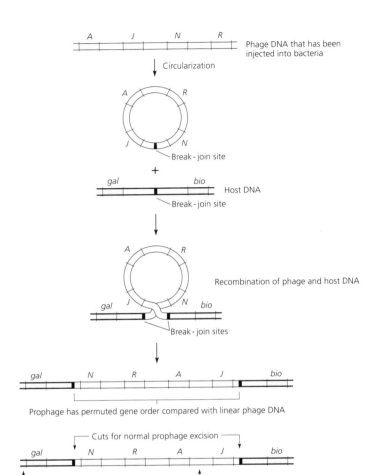

wrong place), how λ could escape from the chromosome (reversing integration), how the F factors integrate into the chromosome (it happens just like λ integration), and a host of other important genetic switching phenomena.

Like most great ideas, the integration concept is a deceptively simple one that seems obvious in hindsight. "How else might it have been?" say our current generation of molecular biology students, who have been fed the answer without any of the prior confusion and therefore start out thinking that science proceeds by logical steps from A to Z. In fact, I think that Allan Campbell came to his model at that time only because he has a superbly "illogical" mind, as judged by stepwise "logical" thinking. After all, we knew that λ DNA was linear, so why would it become a circle? Whereas most of us do think in stepwise, linear terms, except when jarred sideways by an unexpected experimental result, Campbell seemed to be coming in sideways much of the time. I could always tell when he served as an "anonymous" reviewer of one of my genetic papers because he invariably suggested an explanation for some of my data that neither I nor anyone else had ever thought of, and he was often right.

LAMBDOLOGY: THE DEVELOPMENTAL BIOLOGY OF A PHAGE

As was true for most lambdologists, I did not grow up with the subject but somehow arrived there. At Wisconsin in 1962, Julius Adler and I were looking for a good way to study the biochemistry of operon regulation and decided on *gal* because this DNA was available on the λ*gal* phage. The study of λ itself was initially a backup project, initiated by Bill Sly, a postdoctoral fellow with Adler. With every experiment by Sly, λ looked better and better, and *gal* was soon forgotten.

Sly and I plunged into the genetics and cellular biochemistry of λ, joined by my graduate students Lenny Isaacs and Ann Joyner. We found that production of λ RNA had two distinct phases during phage growth: an early phase having a moderate amount of RNA, followed by a later phase having a larger increase in RNA (Figure 5–7a). This observation suggested that we could follow λ development by measuring RNA, short-circuiting the need to identify and develop assays for each developmentally regulated protein. To identify the regulatory genes controlling lytic development, we proposed to examine the pattern of RNA synthesis in various mutant phage. We had at hand Campbell's collection of nonsense mutations in 18 genes that were essential for phage growth. But, which mutants to examine? We reasoned that regulatory mutants should exhibit a major failure in lytic growth, such as preventing lysis of host cells. Leaving out the gene that encoded the lysis protein itself, we were convinced that the lytic regulators should be among the four genes having mutations that prevented lysis. These genes were named *N, O, P,* and *Q.*

We quickly found that *N, O, P,* and *Q* mutants were all defective in producing λ mRNA at late times (Figure 5–7b). The *Q*⁻ mutants were the easiest to interpret. Since *Q* mutants replicated normally, their early functions were probably intact. The most straightforward interpretation of the data for the *Q* gene was that *Q* was a positive regulator of late gene RNA. At this time, Bill Dove, a new

FIGURE 5–7. During lytic development of λ, mRNA synthesis has two phases; the first phase is early transcription and the second phase is late transcription. Early and late gene transcription are differentially assayed by their ability to bind to appropriate pieces of λ DNA containing either predominantly "early" or predominantly "late" genes. Binding can be observed because the RNA is prepared from infected cells in which the RNA had been radioactively labeled. The DNA/RNA hybridization method is described in Figure 5–4. (a). The N and Q genes act sequentially to regulate development; N is required for early RNA transcription and Q is required for late RNA transcription (b).

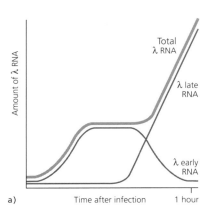

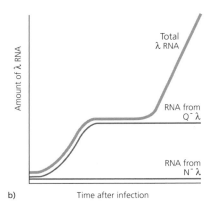

lambdologist, arrived in town. Initially a physical chemist, Dove had begun to study λ development at Cambridge University in England by using protein assays for two λ late proteins. Because Q⁻ mutants were defective in production of both late proteins, Dove had independently concluded that Q was a positive regulator of late events. Bill was friendly, enthusiastic, and communicative. We decided to submit our papers with these results together, and also to present a combined talk at the 1965 Phage Meetings.

The interpretation of the N, O, and P mutants was less clear. As they were defective in DNA replication, they would make very few copies of λ DNA. This could explain their defect in lytic growth—with fewer copies of DNA to use as templates for transcription, they would be unable to make late RNA at a high rate. Or, they might be defective in a function that directly contributed to their inability to make late RNA. To define their roles, we realized that we needed to measure the transcription of early and late genes separately. Fortunately, the biological materials that separated early and late genes, at least crudely, were available. Jean Weigle had just found that the late genes were on the "left" end of λ DNA. Since this end of λ is replaced by bacterial DNA in λgal (see Figure 5–6), λgal could serve as our source of "early" DNA. "Late" DNA was also available. After breaking λ DNA by shearing, Ann Skalka and Al Hershey had learned to isolate the head and tail gene "late" DNA as a separate piece. We joined forces with Skalka to figure out the transcription pattern of the early and late genes using the DNA-RNA hybridization assay. Working with her was fun, even though we did most of our collaboration by phone. Coming from Hurwitz's group, Skalka

Timing in phage λ development is controlled by just a few genes

Genotype	Phenotypes of λ mutants: DNA replication	Early gene expression	Late gene expression	Interpretation
λ^+	+	+	+	Wild type expression
λN^-	low	very low	–	N activates early genes
λQ^-	+	+	low	Q activates late genes
λO^- or P^-	–	+	low	O, P replicate λ DNA

FIGURE 5–8.

had a good sense for what was biochemically reasonable as well as abundant energy and good humor.

We soon had a picture of early and late gene transcription in normal λ. The *N, O, P* gene region was transcribed early in infection; transcription switched to the late (head and tail) gene region about halfway through the growth period (see Figure 5–7a). We could now interpret the transcription patterns in the mutant strains. The N^- mutant made very little early gene RNA and no late RNA. Most likely, *N* protein was a positive regulator of early gene transcription (Figure 5–7b). The O^- and P^- mutants had a normal amount of early gene RNA and showed a normal transition to late gene transcription but had little late gene RNA. Most likely, their primary defect was in replication. The Q^- mutant was completely normal for early gene RNA but defective in transcription of the late (head and tail) genes. Therefore, Q protein was a positive regulator of late gene transcription (Figure 5–7b). The most direct interpretation of our data was sequential positive regulation: N protein turns on most early gene RNA, including Q; Q protein then activates late gene RNA.

At this point, a large number of experimental results converged to the same conclusion. René Thomas had independently done some very clever genetic experiments arguing that N protein was needed to turn on the *P* and Q genes, and Charles Radding had found that N^- mutants were defective in production of two other early λ proteins. Taking all the work together, we were completely convinced that timed expression of genes in lytic development depended on the sequential action of two positive regulators (Figure 5–8).

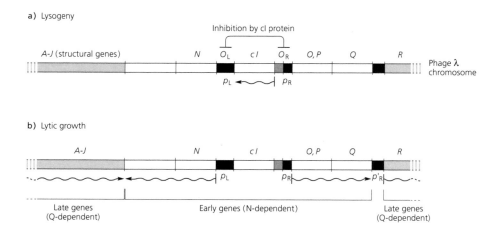

FIGURE 5–9. In the lysogenic state, cI repressor permits its own synthesis, but inhibits transcription from the p_L and p_R promoters by acting at the O_L and O_R operators (a). When the cI repressor is deactivated, lytic growth ensues (b). First, N protein is produced from the p_L promoter. N protein then promotes O, P, and Q transcription from the p_R promoter. The Q protein, in turn, promotes transcription of a very long mRNA originating from the $p_{R'}$ promoter. This transcript encodes all of the structural genes necessary for the production of new phage.

Sequential positive regulation of λ lytic development meant that the cI protein could repress all of the genes required for growth without directly repressing all of its transcription units. The operator sites for cI action had previously been identified by virulent mutants of λ, which had defective operators unable to bind cI. (As a consequence, virulent mutants grow lytically even when they infected a lysogenic cell where cI was present to repress transcription from incoming phage.) Only two operator sites for cI, o_L and o_R had been inferred from the location of the virulent mutations in λ. By acting at these sites, cI could repress expression of the N and Q positive regulators, thus indirectly blocking lytic development (Figure 5–9a). The interaction between positive and negative regulation by λ is an interesting general point for development, but this feature is also historically amusing because Jacob originally came to the operon idea through the belief that cI somehow directly turned off all of the λ genes.

The pathway of lytic development was rapidly filled in. Waclaw Szybalski and his colleagues had learned to prepare each strand of λ duplex DNA separately. By hybridizing λ RNA to the separated strands of DNA, they could decide which of the two DNA strands was copied. In turn, this information indicated the direction of transcription. Two early transcription units originated from the p_L and p_R promoter sites, and proceeded away from the cI gene; late transcription moved rightward. Ethan Signer and his student Ira Herskowitz showed by genetics that a single site to the right of the

Q gene was needed for all late gene transcription; this site was presumably the rightward promoter, $p_{R'}$, at which Q activates transcription of all late genes. This is possible because when λ DNA is circular (as it is during late lytic growth) all of the late genes (R to A) are contiguous (Figure 5–9b).

LAMBDOLOGY AS A COMMUNITY

In retrospect, the rapid pace of new information obtained about λ between 1964 and 1970 was astounding. The developmental biology of an organism (albeit a small parasitic one) was defined in considerable detail: how its genes are regulated; how it replicates and recombines; how it integrates and excises from its host; how it packages its DNA. To describe a partial cast of characters, I liked the molecular approach with simple assumptions; Bill Dove loved genetic complexity and comparative biology; René Thomas and Harvey Eisen went for the clever genetic experiment and the intuitive leap; Ethan Signer was the genetic logician. The time, the people, and the atmosphere were right. We possessed big egos, but we did not let it get in the way of communicating, combining our diverse skills, and having fun at it. Out of chaos came understanding.

The early era of lambdology ended with a meeting held at Cold Spring Harbor in 1970, which gave rise to the old testament bible. The scientific meeting is an interesting institution. The format is designed to exchange information and ideas through formal talks and informal discussions. The meeting is also a grown-up version of show and tell, in which scientists try to convince everyone else that they are doing the most interesting and exciting work and have the best ideas about how it all works. Each meeting also has a certain ambiance, which changes with the time and the particular cast of characters.

The 1970 meeting on temperate phage was characterized by a mixture of euphoria and anxiety—we had come so far so fast, but where were we going? The same emotional mix characterized the political and social atmosphere of the time, and of course science does not proceed independently of its social milieu. The year 1970 was still part of that era known as "the 60's." I had left Madison in 1969 for Berkeley, exchanging one tear-gassed revolutionary center for another. I was very much involved in politics, as were many other scientists. For example, Signer, a member of a peace group, Medical Aid for Indochina, visited Vietnam and China in 1971, one of the first Americans to do so. For a brief moment in 1970, many of us hoped, and many others feared, that a major social and political change was going to happen. My children and I joined some 300,000 people at a political rally in Golden Gate Park in San Francisco and cheered a mixture of speakers dedicated not only to peace but also to the creation of caring communities in which money and power would no longer be the prevailing social force, and racism and sexism would be eliminated.

Amid this political ferment, I decided that perhaps we should consider lambdology. The field was changing—more arguments about who was first and less communication and cooperation. And lambdology was paradise compared to most other fields that I saw. Competition in science is a complex issue. One point that is not widely appreciated is that competition occurs at three levels, one of which I believe to be valuable, the second counterproductive (but probably unavoidable), and the third destructive.

At the first level of competition, all scientists want to think of the best ideas and the best experiments and to advance their field through telling others of our findings. In an atmosphere of open exchange of ideas and mutual trust and appreciation, the competition to be creative accelerates new understanding; the situation is somewhat analogous to trying to be the most valuable member of an athletic team. I believe that this environment characterized early lambdology.

At the second level of competition, the issue is recognition outside of our field. This level of visibility depends on publishing in a "prestige" journal that is widely read. For publication in such journals, it is important to be the first to describe a result, and so the situation becomes more complicated. Often new advances are made independently in two or more places, and there can be a "race into print" in which the earliest discovery or the most complete account may not be properly credited. Worse yet, there can be deliberate noncommunication because two or more scientists see a clear goal and decide that getting there first will be more likely if the other folks do not hear about their ideas and experiments. As a result, there is much needless duplication of experiments and no stimulating exchanges of ideas. The pace of new understanding can be substantially decelerated, and science becomes much less fun. The scientific field becomes somewhat like a street of small shopkeepers each selling the same thing and proclaiming that theirs is the best.

At the third level of competition, there is a quest for complete dominance of a field in the eyes of the outside world. The goal is to have a field uniquely identified with their name; other contributors are not acknowledged. There can be no useful exchange of ideas or cooperation with scientists for whom all concepts and experimental advances are by definition their own. The scientific environment becomes one of warring camps.

Before the 1970 meeting, I proposed a possible remedy to emerging problems, which I called "The New Lambdology." The meeting organizers, Dale Kaiser and Allan Campbell, graciously set aside the final afternoon of the meeting for what proved to be a fascinating discussion. The essence of the idea was to rededicate ourselves to lambdology as a community effort. We would avoid the competition for outside glory by avoiding publication in the old way. We would publish a "Lambdology" book every six months or so that would present all the recent advances in the field, with an introduction by an elected representative of each subfield to make the work accessible to people in unrelated areas of molecular biology. So all the work would appear in the same place at the same time—no priority race, and everyone would know where to find it. In my most extreme proposal, we would even avoid individual authorship; everyone who contributed would be listed as an author for the entire book. (I no longer think that this would work, but I would still bet on the rest of the plan.) Extended to all of molecular biology, I supposed that each area of molecular biology might carry out a similar scheme, generating organized, coherent reports for other communities. Of course, some people would be members of more than one group. In this scheme, unproductive competition might be avoided, and scientific communication improved.

The meeting about the plan drew a much larger crowd than I expected. Generally people begin to disappear toward the end of a long meeting; in this case nearly everyone stayed. There was considerable commentary about competition in science. The discussion also identified some technical concerns that were very important to the younger scientists—for example, how to establish their contributions so that they could compete for research grants and jobs. I did not see how my plan would affect these concerns. Research was still being published, and individual contributions could be spelled out for granting agencies. Besides, jobs often depended mostly on personal recommendations from people in the research field.

For the lambdology plan to work, nearly unanimous agreement was needed. At the end of the discussion, we took a straw vote, and about two-thirds were in favor—not enough. At the time, I was very disappointed; in retrospect, the vote was a remarkable expression of interest in such a radical scheme.

Like most of the idealism of the 1960's, the New Lambdology now seems far away. The individual research groups are typically larger, and less cooperation and communication exists between groups. Some laboratories even function as a large corporate enterprise with a multitude of researchers, working under an executive staff of senior postdoctorals, distantly supervised by the senior scientist in charge. However, I have always believed that new understanding in science is limited by the need for good new ideas and that these arise most readily in a small, interactive community like early lambdology or the early days of the coding problem. The joy and the triumphs of molecular biology derive from creative vision. The most successful scientific community will be the one most supportive of the creativity and communication that accelerate new insights.

Wally Gilbert

THE LAC AND λ REPRESSORS BIND TO OPERATOR DNA

It is time now to return to our story of transcriptional regulation and recount how repressors moved from the genetic construct of Jacob and Monod to their current biochemical reality. By 1966, nearly everyone believed that repressors were proteins that controlled transcription, but until Wally Gilbert and his postdoctoral fellow Benno Müller-Hill captured Lac repressor, the product of the *lacI* gene, nobody had found one to prove it.

Gilbert began his scientific career as a theoretical physicist. He was an assistant professor of physics at Harvard when he migrated to Jim Watson's group during the early days of mRNA. Despite his background in theory, Gilbert was the quintessential experimental innovator at heart. He had an extraordinary sense for the appropriate biochemical approaches to a question and for the development of rigorous new techniques. In the late 1960's and early 1970's, a steady stream of new techniques that revolutionized analysis of the base sequence of DNA and the interactions between DNA and proteins emanated from Gilbert's research group. Surprisingly, Gilbert never seemed truly interested in the detailed molecular interactions that his techniques were designed to probe. In that sense, his eventual departure for biotechnology was preordained.

To follow the Lac repressor during its purification, a biochemical assay was needed. Gilbert and Müller-Hill decided to base their biochemical assay for Lac repressor on the one aspect of its activity that seemed certain—that repressor binds to the inducer. They chose to measure repressor binding to the gratuitous (nonmetabolized) inducer, IPTG, because this compound worked at very low concentrations and so was likely to bind tightly. Moreover, since IPTG was not changed to other compounds by the action of cellular enzymes, IPTG itself should interact with the repressor. It turned out later that this was a very wise choice. Lactose, the natural inducer, binds to repressor only after it is converted to a related compound, called allolactose, by the action of β-galactosidase.

Before starting these experiments, Gilbert realized that experimental limitations might keep him from observing the Lac repressor binding to IPTG. Lac repressor was likely to be present at very low concentrations in the cell. Moreover, the binding of Lac repressor to IPTG might not be tight enough to observe. So before embarking on biochemistry, Gilbert and Müller-Hill took two steps to deal with this problem. They selected and then used a variant of the Lac repressor that bound inducer more tightly, and they increased the amount of repressor per cell, by using a cell with several copies of the Lac repressor gene. When scientists perform experiments at the very edge of feasibility, only the most savvy experimentalists will succeed.

To show the binding of Lac repressor to IPTG, Gilbert and Müller-Hill used equilibrium dialysis, a classical biochemical approach for detecting the binding of proteins to small molecules. In this method, the protein solution is tied inside a sack made of a cellophane membrane that is permeable to small molecules but not to proteins. The membrane, or dialysis sack, is placed in a solution containing the small molecule, in this case radioactive IPTG. If a protein inside the sack binds IPTG, more IPTG winds up inside than outside the sack. Starting with a crude extract of E. coli proteins, Gilbert and Müller-Hill detected the presence of an IPTG-binding protein. This binding activity was not present in extracts from constitutive lacI⁻ mutants that lacked repressor, and so the binding protein in the lacI⁺ extract looked like the real repressor (Figure 5–10).

The question now was what the repressor really did to control transcription. Assuming that the Jacob-Monod proposal—that repressors inhibit mRNA synthesis—was correct, there were two possible answers: specific binding to operator DNA, or specific inhibition of RNA synthesis. Since it was most feasible to test the first idea, initial experiments tested association of repressor with

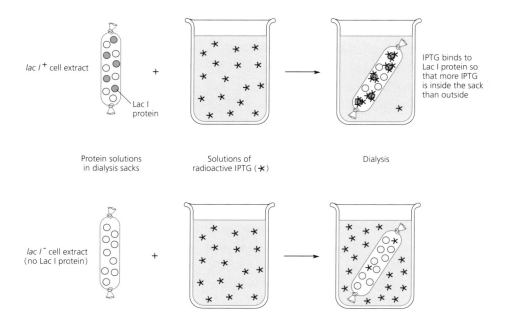

lac I⁺ cell extract

Lac I protein

lac I⁻ cell extract (no Lac I protein)

Protein solutions in dialysis sacks

Solutions of radioactive IPTG (★)

Dialysis

IPTG binds to Lac I protein so that more IPTG is inside the sack than outside

FIGURE 5–10. To see if the *lac* repressor protein could be detected by its ability to bind IPTG, Gilbert and Müller-Hill filled a small sack with a crude extract of either *I*+ or *I*− cells. The protein solution containing *lac* repressor (from *I*+ cells) but not the one lacking repressor (from *I*− cells) bound IPTG, resulting in a higher concentration of radioactivity inside than outside the dialysis sack. This assay was used to purify Lac repressor protein.

DNA, although finding such association was not a conclusive proof of mechanism. At the time, the most straightforward way to determine whether repressor binds to DNA was a centrifuge experiment. In a centrifugal field, the large DNA molecule sediments rapidly and the smaller protein sediments slowly; if repressor bound to DNA, then the protein should move together with the heavier DNA in the centrifugal field. Detection of the protein required either a radioactively labeled repressor that was "radioactively pure" (the only radioactively labeled band in the protein preparation) or a completely pure repressor protein preparation. Gilbert did not have either one. However, Mark Ptashne was in a position to do this experiment because he had prepared radioactively pure λ cI repressor protein.

I first met Ptashne in 1964, when I talked about our results on regulation of λ at an annual meeting of molecular biologists—the Gordon Conference on Nucleic Acids. I was discouraged because there seemed to be little interest in gene regulation from an audience that consisted mostly of biochemists. Ptashne had come to molecular biology with the express purpose of figuring out gene regulation; he was then a graduate student with Matt Meselson, studying λ DNA replication. Mark was supremely self-confident, brash, determined, and unawed by the collection of famous folk. He commented on my talk, "Don't feel bad, those guys just weren't smart enough to understand." A listener to our conversation summarized, "That fellow is the prototype of the derepressed mutant."

On his own at Harvard as a Junior Fellow in 1966, Ptashne set out to identify the λ repressor. He had a fierce determination to solve the mechanism of repression, and it paid off. In a sense, he was the complement to Gilbert, his fellow Harvard biologist and competitor. Ptashne was driven by his interest in molecular mechanisms; he was good at choosing the right experimental techniques but was not at that time a technical innovator. Gilbert and Ptashne would have been ideal collaborators, except that each was intensely competitive and wanted to be Number One. However, by virtue of a common location, they wound up contributing to each other's research in a major way.

Mark Ptashne

Ptashne's idea was to label cI repressor with radioactive amino acids so that he could "track" whether it bound to DNA, but this was not a simple task. The phage side of the experiment was straightforward. When λ infected a cell with a quiescent λ prophage, the cI repressor present in the cell would repress all other λ proteins, so cI should be the only phage protein made (see Figure 5–9a). If many phages infected each cell, quite a bit of cI should be produced. When radioactive amino acids were added to label the new proteins synthesized, cI should be one of the few phage proteins that became radioactively labeled. However, the vast array of host proteins would also appear as radioactive proteins, making the quest for cI extremely difficult. So host protein synthesis had to be severely reduced to make the identification of cI a practical endeavor. A method to do this came from techniques developed for phage T4 by Junko Hosada and Cyrus Levinthal, who found that synthesis of host proteins could be eliminated by irradiating the host with ultraviolet light prior to phage infection. At a high UV dose, host DNA is sufficiently damaged that it can no longer make mRNA, causing host protein synthesis to stop. However, the host cell will still be able to make mRNA and proteins from undamaged phage DNA added after the irradiation.

Irradiating host cells containing the λ prophage with UV introduced another complication: DNA-damaging agents such as UV light will induce the λ prophage to lytic growth. Fortunately, Jacob and Campbell had identified a mutant cI protein that did not

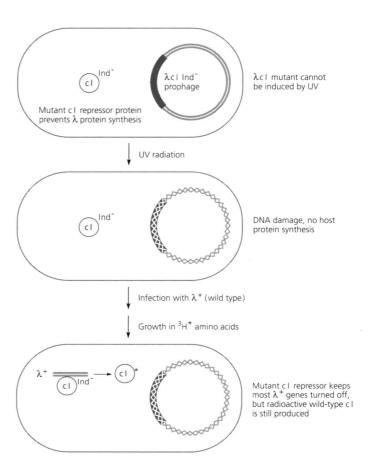

FIGURE 5–11. Ptashne used a clever scheme to trick infected cells into synthesizing the cI protein as one of the few radiolabeled proteins in the cell. Cells lysogenic for a prophage mutant that was not induced by UV irradiation (λcI Ind⁻) were irradiated with UV to prevent synthesis of host proteins. These cells were then infected with high concentrations of λ⁺ (wild type) phage and grown with radioactive (*) amino acids. The λcI Ind⁻ mutant repressor present in the irradiated cells prevented synthesis of most phage proteins, so, under these conditions, radioactive, wild-type cI repressor was one of the few proteins made. It was easy to use standard purification procedures to get rid of the other radioactive proteins present so that cI repressor was the only radioactive protein remaining.

undergo UV induction, called λ*cI* Ind⁻ (for noninducible). This mutant permitted Ptashne to perform his experiment. By irradiating a host cell carrying a *cI* Ind⁻ prophage, Ptashne could prevent host protein synthesis while retaining active λ repressor to repress most transcription from the incoming λ*cI*⁺ phage. λcI protein should be one of the predominant proteins labeled by the radioactive amino acids added after infection. Indeed, Ptashne found a radioactive protein band that was produced after infection by normal λ but not after infection by a *cI*⁻ mutant. By using standard protocols for purification of proteins, cI could be separated from other radioactive proteins, so that it was the only radioactive protein present in the protein preparation (Figure 5–11).

cI repressor protein binds to the λ operator

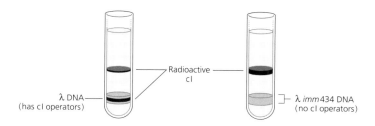

λ DNA
(has cI operators)

Radioactive
cI

λ imm434 DNA
(no cI operators)

FIGURE 5–12. To find out whether cI protein bound to operator DNA, Ptashne mixed radioactively pure cI repressor with either λ DNA or λimm434 DNA, a closely related phage identical to λ except for a different operator region. He then centrifuged these mixtures to separate the large DNA molecules that sedimented rapidly from the small protein molecules that sedimented slowly. He saw that radioactive cI sedimented with λ DNA (which has cI operator sites) but not with λimm434 DNA (which lacks these sites), showing that cI binds operator DNA.

With radioactively pure cI in hand, Ptashne could do the critical experiment. Does cI bind to λ DNA but not to DNA from the related λ*imm*434 phage, which has almost all of λ except for the λ operator region? Ptashne mixed DNA with the radioactive cI protein and centrifuged the mixture to separate the large DNA molecules from the small protein molecules. The samples collected from the centrifuge tube were analyzed for DNA (detected because it absorbed UV light) and for repressor (detected by radioactivity). The first experiment worked—much of the cI repressor sedimented with λ DNA, but none sedimented with λ*imm*434 DNA (Figure 5–12).

Soon after the experiment with λ cI, Gilbert and Müller-Hill succeeded in purifying sufficient amounts of Lac repressor to do the analogous DNA binding experiment. Lac repressor sedimented with φ80 *lac* DNA (which had the *lac* operator) but not with φ80 DNA (which lacked the *lac* operator). Thus, Lac repressor bound specifically to *lac* operator DNA.

These experiments, performed in 1967, yielded the clear conclusion that the λ cI and LacI repressors worked by associating with operator DNA. An interesting historical point is that cI and LacI were most fortunate regulatory proteins to use in this experiment. For this experiment to work, the binding of the protein to DNA must be extremely tight. Most other DNA-binding regulatory proteins do not bind tightly enough to their DNA targets to be detected in this assay.

The question now became how repressors worked. Of course the odds were highly in favor of their inhibiting the RNA polymerase reaction, but this needed to be demonstrated *in vitro* so the repression mechanism could be studied. In 1967, Linda Pilarski and I showed that a cell extract made from *E. coli* cells containing λ prophage producing extra cI could inhibit transcription programmed by λ DNA but not that programmed by φ*imm*434 DNA. Although this result was consistent with cI repressing transcription *in vitro*, transcription was inhibited only twofold, far less than the amount of repression observed *in vivo*. In order to clarify the situation, we needed a clean system with purified repressor, and we needed to know more about the RNA polymerase reaction. The solution of these problems took until 1970 and required the efforts of many people.

THE BIOCHEMISTRY OF RNA POLYMERASE

By 1967, it was clear that RNA polymerase made RNA, but very little was known about how it started and stopped RNA chains. We needed defined templates and precise assays for correctly initiated and terminated transcripts. The DNA of phages λ, T4, and T7 served as the templates that eventually provided this information. Peter Geidusheck and his colleagues gave this approach an early boost with the demonstration that purified *E. coli* RNA polymerase copied only one of the two DNA strands of phage T4 and transcribed largely in the early gene region, as expected for the host enzyme in the absence of phage intervention.

Between 1967 and 1970, experiments from several research groups defined the general mechanism for how RNA polymerase selected promoters, the specific binding sites for RNA polymerase on the DNA. An analysis of the structure of RNA polymerase, mostly by Wolfram Zillig and his associates in Munich, revealed that the active enzyme was a composite of several polypeptide chains (subunits). Dick Burgess and Andrew Travers, working at Harvard with Watson, found that one of these subunits, sigma (σ), was easily dissociated. Without σ, the enzyme started RNA chains from an intact double-strand DNA very poorly. Similar results were obtained by John Dunn and Ekhardt Bautz at Rutgers. RNA polymerase could thus be divided into a "core" complex lacking σ and a "holoenzyme" containing σ. The core enzyme carried out normal chain elongation, but only the holoenzyme recognized promoters. The σ subunit must be the key ingredient for promoter selection. Burgess and Travers then showed that the σ subunit was actually dissociated from RNA polymerase during the chain elongation process. This switch from holoenzyme to core is a clever way to separate the two major tasks of the enzyme. Holoenzyme precisely locates

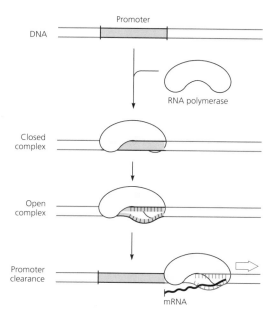

FIGURE 5–13.   With the help of the sigma subunit, RNA polymerase recognizes and binds to the promoter in a "closed complex," so called because the DNA strands at the promoter region remain base-paired. RNA polymerase then separates the DNA strands around the start site of transcription, creating the "open complex" primed for RNA synthesis. In the "promoter clearance" step, RNA polymerase initiates transcription, moves off the promoter, and releases sigma. RNA polymerase is now ready to finish traveling down the DNA.

the promoter site by efficient binding; the core enzyme moves rapidly along the DNA during the process of copying one of the two strands of DNA into RNA.

How does RNA polymerase locate the promoter and begin RNA synthesis? Mike Chamberlin and colleagues at Berkeley, as well as Zillig and his associates, found that RNA polymerase bound extremely tightly at the promoter site and then rapidly initiated RNA chains. However, this rapid-firing complex of enzyme and DNA was not formed efficiently when the duplex base-pairing of the DNA was extremely strong (say at low temperature). Chamberlin used these and other experiments to define a multi-stage process of chain initiation. RNA polymerase first recognizes the base sequence of the promoter site from the "outside" of the duplex, forming a "closed" complex. In a second stage, the promoter DNA is then partially unwound, yielding an "open" complex that is primed to start the RNA chain. The enzyme initiates RNA synthesis by formation of the first dinucleotide linkage and proceeds down the DNA, locally unwinding the duplex to copy the DNA into RNA (Figure 5–13).

Jeff Roberts

How does RNA polymerase know when to terminate the transcript? RNA chains are terminated at two different types of terminators. Intrinsic terminators exhibit a characteristic RNA structure that allows RNA polymerase to terminate on its own. Rho-dependent terminators require a special termination protein, called rho, for termination. For a time, this duality of mechanism caused some confusion. Bob Millette found that RNA polymerase alone could terminate transcription in the early gene region of the phage T7, but Jeff Roberts, a student with Gilbert at Harvard, found that it could not terminate transcription from the p_L and p_R promoters of phage λ. However, extracts of *E. coli* contained a protein that allowed RNA polymerase to produce two shorter RNAs whose sizes were consistent with the earliest stage of transcription in cells infected with λ (a stage before N acted, one that would include transcription of N). Later studies showed that this protein was rho. Early gene expression in phage T7 is controlled by an intrinsic terminator; that of λ is controlled by the rho-dependent terminators t_L and t_R.

Roberts' work on termination in λ was extremely important in three ways. First, he defined a special mechanism for ending many RNA chains in λ and *E. coli*. Second, analysis of the size of the λ early transcripts explicitly defined the two RNAs that were the target of cI-mediated repression. Third, the existence of rho-mediated termination for λ RNA led Roberts to propose that the λ N protein might act as a positive regulator by preventing the termination events, allowing the p_L and p_R transcripts to proceed into the downstream genes. Although not embraced immediately, as described later, this proposal for regulation by anti-termination turned out to be correct.

Increased biochemical understanding of RNA polymerase allowed a more accurate study of the regulation of λ transcription to be performed. "Correct" transcription from p_L and p_R could be increased by selecting RNA polymerases in the "open" (ready-to-start) complex. RNA polymerases bound at other nonpromoter sites (such as DNA ends) could sometimes start RNA chains, but these relatively slow reactions were selectively removed by agents that dissociated or inactivated nontranscribing enzymes. Moreover, the correct RNA chains could be recognized by their specific size, defined by starting at p_L and p_R and terminating at t_L and t_R in the presence of the termination protein ρ. Using these two approaches, selective initiation of RNA synthesis and selective analysis of RNA chains, along with purified repressor, our group and the Ptashne

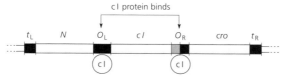

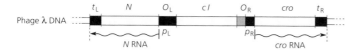

FIGURE 5–14.   In the absence of cI repressor protein, RNA polymerase initiates transcription from the strong p_L and p_R promoters and terminate transcription at the t_L and t_R terminators to give two transcripts each having a defined size. Binding of cI repressor at the O_L and O_R operators inhibits transcription from these promoters, even though RNA polymerase is present in the cell.

group were able to show that λ repressor specifically inhibited transcription from the p_L and p_R promoters. From these experiments, we knew that the cI protein turns off RNA polymerase to provide the dramatic transcriptional shutoff of the entire genome that Jacob envisioned in 1959 (Figure 5–14).

POSITIVE AND NEGATIVE REGULATORS OF THE *lac* OPERON

Just when everything seemed to be working out as predicted by the prophets Jacob and Monod, the *lac* operon turned up with a new twist. Although purified Lac repressor, active in binding to its operator, was available in 1967, all efforts to show that it negatively regulated transcription from the *lac* promoter were unsuccessful. Resolving this issue led to the discovery that something was missing from the purified system—a positive regulator to get specific transcription going in the first place.

Since the purified system used to study *lac* repression seemed not to be working, Geoff Zubay decided to study *lac* regulation in a crude cell extract containing most cellular proteins. If negative regulation could be demonstrated, he could then use the extract system to purify the missing proteins. Disappointingly, when Zubay added *lac* DNA to his extract, only a small amount of β-galactosidase activity was produced. However, a magic ingredient to boost β-galactosidase synthesis was soon uncovered by the seemingly unrelated studies of Ira Pastan and Robert Perlman.

Pastan and Perlman were trying to understand why adding glucose to cells inhibited β-galactosidase production. They found that

Ira Pastan

Jon Beckwith

they could "cure" glucose repression by adding the cyclic AMP nucleotide (cAMP for short) to cells. So cAMP was clearly involved in the regulation of the *lac* operon. But one of the steps in Zubay's extract preparation would have removed cAMP and other small molecules. Zubay added cAMP to his extract and found that β-galactosidase synthesis was greatly increased. Suddenly there was an unexpected new regulatory phenomenon, probably involving positive regulation, and a way to study it.

What was the target for cAMP, and how could it be identified? At this point, genetics came to the rescue. Jon Beckwith figured that a mutant defective in the presumed cAMP activating system would have a most unusual phenotype. Since glucose inhibited the expression of many different operons involved in utilizing sugars, cAMP was likely to have an effect on multiple operons. If this was a positive regulatory system, then mutations inactivating this system should prevent utilization of many different sugars. Beckwith and Daniele Schwartz searched for and found mutants simultaneously unable to grow on two sugars—arabinose and maltose. Spectacularly, the mutants were also unable to grow on lactose, as would be expected if they defined the positive regulatory system involved in the glucose effect. Similar studies were carried out by Pastan and Perlman. These mutants defined two different genes. Why had neither gene been found in previous genetic studies of individual sugar operons? They undoubtedly had been isolated but had been put aside as too hard to understand.

Zubay's extracts now provided an *in vitro* system to clarify the nature of the two genes defined by the mutations. Extracts from each mutant strain were defective in production of β-galactosidase. Those from one mutant strain were restored to activity by adding cAMP. This strain is now known to have a defect in the *cya* gene, coding for the enzyme that produced cAMP. Added cAMP did not restore production of β-galactosidase to the extract from the second mutant strain. With this extract, Zubay used *in vitro* complementation to identify the factor that was missing for production of β-galactosidase just as had been done previously to identify DNA replication proteins (see Figure 4–14). Simultaneously, Pastan and colleagues isolated the cAMP receptor protein (variously called CRP or CAP) by its capacity to bind cAMP, following the same procedure that Gilbert had used to isolate the Lac repressor (see Figure 5–10). Both groups then demonstrated that RNA polymerase required both CRP and cAMP to transcribe *lac* well. Beyond any doubt, transcrip-

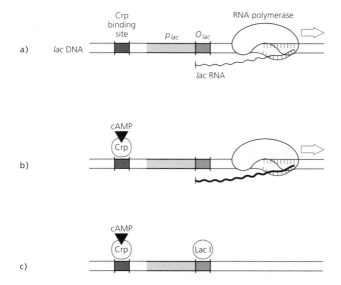

FIGURE 5–15. Without any regulators, RNA polymerase transcribes the *lac* promoter very poorly (a). When the positive regulatory protein Crp is added along with its small molecule effector cAMP, transcription is stimulated about 20-fold (b). Transcription of the *lac* operon can be completely shut off by adding the LacI repressor protein, even though RNA polymerase and Crp-cAMP are present (c).

tion of the Lac operon depended on a positive regulator (Figure 5–15a, b).

With the newly found ability to transcribe the *lac* promoter in a purified system, the Beckwith and Pastan groups could then ask whether addition of the *lac* repressor inhibited transcription. Using approaches conceptually similar to those described for examining λ repressor, they demonstrated clearly that Lac repressor was indeed an inhibitor of transcription by RNA polymerase (Figure 5–15c). Both positive and negative regulators acted to control the RNA polymerase reaction.

THE GENERAL VIEW OF REGULATORY PROTEINS: 1972

The Jacob-Monod model for negative regulation posed three general biochemical questions: What is a repressor? Where does it act? How does it control gene expression? Between 1967 and 1972, these questions were answered for Lac and λ. The Lac and λ repressors are proteins that bind to operator DNA adjacent to the promoter and inhibit the capacity of RNA polymerase to transcribe the operon. Although the available technology did not allow a definitive conclusion, the transcription experiments indicated strongly that the regulatory protein blocked the initiation step of RNA synthesis. The mechanism for positive regulation by Crp was less clear. We now know that Crp does bind specifically to its DNA site; however, the assays available at that time showed only that Crp associated with DNA in a cAMP-dependent reaction. In spite of limited knowledge, a general molecular model emerged for operon regulation: positive and negative regulators are site-specific DNA-binding proteins that control access of RNA polymerase to the promoter site.

This general model posed three new, more specific biochemical questions: How does the regulatory protein recognize its operator site in DNA? How does RNA polymerase recognize its promoter region? How does the regulatory protein direct RNA polymerase to transcribe or not to transcribe? The answer to these questions awaited the development of the technology for determining the sequence of nucleic acids. But before this technology was developed, our ideas about regulation had been considerably broadened.

THE IDEA OF ANTITERMINATION REGULATION: PROBLEMS IN PARADISE

The Jacob-Monod gospel that regulators worked by altering transcriptional initiation was remarkably successful. Initially all regulation was considered to be negative; later the model was reluctantly expanded to include activation. However, some nagging problems remained in extrapolating this model to all cases of regulation. Explication of these problematic systems unveiled a mode of regulation that was completely different from the one envisioned by Jacob and Monod—control over transcription termination.

The genes specifying the synthesis of the amino acid histidine (*his* operon) were repressed by the presence of added histidine. To search for the repressor, Bruce Ames and collaborators isolated and studied many constitutive mutants that had lost repression. But none of the constitutive mutants fit the proper mold of defining a repressor gene; instead, they all affected the production or properties of the histidine·tRNA complex. Intrigued by the apparent involvement of protein synthesis in repression, Ames and Phil Hartman suggested an alternative to the idea of the DNA-binding repressor. In 1963, they proposed that negative regulation might really work by blocking protein synthesis, only indirectly affecting transcription by stopping RNA synthesis. This idea ("transcription-translation coupling") was not generally taken seriously because the Jacob-

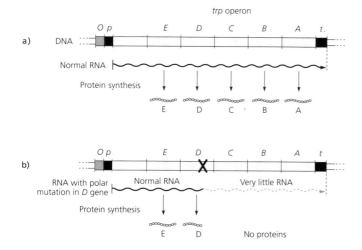

Polar nonsense mutations decrease synthesis
of RNA located downstream of the mutation

FIGURE 5–16. When the *trp* operon is induced, its long mRNA is translated into the five proteins of the operon (*a*). The presence of a nonsense mutation in gene D prematurely terminates translation of D protein. This polar nonsense mutation also inhibits transcription of the downstream *C*, *B*, and *A* genes, thus preventing expression of the C, B, and A proteins (*b*).

Monod model could be accommodated by the alternative proposal that histidine·tRNA (rather than free histidine) was the co-repressor that bound to the repressor, activating it so that it could bind to operator DNA. Even so, it was disturbing that all constitutive mutations affected the co-repressor and none could be isolated in the actual repressor gene. Interestingly, transcription-translation coupling would later be shown to be a major mode of regulation for certain operons, including *his*.

Although one heresy was quelled, there was a new outbreak in 1970. Fumio Imamoto in Osaka was studying a puzzling phenomenon called polarity. Since nonsense mutants terminated polypeptide chains in the middle of a protein, it was expected that the enzyme activity from the mutant gene was lost. However, inexplicably, in many cases the enzyme activities encoded by the downstream genes in an operon were reduced as well. Because downstream but not upstream genes were affected, the nonsense mutations were termed polar, meaning having direction as does an electric field. Moreover, loss of downstream gene expression was correlated with an apparent failure to synthesize downstream RNA. These experiments showed that, as postulated by Ames and Hartman, a failure of protein synthesis could lead to premature termination of RNA chains (Figure 5–16).

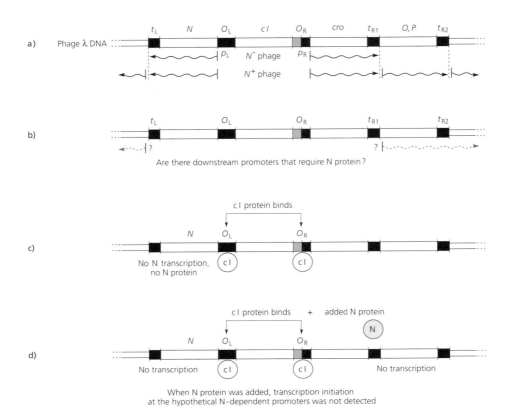

FIGURE 5–17. N protein allows the early genes in λ to be transcribed. By comparing the transcription patterns in N^+ and N^- λ phage, it can be seen that N protein allows transcription downstream of the t_L and t_R terminators (a). This led to the suggestion that N activated transcription from promoters downstream of these terminators (b). In this model, in a lysogenic cell, cI repressor protein indirectly prevents N function at the proposed promoters by shutting off transcription of p_L and thereby synthesis of N protein (c). However, supplying N protein to a lysogenic cell did not activate transcription at the promoters proposed to be downstream of t_L and t_R, thus disproving this model (d).

The λ N protein, which positively regulated early gene expression of λ, provided the most severe crisis for the model that regulators work solely to affect transcription initiation. N activated transcription of the recombination and replication genes located downstream of the t_L and t_R terminators respectively (Figure 5–17a). So, N might be expected to activate additional promoter sites downstream of these terminators (Figure 5–17b). If N works at downstream promoters to allow them to initiate transcription, there is no simple mechanism by which cI could directly block the activity of N when λ is a prophage. But, cI could accomplish this indirectly. By blocking transcription from p_L, it would prevent synthesis of N (Figure 5–17c). This model makes a clear prediction: providing N to a lysogen should activate these proposed N-dependent promoters. Denise Luzzati tested this prediction. She provided N pro-

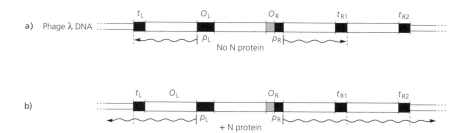

λ N protein is an antiterminator

a) Phage λ DNA

t_L O_L O_R t_{R1} t_{R2}

p_L p_R

No N protein

b)

t_L O_L O_R t_{R1} t_{R2}

p_L p_R

+ N protein

FIGURE 5–18. Transcription can be initiated from the p_L and p_R promoters in the absence of N protein, but it is quickly stopped at the transcriptional termination signals t_L and t_R (a). In the correct hypothesis for N action, Jeff Roberts proposed that the presence of N protein prevents the p_L and p_R transcript from terminating at normal termination sites (b). Since N action requires transcription from the p_L and p_R promoters, which are repressed in the lysogen, providing N protein to a lysogen is not sufficient to restore N function.

tein to a repressed λ prophage by using phage λ*imm*434, which is insensitive to cI repressor. In contrast to expectation, Luzzati found that N protein produced in this way failed to activate transcription from a λ prophage whose operator sites were occupied by cI protein (Figure 5–17d).

Most of us were conservative and sought to accommodate these results within the framework of the Jacob-Monod model of regulation solely at initiation. We proposed that activation of the hypothetical downstream promoters needed transcription from p_L and p_R as well as N protein to be activated. This mechanism seemed plausible because Dove had provided good evidence that transcription somehow activated the replication origin of phage λ. However, Roberts provided the correct idea in 1970 based on his biochemical work with termination protein rho. Roberts proposed that N protein blocked the action of rho, which was required for termination at the t_L and t_{R1} terminators, thereby allowing the p_L and p_R transcripts to extend into the rest of the early gene region. This explained why N could not work when added to a lysogen. Since cI blocks initiation at the p_L and p_R promoter sites, there would be no RNA chains to antiterminate (Figure 5–18).

The idea of N-mediated antitermination gave a simple and clear picture for how N protein might work. The concept also broadened the ways in which people thought about regulation. However, an attractive idea does not prove a mechanism. The definitive evidence that termination could be regulated was first provided by work with the *trp* operon. I will describe this work and then return later to the complicated (and still incompletely understood) interactions involved in the action of N protein as an antiterminator.

Charley Yanofsky

The *trp* operon of *E. coli* consists of five genes encoding the enzymes that catalyze the synthesis of the amino acid tryptophan. However, most of us think of it as the Charley Yanofsky operon because of his 40 years of work on the molecular genetics of tryptophan biosynthesis. Yanofsky began his research in the late 1940's, following the steps of Beadle and Tatum with the mold *Neurospora*. After receiving his Ph.D., he began to work with *E. coli*. Yanofsky worked out the gene-enzyme relationships responsible for tryptophan production in the 1950's. As noted in Chapter 2, Yanofsky's studies of the *trpA* gene product most clearly established the colinearity of gene and protein (Figure 2–4).

By the late 1960's, the coding problem was solved and the Jacob-Monod model seemed to explain most of regulation, even at a molecular level. Nearly everyone from the early coding group moved into new fields, sometimes accompanied by pronouncements that everything fundamental was known in molecular genetics (at least for prokaryotes). Gunther Stent wrote a book to make the general point, and he extended the analysis to proclaim that everything had been done in art and music as well. However, many of us were convinced that the fun was just beginning and that the molecular genetics of bacteria and phage would remain a source of exciting new principles about the expression, duplication, and recombination of genomes. We were right. Yanofsky was a member of the latter school. He was rewarded by the discovery of a brand new regulatory element, the attenuator.

Someone once described Charley Yanofsky to me as the ultimate "blue collar" scientist. He meant Yanofsky's dig-it-out, data-based approach in which the biological problem was studied in detail from all possible experimental angles and then analyzed for mechanism—no pyramid of hypotheses, spectacular guesses, or intuitive leaps. In the early 1970's, all experimental roads in Yanofsky's lab led to regulated transcription termination—a new regulatory concept with a completely unanticipated mechanism.

In early studies of *trp* operon regulation, everything seemed to follow the Jacob-Monod model. A repressor gene had been identified by constitutive (*trpR⁻*) mutations and *trpR⁻* cells produced high levels of *trp* operon RNA as expected. Purification of the Trp repressor was under way. However, Yanofsky's broad program on *trp* repression began to turn up some complications. First, Ron Baker

a) Attenuation with repression

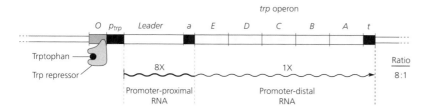

b) Attenuation in the absence of repression

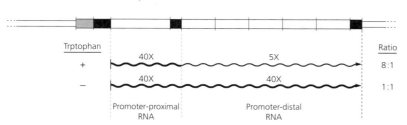

FIGURE 5–19.   By examining the relative production of RNA close to the promoter ("promoter proximal") and RNA downstream from the promoter ("promoter distal"), Yanofsky found that the *trp* operon is regulated by an attenuator ("*a*") site as well as by the Trp repressor. The attenuator is located within the transcribed RNA, immediately downstream of the "leader" region of the RNA and promotes termination (*a*). The attenuator works independently of Trp repressor: it decreases transcription about eight-fold, either in the presence or absence of Trp repressor. Tryptophan is required both for attenuation and repression control; in its absence, RNA expression is high all along the operon (*b*).

found that, even when the cells lacked the Trp repressor, production of *trp* mRNA increased in the absence of tryptophan. This was puzzling. Tryptophan was considered to function as a co-repressor, binding to the TrpR protein and converting it to an active repressor. When TrpR was inactivated by mutation, tryptophan should no longer be able to repress transcription of *trp* mRNA. Then, Ethel Jackson found that deletions of DNA downstream from the operator-promoter region increased expression of the *trp* operon, indicating that some other regulatory element might be present. Finally, Frank Lee and Cathy Squires found that transcription of the normal *trp* operon by purified RNA polymerase yielded a most surprising result: almost all of the RNA terminated after 140 bases, before transcription of the first gene of the operon had even begun.

In an attempt to figure out what was happening, Yanofsky and Kevin Bertrand turned to measuring the amount of RNA synthesized at various parts of the *trp* operon, comparing the amount of RNA made from DNA adjacent to the promoter (promoter proximal RNA) with that from DNA further downstream (promoter distal RNA) (Figure 5–19). Under repressed conditions (when tryptophan was present), the cell produced considerably more promoter proximal than promoter distal RNA. These results clearly identified a regulatory site located downstream of the *trp* operator-promoter, but preceding the first of the genes for the tryptophan enzymes. This new regulatory element was termed the attenuator site "*a*" because RNA synthesis dropped abruptly there in the presence of tryptophan (Figure 5–19a). The attenuator site functioned independently of the operator. Constitutive *trpR⁻* mutants, which lacked the Trp repressor showed the same relative decrease in RNA downstream of the attenuator site as *trpR⁺* cells, which had the repressor (compare Figures 5–19a and b). In the absence of tryptophan, regulation at the attenuator was abolished, completely derepressing the operon (Figure 5–19b). Thus, tryptophan is required for attenuation control as well as for repression control. Control of attenuation by tryptophan explains Ron Baker's observation that tryptophan reduces *trp* gene expression even in a *trpR⁻* cell.

The attenuator site identified in these experiments was in the same region of the *trp* operon as the site where purified RNA polymerase terminated transcription of *trp in vitro*. So, regulation at the attenuator was almost surely by an antitermination mechanism. Understanding how this antitermination system worked required insight into how RNA chains are normally terminated and a knowledge of events occurring in the region upstream of the attenuator, called the "leader region," where tryptophan must somehow intervene to alter termination. Fortunately, the rapid development of methods for determining nucleic acid sequences allowed these questions to be answered.

The "clover-leaf" model for tRNA developed by Bob Holley had focused attention on the likely functional importance of regions in RNA that could base-pair with each other and create ordered structures in RNA (see Figure 6–2). So, when Frank Lee and Craig Squires obtained the RNA sequence up to the attenuator site, it was examined for possible base-paired secondary structures. Several such "stem and loop" structures were found. *trp* RNA could be folded into a stem-and-loop secondary structure (the 3–4 stem and loop in Figure 5–20a) followed by a terminating string of U nucleotides at the attenuator site. This structure was also found at other intrinsic termination sites, explaining why RNA synthesis terminated here *in vitro*. By itself, this was not especially interesting. However, two other features of the sequence were fascinating. First, the sequence had a less obvious additional stem-and-loop secondary structure lacking a

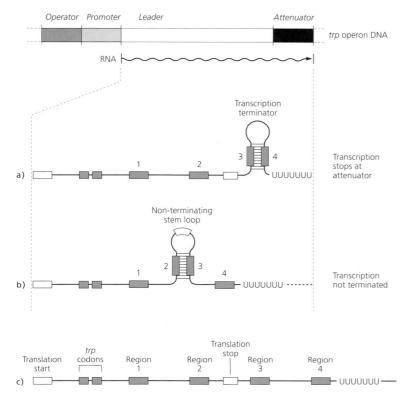

FIGURE 5–20.   When the leader region was sequenced, Yanofsky found that the attenuator encoded a classical terminator: a stem and loop structure (called the 3–4 structure in the diagram) followed by a string of U's, explaining why transcription decreased eight-fold at the attenuator (*a*). He also found another possible stem and loop structure (called the 2–3 structure in the diagram), whose formation might compete with formation of the 3–4 terminator (*b*). Finally, he found translation start and stop signals for a small peptide. Since the RNA transcript contained two *trp* codons, this peptide would include two *trp* residues (*c*). Yanofsky suspected that the presence of tryptophan in this peptide might be involved in regulation of the attenuator.

terminating string of U nucleotides (the 2–3 stem and loop in Figure 5–20b). Interestingly, if this alternative stem-and-loop structure formed, it would capture the 3-segment of the 3–4 stem and loop, thus disrupting the terminating stem-and-loop structure at the attenuator. Second, there was a start and stop sequence for protein synthesis—perhaps the leader RNA was translated. The possible translation product had two tryptophan amino acids in a row, indicating a potential entry point for regulation by tryptophan (Figure 5–20c). After all, first Ames and Hartman and later Imamoto had provided evidence for coupling of transcription and translation.

Peering at the alternative RNA structures on his blackboard, Yanofsky saw a way to couple translation of the leader peptide to termination of RNA synthesis. When the small peptide in the leader region (the "leader" peptide) was translated in the presence of tryptophan, ribosomes would move across the complete leader message to

FIGURE 5–21. When tryptophan is present, translation of the leader peptide continues past the two tryptophan codons to the translation stop site. Translation of the entire leader peptide allows the ribosome to block formation of the 2–3 stem-loop structure, thereby promoting formation of the competing 3–4 terminator structure and termination of RNA synthesis (a). When tryptophan is absent, translation of the leader peptide stalls at the *trp* codons, allowing formation of the 2–3 stem loop and preventing formation of the competing 3–4 terminator structure. RNA synthesis continues into the *trp* operon genes, and ribosomes bind to the exposed translation start sites allowing expression of the *E, D, C, B,* and *A* gene products.

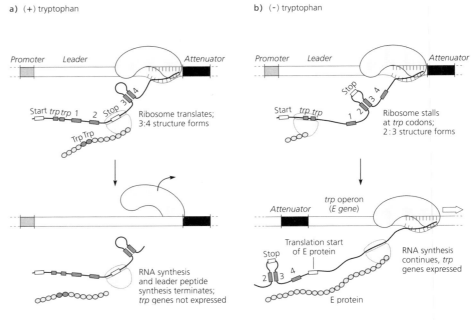

a) (+) tryptophan

b) (−) tryptophan

the stop codon and prevent formation of the 2–3 stem-and-loop structure. Under these conditions, the 3–4 stem-and-loop structure should form efficiently, and RNA polymerase would terminate (Figure 5–21a). In contrast, when the leader peptide was translated in the absence of tryptophan, ribosomes would stall at the Trp codon because there was no tryptophan·tRNA. A ribosome at this position would allow the 2–3 stem-and-loop structure to form, delaying formation of the 3–4 terminator stem-and-loop structure. Since RNA polymerase sees no terminator sequence, the enzyme will continue RNA synthesis into the downstream structural genes for the tryptophan enzymes (Figure 5–21b). The presence of tandem Trp codons, a key in pointing to the mechanism, ensures an efficient and sensitive response to a shortage of tryptophan in the cell (Figure 5–21).

This picture of the attenuator structure and its translational regulation, developed in 1976, has served to explain the transcriptional control of the *trp* operon and all other operons for amino acid synthesis in *E. coli* examined so far. When Yanofsky first told me of his scheme for regulation, I commented that the idea was wonderful but the two Trp codons could be a fortuitous property of the leader peptide. The next time I saw Yanofsky, he said, "What about seven histidine codons in a row in the leader peptide for the histidine operon?" I agreed that fortuity went only so far. Whereas the *trp* operon is regulated both by repression and attenuation, the *his* operon is regulated only by attenuation, explaining why all of the constitutive mutations in *his* isolated by Ames altered histidine tRNA.

One general feature of *trp* and other amino acid operons not used in the very first model is a third secondary structure, the 1–2 stem and loop, present earlier in the RNA sequence (see Figure 5–20). In the absence of any protein synthesis, the 1–2 structure will form, preventing formation of the 2–3 stem-and-loop structure, thus ensuring that the terminating 3–4 stem-and-loop structure will form. There is no point in making RNA for amino acid operons if protein synthesis is not happening.

Three critical predictions follow from the Yanofsky attenuator model. First, synthesis of the leader peptide should be required for *trp* operon transcription. Second, mutations changing the codon for the limiting amino acid should change the amino acid to which the operon responds. Third, mutations disrupting the 2–3 structure should prevent RNA synthesis from the operon. These predictions have been verified for *trp* and other amino acid operons.

The understanding of termination control in the *trp* operon provided a way to think about Imamoto's finding that nonsense mutations in *trp* reduced RNA synthesis of downstream genes (see Figure 5–16). Perhaps protein synthesis prevents transcription termination by covering the RNA sequences that signal termination. When protein synthesis is stopped prematurely by a nonsense mutation in a structural gene, the structures or sequences that provoke premature termination of the multi-gene RNA chain will be unmasked and diminish downstream gene expression. In support of this idea, John Richardson showed that a mutation reducing the activity of the termination protein rho prevents the reduction in downstream RNA synthesis caused by nonsense mutations. Thus, the "polarity" of many nonsense mutations probably arises from unwanted rho-mediated termination events.

The attenuator mechanism established a mode of regulation for bacterial operons that was not even suspected in the early regulatory world of Jacob and Monod. Parallel work established antitermination as the major regulatory principle for lytic development by phage λ. There were clearly multiple pathways in the regulatory universe.

Armed with an expanded regulatory framework, we return now to the story of how the λ N and Q proteins positively regulate λ development. Jeff Roberts was the first to suggest that N worked by preventing termination. We now know that both N and Q carry out their regulatory roles by eliminating the normal response of RNA polymerase to all terminators—both rho dependent and intrinsic. However, the mechanism by which N and Q act remains somewhat murky even at the present.

Evidence that N was an antiterminator was first provided in 1974 by a clever genetic experiment using polar nonsense mutants that cause RNA synthesis of downstream genes to terminate (shown for the *trp* operon in Figure 5–16). If N was an antiterminator protein, it might override this effect, thereby restoring expression of previously silent downstream genes. Naomi Franklin and Sankar Adhya used special genetic constructions to show that the N protein reversed the effect of polar mutations in the *E. coli trp* operon and the *gal* operon, respectively. Later, Douglass Forbes and Ira Herskowitz provided a similar genetic argument for antitermination by λ Q protein, the positive regulator of late gene expression in λ. A curious historical footnote is that, because the N and Q proteins prevent polarity in λ, nonsense mutations in λ, unlike those in *E. coli*, do not affect downstream gene expression. Because λ does not exhibit polarity, the Campbell nonsense mutations affected expression only of the gene in which they resided. This accounts for the utility of Campbell's collection of nonsense mutations that were so instrumental in understanding λ development.

These genetic experiments provided a strong indication for regulation by antitermination but did not reveal the biochemical mechanism for accomplishing this. The quest for the mechanism of N-mediated antitermination has been important not only for its own sake but also for the valuable new insights into the transcription process itself. Transcription has turned out to be a more dynamic enterprise than was initially supposed. Additional proteins interact with RNA polymerase during elongation to influence the termination response of the enzyme. The genes for these host proteins were identified as *E. coli* mutations that prevented N from working. David Friedman, who carried out most of the genetic work, termed these genes *nus* (for N-utilization substance) and found three new genes:

Purification of λ N protein was possible because it promoted expression of endolysin

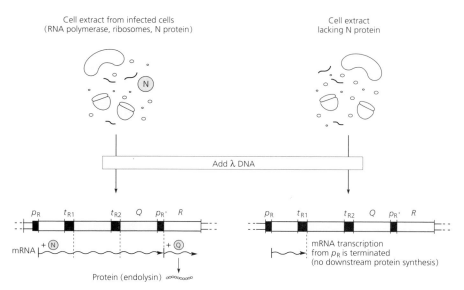

Cell extract from infected cells
(RNA polymerase, ribosomes, N protein)

Cell extract
lacking N protein

Add λ DNA

p_R t_{R1} t_{R2} Q $p_{R'}$ R

mRNA $+$ N

Protein (endolysin)

p_R t_{R1} t_{R2} Q $p_{R'}$ R

mRNA transcription
from p_R is terminated
(no downstream protein synthesis)

FIGURE 5–22. Using a crude protein extract that carries out both transcription and translation when programmed with added DNA, Greenblatt developed an assay for N function. An extract from λ infected cells (*left*) but not one from uninfected cells (*right*) makes endolysin, the product of the R gene when programmed with λ DNA. Since several termination sites (t_{R1}, t_{R2}) intervened between the promoter and the R gene, the infected cell extract was carrying out antitermination. To purify N, Greenblatt fractionated the infected cell extract and identified the fraction that allowed an uninfected extract to produce endolysin. Endolysin is actually a "late gene" and requires λQ protein for expression. The assay worked because when N was present, the extract first made Q protein, which then allowed expression of endolysin. This same type of assay was later used to purify the Q protein.

nusA, *nusB*, and *nusE*. Remarkably, the special multi-protein complex assembled in the presence of N protein, which includes three to five host proteins as well as N, does not respond to any known specific termination signals. According to these ideas, transcription would be eternal. Why then, does N not suffice for late gene transcription? The answer to this question is not totally clear. The need for a second antitermination system probably results from the fact that multiprotein complex assembled by N is not stable—it decays over time. The Q antitermination modification to RNA polymerase is very stable, allowing RNA polymerase to bypass all of the termination signals in the very long late gene transcript.

The initial biochemical insights into N-mediated antitermination were provided by Jack Greenblatt, who used an extract system to identify an activity for N protein. Greenblatt found that an extract made from λ-infected cells could make the *R* gene product, endolysin, even though this gene was separated from the p_R promoter by several termination sites; an extract from uninfected cells cannot make endolysin. Greenblatt used this assay for N function to purify the protein (Figure 5–22).

The next task was to identify the host *nus* proteins needed for

Jack Greenblatt

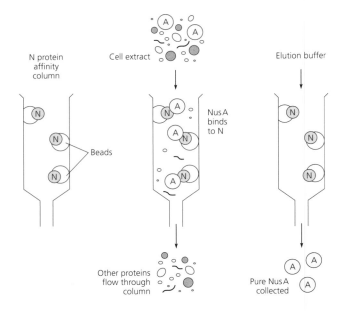

FIGURE 5–23.   In order to detect proteins that associated with the λN protein, N was bound to beads that were packed into tube, or "column." A cell extract containing many proteins was passed through the column, and one of these proteins (NusA) had an "affinity" for and bound specifically to N. After all of the other extract proteins had passed through the column, NusA was made to release its grip on N with a special buffer and then collected as it came out of the column.

antitermination. Because Friedman's genetic strategy suggested that *nus* proteins interacted with N, to search for Nus proteins Greenblatt used a powerful biochemical method, called "affinity chromatography," which identifies proteins that interact. One protein (e.g., N) is linked to insoluble beads, and a mixture of other proteins in solution is passed through the beads. A soluble protein (e.g., Nus) binding tightly to N will be retained on the column and only released ("eluted") by a new solution that strongly dissociates such protein-protein interactions. In this way, Greenblatt identified NusA protein by its ability to associate with N (Figure 5–23). Greenblatt's work showed that N, NusA, and RNA polymerase are probably all present in one multi-protein complex.

More recently, Greenblatt, Asis Das, Max Gottesman, and their colleagues showed that the NusB and NusE proteins, as well as a newly identified NusG protein, also join N and NusA in the multi-protein antitermination structure. So the form of RNA polymerase that meets termination signals is preprogrammed to escape termination by the prior addition of various signaling proteins. The N protein joins RNA polymerase neither at the promoter nor at the terminator but at a special site in between—the *nut* site (which stands for N-utilization and not for the people who work in this difficult field).

In an important experiment, Greenblatt showed that NusA binds to RNA polymerase even in the absence of N, indicating that NusA is a normal component of the elongating form of RNA polymerase. Mike Chamberlin showed that NusA altered the elongation rate and influenced the capacity of RNA polymerase to terminate. These experiments were the first to show that transcription elongation can be regulated.

The mechanism of action of the Q protein is similar to that of N, but the machinery is simpler. Q protein was isolated by Jeff Roberts, using the same extract approach developed for N. In the presence of Q, RNA polymerase overrides a rho-independent termination signal just downstream of the $p_{R'}$ promoter (see Figure 5–22). RNA polymerase then transcribes all of the late-gene region. In test-tube reactions, the ability of Q protein to direct antitermination depends on NusA, but not NusB and NusE.

How do termination and antitermination really work? Based on experiments by many research groups, termination involves two events: transcription by RNA polymerase is halted, and then RNA polymerase and the completed RNA chain are released from the DNA. The RNA itself appears to play an active role in this process. For intrinsic terminators, both of these events are mediated by the terminator structure in the RNA—the stem-and-loop and string of U's described for the *trp* attenuator. For rho-dependent terminators, RNA polymerase pauses in the absence of rho. However, the enzyme will eventually continue unless rho is present to provide for release of the polymerase and RNA. Rho probably binds to RNA near the termination point and then moves along the RNA, contacting and freeing the polymerase and the RNA. The antitermination proteins may do the job by blocking the pause response of RNA polymerase—the enzyme behaves as a train that is unable to respond to stop signals because it has lost its braking system.

Control of termination is an important and widespread regulatory mechanism, operating in both prokaryotic and eukaryotic cells. Much has been learned about it, but many aspects remain for further exploration. Having reached the level of understanding about regulated termination of RNA synthesis that prevailed as of the writing of this book, we return now to the story of how initiation of transcription is regulated.

FIGURE 5–24.   To identify the *lac* operator, Gilbert used an enzyme that chops up DNA (DNase I) to degrade either free DNA (*left*) or DNA with Lac repressor bound (*right*). When repressor was bound, a 24-base-paired region of the DNA was protected from DNase I digestion. This region was the *lac* operator.

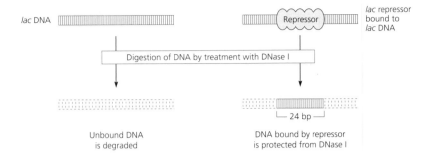

THE REGULATOR-OPERATOR INTERACTION: HOW TO LOCATE A SITE ON DNA

By 1972, a general model to account for the action of regulatory proteins that control transcription initiation had been developed: the regulator binds to a specific operator and either inhibits or facilitates a productive interaction by RNA polymerase with its promoter site. The critical problem then became understanding the nature of the operator and how it is selected from among the five million base-pairs of the *E. coli* genome.

The study of site-specific DNA-protein interactions required new technology. Many of the new approaches were developed by Wally Gilbert and his colleagues in a brilliant flurry of technological achievement. Gilbert began his study of the Lac repressor-operator interaction in an obvious way. If repressor bound tightly to operator, this DNA should be protected from degradation by the well-known DNA-degrading enzyme, pancreatic DNase, or DNaseI. DNaseI is "nonspecific"; it makes more or less random cuts within the sugar-phosphate backbone of DNA to release double-strand fragments. As a consequence, all of a *lac* DNA segment should be munched up except for the protein-protected operator site (Figure 5–24).

Gilbert and Allan Maxam found that Lac repressor protected a segment of DNA some 24 base-pairs in length from DNase degradation. Now the problem was to determine the DNA sequence of the operator region. Sequencing a relatively short RNA was routine using RNA-cleaving enzymes (RNases) that cut at specific bases in the polynucleotide chain. But such enzymes were not available for DNA. One possible solution was to transcribe the DNA segment into RNA and then sequence the RNA. Nancy Maizels accom-

a)

5' — T G G A A T T G T G A G C G G A T A A C A A T T — 3'
3' — A C C T T A A C A C T C G C C T A T T G T T A A — 5'

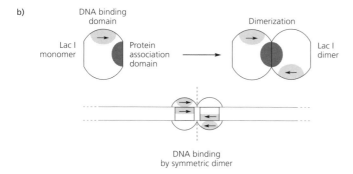

b)

DNA binding domain

Dimerization

Lac I monomer

Protein association domain

Lac I dimer

DNA binding by symmetric dimer

FIGURE 5–25.   Sequencing the *lac* operator revealed that it was partially symmetric; the upper shaded segment can be converted to the lower shaded segment by rotation of 180° around the central GC base-pair (*a*). Such sites are recognized by symmetric dimers or even tetramers of the regulatory protein (*b*). This doubles the size of the region covered by a regulatory protein, ensuring that the operator is unique in the genome.

plished this task, revealing for the first time the base sequence of an operator site (Figure 5–25a). Although not obvious at the time, transcription of the operator segment depended on the ability of RNA polymerase to start synthesis (reluctantly) from DNA ends—the operator segment did not include the essential promoter sequences.

To the casual observer, the base sequence of the *lac* operator is not especially inspiring—we already knew that the operator would contain bases. However, one very important feature is the symmetry of the sequence. The upper shaded sequence can be converted into the lower shaded sequence by rotation of 180° about the central G-C base-pair (see Figure 5–25a). The symmetric sequence of the operator had been guessed previously by John Sadler, based on his genetic study of *lac* operator-constitutive mutations.

What use is the symmetric sequence? The most obvious (and correct) possibility was that symmetry allowed a repressor protein to be used twice, doubling its recognition capacity. A symmetric dimer could match its DNA-binding regions to a symmetric operator. In this way, a protein large enough for specific recognition of 10 bases could actually recognize 20 (Figure 5–25b). Specific recognition of an operator DNA segment this long would clearly provide for binding to a unique site in the *E. coli* genome. (This calculation is actually rather easy to do. Since four bases [A,T,G,C] can occur at each position, the number of bases of DNA necessary for the sequence to appear once by random chance is 4^x [where x = the length of the particular sequence]. If the sequence were 10 bases long, it would appear at random five times in the genome [4^{10} = 1 million; total *E. coli* genome size about 5 million]; the likelihood that a sequence 20 long would appear at random is vanishingly small [4^{20} = 1 trillion]). Recognition of symmetric sequences by symmetric dimers has turned out to be a general feature of the binding of regulatory proteins to their operator sites.

How does the repressor monomer recognize its specific target sequence? There were three general possibilities: unwinding the DNA to look at each base; making each operator a special DNA structure that differed from the standard DNA double helix; or recognizing the base sequence in normal duplex DNA from the "outside" by using structural features of the bases distinct from those involved in the Watson-Crick base pairs. The first two possibilities seemed rather unlikely from the outset. Repressors found their sites too rapidly for a lengthy search by DNA-unwinding. A notably different DNA structure for each of the multitude of operator sites was hard to imagine. The third possibility made good sense; the DNA duplex has a structural feature termed the "major groove" where side groups on the DNA bases were exposed to solution and available for contact with reactive groups in proteins.

The Lac repressor specifically recognized a duplex DNA operator and not its single-stranded form, ruling out single-strand recognition. Although the base sequence did not give any indication of a deviant duplex structure, one possible variant remained in the running for a while. Alfred Gierer had pointed out that each strand of a symmetric DNA sequence could pair with itself to give a "hairpin" structure (like the RNA stem-and-loop structure involved in signaling termination). A variety of evidence argued strongly against the hairpin model, but it was only completely ruled out by later studies that used X-ray crystallography to reveal the atomic interactions involved in operator-repressor recognition.

A NEW APPROACH TO DNA SEQUENCE AND ITS RECOGNITION: SINGLE-CLEAVAGE ANALYSIS

By 1975, a new technological advance had vastly simplified the preparation of operator-containing DNA segments. Werner Arber, Ham Smith, Dan Nathans, and their colleagues had all demonstrated the existence and utility of site-specific DNA endonucleases. These

Restriction enzymes allowed the preparation of small fragments containing the *lac* operator

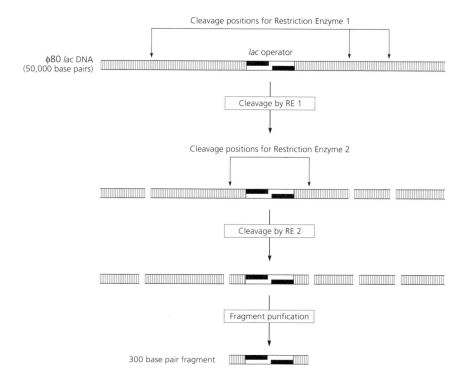

FIGURE 5–26. Previous studies of bacterial genes had to rely on phage variants that had incorporated pieces of bacterial DNA. Because phage were so large (50,000 bp), the region of interest was only a small fraction of the total phage DNA. Restriction enzymes cut DNA in specific spots, so they could be used to chop out a tiny piece of DNA that contained the *lac* operator (or any other sequence of interest). The pieces of DNA could then be purified using gel electrophoresis to separate them according to size.

enzymes were called "restriction" enzymes because their biological role is to digest foreign DNA that enters a bacterium, thus "restricting" its influence on that host cell. This work is described in detail in Chapter 10. With restriction enzymes, substantial amounts of DNA fragments of relatively small size and specific sequence could be prepared (Figure 5–26). The availability of small fragments with specific ends resulted in an explosion of new techniques for DNA–protein studies and also gave rise to the technology of genetic engineering.

Using the new restriction enzyme technology, Maxam and Gilbert could prepare the *lac* operator on a small fragment, bind Lac repressor to this fragment, and probe the molecular interactions resulting in repressor–operator binding. But, first, they needed to develop a new technology able to do two things: identify and nick the DNA at specific bases, and then detect different sized bands of DNA that would correspond to DNA nicked at different bases. To accomplish the second, they labeled their *lac* operator DNA fragment at one end with radioactive phosphate (^{32}P$\star$), separated the two strands of DNA by heating, and then used "gel electrophoresis" (an

DNA fragments can be separated by size using electrophoresis and visualized by autoradiography

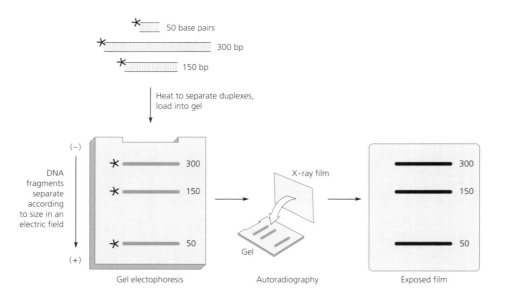

FIGURE 5–27. DNA has a negative electrical charge, and it migrates in an electric field. This property of DNA has been used to devise a procedure to separate pieces of DNA by size. Radioactive DNA fragments are loaded onto a polyacrylamide gel and subjected to an electric field (electrophoresis), which causes the DNA to move. The smaller fragments move easily through the gel and travel long distances, whereas the larger fragments move slowly. Following electrophoresis, the gel is covered with a piece of X-ray film. Radiation from the DNA locally fogs the film, allowing visualization of the DNA fragments when the film is developed. In the example shown, the double-strand DNA is first heated to separate the two strands. Only the radioactively labeled strand is visualized.

electric field in a supporting gel of polyacrylamide) to separate the single-stranded DNA molecules by size. Small DNA molecules moved rapidly through the gel whereas the larger DNA molecules moved slowly. The separated DNA fragments were visualized by "autoradiography," the blackening on an X-ray film caused by the decay of the radioactive phosphorus (Figure 5–27).

To identify and nick DNA at specific bases, Gilbert and Maxam set out to find chemical treatments that would modify or destroy each base. The first reagent they found was DMS. DMS adds a methyl group to a nitrogen of G; because of this chemical change, the base is cleaved from its sugar in an alkaline solution, and the sugar-phosphate linkage is severed leaving a single base gap in the duplex DNA. They could use this reagent to determine which G residues in the operator were protected from methylation when repressor bound. Maxam and Gilbert bound repressor to the ^{32}P★ "end-labeled" operator DNA, treated with DMS, exposed to alkali to nick the DNA, heated to ensure that the nicked fragments were single-stranded, separated the fragments with electrophoresis and visualized them with autoradiography. Any G-bases touched by the Lac repressor should be protected from modification by DMS and would therefore not be cleaved in alkali solution. However, these bases would be present in the control lane where free end labeled

FIGURE 5–28.   DNA methylated at G
residues can be cleaved in alkaline
solution (*center and right*), whereas
unmethylated DNA is not cleaved under
these conditions (*left*). This property was
used to devise a procedure to determine
the sequence of G residues in a DNA
molecule. End-labeled DNA was
methylated at G residues, treated with
alkali to cleave the methylated G's,
heated to separate DNA strands, and
subjected to gel electrophoresis and
autoradiography to identify all cleaved
bands that still contained the radioactive
label at the end of the DNA. If the DNA is
heavily methylated, then cleavage will
occur at every G base, but only the band
having the G residue closest to the end-
label will be radioactive (*center*). If the
DNA is lightly methylated (about one site
per DNA molecule), then each DNA
molecule will be cleaved at only a single G
residue. However, among the entire
population of DNA molecules, cleavage
will occur at every G in the sequence.
Since every cleaved molecule retains its
end-label, they can all be visualized by
autoradiography after size separation. By
developing specific cleavage strategies for
the other three bases, Maxam and Gilbert
were able to determine the nucleotide
sequence of any DNA that could be
isolated as a restriction fragment.

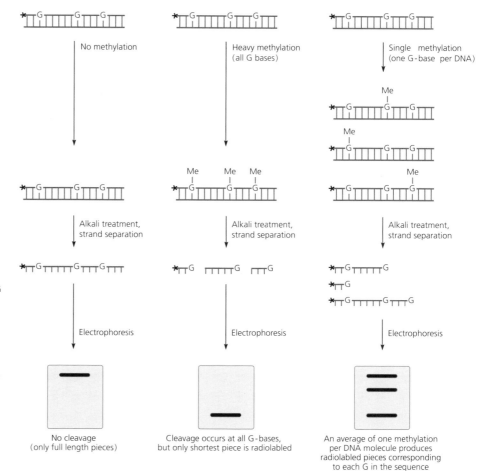

DNA (no Lac repressor present) was treated with DMS. Thus, the
repressor binding site would be identified by the absence of particular size classes on the X-ray film.

For this experiment to be successful, the extent of DMS modification was important. If the DNA were treated exhaustively with
DMS, every G residue would be modified and the DNA would be
uniformly cleaved at the first G following the labeled end—the operator would never be seen (Figure 5–28, center). To avoid this, each
DNA duplex was treated only briefly so that only one or very few
G residues per DNA molecule were modified with DMS (Figure
5–28, right).

The control experiment for this repressor protection experiment was to treat the free DNA with DMS. Generally, control experiments are boring necessities. In this case the control was spectacular—each G base appeared on the autoradiograph of the gel as a DNA fragment of specific length (Figure 5–28, right). Maxam and Gilbert realized that they had just developed an effective method for determining the base sequence of DNA! If specific chemical cleavage techniques could be developed for the other bases, the DNA sequence could be "read" from the lengths of DNA fragments. The development of such methods led to the Maxam-Gilbert sequencing technique.

The ability to sequence DNA produced a phenomenal leap in the capacity to learn about genes at a molecular level. The approach was also an interesting reversal of classical sequencing techniques used for RNA and proteins. In the earlier technology, the first step was to obtain pure separated fragments; sequences within the fragment were then determined using a secondary degradative analysis. In this new technology, the order was reversed: a single cleavage per DNA molecule was followed by a separation technology that allowed the rapid and efficient determination of DNA sequence for several hundred bases.

In addition to its utility for DNA sequencing, the single-cleavage chemical approach was also useful for its original purpose of studying protein-DNA interactions. When the Lac repressor was added to a *lac* operator segment, some G residues were protected from DMS attack; these bases were in contact with the protein. This gap, termed a "footprint," identified the target sequence for the DNA-binding protein. Of the various chemical cleavage reagents developed for DNA sequencing, the original DMS reaction is the most widely used for the study of protein-DNA interactions (Figure 5–29). The harsh conditions required for the other cleavage reactions destroy most protein-DNA interactions, limiting their use in studying such interactions.

Several years later, David Galas and Albert Schmitz developed the most generally useful way to identify DNA sequences bound by regulatory proteins. They adapted the original DNase digestion method of Maxam and Gilbert to the single-cleavage approach. Instead of adding large amounts of DNase to degrade all of the DNA except the protected segment, Galas and Schmitz added only enough DNase to give one cleavage per end-labeled fragment of DNA and then processed the samples just as one does in a sequencing reaction. Each of the roughly random DNase cleavages could be

Protein-DNA interaction can be detected by "footprinting"

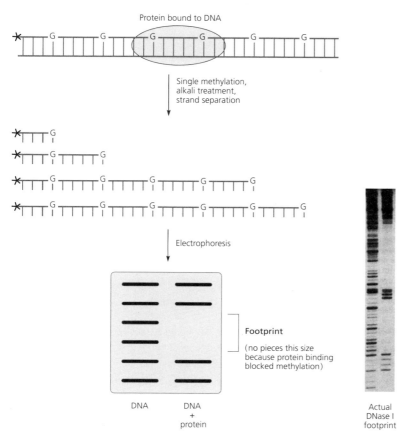

Protein bound to DNA

Single methylation,
alkali treatment,
strand separation

Electrophoresis

Footprint

(no pieces this size
because protein binding
blocked methylation)

DNA

DNA
+
protein

Actual
DNase I
footprint

FIGURE 5–29. When an end-labeled DNA molecule with bound protein is singly methylated at G residues and then cleaved by alkali, all G's except those within the protein binding region can be cleaved (*top and middle*). Following size separation and autoradiography, the bands missing from digestion of the protein-protected DNA fragment but present when DNA alone is digested can be observed. The missing bands are called the "footprint" of the protein (*bottom*). Most generally, footprints are obtained by using low concentrations of the nonspecific endonuclease DNase I to cut DNA approximately once per molecule. This reagent has the advantage of cutting after all four bases rather than exclusively at G's. An autoradiograph of a DNase I footprint of RNA polymerase binding to its promoter region on DNA is shown (*right*). (Autoradiogram courtesy of Carol Gross.)

visualized as a separate "band" on the autoradiograph. As for DMS cleavage, the gap in the band pattern, or footprint, identifies the target sequence for the DNA-binding protein (see Figure 5–29, right).

DETERMINING DNA SEQUENCE BY DNA REPLICATION

In work going on at the same time as that of Maxam and Gilbert, Fred Sanger and his colleagues developed a completely different approach to determining the base sequence of DNA. The Sanger technique also depended on specific fragments of DNA prepared by restriction enzymes. However, in this case the sequence of bases was determined by controlled DNA replication. A small single-stranded DNA fragment paired with a longer DNA segment

Fred Sanger

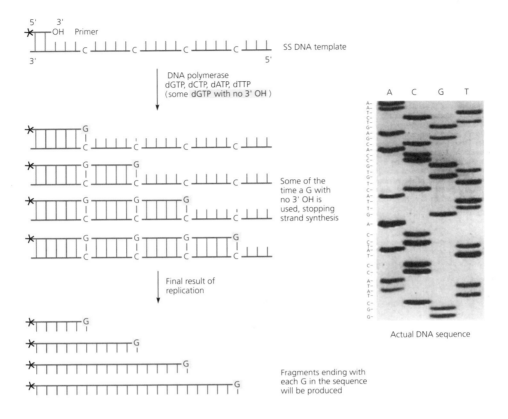

FIGURE 5–30. By starting replication of a single-strand DNA template with a radioactive primer, radioactive strands of DNA are produced. In the presence of a small amount of a nucleotide that causes replication to terminate (a nucleotide without a 3' OH group), some of the replicated strands will be terminated every time that base appears in the sequence. In the example shown, a mixture of dGTP and a chain-terminating dGTP produce a large collection of different fragments, some of which will be terminated at each G in replicating chain (*C in the template*). Four separate reactions, each of which has a different chain terminating nucleotide (A, C, T, or G) will produce a collection of radioactive strands that end at each position of the sequence. Following size separation and autoradiography, the DNA sequence can be read from the four reactions. An autoradiogram that allowed an actual DNA sequence to be determined by this procedure is shown (*right*). (Autoradiogram courtesy of Lloyd M. Smith from "DNA sequence and analysis: Past, present, and future," *American Biotechnology Laboratory* 7, no. 5 [1989].)

served as a primer for DNA replication; replication was started by adding DNA polymerase, and DNA chain elongation was stopped at a designated base by eliminating that triphosphate from the substrate mixture. Later the technique was modified to its current, more efficient form by adding small amounts of one nucleotide that terminated replication (because it lacked the 3'-OH group necessary for addition of the next nucleotide) to the polymerase reaction. This would cause a small fraction of the replicating chains to terminate each time that nucleotide was reached. The replicated DNA chains were separated by gel electrophoresis and identified by autoradiography. The modified Sanger approach is now the principal technique used for DNA sequencing work. Sanger's unique dedication to and genius for understanding primary structure has given us the way to determine the amino acid sequence of proteins and the base sequence of both DNA and RNA (Figure 5–30).

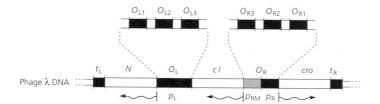

FIGURE 5–31. Sequence analysis of the O_L and O_R operators revealed that each operator region is composed of three binding sites. Two different negative regulators, cI repressor and Cro, bind to these sites. Multiple operator sites were necessary to mediate the complex interactions that govern the life-style decisions of λ.

THE MOLECULAR STUDY OF PROTEIN-DNA INTERACTIONS

The rapid development of restriction enzyme technology, DNA sequencing, and a series of reactions to examine protein–DNA interactions opened the floodgates to the molecular study of the many interesting biological reactions that had heretofore been examined only at the genetic and biochemical levels. Among these, the study of two systems had the most far-reaching consequences: the interaction of cI repressor with its operator, and the interaction of RNA polymerase with its promoter.

The interaction between λ repressor and its operator was the special focus of work by Mark Ptashne and his colleagues for a number of years. The first big news from the λ work was the discovery of multiple operators. In early experiments, Vince Pirrotta, Tom Maniatis, and Ptashne found that the size of the DNA fragment protected by cI repressor against complete digestion by DNase varied with the amount of cI added to the reaction. The maximum size was much too long for a respectable single operator, based both on the *lac* work and the rather small size of the cI protein. Understanding this finding depended on the newly found ability to sequence DNA. Maniatis sequenced the o_L and o_R operator regions. At first, it was hard to reconcile the sequence of the operators with the DNase protection experiments because only the strongest operator site was apparent. But, by staring at the sequence for many days, Keith Backman figured out that o_L and o_R really consisted of three contiguous operator sites: o_{L1}, o_{L2}, and o_{L3}; and o_{R1}, o_{R2}, and o_{R3}. Each site diverged sufficiently from the others that it had initially been difficult to "see" the individual operator sites. The sequence differences of each cI binding site explained why cI had differential affinity to these sites in the DNase protection experiments (Figure 5–31).

Why three operator sites? An enormous amount of regulation occurs in the o_R region, which influences the decision between lysis and lysogeny and controls induction of λ. Genetic and biochemical studies indicated that cI binding at o_R regulates two different promoter sites: the p_R promoter for lytic growth is repressed; and the p_{RM} promoter for expression of *cI* gene is activated. In addition, another regulatory protein, the Cro repressor, binds at o_R negatively regulating both p_R and p_{RM}. Indeed, later studies (described in Chapter 8) showed that the differential affinity of the three o_R sites for Cro and cI repressors provides for a finely tuned system that makes life-determining decisions for the λ phage.

The need for so many regulatory sequences at o_L is less clear. Part of the answer is probably that the two repressor proteins bind to each other as well as to the DNA, thereby providing a more stable association of each cI dimer ("cooperative binding"). Cooperative binding of the two dimers requires two operator sites. For many years, multiple operators were considered to be a freak of λ and its relatives. Later, multiple noncontiguous operators showed up as a common feature of bacterial regulatory circuits. Even *lac* has additional operators, which contribute to its very tight regulation. In eukaryotes, multiple operator sites are clearly the rule.

The second key question was how RNA polymerase was instructed to start transcription at the promoter. Earlier work had defined a two-stage reaction for transcription initiation: formation of the weakly bound closed complex (where the DNA remains base-paired), followed by the transition into the open complex (where strands separate to expose the transcription start site). The enzyme initiates RNA chains very rapidly (less than a second) from the open complex. The task for the new sequencing technology was to define the base recognition elements of the promoter. Once the DNA sequence of the promoter region was known, the start point for RNA synthesis could be identified by RNA-sequencing, and the DNA region examined for general recognition features.

The initial data from several promoter sequences in *lac*, λ, and phage T7 were disappointing—no obvious "operator"-like sequence was present. In the open complex, RNA polymerase protected about 60 base-pairs (bp) of DNA from cleavage by DNase—from 40 bp upstream to 20 bp downstream of the start point of RNA synthesis. But the base sequences in this segment of DNA looked disturbingly like any old DNA. David Pribnow, a student with Gilbert, provided the critical insight. Pribnow noticed a sequence similarity (but not identity) of six bases located about ten

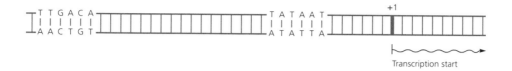

Transcription start

bp upstream from the RNA start. Perhaps RNA polymerase could recognize a variety of similar sequences. Because Pribnow marked his proposed recognition element by drawing a box around it, the sequence was initially called the "Pribnow box" (now it is usually called the "–10 region").

The Pribnow box helped, but it also raised a major question: how could RNA polymerase bind to a promoter sequence with an enormous specificity and affinity if only six bases were recognized and variety was tolerated? The answer turned out to be a second recognition element about 35 bp upstream from the RNA start—so far away that it was overlooked in the earliest work. RNA polymerase recognizes two sequences, the –10 and –35 regions, and also demands a relatively fixed spacing (16–18 bp) between the two recognition elements. Although variation in some of the bases of these recognition sequences is tolerated, there is a single sequence that is preferred in terms of the efficiency with which the promoter is used by RNA polymerase (Figure 5–32).

Additional experiments using chemical protection methods, mostly by Gilbert and colleagues, showed that bases in both the –10 and –35 regions are contacted by RNA polymerase. Moreover, in the open complex, RNA polymerase was found to unwind about 12 bp of the DNA duplex from –10 to +2. The precise relationship between the closed and open complexes has not been defined, mostly because the closed complex is a weakly bound, unstable intermediate that normally moves rapidly into the very stable open association or into enzyme dissociated from DNA.

By 1979, the major features of the polymerase-promoter interaction had been identified. The enzyme recognizes the two correctly spaced –10 and –35 region sequences and then forms a highly stable association in which the duplex DNA is opened around the start site for RNA synthesis. This information opened the door to a study of how regulator proteins alter RNA polymerase interactions with the promoter to control transcription initiation. Although the promoter sequences are different and more than 50 proteins are

FIGURE 5–32.   Sequence comparisons from known promoters did not show an "operator-like" sequence that might be used as a recognition element. Finally, Pribnow identified a relatively conserved region (TATAAT) present in strong promoters, centered about –10 base-pairs from the start site of transcription. But this was not enough to specify a promoter. Additional work, including chemical protection experiments, identified a second conserved region (TTGACA) centered about –35 bp from the start site of transcription. In strong promoters, the two conserved regions are separated by about 17 base-pairs.

involved, eukaryotic RNA polymerases follow this same sequence of reactions to start transcription.

BUILDING A REGULATORY PROTEIN: A DNA-BINDING MOTIF

The knowledge that regulatory proteins locate a sequence of DNA bases led to another question: what chemical bonds between protein and DNA are responsible for this recognition? A second related question was the eternal universal: are there any general rules for building a regulatory protein? The ultimate weapon for this struggle was X-ray crystallography, a technique capable of visualizing the individual atoms in a complicated molecule such as a protein.

X-ray crystallography depends on the scattering of X-rays into a pattern dependent on the atomic arrangement within the crystal. The fundamental scattering units are the electrons of the individual atoms. The pattern of X-rays produced from the crystal (called a "diffraction pattern") depends on the reinforcement or cancellation of scattering from individual electrons. The protein structure is determined from the X-ray diffraction pattern by a complicated mathematical treatment (called Fourier analysis), which converts the intensity of scattered X-rays into an "electron density map" that defines the atoms within a protein.

The first X-ray structures of proteins, those of myoglobin and hemoglobin, were solved in the early 1960's by John Kendrew, Max Perutz, and their colleagues. By the 1970's, protein crystallography had become a worldwide enterprise and many structures had been solved. But it was not very easy. X-ray crystallography was very slow and erratic because large crystals of the protein had to be grown from solution, and many proteins were reluctant to cooperate. Moreover, a lot of protein was required. This was a big problem for regulatory proteins, which are present in very small amounts in the cell. Because of these limitations, no atomic structure was yet available for any regulatory protein.

In 1977, I found myself in the X-ray business. Howard Schachman, a colleague at Berkeley, asked me if I was interested in providing enough Cro protein so that Brian Matthews, a crystallographer at Oregon, could try for a structure. Although my Ph.D. in Physics involved low resolution studies of proteins by X-ray scattering, and I got my first faculty job in the Biochemistry Department at Wisconsin as the "department physical chemist," this was the first time since my graduate work that I collaborated on a physical study of a protein. The Cro protein was enticing for X-ray work: its small

Brian Matthews

size would reduce the complexity of the task yet would provide structural information on a negative regulatory protein.

Matthews turned out to be the right guy, and Cro turned out to be the right protein. Matthews had all the right credentials, having been trained at the MRC Lab in Cambridge, England; in addition, he had a keen interest in how structures have evolved to solve biological problems. At that time, he was working on a lysis protein ("lysozyme") produced by phage T4 because many mutants were available that might help in learning about rules for the folding of amino acid chains into globular proteins. Having already worked with a limited quantity of protein, Matthews appreciated the problems in obtaining large amounts of Cro, and he optimistically proposed to do the structure with about 2 milligrams of Cro (a minuscule amount by traditional crystallographic standards).

Yoshi Takeda set out to purify the necessary Cro with a specially designed λ strain, first while a postdoctoral fellow in my laboratory and later as a beginning faculty member at the University of Maryland. He sent the precious sample to Matthews without a great deal of optimism for a rapid success story. Tom Steitz had been working on Lac repressor for some time, as had Mark Ptashne and colleagues with λ cI. And, Steitz and David McKay were working on Crp. However, Cro crystallized easily; Matthews and Wayne Anderson soon had an excellent diffraction pattern, and the structure was solved in 1981, after about two years of full-scale effort (extremely rapidly by standards of the time). The structures of Cro and Crp were published simultaneously in 1981.

The actual X-ray diffraction pattern of Cro protein is shown in Figure 5–33a. The intensity of spots on the photographic film was then converted to a series of electron density maps, revealing the contours of atoms. One of these maps is shown in Figure 5–33b, identifying a segment of α-helix (top left of figure) and one of β-sheet (bottom right of figure). A large number of these maps stacked on top of each other revealed the complete structure of the protein.

Most biochemical and genetic research has experimental results and new things to think about on a daily basis. In contrast, the time scale for solution of a protein structure by X-ray diffraction is very long. Crystallographers can only pray that their structure will be interpretable and interesting. For a DNA-binding protein, the anxiety was even greater because the DNA was not in the crystal, and the big news was clearly how the protein recognized DNA. Fortunately, Cro turned out to have a great deal to say about how proteins find sites on DNA.

Crystallographic data are used to solve protein structures

a)

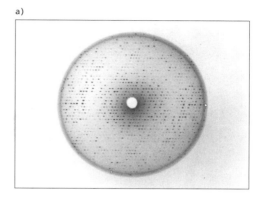

b)

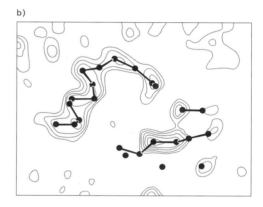

FIGURE 5–33. X-rays scattered by diffraction from a crystal of Cro repressor protein form a specific pattern on photographic film (a). This diffraction pattern is used by a computer to generate a map of the electron densities in the crystal. The carbon backbone of the protein (α-carbons only) is superimposed on one portion of the density and shown as connected black dots (b). This map is then used to build a three-dimensional model of the structure of Cro. (Image and illustration courtesy of Brian Matthews.)

When Matthews and colleagues matched the symmetry of the dimeric cro protein to a DNA model of the symmetric operator, they found an α-helix in each monomer perfectly positioned to touch the bases of the operator and an adjacent α-helix positioned at a perfect angle to lie along the sugar-phosphate backbone. Cro recognized its operator site by a "helix-turn-helix motif." The interaction can be visualized as a thumb-and-forefinger grip by the protein on a standard Watson-Crick (B-DNA) double helix. Two views of this "helix-turn-helix" recognition are shown in Figure 5–34. In the schematic "ribbon diagram," the helices of Cro and DNA are easy visualize (Figure 5–34a). In the "space-filling model," which accurately represents the complex, the amino acids of Cro and the phosphates of DNA are shown with the approximate sizes represented by spheres (Figure 5–34b).

The helix-turn-helix soon became a theme for operator recognition. When the structure of the operator-binding part of cI was crystallized by Carl Pabo and Mitch Lewis, it revealed a helix-turn-helix that recognized the symmetric operator in the same way as Cro protein. The third early member of the helix-turn-helix club was the Crp protein.

Armed with three examples, Matthews, Bob Sauer, and colleagues looked for similarities to the helix-turn-helix structure in other DNA-binding proteins whose sequence was known. A number of known regulatory proteins, including Lac repressor, revealed significant similarity to the helix-turn-helix region of Cro, cI, and Crp. From these and later studies, the helix-turn-helix has emerged as a widespread structural motif by which transcription repressors and activators recognize their DNA target. More generally, the approach of using the computer to search for sequence "motifs" known to carry out a particular function has emerged as an extremely powerful way to discern the function of an unknown protein.

How does the helix-turn-helix carry out specific recognition of DNA bases? The computer modeling of Cro protein bound to operator DNA provided a remarkably good guess. The definitive answer to the recognition question resulted from determining the structure of a regulatory protein bound to its operator site in a crystal containing both DNA and protein; this approach provided a direct visualization of interacting chemical groups. These structural solutions were carried out for λ cI by Pabo and colleagues and for phage 434 cI by Steve Harrison and associates. The general picture involves mainly two types of interactions. The "recognition" helix of

Cro repressor binds specifically to DNA

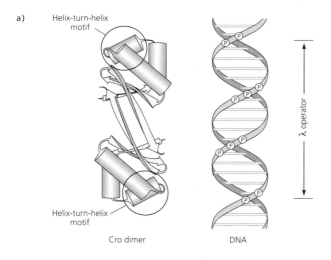

a)

Helix-turn-helix
motif

Helix-turn-helix
motif

Cro dimer

DNA

λ operator

FIGURE 5–34.   A dimer of λ Cro repressor protein binds to DNA at the λ operator. Cartoons of a Cro dimer and the λ operator DNA are shown in (a). Each Cro monomer uses a helix-turn-helix motif to make contact with DNA. An accurate representation (using a space-filling model) of the Cro dimer bound to DNA is shown in (b). The dark spheres denote the phosphate backbone of the DNA, making it easy to see the helical form of the DNA. (Illustrations adapted from materials courtesy of Brian Matthews.)

b)

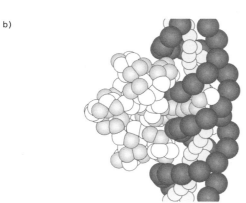

the regulatory protein makes specific hydrogen-bonds between its amino acid side chains and the bases in the operator; the other helix makes less specific electrostatic interactions between positively charged amino acids and the negatively charged phosphates of the DNA backbone.

One of the questions we began with was whether there were rules for binding of a regulatory protein to its target DNA. As I have just recounted, many proteins use the helix-turn-helix motif to recognize DNA. Although we now know that there are several different protein motifs that carry out the job of recognizing DNA, commonalties have emerged. Just as was the case for the helix-turn-helix proteins, many DNA binding proteins use an α-helix to specifically recognize DNA. When the two antiparallel DNA chains wind around each other, they have a large ("major") groove on one side of the helix and a smaller ("minor") groove on the other side. An α-helix can nestle inside the major groove and interact with the bases. So, specific recognition most often involves contacts between bases in the major groove and amino acids in an α-helix. And, just as was seen for the helix-turn-helix interaction, there is always a further interaction between the protein and the DNA backbone. However, there is no simple "code" for DNA-protein recognition. The same amino acid can recognize different bases, depending upon the orientation of the α-helix in the major groove, and on the surrounding sequence of both DNA and protein. Finally, other motifs in a protein (β-sheets) can carry out DNA recognition and often the minor groove of DNA is used for recognition.

One overall similarity of DNA-binding regulatory proteins has been noted. They all have a general capacity to associate weakly with any double-strand DNA (termed "nonspecific" binding). Less specific interactions are probably involved in this association, providing a "base line" energy on which the highly specific bonds are superimposed to give the very tight and specific operator binding. Peter von Hippel has pointed out that the weak nonspecific association also serves to allow the DNA-binding proteins to "slide" along the DNA in quest of their operator site, a more efficient search process than roaming three-dimensionally through solution.

THE REGULATOR-POLYMERASE INTERACTION: GENERAL RULES?

How do regulatory proteins communicate with RNA polymerase? In the early 1980's, the prevailing general model for bacterial gene regulation envisioned protein-protein interactions on adjacent or overlapping sites on linear DNA. When many eukaryotic regulatory sites turned out to be much too far away for this notion, there was a quest for a new mechanism. Perhaps eukaryotes had thought up a different approach.

Initially, the ideas for how far-away proteins could communicate with polymerase were constrained by the notion that regulatory proteins were bound in a linear array. Within this framework, several ideas came forward. First, the distant binding site might define an "entry point" for the transcription activator. RNA polymerase would bind to the DNA with the help of the activator and migrate along the DNA to the promoter. In a modified version of the same notion, the activator itself migrated to the promoter site to nudge RNA polymerase

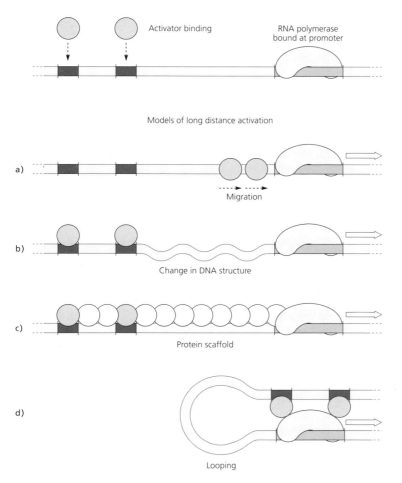

Several models can explain how activators binding far from RNA polymerase enhance transcription initiation

Activator binding

RNA polymerase bound at promoter

Models of long distance activation

a)

Migration

b)

Change in DNA structure

c)

Protein scaffold

d)

Looping

FIGURE 5–35.   Activators may first bind to their sites and then migrate along the DNA until they reach RNA polymerase (a). Activators may remain at their site but transmit a signal by altering the DNA structure between the bound activator and polymerase (b). Activators may remain at their site but recruit additional proteins that build a scaffold to reach RNA polymerase (c). Finally, the bound activators, regardless of their position on the DNA, may simply make contact with RNA polymerase at the promoter by looping out the DNA between activator and RNA polymerase (d).

into activity (Figure 5–35a). A second general idea supposed that the transcription activator altered the structure of the DNA, perhaps generating local supercoiling, so that RNA polymerase could interact productively with the promoter site (Figure 5–35b). In a third view, the known activator protein might associate with other proteins to form a continuous protein scaffold reaching out to RNA polymerase (Figure 5–35c). A fourth idea posited that activator proteins do not bind and act in a linear array; DNA-bound proteins associate to form a reactive multi-protein structure at the promoter, looping or winding the DNA in between. The activation assembly could be termed a nucleoprotein complex because the DNA and protein are both part of the reactive structure (Figure 5–35d).

The principal initial reason for my belief in the nucleoprotein or DNA-looping model came from our demonstration that nucleoprotein structures are important in λ site-specific recombination and DNA replication. If the association of DNA-bound proteins is a general feature of these DNA transactions, why not transcription? Sankar Adhya had also argued that DNA-looping provided the only likely explanation for why two operators and their associated repressors, one on either side of the promoter, were required to repress the *gal* operon of *E. coli*. The association of these two repressor dimers into a regulatory complex seemed a highly plausible mechanism for the repressors to control transcription. In addition, Bob Schleif noted that the association of DNA-bound regulatory proteins made sense in understanding distant regulatory sites in the *ara* operon.

The concept of looping interactions has broadened the way we think about communication between proteins. Bringing activators and RNA polymerase into close contact by looping interactions is a dominant theme in eukaryotic transcription and is used in regulating many bacterial operons. A variety of evidence now supports DNA-looping as the regulatory mechanism for the *gal* and *ara* operons, and even the *lac* operon turns out to have more than one operator to provide the last bit of extremely tight regulation. The association of DNA-bound regulatory proteins in "looping" reactions has been visualized directly by electron microscopy. One such example is shown in Figure 5–36, in which the linear DNA is looped by the binding of the activator protein NtrC to RNA polymerase. The elegant work of Sydney Kustu has indicated this looping reaction brings NtrC into contact with polymerase, which facilitates formation of the open complex necessary for transcription initiation.

Interestingly, all of the mechanisms described above, or variants of them, are used to facilitate communication between regulators and RNA polymerase over long distances. In some bacterial operons, regulators binding far from the promoter use a scaffold of proteins that reaches from the operator to the promoter to communicate with RNA polymerase. Peter Geiduschek has documented the case of an activator that binds at a distant site and migrates, together with replication clamp proteins, to the promoter. Finally, a major emerging mode of transcriptional regulation in eukaryotic cells is altering the accessibility of the DNA in chromosomes. In eukaryotic cells, DNA is condensed around a core of different but related proteins called histones. Together this complex

The activator, NtrC, makes contact with RNA
polymerase by looping intervening DNA

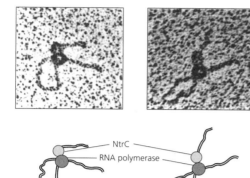

NtrC
RNA polymerase
DNA

FIGURE 5–36.   When RNA polymerase and NtrC bind to
DNA, a looped out piece of DNA can be seen between them
in the electron micrograph. Because RNA polymerase
(*bottom*) stains darker than NtrC (top), the two proteins can
be distinguished and seen to be touching. (Micrographs
courtesy of Sydney Kustu.)

is called chromatin. Some very tightly wound portions of chromatin
are inaccessible to binding by most transcriptional activators. Some
transcriptional activators can alter the state of the histones so that
the DNA is now accessible for binding by other transcriptional regulators. This kind of regulation can change the transcriptional potential of DNA over very long distances.

FURTHER READING

Edgar, R. S. (1966) Conditional lethals, in J. Cairns, G. S. Stent, and
J. D. Watson, eds., *Phage and the Origins of Molecular Biology.* Cold
Spring Harbor, N.Y.: Cold Spring Harbor Laboratory Press.

Maniatis, T., and M. Ptashne (1976) A DNA operator-repressor system. Sci. Am. 234, 64–76.

Ptashne, M. (1973/74) Repressor, operators, and promoters in bacteriophage lambda. Harvey Lectures Series 1973–74, vol. 69,
143–71.

6

THE RNA WORLD:

NEW PROTEINS AND REVISED RNAS

The identification of the code from gene to protein described in Chapter 2 was achieved at such a marvelous pace in part because the critical experiments could proceed with almost no knowledge of the mechanism by which the code was actually translated—the process of protein synthesis. The ribosome not only exhibited heroic stability but also clutched to its bosom much of the protein machinery necessary for translation. The decoders could therefore ascertain the identity of the RNA codons simply by adding synthetic mRNA to either the cellular "soup" or crudely prepared ribosomes resulting from the biochemical fractionation scheme of Paul Zamecnik and Alfred Tissières, and then reading out the selected amino acid.

A clear set of questions resulted from unveiling the code. What are the rules for tRNA recognition of its specific codon in mRNA? What is special about starting and ending translation of mRNA? How is mRNA used as a message tape to recognize the incoming tRNA, and how does it move on after the new amino acid is added to the growing polypeptide chain? And finally, what do ribosomes really do? In principle, these questions could be answered by using the knowledge about mRNA decoding to figure out the players that did this job—just the reverse of the approach used to solve the code. The first sections of this chapter describe the experiments that answered these questions.

In the latter sections of this chapter, we will consider a fascinating property of RNA itself: the production of "revised" RNAs by the programmed cutting and splicing of the transcribed gene product. This work has led to two spectacular surprises of recent molecular biology: the informational discontinuity of many eukaryotic genes that is remedied by RNA splicing, and the ability of RNA itself to function as a catalyst ("ribozyme") in RNA processing reactions. These findings have led to our current view that the "pre-biological" world was an "RNA world" populated by RNAs that performed both the informational function of DNA and the catalytic function of proteins.

THE ANTICODON: BASE-PAIRS CAN WOBBLE

Transfer RNA was clearly the decoding element in protein synthesis—the adapter that converted the nucleic acid language of mRNA into the amino acid language of proteins. But how did it recognize its mRNA codon? At first thought, a reasonable scheme would be one specific decoding tRNA for each of the 61 sense codons, allowing the three base "anticodon" in each tRNA to recognize one mRNA codon with Watson-Crick base-pairing. Efforts to fractionate tRNAs by various schemes indicated that the number of tRNAs was certainly less than 61, so another rule had to be found.

A clue to this rule came from studies identifying the RNA triplets recognized by individual tRNAs. Of special importance was the work of Dieter Söll and Gobind Khorana showing that a single tRNA could recognize more than one codon. For example, one *E. coli* serine tRNA recognizes both the UCG and UCA codons; a second *E. coli* serine tRNA recognizes both the AGC and AGU codons. Notably, the two codons recognized by a single tRNA always differed only in the third coding position. When coding differences involved other bases, a different tRNA provides the recognition. There is one exception to this rule. As we will see later, the initiation-specific tRNA recognizes both AUG and GUG.

Francis Crick eventually inferred the correct recognition principle, "base-pair wobble," from the nature of the code and the emerging biochemistry of tRNA. The mechanism involved a cute twist on the Watson-Crick rules for base-pairing in DNA. In tRNA-mRNA recognition, the geometrical identity of the RNA base-pairs in the third position of the codon is relaxed. In addition to the Watson-Crick base-pairs of G-C and A-U (replacing the A-T pairing in DNA), "wobble" base-pairs are also allowed in this position. At the third base in the codon, U and G can pair with each other. In addi-

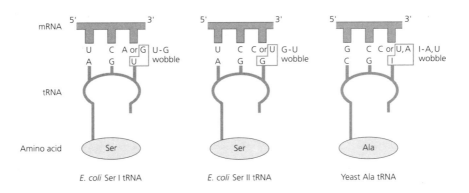

FIGURE 6–1. The three base codon in mRNA is recognized by base-pairing with the three base anticodon in tRNA. However, base-pairing rules are relaxed for the third position of the codon. In that position, the "wobble" G-U and I-U/A base-pairs are allowed in addition to the Watson-Crick A-T/U and C-G/I base-pairs. Base-pair wobble, proposed by Crick, explains how 61 different codons can be recognized by about half that number of tRNA molecules.

tion, inosine, a G–like base found only in tRNA, can pair with U or A as well as with C (Figure 6–1). As a result of wobble recognition, only about 35 tRNAs are required to do the job, rather than the 61 tRNAs necessary if the system were restricted to Watson–Crick pairing. Why do wobble base pairs not appear in DNA? To ensure genome fidelity, the replication machinery goes to a great deal of trouble to demand that only the geometrically identical Watson-Crick A-T and G-C base pairs are maintained in DNA.

BASE SEQUENCE OF tRNA: A THREE-LEAF CLOVER FOR HOLLEY

Understanding codon-anticodon recognition required some knowledge of the structure of tRNA. Bob Holley provided the initial critical insight by determining the base sequence of alanine tRNA from yeast. Holley's remarkable technical achievement was almost immediately rewarded by a spectacular vision of a general secondary structure for tRNAs—a cloverleaf with an anticodon loop.

Determining the sequence of a tRNA was a major goal for several research groups. Holley was a long shot in the race; he had only limited experience with nucleic acid chemistry, and he worked in an unlikely place far from the scientific main stream—the U.S. Department of Agriculture Lab in Ithaca, New York. A rigorous, experienced biochemist with an abiding passion for protein synthesis, Holley possessed definite assets, however, and he thought big. In the mid-1950's, Holley became interested in the "activation" of amino acids for protein synthesis, the research area pioneered by Zamecnik

Bob Holley

(see Figure 2–6). By pursuing this problem, he obtained independent evidence for the transfer of amino acids to tRNA. The next goal for him was the structure of tRNA.

In 1959, determining the base sequence of a tRNA was a formidable task. Because tRNAs are a collection of closely similar nucleic acid molecules, no obvious preparative route existed to obtain a single pure tRNA species. Holley chose a classical, largely forgotten procedure called counter-current distribution to separate the tRNAs. This procedure partitions RNA between a water phase and an oily (organic) phase over and over again, thereby amplifying small chemical differences between different tRNAs and allowing them to be separated from each other. Holley chose to purify alanine tRNA because it was the easiest to fractionate away from the others in the counter-current distribution apparatus, and he ran his purification procedure on a grand scale, using 50 grams of tRNA purified from 100-pound batches of yeast. The elaborate glass machine to carry out the counter-current distribution was huge—it took up an entire small room.

To obtain the base sequence of the tRNA, Holley and his associates adapted the strategy developed by Fred Sanger for determining the amino acid sequence of proteins. The purified alanine tRNA was cut into pieces ("oligonucleotides") by enzymes ("nucleases") that cut at specific bases. These small pieces were then separated from each other. Good separation of the oligonucleotides required them to be passed through a 25-foot-long chromatographic column. This column was located in a stairwell and spanned several floors so that the mixture of oligonucleotides could be loaded easily from an upper floor. The separated oligonucleotides were then collected from the bottom of the column. The individual purified oligonucleotides were small enough that their base sequence could be determined by further degradative procedures. In the last step, the oligonucleotides were arranged in the correct order by obtaining larger fragments of the tRNA and then determining which of the oligonucleotides were contained in each larger fragment. By 1965, the task was finished.

The successful determination of the base sequence of alanine tRNA, the first primary structure of a nucleic acid, was an obvious technical milestone. Holley did not really expect the base sequence to illuminate the function of the tRNA. However, because physical studies of tRNAs had demonstrated that they were likely to have a base-paired secondary structure, Holley thought that he should

FIGURE 6–2. When the sequence of alanine tRNA molecule was analyzed for its secondary structure (regions of the RNA that could base-pair with other regions), one model showed the primary sequence folding into a dramatic cloverleaf shape. The "stem" of the cloverleaf carries the amino acid, whereas the loop opposite the stem is the anticodon that recognizes mRNA. In this model, about half of the bases are base-paired (indicated by dashes between bases) (*left*). The tertiary structure of the molecule (determined by X-ray crystallography) looks like an L (*right*). Some of the bases of the tRNA (indicated by shading in both structures) form connections that twist the cloverleaf structure of the tRNA into an L-shape.

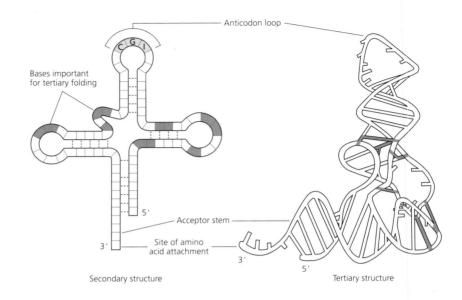

include some possible secondary structures with publication of the sequence. James Penswick, a student with the Holley group, drew some possibilities with Elizabeth Keller, a professor at Cornell who had earlier worked with Zamecnik. Keller looked at one that resembled a clover leaf and announced, "That looks so beautiful, it must be the structure"—and it was (Figure 6–2, left).

By itself, the clover-leaf picture of alanine tRNA paid homage to simplicity and elegance but initially had no direct evidence to recommend it. The structure did display a likely anticodon, CGI, in an unpaired region that would be available for interaction with the alanine codons GCU, GCC, or GCA. The limitation of one example did not last long. Within six months, Holley's colleague at Cornell, James Madison, had progressed far enough with yeast tyrosine tRNA to see that the clover-leaf model would fit. Within a year, Hans Zachau and his colleagues in Germany had added two yeast serine tRNAs to the clover-leaf collection. In this model, the "anticodon loop" occupies its characteristic spot opposite the "acceptor stem" that receives the amino acid.

All tRNAs currently known exhibit a secondary structure that fits the clover-leaf pattern. This generality itself provided a compelling argument for the accuracy of the clover-leaf model. Later, an

X-ray crystal structure of alanine tRNA was obtained by Sung-Hou Kim and Alex Rich, and by Brian Clark, Aaron Klug, and their colleagues. The crystal structure confirms the clover-leaf secondary structure and reveals a tertiary fold that gives the molecule more of an L-shape (Figure 6–2, right). The remarkable structural similarity of tRNAs poses a problem for understanding the very specific recognition of a given tRNA by its "charging" enzyme (the enzyme that adds the appropriate amino acid to that tRNA). Work by John Abelson, Paul Schimmel, William McClain, and La Donne Schulman indicated that the specific recognition of tRNA by its charging enzyme generally depends on very subtle and highly localized structural differences among tRNAs. Most frequently, these differences are localized to the anticodon. High resolution structures of charging enzymes with their tRNAs now allow us to "see" the subtle differences that underlie specific recognition.

MESSAGE DECODING: METHODS AND PROTAGONISTS

By 1964, the problems of message decoding were clear: How does a polypeptide chain start? How is the message translated sequentially into the growing protein? How is the completed polypeptide chain terminated? The appropriate approach was also clear: start with a unique mRNA and analyze the biochemical requirements for each stage. The choice of mRNAs was limited. The only natural RNA available with sufficient abundance and purity was the viral genome of a recently discovered small phage that used RNA as its genetic material. The alternative choice was a "fake" (synthetic) mRNA with a defined sequence. Both types of RNA were used with success by a number of research groups. As is often the case when the problem is important and the approach is evident, there followed an overlapping and highly competitive period of research that cannot be recounted in a few pages with fair credit to all.

The RNA phage, f2, emerged from the New York sewers in 1961, with a timing as if sent by God for the study of protein synthesis. The phage was initially isolated from sewage (the traditional source of *E. coli* phages) by Tim Loeb at the Rockefeller University for another intended purpose—its ability to grow only on male (F+) strains of *E. coli*. When Loeb and Norton Zinder purified f2, they learned to their surprise that the genome of this virus was RNA instead of DNA. Plant and animal viruses with RNA genomes were known, but no RNA phage had been previously isolated. The RNA genome

Norton Zinder

of f2 was small, coding for only three proteins. Suddenly, a packaged, easily purified mRNA was available for protein synthesis studies. The previous (and current) tradition in molecular biology was to give biological materials to other scientists after the initial characterization had been published. Zinder refused to do this with the f2 phage. As a result, closely similar phage were isolated around the world and given different names. The irrepressible Sol Spiegelman claimed that he got the virus by obtaining phage plaques from the paper of Zinder's letter of refusal, reasoning that Zinder's lab must be filled with the phage. For everyone but the purists, all of these related viruses were called "RNA phage," and that term will be used here.

Jim Watson emerged as Zinder's principal rival for studying translational punctuation questions. Watson had taken a faculty position at Harvard in 1955, with research initially focused on the ribosome. His group carried out the critical early experiments demonstrating that *E. coli* has mRNA (Chapter 3), and Watson quickly realized the importance of an abundant natural mRNA for understanding the decoding process. Watson's exceptional capacity for recognizing important problems and how to approach them did not end with the structure of DNA. Nor did his competitive drive and yearning for the quick, spectacular payoff. Watson did not acknowledge the experimental and intellectual boundaries of prior experience that constrain most scientists to limit the scope of their research, and he imbued his students and associates with a similar spirit, mostly with great success. Interestingly, the big payoffs did not happen fast enough for Watson in the translation-initiation field, and he had little patience with the lengthy struggle often needed for major new insights. In the late 1960's, he moved his research from translational decoding to transcriptional regulation, with diminishing returns in terms of big news results. In the early 1970's, Watson transferred his abundant energy entirely into directing the Cold Spring Harbor Laboratory, making it the ever-expanding high visibility research enterprise and meeting center that it is today.

THE PROBLEM OF STARTING A PROTEIN

A protein is a chain of amino acids with a direction: it starts with an amino (N) end and finishes with a carboxyl (C) end. This distinction occurs because each amino acid subunit has an amino and a carboxyl end, and in a protein the amino acids are joined end to end in a linear array (see Figure 1–1). Howard Dintzis demonstrated in

1961 that protein synthesis in cells begins at the N-terminal end and proceeds sequentially to the C-terminal end. But how is the precise N-terminal start chosen from a long array of potential coding triplets? This question defines the most fundamental decoding problem in converting a string of codons into a real protein.

The first clue about the nature of the start site for protein synthesis came from work by Kjeld Marcker in Fred Sanger's group at Cambridge. Sanger had turned his sequencing skills from proteins to nucleic acids, and, before devising DNA sequencing methods (see Figure 5–30), his group had developed a rapid method for determining RNA sequences starting with small amounts of a radioactive RNA. In the course of separating tRNAs for sequencing, Marcker found an odd tRNA carrying a variant of the amino acid methionine. The N-terminal end of this methionine had an additional chemical group termed formyl. This formyl group prevents the methionine from joining to the C-terminal end of the preceding amino acid in the chain. Clearly, formyl methionine could not be used for elongating an amino acid chain. Perhaps "fmet-tRNA$_f$" specialized in starting amino acid chains (Figure 6–3 on page 174).

Marcker and Brian Clark examined the decoding properties of this special tRNA (fmet-tRNA$_f$). In the triplet binding assay of Phil Leder and Marshall Nirenberg (see Figure 2–10), the "standard" met tRNA (met-tRNA$_m$) recognized only AUG (the previously defined codon for methionine). In contrast, fmet-tRNA$_f$ recognized the triplet AUG best, but also GUG and UUG, a somewhat ambiguous result. The real clue for fmet-tRNA$_f$ function came from a synthetic mRNA assay. Here, fmet-tRNA$_f$ could start chains but refused to insert methionine internally into a polypeptide chain. Thus fmet-tRNA$_f$ appeared to be the initiator tRNA. In a remarkable success for the crude triplet assay, AUG, GUG, and UUG have all been found to initiate proteins in cells, with AUG by far the most frequent initiator (Figure 6–4 on page 175).

Although the biochemical data looked impressive for initiation by fmet-tRNA$_f$ in the test tube, at first the case from *in vivo* work in the living cell was not nearly as strong. There were two problems. Jean-Pierre Waller had found earlier that only about half of the proteins in *E. coli* begin with an N-terminal methionine, leaving more than half starting with other amino acids. In addition, formyl methionine was not present in proteins isolated from cells. Although the formyl group might have been clipped off *in vivo*, the nonmethionine ends were still unexplained.

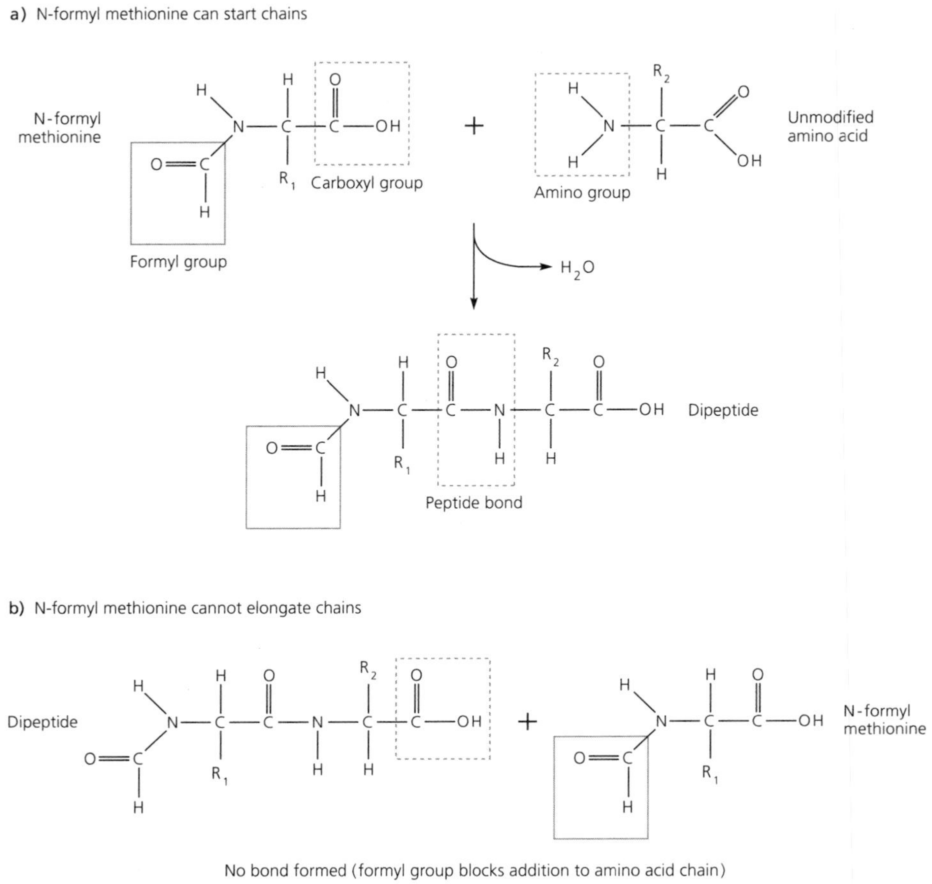

a) N-formyl methionine can start chains

FIGURE 6–3.   The modified amino acid,
N-formyl methionine, can be used to start
chains (a) but cannot be used at an
internal position of a protein (b). The
formyl group blocks formation of the
peptide bond that adds amino acids to the
growing chain.

b) N-formyl methionine cannot elongate chains

No bond formed (formyl group blocks addition to amino acid chain)

The gap between these discrepant results was bridged through concurrent work by the Watson and Zinder groups comparing the *in vivo* and *in vitro* translation products from the natural mRNA of the RNA phage. An especially complete analysis came from Jerry Adams and Mario Capecchi in the Watson group. When cell extracts were programmed with phage RNA, they mostly produced the coat protein of the virus particle, with detectable amounts of the viral RNA synthetase that replicates the phage RNA. When the extract was supplemented with radioactive formyl methionine, both proteins were found to begin with formyl methionine. Coat protein isolated from phage infected cells, however, began with alanine, which was the second amino acid of coat protein made in extracts.

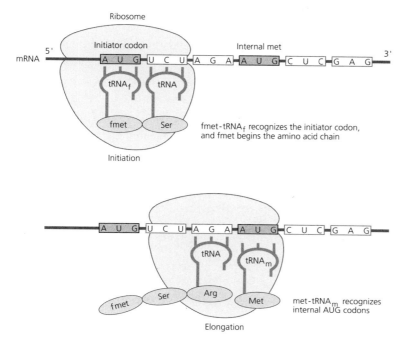

FIGURE 6–4. The initiating AUG codon is always recognized by formyl methionine-tRNA$_f$ (fmet-tRNA$_f$), thus setting the reading frame for the protein. This occurs because fmet-tRNA$_f$ is brought to the ribosome by an initiation factor that does not recognize methionine-tRNA$_m$ (met-tRNA$_m$) (see Figure 6–19). In contrast, internal AUG codons are always recognized by met-tRNA$_m$. This occurs because the elongation factors that escort tRNAs to internal codons recognize met-tRNA$_m$ but not fmet-tRNA$_f$. Thus, even when fmet-tRNA$_f$ carries normal (nonformylated) methionine, it cannot be used at internal AUG codons (see Figure 6–12).

Presumably, *E. coli* proteins always initiate with formyl methionine. The formyl group is always removed, and sometimes the terminal methionine is removed as well. This conclusion was substantiated through additional work by Capecchi with mRNA from *E. coli*.

The existence of a special tRNA for initiation helped to elucidate correct initiation: fmet-tRNA$_f$ establishes the "reading frame" for the coding sequence of a specific protein. How does the fmet-tRNA$_f$ know to select the "initiator AUG" from all other AUGs coding for the internal methionines in proteins? As will be discussed later, selecting the initiator AUG is one of the activities of a ribosome.

ENDING A PROTEIN: THE VALUE OF NONSENSE

Once started, synthesis of a protein uses a succession of tRNAs to translate consecutive RNA triplets and thereby insert amino acids sequentially into the growing chain of the protein. How does this sequential translation process terminate to give a finished protein?

The general answer to this question preceded the detailed analysis of coding: nonsense codons do the job. There are three of these—UAG, UAA, and UGA—all of which are used in cells.

The notion of nonsense in the genetic code began with the information theorists. The first experimental evidence for this concept came from work on strains having "suppressor" mutations. These "suppressor" strains were able to restore wild-type function to certain mutations. Seymour Benzer and Allan Garen inferred that the major class of mutations that could be suppressed were of the nonsense type, but the argument was indirect: in a suppressing host, the mutants made a full-length protein, but in the nonsuppressing host (where the mutant phenotype was expressed), full-length protein could not be found. Rigorous proof required showing that the presumed nonsense mutants actually produced a shortened protein in the nonsuppressing host. Sydney Brenner and colleagues quickly provided the necessary evidence using the major head protein of phage T4. Using the head protein, Brenner's group had been struggling to identify the amino acid changes caused by missense mutations (those mutations that change one amino acid into another) to break the genetic code (see Figure 2–11). This work was largely without success. Nonsense mutations were a much-needed piece of cake. Because the head protein was so abundant in T4 infected cells, the predicted shortened protein encoded by nonsense mutations in the head protein was readily found in nonsuppressing hosts.

Brenner and colleagues then used the nonsense mutations in the phage head protein in a series of elegant experiments. Besides verifying the colinear relationship between gene and protein *in vivo* (see Figure 2–4), they deduced the sequences of two of the nonsense codons themselves. By comparing the codons for the amino acid inserted by each suppressor tRNA with the codons for the mutated amino acid, Brenner was able to infer that UAG and UAA were nonsense codons. Garen and associates provided a similar analysis for alkaline phosphatase.

In the meantime, biochemical experiments with cell-free systems were yielding the same conclusions about "nonsense" as the *in vivo* suppressor analysis. As expected for nonsense codons, the UAA, UAG and UGA triplets failed to give a tRNA binding response in the trinucleotide binding assay. Further support for the idea that "nonsense" codons caused translation termination came from work with phage RNA by the Zinder and Watson groups. Their studies revealed that the amino acid chain terminated just at

a)

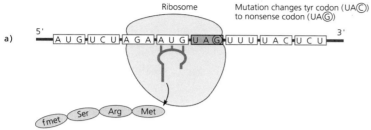

b)

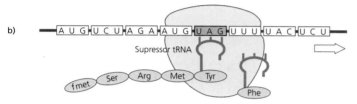

c)

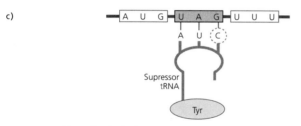

FIGURE 6–5. A "nonsense" mutation is created when a codon specifying an amino acid is mutated to a codon that specifies no amino acid. Translation terminates at the nonsense mutation, producing a shortened protein (*a*). Suppressor tRNAs read nonsense codons as codons for an amino acid, allowing protein synthesis to continue (*b*). This occurs because the suppressor tRNA has a mutated anticodon that recognizes the nonsense codon (*c*). These principles were inferred from *in vivo* experiments and directly demonstrated in a cell-free translation system programmed with RNA phage.

the position of the presumptive nonsense codon in the viral coat protein (Figure 6–5a). Moreover, synthesis of full-length coat protein was restored by the addition of a tRNA fraction from suppressing cells (Figure 6–5b). Thus, suppression depended on the capacity of a mutant tRNA to read a nonsense codon as a sense codon and insert an amino acid at that position. These experiments demonstrated directly that nonsense mutations acted during protein synthesis and could be corrected by an altered tRNA. Later work by Brenner, John Smith, and colleagues revealed that a particular suppressor tRNA carried a single mutation in the anticodon, allowing the tRNA to translate the nonsense codon UAG as the amino acid tyrosine (Figure 6–5c). An interesting historical

FIGURE 6–6.   Sequencing the end of the coat protein gene of RNA phage revealed that two nonsense codons immediately follow the codon for the last amino acid in the protein, indicating that the normal role of nonsense codons is to terminate protein synthesis.

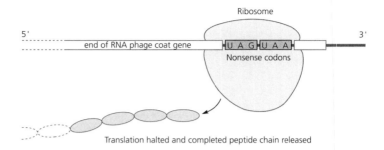

Translation halted and completed peptide chain released

footnote is that this mutation was the very first to be sequenced at the nucleic acid level.

The experiments with nonsense mutations demonstrated that UAG and UAA were chain-terminating codons. The only major missing point was showing that nonsense codons were used at the ends of genes to terminate normal proteins. The initial verification was provided by John Nichols of the Sanger group, who showed that an RNA oligonucleotide corresponding to the end of the coat protein gene carried both UAG and UAA at the end of the coding sequence (Figure 6–6). The use of nonsense codons to terminate natural proteins has been abundantly demonstrated since. Generally, a single nonsense codon ends the protein.

PROTEIN TRANSLATION FACTORS: TURNING THE GEARS IN PROTEIN SYNTHESIS

So far, we have considered solely the mRNA-tRNA interaction, ignoring the ribosome. This approach largely defined the early decoding studies of the translational apparatus and achieved notable success. It correctly identified both the nature of the codon-anticodon interaction and the punctuation in the mRNA message, establishing that protein synthesis begins with an initiating AUG and ends with a nonsense codon.

Work in the late 1960's switched to the mechanics of protein synthesis—the way in which tRNAs were brought into the ribosome and the way amino acids sequentially joined to the growing protein. This process is simple in an informational sense, although highly complex in its biochemical execution. The mRNA tape does not select the free tRNAs from solution solely by the three base codon-anticodon interaction; this would not provide a stable association at the growth temperatures of organisms. Protein synthesis requires the catalytic activities of proteins that act sequentially to direct the initiation, elongation, and termination of the amino acid chain, as was true for DNA replication and transcription. The ribosomal RNA itself also participates in these reactions.

The identity of the proteins that turn the translational gears emerged from failed efforts to carry out protein synthesis solely with ribosomes, mRNA, and tRNA. These additional proteins were originally termed "factors"—the biochemist's name for mysterious somethings that are needed for reactions, but no one knows why. The experiments described below show how the many proteins, which work sequentially to direct initiation, elongation, and termination of translation, were identified. These experiments also reveal both the pleasures and perils of biochemistry—a touch of salt, a pinch of magnesium, and, voilà, proteins made in a test tube!

The first "factors" to be isolated were those required for elongation. Protein synthesis was originally achieved in a crude extract of cells with all of the intracellular constituents; this extract was termed an "S-30" because the cell debris had been removed by centrifugation at 30,000 times gravity, to generate a soluble supernatant (Figure 6–7a). Interestingly, the ribosomes used for the triplet RNA binding assay (prepared by centrifuging the S-30 extract at a higher speed [100,000 times gravity]) were unable to carry out protein synthesis directed either by the synthetic polyU message or by natural mRNAs. Clearly, something was missing from the squishy ribosome "pellet" at the bottom of the centrifuge tube. When the proteins remaining in solution (called an "S-100" extract) were added to the ribosome fraction, protein synthesis directed by polyU or by natural mRNA was restored. Thus began the isolation of the first protein "factors" turning the translational gears (Figure 6–7b). These "elongation factors," designated EF-Tu, EF-Ts, and EF-G, were purified principally through the efforts of Jean Lucas-Lenard in Fritz Lipmann's group at Rockefeller University (the EF is a later addition to distinguish elongation factors from other minions of protein synthesis).

As work on elongation factors got under way, another line of research by several groups turned up the initiation factors required for the first step in protein synthesis. The crude ribosome pellet obtained after high-speed centrifugation could be "purified" by

FIGURE 6–7. When a cell extract is spun at 30,000 times gravity to remove cell walls, the supernatant, called an "S-30" extract, can carry out protein synthesis (a). Further centrifugation of the S-30 at 100,000 times gravity pellets the ribosomes. This crude ribosome pellet cannot carry out protein synthesis from either poly-U RNA or natural mRNA. However, adding back the supernatant (called an "S-100" extract) to the crude ribosomal pellet restores protein synthesis (b). This assay was used to purify the elongation factors (EF-Tu, Ef-Ts, and EF-G) from the S-100 extract.

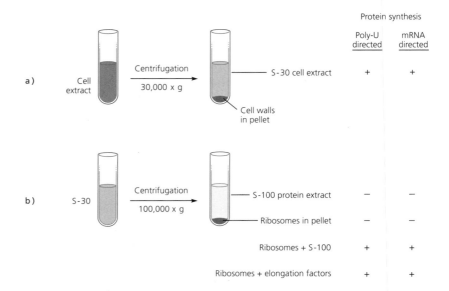

treatment with a high–salt solution ("washing") to remove loosely associated proteins. When supplemented with elongation factors, the pure ribosomes worked for polyU–directed protein synthesis (which does not use a normal initiation mechanism) but failed for a natural mRNA such as that from the RNA phage. Adding back the proteins released by the salt wash restored protein synthesis from a natural mRNA (Figure 6–8). The existence of "initiation factors" required to start protein synthesis from natural mRNA was inferred from these experiments, which were mainly carried out by Michel Revel with François Gros in Paris and by the group of Severo Ochoa at New York University. Isolation of the initiation factors, designated IF1, IF2, and IF3, occurred principally through the efforts of these two research groups and that of Bob Thach and colleagues at Harvard. Not surprisingly, as described below, termination of protein synthesis also turned out to depend on specific protein factors, termed "release factors," which catalyzed the release of the amino acid chain from tRNA and the ribosome.

In this same period, ribosome dissociation turned up as another key element in the pathway of protein synthesis. Isolated ribosomes had long been known to undergo a reversible transition from the complete RNA-protein complex into two smaller subunits. This

Ribosomes need initiation factors to start protein synthesis from natural mRNAs

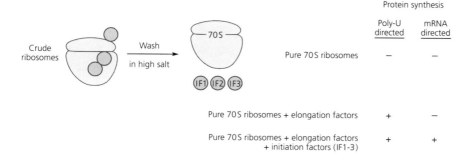

	Protein synthesis	
	Poly-U directed	mRNA directed
Pure 70S ribosomes	−	−
Pure 70S ribosomes + elongation factors	+	−
Pure 70S ribosomes + elongation factors + initiation factors (IF1-3)	+	+

FIGURE 6–8. When the crude ribosome pellet isolated by centrifugation (described in Figure 6–7) is washed with high salt, these "purified ribosomes" can no longer carry out protein synthesis. When supplemented with elongation factors, they can carry out poly-U directed protein synthesis but cannot make proteins from natural mRNAs. Adding back the high salt wash to purified ribosomes supplemented with elongation factors restores protein synthesis from natural mRNAs. This assay was used to purify the initiation factors (IF-1, IF-2, and IF-3), which are necessary for initiating translation of natural mRNAs.

Ribosomes are composed of smaller subunits

FIGURE 6–9. When intact "70S" ribosomes are placed in a solution with very low magnesium, they separate into two smaller subunits, called "30S" and "50S" subunits. The names of these particles refer to how rapidly they move during centrifugation.

transition depended on the amount of magnesium in the solution. The complete single ribosome used in protein synthesis was termed "70S" because of its rate of sedimentation in the centrifugal field of a high-speed centrifuge. The ribosome subunits found in low magnesium were termed "50S" and "30S" based on this same measurement (Figure 6–9). Interestingly, it was test-tube conditions with rather low amounts of magnesium that favored recognition of triplet AUG by the ribosome and protein synthesis from natural mRNA. Masayasu Nomura, who was studying ribosome structure at Wisconsin, wondered whether the preference for low magnesium might indicate a special role for the 30S and 50S ribosome subunits in initiation. A series of experiments with Chuck Lowry and Christine Guthrie verified this idea.

In the first experiment, Lowry compared binding of the initiator tRNA to the 30S ribosome subunit and to the 70S ribosome. He found that the 30S subunit with AUG and initiation factors bound the initiator tRNA better than the traditional 70S complex. In the

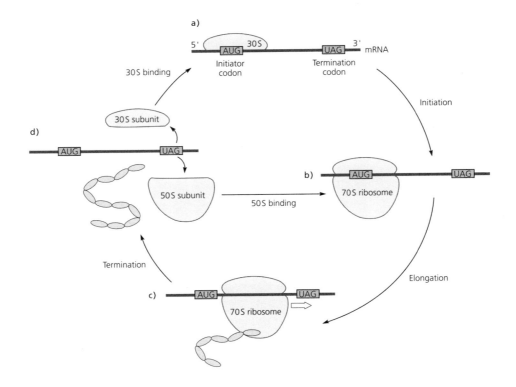

FIGURE 6–10. To begin protein synthesis, the ribosomal 30S subunit (along with initiation factors and fmet-tRNA$_f$) binds to the region of mRNA surrounding the AUG initiation codon (a). The 50S subunit joins the 30S subunit, forming the 70S ribosome (b), and the elongation phase of protein synthesis occurs (c). When the ribosome encounters a termination (nonsense) codon, protein synthesis stops, the 70S ribosome dissociates into the 50S and 30S subunits (d), and the cycle begins again (a). Initiation factors and tRNAs are omitted for clarity.

second experiment, Guthrie used natural mRNA from an RNA phage to show that the 30S subunit first complexed with mRNA and fmet-tRNA$_f$, and then joined to the 50S subunit to form the 70S ribosome, which carried out protein synthesis. These experiments argued for a ribosome cycle: protein synthesis begins with a 30S subunit, which is produced by dissociation of a 70S ribosome into its 50S and 30S constituents at the conclusion of protein synthesis (Figure 6–10). The existence of a dissociation cycle was also supported by other experiments *in vivo* by David Schlessinger and associates and by Raymond Kaempfer and Matt Meselson.

USING THE RIBOSOME: GTP AND TRANSLOCATION

Translation factors clearly guided protein synthesis, but how? What does protein synthesis really look like on a ribosome? The further development of the field depended on a conceptual framework and a biochemical clue, both of which had been around for some time.

Watson provided the conceptual framework. Although the role of the ribosome had been a matter of general discussion since the early days of protein synthesis, Watson led the development of the post-mRNA view of protein synthesis, one that acquired biochemical reality through the study of translation factors. For mRNA to be translated on the surface of a ribosome, there must be communication between two mRNA codons: the triplet holding the growing amino acid chain (presumably via a tRNA), and the adjacent triplet in contact with the incoming tRNA carrying the next amino acid. Thus, the ribosome must carry at least two sites, a P site (for peptide) and an A site (for accepting the next tRNA). We now know there is also an E site for exit of the tRNA after it has given its amino acid to the growing polypeptide chain (Figure 6–11a, b). During elongation, the growing chain in the P site is joined to the incoming amino acid in the A site, and the freed tRNA moves from the P site to the E site. For translation to proceed, the mRNA must move with respect to the ribosome. In this step, the tRNA carrying the growing protein in the A site moves with its mRNA codon to the P site. After this somewhat mystical step, termed translocation, the A site is available for the next amino acid (Figure 6–11c, d). Subsequent chain elongation proceeds through a reiteration of the same events. In this scheme, the only free amino acid tRNA ever to occupy the P site would be the initiator fmet-tRNA$_f$.

Zamecnik provided the biochemical clue. In the 1950's, he found that the ribonucleotide GTP was a required "cofactor" for protein synthesis. Somehow the binding and cleavage of GTP (the triphosphate) to GDP (the diphosphate) must be involved in turning the gears of protein synthesis. Early thoughts about GTP considered it likely to be a specialized energy source to drive protein synthesis, a role often carried out for other processes by its sister nucleotide ATP. This turned out not to be the case. Instead, GTP mediates conformational changes in translation factors. Promoting conformational transitions in proteins is also a common function of ATP. GTP is needed as a cofactor for three different translation factors to function: two elongation factors and an initiation factor. In their GTP-bound form, these translation factors bind to the ribosome; when GTP is converted to GDP, conformational changes in the translation factors release them from the ribosome. The remainder of this section presents a synopsis of how translation elongation, initiation, and termination are accomplished.

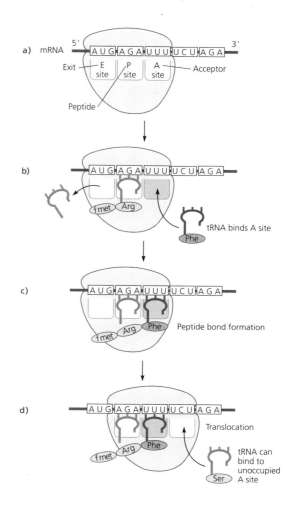

FIGURE 6–11. To carry out elongation, the 70S ribosome has a binding site for the tRNA holding onto the growing peptide (the "P" or peptide site), a binding site for the tRNA delivering the next amino acid (the "A" or acceptor site), and a binding site for the departing tRNA (the "E" or exit site) (a). When the empty A site accepts a tRNA, the E site tRNA dissociates (b). When both the P and A sites are occupied, a peptide bond forms, adding a new amino acid to the growing chain and presumably temporarily transferring the amino acid chain to the A site (c). During translocation, the mRNA and tRNAs move relative to the ribosome. The P site tRNA, free of its amino acid, moves into the E site, and the A site tRNA (with its peptide chain) moves to the P site, allowing the cycle to begin again (d).

FIGURE 6–12. A tRNA molecule carrying an amino acid ("aa-tRNA") binds to the elongation factor Tu, which carries GTP (a). This complex then binds to the A site of the ribosome (b). Converting GTP to GDP by the loss of a phosphate causes EF-Tu to dissociate from the tRNA and ribosome (c). The freed EF-Tu•GDP complex binds a second elongation factor, Ts, displacing GDP (d). In turn, Ts is displaced when Tu binds a new GTP (e) and is ready to associate with a new aa-tRNA (a). This complex pathway was worked out by a number of research groups, with especially definitive contributions by Jean Lucas-Lenard, Joanne Ravel, Peter Lengyel, Herb Weissbach, and their associates.

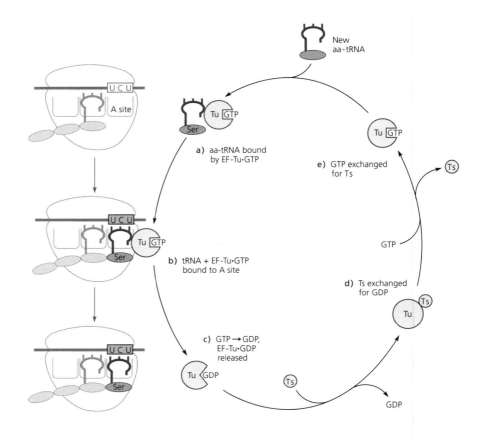

How is the amino acid brought to the ribosome during translation elongation? The first explicit role determined for GTP was in the reaction that brings the tRNA bearing an amino acid (aa-tRNA) to the ribosome. The aa-tRNA forms a three-component complex with EF-Tu and GTP (Figure 6–12a). This complex binds to the available codon in the A site of the ribosome (Figure 6–12b). Following binding, GTP is converted to GDP by loss of a phosphate, a transition that results in the release of EF-Tu·GDP from the ribosome (Figure 6–12c). The EF-Tu·GDP is inactive in further tRNA binding but is recycled into an active conformation by association with EF-Ts, losing its GDP and reacquiring the ability to bind the GTP and tRNA (Figure 6–12d, e). Peter Lengyel was the first to point out that the GTP to GDP conversion most likely changes EF-Tu from an active to inactive conformation so that it can be released from the ribosome.

After the incoming amino acid arrives in the A site of the ribosome, the joining reaction (peptide bond formation) occurs. In this reaction, the growing amino acid chain in the P site is added to the new amino acid in the A site, transferring the growing amino acid chain to the A site (see Figure 6–11c). This reaction is catalyzed by a ribosome activity termed peptidyl transferase. This joining activity, discovered and characterized by Robin Monro, can occur with isolated 50S subunits under special conditions. The peptidyl transferase reaction is the sole catalytic activity of the ribosome itself. All efforts to associate peptide bond formation with a specific 50S ribosomal protein have failed. Based initially on other much later evidence of catalysis by RNA, the rather heretical notion that ribosomal RNA is the catalytic center of the ribosome was explored. Amazingly, this idea is almost certainly correct.

For protein synthesis to proceed, a translocation step must move the tRNA with its attached growing chain from the A site to the P site, creating a vacant A site able to bind the next aa-tRNA. This translocation step is mediated by the EF-G protein and GTP. Phil Leder and associates carried out a particularly clear experiment establishing the role of EF-G, using the synthetic mRNA, AUGUUUUUU, especially designed for observing sequential events. When only initiation factors are present, fmet-tRNA$_f$ binds to the initiator codon, AUG. Addition of EF-Tu and GTP allows the entry of phe-tRNA and formation of the fmet-phe-tRNA product (Figure 6–13a). However, the second phe cannot be added in response to the second UUU triplet, presumably because the A site on the ribosome is blocked by the continued presence of fmet-phe-tRNA. Addition of EF-G and GTP allows joining of the second phe to form the fmet-phe-phe-tRNA product (Figure 6–13b). Thus, binding of phe tRNA at the second UUU triplet depended on the action of EF-G and GTP, most likely because they moved the growing polypeptide chain, fmet-phe-tRNA, to the P site.

Lengyel and coworkers provided the eventual direct demonstration of physical translocation mediated by EF-G. The translocation reaction should move the mRNA one triplet with respect to the ribosome. Because the ribosome protects mRNA from the attack of nucleases, each translocation should result in protecting a new sequence of RNA further along the message. Using mRNA from the RNA phage, Lengyel generated an amino acid chain joined to tRNA in the A site in the absence of EF-G. Addition of EF-G moved the ribosome-protected RNA three bases, showing that EF-G indeed promotes translocation, acting as a sort of ribosomal

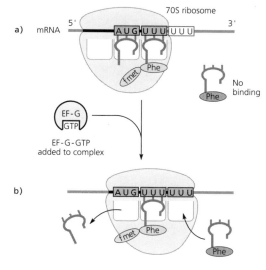

Elongation factor G allows mRNA and tRNAs to move on the ribosome

FIGURE 6–13. Using a short RNA that coded for the peptide f-met-phe-phe, Leder showed that only fmet and phe were joined in the absence of EF-G and GTP (a). When EF-G and GTP were provided, the second phenylanine was added to this mini-protein (b). Since this step requires the mRNA and tRNAs to move, or "translocate" relative to the ribosome, so that the A site is free to bind the second tRNA-phe, this experiment strongly suggested that EF-G and GTP were involved in translocation.

muscle protein. We now know that translocation can occur very, very slowly in the absence of EF-G, but EF-G is required to achieve the rates of translocation necessary for cell growth.

The three initiation factors have two roles: to position the fmet-tRNA$_f$, and to facilitate the binding of mRNA. The initiation factor primarily responsible for positioning the fmet-tRNA$_f$ is IF2. IF2, GTP, and fmet-tRNA$_f$ add to the 30S subunit together (Figure 6–14a). Except that a tight three-component complex is not found in solution, the action of IF2 in initiation is analogous to that of EF-Tu in elongation: both factors bring tRNAs to the ribosome (compare Figure 6–12a, b, and Figure 6–14a). IF1 is an adjunct protein for this reaction, enhancing the capacity of IF2 to bring fmet-tRNA$_f$ to the ribosome. IF3 carries out the second function of initiation factors: it is needed for effective addition of a natural mRNA to the 30S subunit but is not stably retained after the mRNA is bound (Figure 6–14b). Departure of IF3 allows the 50S subunit to bind. The union of the 30S and 50S ribosome subunits occurs with the cleavage of GTP to GDP, facilitating release of IF2. The resultant 70S structure now has fmet-tRNA$_f$ in the P site, and an open A site that is ready for the elongation steps that produce the new protein (Figure 6–14c).

The termination of the amino acid chain involves fooling the elongation machinery into joining the final (C-terminal) amino acid of the protein to water instead of the next amino acid. Simply halting protein synthesis will not free the polypeptide chain because the terminal amino acid would still be attached to the mRNA at the P site of the ribosome by the terminal tRNA. One might have guessed that special tRNA would be used to facilitate release, but early experiments by Mark Bretscher showed that this was not true. The necessary ruse is accomplished by special proteins called release factors, which work in concert with a nonsense codon in the mRNA to free the terminal amino acid from the grip of its tRNA.

The release proteins were identified by Capecchi and Tom Caskey, using two different templates and similar assays. Capecchi started with mRNA from an RNA phage that had a nonsense mutation after the first six triplet codons of the gene for the viral coat protein. Caskey used the minimalist approach of the Nirenberg group, employing a six-base "micromessage" mRNA consisting of an initiator triplet followed by a nonsense triplet. Both groups studied chain termination by following the release of the amino acid chain—comprising six-amino acids in the case of the phage RNA, or only fmet in the case of the micromessage. Two release factors

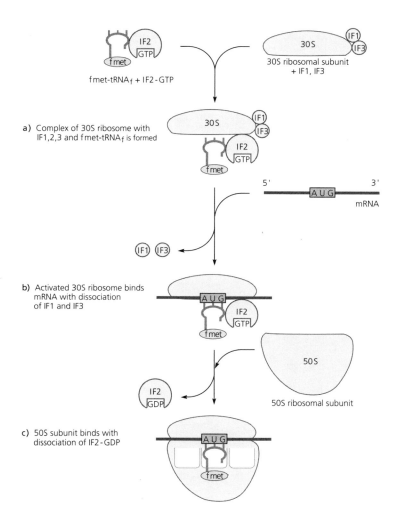

a) Complex of 30S ribosome with IF1,2,3 and fmet-tRNAf is formed

b) Activated 30S ribosome binds mRNA with dissociation of IF1 and IF3

c) 50S subunit binds with dissociation of IF2-GDP

FIGURE 6–14.   Initiation begins when IF2 carrying GTP escorts f-met tRNAf to the 30S subunit of the ribosome, assisted by IF1. IF3 holds the 30S subunit in an active conformation for binding to mRNA (a). IF3 and IF1 dissociate when mRNA binds (b), allowing the 50S subunit to join this complex. Binding of the 50S subunit triggers conversion of GTP to GDP, facilitating the loss of IF2 (c). The ribosome is now ready to begin translation.

were isolated, designated RF1 and RF2. RF1 works at UAG or UAA nonsense codons, whereas RF2 works at UAA or UGA nonsense codons. Since RF1 and RF2 respond to different nonsense codons, they must recognize to the nonsense codons themselves rather than simply speeding up a ribosome-directed reaction. Based on modeling studies, it now seems likely that release factors accomplish this by looking like tRNAs themselves! Another protein, RF3, speeds up the release reaction. The peptidyl transferase reaction also participates by transferring the terminal amino acid from tRNA to water (Figure 6–15).

The identification and characterization of the translation factors,

FIGURE 6–15.   When the A site of a translating ribosome encounters a termination or nonsense codon, the A site remains empty of tRNAs (a). Instead, a release factor binds to the A site. RF1 binds when UAG or UAA is present, and RF2 binds when UAA or UGA is present (b). The release factors facilitate joining the last amino acid of the peptide chain to water. The addition of RF3 accelerates the detachment of the amino acid chain from the grip of the tRNA in the P site (c). Following chain release, the translation complex dissociates (d).

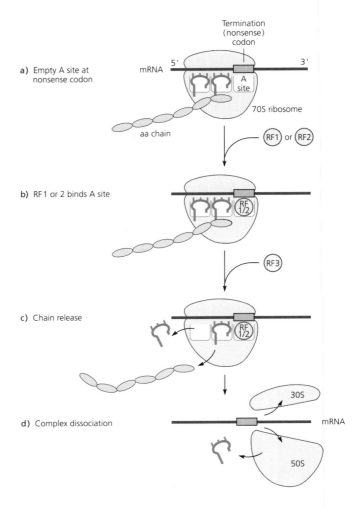

largely completed by 1970, led to the view that protein synthesis was a sequential, protein-catalyzed reaction. The genetic information resides in the mRNA triplet and the tRNA anticodon, but the execution of the translation process depended on the ability of proteins to get the job done at a pace that allows competitive growth rates and thus survival of an organism. At this time, the ribosome was a murky entity lurking in the background, and its role in translation was very poorly understood. This view was to change radically in the next two decades.

In teaching molecular biology, one approach is to present contemporary concepts as a "done deal"; the other is to emphasize the historical struggles responsible for these ideas. The former approach promotes the illusion of "knowing" molecular biology. I prefer the latter approach because it teaches how molecular biology is really done and points out the fragility of many contemporary concepts—a textbook conclusion of one year can be gone the next. The ribosome constitutes a wonderful example of conceptual fragility. Our mental portrait of a ribosome has changed with each decade of molecular biology. The ribosome is a chimera with a new face for each era.

In the late 1950's, the ribosome was viewed as an RNA virus, with its RNA surrounded by an inert coat composed of multiple copies of one or two proteins. By the late 1960's, the ribosome appeared as an RNA structural rack holding a beehive of diverse proteins, each with a specialized activity for protein synthesis. A decade later the ribosome exhibited a highly specialized RNA and protein topography that served to provide critical structural niches for the translation machine. In the contemporary view, the specific three-dimensional structure of the RNA provides both the essential structural features and the catalytic activity for amino acid joining, whereas ribosomal proteins simply stabilize the structure and speed up the operation.

The virus model, suggested by Watson, was initially appealing when ribosomal RNA was thought to be the informational message that must become accessible to the protein synthesis apparatus. Cellular mRNA, masquerading as ribosomal RNA, would emerge from its coat just as the genetic RNA emerged from the virus. The model persisted after mRNA was discovered, because the ribosomal RNA was considered to be a structural surface in the ribosome that held the mRNA. In the early 1960's, Jean-Pierre Waller concluded that there were multiple ribosomal proteins, based on his observation of distinctly migrating proteins after separation of ribosomal constituents in an electric field ("electrophoresis") carried out on a starch gel matrix. Joel Flaks also observed multiple protein species using electrophoresis in a polyacrylamide gel, the procedure that soon dominated the practice of cataloging the protein constituents in a mixture. However, so strongly ingrained was the notion of ribosome simplicity that these studies were greeted with skepticism. In the mid-1960's, I remember a dramatic announcement at a major

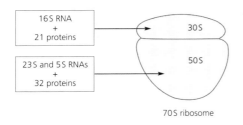

The 70S ribosome is composed of
three RNAs and many proteins

16S RNA
+
21 proteins → 30S

23S and 5S RNAs
+
32 proteins → 50S

70S ribosome

FIGURE 6–16.

Joan Steitz

molecular biology meeting by two different research groups who claimed that Waller and Flaks were wrong—after all, the ribosome contained multiple copies of a single ribosomal protein. Waller's finding that most of the presumably distinct proteins had methionine as the N-terminal amino acid strengthened the doubts of the skeptics. This particular problem was removed a few years later by the realization that all proteins begin with methionine.

By 1967, Waller and Flaks were right again, and the virus view of the ribosome was gone forever. Two research groups separated the ribosomal proteins and obtained large quantities of multiple individual proteins with clearly distinct physical properties and amino acid composition. Harry Noller, Rob Traut, and Peter Moore accomplished this task with Alfred Tissières at Geneva, and Chuck Kurland with Gary Craven, Masayasu Nomura, and collaborators achieved similar results at Wisconsin. The smaller 30S ribosome subunit carried 21 different proteins along with the single "16S RNA" that had previously been identified by Kurland when he was a graduate student with Watson. The Tissières group argued that each ribosome subunit in the cell probably carries one copy of all 21 proteins. Although purification of this ideal population was not achieved, the numbers were close. Later work, mainly from the group of Heinz-Günter Wittmann in Berlin, defined 32 proteins in the 50S ribosomal subunit, to go with the previously identified "23S" and "5S" RNA species (Figure 6–16). By 1967, the ribosome was clearly a complicated beast with a large array of distinct proteins. What did all of the proteins do? For what was the RNA used?

The subsequent story of the ribosome mirrors the changing view of RNA by biologists and chemists over time. As we have just recounted, an understanding of the biochemical complexity of the translation process emerged in the late 1960's. But the most important conceptual transition in the route to protein synthesis has been the realization of the catalytic capacity and functional diversity of RNA. In the recognition events responsible for translation of the genetic code, as well as in the formation of the peptide bond per se, RNA is not a passive voyeur but an active participant.

Joan Steitz has been at the center of the "RNA world" since the late 1960's. In 1962, she began graduate work at Harvard (as Joan Argetsinger) already an experienced molecular biologist from sojourns at MIT and Tübingen as part of the work-study academic program at Antioch College. Argetsinger became a student with

Watson after her initial first choice for a thesis adviser asked her why she wanted to bother with research as she would not continue in science once she got married. After a successful graduate project identifying A protein, the attachment protein for the RNA phage, Joan married Tom Steitz (an X-ray crystallographer), and the two left for postdoctoral work in the MRC Lab at Cambridge. On arriving in Cambridge, they found that there was lab space only for Tom, and Francis Crick suggested that Joan do a theoretical project in the library.

Undeterred, Joan begged some space from Mark Bretscher and Brian Clark, who were working on initiation of protein synthesis, and she set out on a difficult project that had scared everyone else away—directly identifying the mRNA sequence that is bound by the ribosome at initiation. Mitsuru Takanami had demonstrated some years earlier that the ribosome could protect mRNA from attack by degradative nucleases, and Peter Lengyel had worked out conditions where mRNA binding depended on initiator tRNA. But obtaining specific protected pieces of mRNA in sufficient quantity to determine their sequences was extremely difficult. After a year of trying and failing to define the parts of RNA phage that were protected by ribosomes, Steitz did "one last experiment" that captured three protected fragments, one for each gene of the phage. The base sequences of the radioactive pieces were obtained by the Sanger methodology, a local product at the MRC Lab, and the first initiator sequences of a real mRNA had been identified. Each of the three protected fragments carried an AUG initiator codon followed by codons for an amino acid sequence that was consistent with the three known gene products of the viral RNA–coat protein, synthetase, and A protein (Figure 6–17). AUG was an initiator codon *in vivo*, and ribosomes must somehow know how to pick the right AUG from a very long RNA chain.

The initiator RNA experiment defined the Joan Steitz approach to molecular biology—an exceptionally successful combination of biological vision and rigorous biochemistry, dedicated to the proposition that the structure of RNA has profound implications. Steitz brought to the RNA field an infectious enthusiasm and passion for clear, definitive experiments that made her an effective missionary as well as a leading participant. She has a love for science that illuminates any conversation.

At the conclusion of their postdoctoral stay at Cambridge, Joan and Tom Steitz faced a problem that has become increasingly

FIGURE 6–17. To identify the mRNA sequences recognized by the ribosome, Joan Steitz bound ribosomes to RNA phage mRNA, isolated the mRNA pieces protected from nuclease attack, and then determined their sequences. Steitz isolated three protected pieces, one for each of the three genes in the phage RNA. Each had the initiator AUG codon for that gene, demonstrating that ribosomes could pick out this special AUG from all other AUGs in the mRNA.

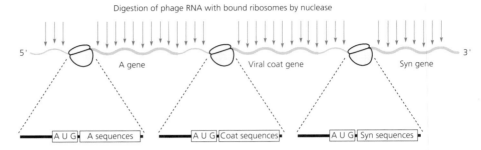

Digestion of phage RNA with bound ribosomes by nuclease

Phage RNA fragments protected from digestion (about 30 bases long)

frequent in molecular biology—acquiring two academic jobs in one place. Tom accepted an assistant professor position at Berkeley, but Joan was offered only another postdoctoral job. Tom's efforts to obtain funding at Berkeley ran into administrative snags, and he was told by two senior faculty members that a woman such as Joan would have difficulty competing for an academic position at Berkeley. In the meantime, Yale offered assistant professor positions to both, and the Steitzes were off to New Haven. Attitudes toward women in science have changed since 1970, but the historical problems are recent enough to be worth remembering. The number of women with tenured faculty positions is still alarmingly small.

Safely ensconced in her position at Yale, Steitz was now ready to tackle a major question: what are the determinants in mRNA required to bind ribosomes and thereby localize initiation? The AUG sequence in this region could not be sufficient by itself for recognition—there were many other AUG triplets in the 3,300 bases of phage RNA. Steitz had looked hopefully for some common sequence upstream from the AUG triplet, but no consistent pattern emerged. Now she decided to expand her sample size by looking at RNA sequences from other RNA phages, and the T7 DNA phage. Other labs contributed ribosome binding site sequences from a few bacterial mRNAs.

The big news on initiator recognition finally came in 1974. Steitz received a visit from Lynn Dalgarno, an Australian unknown in the field, who worked on 16S RNA, the RNA component of the 30S ribosomal subunit. Dalgarno and John Shine had resequenced the

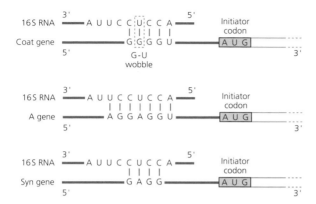

The Shine-Dalgarno hypothesis shows how the ribosome could find the initiator codon in mRNA

FIGURE 6–18. Shine and Dalgarno proposed that the additional interactions necessary for the ribosome to recognize the initiating AUG were provided by base-pairing between the 5'ACCUCCUUA 3' sequence at the end of 16S RNA and mRNA sequences upstream of the initiator codon. They suggested that an exact match was unnecessary so long as some base-pairing was possible. This allowed the 16S sequence to recognize many similar, but not identical mRNA sequences, each located upstream of the initiating AUG.

end of the 16S RNA molecule (correcting an earlier version), and they had identified some base-pairing potential between this end of 16S RNA and the mRNA sequence upstream from the AUG. Based on her own work and experiments from other labs, Steitz reinforced Dalgarno's hypothesis that this additional base-pairing defined the initiator AUG. All of the known initiator sequences could fit his proposal. Dalgarno's spectacular insight was the realization that the base-pairing between the two RNAs was somewhat flexible. The recognition site on 16S RNA consisted of a set of bases available to pair with mRNA sequences upstream of the AUG. Every mRNA would contain some bases able to pair with 16S RNA recognition site. This recognition mechanism involved the concept of a "consensus sequence"—the recognition sequences and spacing of the mRNA site upstream of the AUG were not precisely identical but were very similar.

Other than Steitz, few people were initially impressed with the Shine-Dalgarno hypothesis. Dalgarno had sequenced only the very end of the 16S RNA—it would be tremendous luck to find the crucial pairing sequence in that small stretch. Moreover, some of the sequence pairings were very bad. The problems can be illustrated by looking at pairings of the three initiator regions of the RNA phage with the Shine-Dalgarno recognition sequence (Figure 6–18). Since the coat protein was translated best, it should have the best pairing. Instead, the coat protein RNA required a G–U wobble base-pair to get just five base-pairs in a row, whereas the poorly translated A protein RNA made an impressive seven-base duplex with 16S RNA.

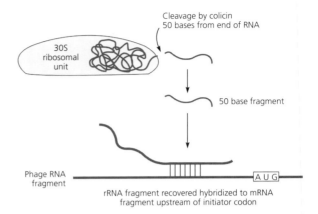

A fragment of 16S rRNA containing the ACCUCCUUA sequence binds the initiator region

FIGURE 6–19.   A small fragment of 16S RNA containing the Shine-Dalgarno sequence is released from the ribosome by cleavage with a colicin. Joan Steitz was able to show that this 50-base fragment was recovered hybridized to an initiator region of phage RNA, proving that the Shine-Dalgarno sequence could recognize mRNA.

However, Steitz had an explanation for this. She had found that, when she rebound ribosomes to her isolated fragments, they associated best with A RNA, as predicted by Shine-Dalgarno pairing. Maybe the intact viral RNA contained a secondary structure that masked the A binding site.

Fortified by this insight, Steitz set out to determine whether the Shine-Dalgarno pairing really occurred. By good fortune, this experiment was possible, thanks to "colicins," proteins used by one *E. coli* strain to war against others. One particular colicin had been shown to cleave off a 50-base segment containing the Shine-Dalgarno sequence from the end of 16S RNA, releasing it from the ribosome. Steitz showed that this 50-base segment exhibited RNA-RNA pairing with the initiator region of phage RNA (Figure 6–19). The mechanism for localization of the initiation site in mRNA was solved. Moreover, the much-maligned ribosomal RNA was providing the critical interaction.

At the time of this experiment, ribosome research was focused on the mysterious collection of ribosomal proteins, so experiments on the specific role of 16S RNA were distinctly off the mainstream. The surprising abundance of ribosomal proteins seemed to imply an array of distinct biochemical functions. Maybe each individual protein was associated with a particular reaction. In 1967, Peter Traub and Masayasu Nomura succeeded in rebuilding a 30S ribosome from the 16S RNA and 21 isolated proteins. The trick in ribosome assembly was to carry out the process at biological growth temperature (37°C), rather than the classical refrigerator temperatures usually used in handling proteins. This stunning technical feat seemed to provide a potential assay for the function of the individual proteins, which could be deleted singly or in groups from the assembly mix to identify different protein functions.

Nomura devoted the first 30 years of his research career to the structure, function, and synthesis of ribosomes, with exceptional success. He came to the United States from Japan in 1960 as a post-doctoral fellow with Spiegelman. He was introduced to analytical studies of ribosomes during a brief visit to Watson's lab and began to study the assembly of ribosomes when he returned to Japan in 1962. The Japanese academic system was not designed to encourage independent research by young investigators, a situation that has not changed notably in the ensuing years. As a result, Nomura returned to the United States in 1963, accepting a faculty job at Wisconsin, where he and I jointly taught a molecular genetics course for several years.

Masayasu Nomura

Nomura immediately evinced an exceptional ability to apply biochemical approaches to complex biological problems. His later work analyzing how production of ribosomes is regulated in cells demonstrated his outstanding capacity to think from both genetic and biological perspectives. Nomura's intense drive to do everything that was clearly interesting in the ribosome field sometimes led to strained relations with other groups, but he achieved consistent success in his areas of focus. He had the important role, but perhaps the historical misfortune, to demonstrate that ribosomal proteins are not all that important in ribosome function. Using the cell-free assembly system, Traub and later members of Nomura's group vigorously pursued the role of ribosomal proteins, with disappointing results. Omission of some proteins lowered the activity of the ribosome for protein synthesis but also interfered with the assembly process. Other proteins could be left out of the assembly mix altogether without a discernible deleterious effect. The ribosomal proteins were clearly not a set of individual enzymes facilitating the translational pathway of initiation, elongation, and termination. The protein beehive view of the ribosome had to be abandoned.

Efforts to find enzymatic functions for the ribosomal proteins shifted to the 50S subunit, which carries the peptidyl transferase activity joining the incoming amino acid to the growing chain. The detailed functional analysis was carried out primarily by Knud Nierhaus and associates in Wittmann's group in Berlin using a 50S subunit that was rebuilt from 23S RNA, 5S RNA, and 34 isolated proteins. Nierhaus and coworkers obtained essentially the same general results with the 50S ribosome as had Nomura with the 30S—no functions were attributable to single proteins. The peptidyl transferase activity was elusive. By 1980, the amino acid joining activity had been "semi-localized" to 23S RNA plus several 50S proteins.

In the meantime, during the 1970's, a major effort was directed toward determining the arrangement of proteins in the ribosome. The ribosome was an awkward structure for physical studies—large, asymmetric, and composed of many proteins as well as RNA. The study of ribosome topography employed a variety of techniques. Several groups used chemical cross-linking agents, which join together proteins that are located very close to each other, providing information about which proteins are neighbors in the ribosome. Peter Moore and Don Engelman figured out how to use neutron scattering to determine distances between proteins; this information has provided powerful constraints for models of the ribosome. The most successful initial approach turned out to be electron microscopy, work carried out mainly by Jim Lake at the University of California–Los Angeles (UCLA) and George Stöffler in Berlin. The ribosome is a reasonable size for visualization by electron microscopy, although the technique has limited ability to see detailed structure because biological molecules must be "shadowed" with a dense material in order to be observed. However, the individual proteins could be "observed" by tagging them with a specific antibody, and their locations could be determined with respect to the asymmetric outline of the ribosome. After some initial disagreement between the two groups, the general features of the Lake view of the 30S ribosome were accepted. This model is also consistent with the protein assignments from the neutron scattering work, eventually completed by Moore and colleagues in 1987. Later experiments by the Lake and Stöffler groups defined the location of many of the 50S proteins.

The 30S ribosome looks something like a mittened hand in repose; in similar vein, the 50S ribosome might qualify as a mittened hand in a ghost pantomime. The 70S forms by a clasping of the two hands. The P site for amino acid tRNA is roughly under the "thumb" of the 30S subunit and the A site is somewhat to the right of that (Figure 6–20).

Overall, the 1970's was not a joyous period in ribosome research. An enormous amount of work had revealed something about struc-

The ribosomal subunits are shaped like mittens

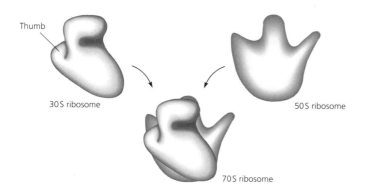

Thumb

30S ribosome

70S ribosome

50S ribosome

FIGURE 6–20. (Ribosomal models courtesy of J. Lake.)

ture, but essentially nothing about the function of the ribosomal proteins. The sole clear experiments on ribosome mechanism had been those defining the importance of the 16S "Shine-Dalgarno sequence" in grabbing the initiator sequence in the mRNA. The initiative started to shift to the "RNA-centric" view of the ribosome.

Harry Noller and Carl Woese were the major initial voices for the RNA ribosome. Two developments in the early 1970's sparked that view. First, Noller found that he could destroy the ability of tRNA to bind to the ribosome by modifying only a few bases in ribosomal RNA. Simultaneously, Woese was comparing sequences of 16S rRNA from different bacterial species as a means to study evolutionary history. Determining base sequences was difficult then, but Woese and coworkers started with short segments of RNA. Sequence comparisons among organisms representing vast evolutionary divergence revealed that some sequences showed a remarkable identity, whereas others differed greatly. Significantly, the rRNA sites that Noller had shown were important for tRNAs to bind to ribosomes were among those universally conserved. In 1979, Woese presented a talk at a large ribosome meeting at Wisconsin, where he argued that the phenomenal sequence conservation he had observed was compatible only with a vital functional role for those conserved regions of rRNA. rRNA must be a participant in protein synthesis, not a coat rack for ribosomal proteins.

By the late 1970's, techniques for molecular cloning and DNA sequencing permitted the determination of DNA sequences for specific genes, including those for rRNA. In 1978, Noller and his colleagues at the University of California–Santa Cruz determined

Harry Noller

Carl Woese

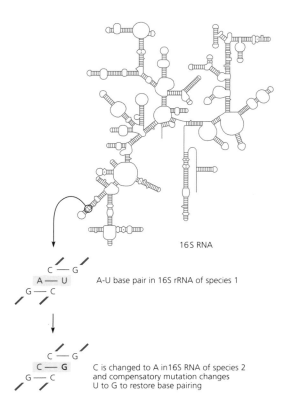

Base-pairing in 16S RNA is important for its function

16S RNA

C — G
A — U   A-U base pair in 16S rRNA of species 1
G — C

C — G
C — G   C is changed to A in16S RNA of species 2
G — C   and compensatory mutation changes
         U to G to restore base pairing

FIGURE 6–21.   Many parts of 16S RNA can base-pair with other parts of 16S RNA to give a complex secondary structure. Woese and Noller used sequence data from different species to assess which base-paired segments were important for function. Functionally important base-pairing should be conserved from species to species, even if the identity of the base-pair changes. In the example shown, an A-U base-pair in species 1 is replaced by a C-G base-pair in species 2, indicating the importance of maintaining base-pairing in this segment. Many comparisons of this nature indicate that much of the secondary structure of 16S RNA is important for its function.

the complete sequence of a 16S RNA from its DNA sequence. Woese and Noller found that regions of structural conservation in 16S RNA included many short regions of RNA that could base-pair with each other (Figure 6–21, top). They realized that sequence comparisons provided a clever way of "proving" the existence of the presumed secondary structures in RNA. If base-pairing is important for function, then organisms will preserve the ability to base-pair even when the sequences change. Therefore, an evolutionary change (mutation) in a base that must base-pair will be accompanied by a compensatory alteration that preserves the base-pairing potential. For example, assume one organism has an A–U base-pair in the RNA; if a second organism has replaced the A with a C, then it will also have changed the base-pairing partner from a U to a G to preserve base-pairing at that position. Examining many sequences proved that compensatory mutations to preserve base-pairing were often present in these segments of RNA (Figure 6–21, bottom). Woese and Noller argued that rRNA has a precise, highly conserved structure that is extremely likely to participate actively in protein synthesis.

The major new intellectual force behind the RNA view of the ribosome has been the discovery of the catalytic capacity of RNA in certain RNA processing reactions, work described in the final section of this chapter. The discovery by Tom Cech in 1981 that RNA could work like an enzyme profoundly influenced the ribosome field. The search for the biochemical activities of the ribosome that carry out the stages of protein synthesis shifted from the proteins to rRNA. The conceptual switch was fostered by new techniques that revealed that specific nucleotides of rRNA directly interact with tRNA and mRNA. A specific base pair between 23S rRNA and tRNA even positions tRNA at the peptide transferase center of the ribosome. Noller's finding that 23S rRNA largely devoid of proteins carried out peptide transfer significantly advanced the idea that rRNA itself catalyzes peptide bond formation and the high resolution crystal structure of the 50S ribosomal particle now seems to prove the case. Tom Steitz and colleagues show that the peptide transferase center of the ribosome is composed solely of RNA and that the many ribosomal proteins are mostly on the outside, helping to organize this complex RNA structure. Although the precise mechanism of this RNA catalyzed reaction is yet to be solved, it now seems clear that the ribosome is a proud representative of the primordial RNA world.

The discovery of mRNA brought with it the realization that RNA is not at all simply the single-strand alter ego of DNA. Whereas DNA is a stable repository of the genes, mRNA carries out its translation function only transiently: it is degraded with an intracellular lifetime of a few minutes. The instability of mRNA means that the translation of unneeded proteins can be rapidly terminated when transcription of that mRNA stops. Thus, turnover of mRNA permits rapid cellular adaptation to new environments.

In the 1960's, mRNA was considered to be a special class of RNA, distinct from the "stable RNAs," tRNA and rRNA. Because there appeared to be no logical reason for it, no one thought that stable RNAs might also be cleaved after transcription. But this logic was not evident to Mother Nature. Both tRNA and rRNA are transcribed from DNA as much longer molecules than the final stable species used for protein synthesis. The RNA slicing events that make the "mature" rRNA and tRNA are collectively termed RNA processing.

Like mRNA, the precursors of stable RNA are not easily noticed because these large RNAs are also unstable—they are rapidly processed to the mature form. The first clear demonstration of a tRNA precursor came ten years after the discovery of mRNA—the product of frustration rather than design. To understand the structure of tRNA, John Smith and Sydney Brenner at the MRC lab in Cambridge were determining the base sequence changes that altered the activity of tyrosine tRNA. Their initial big success had been finding that a change in the anticodon allowed tyrosine tRNA to work as a suppressor at the nonsense codon UAG (see Figure 6–5c).

As a new postdoc at the MRC lab, Sidney Altman carried on earlier work to study other tRNA mutations that led to loss of activity. Unfortunately, most of the mutant tRNAs were not to be found in cell extracts. Suspecting that the altered tRNAs might be degraded, Altman carried out a "last resort," instant tRNA extraction procedure (such as would be used for unstable mRNA). The procedure identified an RNA transcript for the mutant tRNA, but the RNA was about 50% too long. Altman also identified a similarly overlength RNA from the wild-type tRNA gene. Mature tRNA was evidently born as an unstable precursor, which was then processed to the stable tRNA molecules found in cells. This spectacular conclusion shelved work on the mutants (the mutant tRNAs were presumably completely degraded because those molecules failed to fold into a tRNA structure, and so were rudely treated like mRNA). The announcement of the tRNA precursor by Altman and Smith in 1971 initiated the study of tRNA processing by a number of research groups.

Using the MRC lab protocols for RNA sequencing, Altman learned that the precursor tRNA carried 41 extra bases preceding and 3 extra bases following the mature tRNA. (We now know that the true precursor is even larger—the Altman precursor was already partially processed.) Hugh Robertson and Altman found that the 41 bases preceding tRNA could be removed by a single cleavage mediated by an RNA-cutting activity present in extracts of *E. coli*; the presumed enzyme was termed RNase P for processing ribonuclease. Altman moved to Yale in 1971, where he sought to isolate the RNase P enzyme. This long-term effort identified RNase P as an enzyme with both a protein and an RNA component. This ultimately led to the realization in 1983 that the catalytic activity of RNase P resides in the RNA component of the enzyme.

The generality of tRNA processing was initially hard to assess because the precursors did not last long in cells. This impasse ended in 1973 when Paul Schedl and Paul Primakoff, working in Paul Berg's lab at Stanford, isolated a temperature-sensitive mutation in RNase P. A zoo of tRNA precursors accumulated in cells when the RNase P activity was inactivated by elevated temperature; RNase P must be a general processing enzyme for tRNAs. Based on experiments by Schedl and Primakoff and by Haruo Ozeki and coworkers with other RNase P mutants, all of the tRNAs in *E. coli* are probably produced as longer precursor molecules; some precursors have only a single tRNA, whereas others carry two or more tRNAs.

The accumulation of tRNA precursors in the RNase P mutant showed that RNase P was probably responsible for making one mature end (the 5′ end) of all tRNA molecules. Because the cleavage sites of the different tRNAs have different base sequences, RNase P presumably recognizes the specific three dimensional structure of tRNA, which folds into its final shape within the longer precursor RNA. RNase P is termed an endonuclease because the cleavage site is within an RNA molecule. Additional work indicated that the other end of the mature tRNA (the 3′ end) is probably produced by one or more enzymes that remove one base at a time from the end of the precursor RNA transcript; these exonucleases stop their nibbling action at the end of mature tRNA (Figure 6–22).

Close on the heels of tRNA maturation came the discovery that mRNA and rRNA are also processed, in this case by another fascinating RNA slicing enzyme, RNase III. Working at the Brookhaven Laboratory on Long Island, Bill Studier began a detailed study of the transcriptional and translational expression of phage T7, work that he pursued with notable success for many years. Studier knew from the work of Ruth Siegel and William Summers that, when T7 infected cells, five small species of mRNA were produced from the T7 early genes. He found to his surprise that, when RNA polymerase transcribed T7 DNA *in vitro*, only a single large RNA was made. John Dunn then took a position at the lab. With a background in RNA polymerase, he set out to understand why there were so many RNAs in cells. Dunn could not reproduce the multiplicity of RNAs in his purified system, so he fractionated cell extracts to look for a new biochemical activity that might produce the shorter RNAs. An RNA endonuclease emerged that cleaved the long RNA into the five RNAs found in cells. T7 mRNA was processed, a most heretical behavior. Dunn purified the cleaving protein a bit and read up on other RNA nucleases. He concluded that his processing enzyme resembled RNase III. This enzyme had previously been isolated by Hugh Robertson and Norton Zinder, based on its ability to degrade double-strand RNA (nucleases were usually identified without knowledge of function and classified generically as RNase I, RNase II, RNase III, etc.).

At this point, Studier went on a lecture tour in Europe, where he talked with Peter Hofschneider, who had just isolated an *E. coli* mutant deficient in RNase III. This strain was just what Studier needed to test whether RNase III processed T7 RNA *in vivo*. Studier returned with the mutant strain and hit the jackpot with the first experiments. After phage infection, a single long T7 transcript carrying all of the early genes appeared in the mutant cells rather than the five shorter RNAs seen in normal cells. Dunn and Studier also discovered that the RNase III mutant strain made a very long cellular RNA. This turned out to be an RNA transcript containing the ribosomal 16S, 23S, and 5S sequences. RNase III processed rRNA as well as T7 early mRNA! David Schlessinger and coworkers at Washington University independently established the role of RNase III in rRNA processing. RNase III is clearly the central processing enzyme for rRNA but does not appear to be a general processing enzyme for mRNA. Instead, it is used for special regulatory functions with certain mRNA species.

Dunn, Robertson, and others went on to characterize the target sites for RNase III in phage T7 RNA. RNase III cleaves in a region

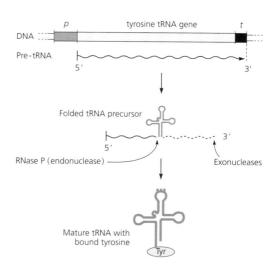

tRNA is cut from a larger transcript

FIGURE 6–22. The tRNA gene is transcribed as part of a larger precursor RNA, beginning at its promoter (*p*) and ending at its terminator (*t*). It assumes a cloverleaf structure while still a part of this transcript. RNase P endonuclease cuts off the unfolded RNA at the 5′ end of the tRNA and exonucleases degrade the unfolded RNA at the 3′ end of the tRNA, releasing mature tRNA. "Endo-" nucleases cut at an internal site, whereas "exo-" nucleases digest from an end of the RNA.

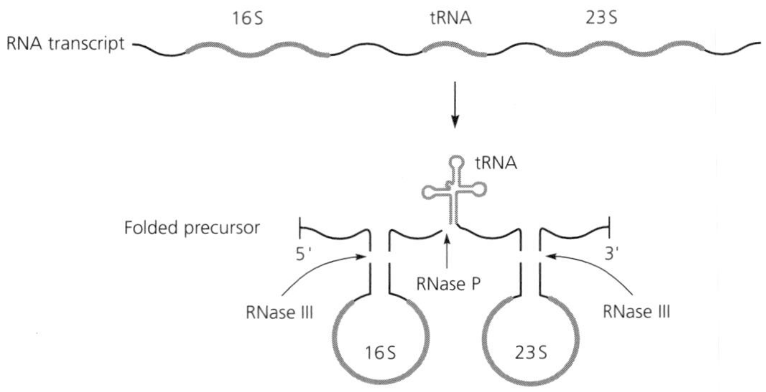

FIGURE 6–23. The 16S, 23S, and 5S RNA genes are transcribed as part of a larger precursor RNA that often includes some tRNAs as well. Nucleases release each RNA species. The 16S and 23S sequences are each freed from the larger transcript by RNase III. The long duplex required for RNase III to cut RNA is formed by base-pairing segments on either side of the 16S and 23S RNAs. The 5' end of the tRNAs is freed by the endonuclease RNase P. Exonucleases finish the job of removing the leftover sequences in each RNA. 5S RNA is not shown.

of RNA with extensive base-pairing—a long "RNA hairpin." Based on the study of T7 sequences and other work, the recognition sequence for RNase III appears to be a duplex RNA region of at least ten base-pairs in length.

The most spectacular example of RNase III activity occurs in ribosomal RNA processing. Earlier experiments had established that the genes for 16S, 23S, and 5S RNA were on the same transcript. Nomura and coworkers found that the genome of *E. coli* had seven "rRNA operons." With the development of DNA sequencing methodology, the detailed organization of each of the rRNA operons became available for study. Nomura, Jim Dahlberg, and their associates at Wisconsin learned that certain tRNA genes were contained within the rRNA transcript, between the 16S and 23S RNAs. The rRNA operon was "RNA processor's heaven."

Rick Young and Joan Steitz wondered how the large rRNAs were cut out of the long transcript. Young determined the sequences adjacent to mature 16S RNA and found that bases on either side of the 1,542-base rRNA had the potential to pair with each other, forming a 26 base-paired RNA duplex. RNase III cuts within this duplex region, leaving some extra RNA on either end of the 16S RNA. Other processing steps finish the job by trimming off the extra bases. Similarly, a long duplex made from bases flanking 23S RNA serves as a target for RNase III, after which additional processing reactions yield mature 23S RNA (Figure 6–23).

The experiments with RNase III unveiled a complex and fascinating pathway for rRNA maturation. In addition, this work proved the biological importance of secondary structure in large RNA molecules. The existence of RNA duplexes had been initially inferred from physical measurements in the late 1950's by Paul Doty and his colleagues at Harvard. The universal clover leaf presented a convincing case for tRNA. The RNA phage workers provided a variety of strong indirect arguments for the regulatory importance of RNA secondary structure in modulating translation of the viral mRNA. The discovery that an RNA duplex was the target of RNase III routed the remaining skeptics.

RNA SPLICING IN EUKARYOTES

The cellular organization of eukaryotic organisms is clearly more complex than that of bacteria. The much larger eukaryotic genome is packaged into an organized chromosomal structure in the nucleus, and the mRNAs made in the nucleus are transported to the cytoplasm for protein synthesis. In general, however, the molecular mechanism for the transfer of genetic information is extremely similar in prokaryotes and eukaryotes. A major special property of genetic expression in eukaryotes emerged in the mid-1970's: the DNA coding sequence of many genes is discontinuous, requiring an RNA splicing reaction to remove the extraneous information interspersed with the useful bits and assemble a functional mRNA capable of specifying a complete protein. RNA splicing was such an unexpected concept that, in order to avoid it, much evidence for splicing was initially interpreted in other, often strained ways. Once the idea was presented in 1977, a flurry of papers followed with earlier experiments reinterpreted in terms of the new concept. RNA splicing occurs by a unique biochemical mechanism, involving a complex cellular machinery (with interesting exceptions).

The first indication that something might be different about eukaryotic transcripts harkened back to work on mRNA synthesis in animal cells carried out by Jim Darnell and coworkers at the Rockefeller University in the 1960's. When they added a radioactive precursor of RNA to animal cells, they detected very long radioactive RNA molecules in the nucleus, which they termed heterogeneous nuclear RNA (hnRNA). Surprisingly, the bulk of this RNA was degraded in the nucleus, never reaching the cytoplasmic ribosomes, the proper target for mRNA. What was hnRNA, and how was it related to mRNA? These questions were difficult to answer without being able to follow the history of individual mRNAs, an experiment that was not technically possible at the time.

Work in the early 1970's introduced a way to identify both the beginning and the end of eukaryotic mRNA, allowing workers to determine which end of the nuclear hnRNA was transported to the cytoplasm. Following up on an initial observation by Mary Edmonds, all mRNA molecules were found to carry a chain of A bases at the very 3′ end of the transcript. Darnell and coworkers found that poly A "tails" were present at the ends of both hnRNA and cytoplasmic mRNA. Based on this observation, they suggested hnRNA was cleaved and the 3′ segment with the polyA tail was transported into the cytoplasm as mRNA, leaving the 5′ segment of hnRNA in the nucleus. But this was soon shown to be only half right. In 1972, Aaron Shatkin at the Roche Institute discovered that the 5′ end of mRNA in animal cells was covered with a protective "cap" made out of an unusual backwards chemical linkage with a G nucleotide (making the RNA unpalatable to degrading nucleases). Bob Perry quickly learned that both hnRNA and cytoplasmic mRNA carried the G cap, suggesting that it was the 5′ end of hnRNA that was transported to the cytoplasm. This presented an apparent conundrum: the long hnRNA seemed to be processed into the shorter mRNA by a reaction that conserved both ends of hnRNA (Figure 6–24). There was no precedent for such a reaction. The resolution of this paradox required a way to look at individual mRNAs. This quest had been under way for some time, largely focused on the problem of defining the eukaryotic gene and its regulatory sequences.

The crucial experiments that led to the splicing concept were carried out with the mRNA transcripts of adenovirus DNA, which encoded few enough abundant RNA transcripts that each could be followed with high resolution. The principal contributors represented two research groups: Rich Roberts, Louise Chow, and their colleagues at the Cold Spring Harbor Laboratory; and Phil Sharp, a Cold Spring Harbor graduate who had recently moved to MIT. The discovery depended on a large body of previous work at Cold Spring Harbor devoted to understanding the transcriptional pattern of adenovirus.

In 1968 Watson, the new director of Cold Spring Harbor, decided that a major focus of the laboratory should be on the DNA animal viruses that were capable of causing cancer, with a view to an eventual understanding of the disease. Joe Sambrook came in 1968 to start the animal virus group, which expanded rapidly with the addition of a number of exceptionally talented young scientists. The

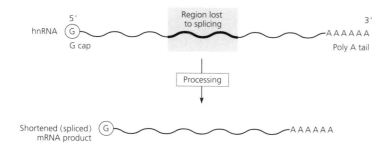

hnRNA

5'

G

G cap

Region lost
to splicing

3'

A A A A A A

Poly A tail

Processing

Shortened (spliced)
mRNA product

G

A A A A A A

FIGURE 6–24.   Both ends of the long RNA transcripts in the nucleus ("hnRNA") are marked: they begin with a modified 5' G base, called a "G cap," and end with a string of A bases, known as a "poly A tail." Without losing either of its distinctive ends, hnRNA is processed into shorter pieces and transported to the cytoplasm to serve as mRNA. This is accomplished by deleting ("splicing" out) sequences in the middle of the long transcript.

focus of the research was on SV40, a small virus with only a few genes, and adenovirus, one cause of the common cold, which is comparable in size and genome complexity to phage λ. The value of DNA animal viruses in understanding cancer emerged only 20 years later. But these viruses immediately provided a superlative insight into the workings of the normal eukaryotic cell, in the same way that we have described earlier for the bacterial DNA phage.

The Cold Spring Harbor virus group became a remarkable enterprise. Watson had exceptional skill in recognizing talented individuals, and he argued persuasively for the future importance of DNA animal viruses in research on eukaryotic molecular biology and cancer. He served both as a talent scout and prophet of a new era, a role very similar to that of Max Delbrück with the phage group 20 years earlier. However, there was one notable difference in style and philosophy. Delbrück believed in a society devoted to shared contribution and mutual trust (at least within his phage group). As he describes in *The Double Helix*, Watson was an individualist devoted to getting there first. He imposed no Delbrückian social goals on Cold Spring Harbor. Whether by accident or design, the social structure of the virus group became an experiment in rapid Darwinian selection. The competition among members of the virus group was intense. Of this exceptional group, a few eventually did very well; a larger number did not emerge as major contributors to molecular biology.

Phil Sharp has been the extraordinary achiever from this early virus group. He arrived at Cold Spring Harbor with the right combination of personality, goals, and scientific skills to derive maximum benefit, and he carried his momentum into the discovery of splicing and a succession of other major scientific achievements at MIT.

Phil Sharp

Sharp had already come a long way from a small college in rural Kentucky, stopping at Illinois for his Ph.D. and at Cal Tech for post-doctoral work with Norman Davidson, an enormously talented physical chemist. At Cal Tech, Sharp had learned the physical chemistry of DNA and how to use the electron microscope to locate particular DNA sequences. The DNA duplexes that result from hybridizing small DNA segments to a longer single-strand DNA molecule can be directly visualized in the electron microscope.

With an intense interest in defining the eukaryotic gene and its expression, Sharp arrived at the right time both for himself and for the virus group. Sharp loved competition, and he possessed an experimental wizardry for developing the ideal technique to solve a given problem. In a research discussion, he pounces on an interesting bit of data like a hungry tiger. Sambrook provided a background in virology and a superlative sense of the attainable goal, always a critical attribute in an emerging area. Working with SV40, Dan Nathans at Johns Hopkins had introduced restriction enzymes as a way to make fragments of viral DNA. By hybridizing the RNA transcripts to the isolated fragments of DNA, Nathans was able to locate the DNA source of RNA transcripts (see Figure 10–8). Sambrook realized the power of this technique as a universal approach to studying gene expression, and Sharp supplied his "golden hands" to make the technology simple and fast. Specific DNA fragments were prepared by gel electrophoresis. They hybridized viral RNA made either early or late in infection to these fragments, thereby locating the DNA regions active in production of SV40 and adenovirus RNA. This technique allowed Sharp, Sambrook, and associates to define a temporal program for SV40 and adenovirus growth, as had been done previously for λ and T4 phage (see Figures 5–5 and 5–7). Adenovirus DNA directed an "early" class of RNA, followed by a "late" population of RNAs devoted to making proteins for viral encapsulation (Figure 6–25).

Sharp moved to MIT in 1974, carrying with him his passion for tracking adeno RNA to its genomic birthplace in the adeno DNA. The late RNAs were promising for detailed study because they were so abundant and because they made known viral proteins. Sue Berget and Sharp learned to isolate specific late RNAs by gel electrophoresis, and they selected one to study. Their immediate goal was to identify the promoter site for this DNA, but there were no direct techniques available for mapping the RNA to its DNA source with the requisite precision. The best available approach was

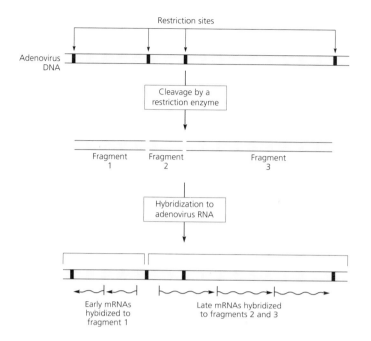

Adenovirus DNA produces two classes of mRNA

Restriction sites

Adenovirus DNA

Cleavage by a restriction enzyme

Fragment 1   Fragment 2   Fragment 3

Hybridization to adenovirus RNA

Early mRNAs hybidized to fragment 1

Late mRNAs hybridized to fragments 2 and 3

FIGURE 6–25. Like phage λ, adenovirus has two classes of genes: those expressed early and those expressed later in the life cycle of the virus. By cutting adenovirus DNA with restriction enzymes and then hybridizing adenovirus RNA made at different times after infection to the separated strands of these specific isolated fragments, the DNA regions responsible for early and late transcripts were identified. Early genes are encoded on the left side and late genes on the right side of the viral genome.

"R-loop mapping," a technique developed by Ron Davis. Under appropriate conditions, a homologous RNA could form a DNA-RNA hybrid, termed an "R-loop," by displacing one DNA strand within otherwise duplex DNA (because DNA-RNA duplexes are more stable than DNA-DNA duplexes). Berget carried out the R-looping protocol with her RNA and adenovirus DNA restriction fragments, and made a most surprising observation—the DNA-RNA hybrid revealed an RNA "tail" at each end, indicating that both ends of the transcript were nonhomologous to the DNA (Figure 6–26a). A tail at the 3' polyA end was expected because the polyA track was not encoded in the DNA; Norman Davidson had already characterized this feature. But why did there also appear to be a tail at the 5' G-cap (promoter) end? The phenomenon of "branch migration" (meaning that the duplex region can move) might be an explanation. At the very ends of the RNA, the DNA-DNA duplex might be more stable than the RNA-DNA hybrid, allowing it to displace the RNA and create the 5' tail.

To avoid the possible RNA displacement phenomenon, Berget carried out electron microscopy with single-stranded DNA so that

The 5' end of an adenovirus late mRNA is not complementary
to the DNA fragment adjacent to the gene

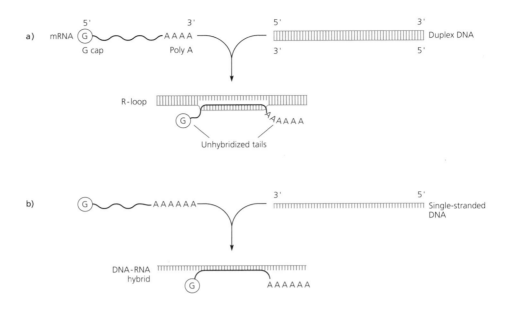

FIGURE 6–26. When an mRNA molecule is mixed with the DNA that encodes it, the mRNA will hybridize to its complementary strand, displacing the other DNA strand. This structure is called an "R-loop." When a purified adenovirus late RNA was hybridized to its complementary DNA, the 5' end of the RNA did not pair and instead extended as a nonhybridized "tail" (a). The same 5' tail is seen even when the late RNA is hybridized to single-stranded DNA, indicating that the tail is not an artifact of the R-loop method (b). Further experiments showed that the 5' tail comes from a part of the adenovirus genome termed the VA region.

only DNA-RNA hybrids could form. The 5′ tail persisted (Figure 6–26b). Berget and Sharp made a "list of possible artifacts," which they ruled out by more experiments. Having exhausted all other possibilities, they asked the definitive experimental question— did the 5′ tail originate from another region of adeno DNA, as judged by hybridization to a DNA fragment representing another portion of the adeno genome? The answer was yes! The adeno mRNA must be produced by a mechanism that eliminated an internal part of the primary RNA transcript from DNA.

A number of scientists at Cold Spring Harbor had also concluded that the late adeno RNAs might provide the pathway to the eukaryotic promoter. To get at the DNA sequence at the RNA start site, Rich Roberts took a very different approach—looking for RNA segments that started the mRNA, each of which should have a 5′ G cap. Roberts came to Cold Spring Harbor with a strong chemical background, and figured that he could exploit the special chemistry of the G cap to grab the RNA sequence adjacent to every promoter. About 80% of the adenoviral genome was devoted to late RNA. Based on the size of typical adenovirus mRNAs, Roberts and his associate Rich Gelinas expected 15–20 RNAs, each presumably made from a distinct promoter. Each RNA should have a distinct sequence adjoining the cap, which they could determine from a nuclease cleavage near the cap. Instead, they found only a single short sequence, indicating that the transcription start of all late mRNA was the same. One possible explanation was that all late mRNAs originated from a single transcript, which was then processed into the individual mRNAs (as for phage T7). If so, only the first RNA of the long transcript would have the 5' cap, so only that RNA would have been grabbed by Roberts and Gelinas. To test this, Gelinas first fractionated the RNAs by size and by hybridization to DNA restriction fragments and sequenced the 5' end of the separated RNA fragments. Each class of RNA appeared to have the same end sequence at the cap, eliminating this hypothesis. Concurrently, Dan Klessig had taken the approach of purifying two defined mRNAs for viral structural proteins. He found that both RNAs carried the same ubiquitous sequence next to the cap.

The hybridization experiments with defined DNA sequences carried out by Roberts and Gelinas showed that every adeno late RNA had the same sequence next to the cap, but an extension of this work seemed to say that the capped sequence came from nowhere! The DNA-RNA duplexes produced by hybridization are very resistant to destruction by the RNA-degrading enzyme RNase, but the capped sequence was highly sensitive to nuclease attack, indicating that this piece of RNA came from a different DNA site than the rest of the mRNA.

What was the source of the cap end of adeno late mRNA? DNA polymerases require a primer for synthesis; by analogy Roberts thought that adenovirus late RNA might use a "primer RNA" produced elsewhere on the virus. Where might the "primer RNA" come from? One guess was that an abundant small viral RNA of unknown function, called VA-RNA, might be the primer. An electron microscope experiment was proposed to determine this. If the VA-RNA really was the "primer RNA" present next to the 5' G cap, then the 5' RNA "tail" of the late RNAs should hybridize to single-strand DNA carrying the VA sequence. Louise Chow and her husband, Tom Broker, had already found 5' tails much earlier in R-looping experiments like those of Berget and Sharp (see Figure 6–26a) but had ascribed them to branch migration. In the new experiments, Chow and Broker observed an RNA tail that hybridized to the VA-region of DNA not once but at three separate locations, none of which corresponded to the sequence of the VA

Richard Roberts

Louise Chow

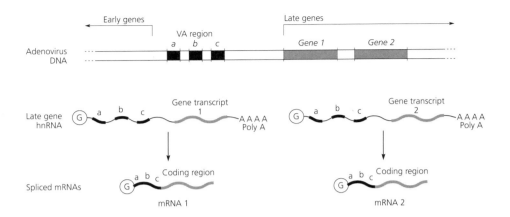

FIGURE 6–27. The 5' tail that showed up in R-loops was found to be transcribed from three short separated sequences within the VA region of the adenovirus genome. Every late RNA begins with these three segments spliced together. This leader region is then spliced to each late gene coding region.

transcript. The conclusion was inescapable—adenovirus late mRNA was created by RNA segments from at least four regions of viral DNA (Figure 6–27). By luck, the single-stranded DNA used to hybridize with the cap sequence happened to contain all three DNA regions represented in all late mRNAs.

Klessig at Cold Spring Harbor completed the case for splicing by demonstrating with DNA-RNA hybridization that the cap end of late mRNA did indeed come from a specific site on adenovirus DNA. Other puzzling data on adenovirus RNA at Cold Spring Harbor also became clear in light of the splicing concept. RNA splicing was signed, sealed, and delivered for publication to the journals *Cell* and *Proceedings of the National Academy of Sciences* in the form of four papers from Cold Spring Harbor and one from MIT. Although there is disagreement about who first thought of the electron microscopy experiment that identified the 5' tail, there seems to be no doubt that the credits for the discovery of splicing belong jointly to Berget and Sharp and to Roberts, Chow, and their associates of the Cold Spring Harbor group.

The remaining question about RNA splicing was whether this processing pathway would be universal. The answer arrived rapidly. Many people had been trying to track eukaryotic mRNAs to their genes by DNA-RNA hybridization approaches. The common result was that mRNAs appeared to hybridize to DNA segments from several different places on the genome, as would be expected if splicing were occurring. Three groups demonstrated splicing in SV40, clarifying several puzzling earlier findings. Richard Flavell provided early evidence that the mRNA for the hemoglobin gene was spliced; he noted an "extra" restriction site in genomic DNA that defined a region of the gene not present in the mRNA. Phil Leder also found evidence for splicing of hemoglobin mRNA based on R-looping. Susumu Tonegawa identified splicing in antibody mRNA. Pierre Chambon discovered a complex splicing pathway for RNA from the ovalbumin gene. We now know that expression of nearly all genes of higher eukaryotes depends on splicing together RNA sequences that are not adjacent in the DNA. The non-informational spliced-out segments are called intervening sequences, or "introns." The informational sequences are called "exons." Splicing is rarer in lower eukaryotes, occurring in only 2–3% of yeast genes.

THE BIOCHEMISTRY OF SPLICING: LARIATS ON A SPLICEOSOME

Once splicing was discovered, the focus of research shifted to how the reaction was accomplished. The favorite initial idea supposed that an RNase III-like cutting reaction took place at an RNA duplex (see Figure 6–23), except that RNA ends were rejoined rather than simply cleaved. The actual process of pre-mRNA splicing turned out to be a big surprise, utilizing unexpected biochemistry and occurring in a huge RNA-protein complex ("spliceosome") analogous to a ribosome. The first splicing reaction reproduced in a pure system turned out to be an even bigger surprise; in 1981, Tom Cech learned that the ribosomal RNA precursor from a protozoan could splice itself—the RNA carried the catalytic activity. This work is described in the next section on catalytic RNA.

The development of an effective cell-free system that would execute the splicing reaction for mRNA was not accomplished until 1983, six years after discovery of the phenomenon. One problem was an adequate supply of an unspliced precursor; another difficulty was the unexpected complexity of the reaction. Rick Padgett and Phil Sharp eventually succeeded in detecting a splicing reaction in an extract system that synthesized new mRNA. Nouria Hernandez and Walter Keller in Heidelberg synthesized a similar RNA under conditions that prevented splicing; they then isolated this RNA and showed that it was spliced by an extract from nuclei. Michael Green and Tom Maniatis at Harvard solved the precursor problem in a simpler way; they hooked their eukaryotic gene to a phage promoter and used purified RNA polymerase to make the unspliced RNA,

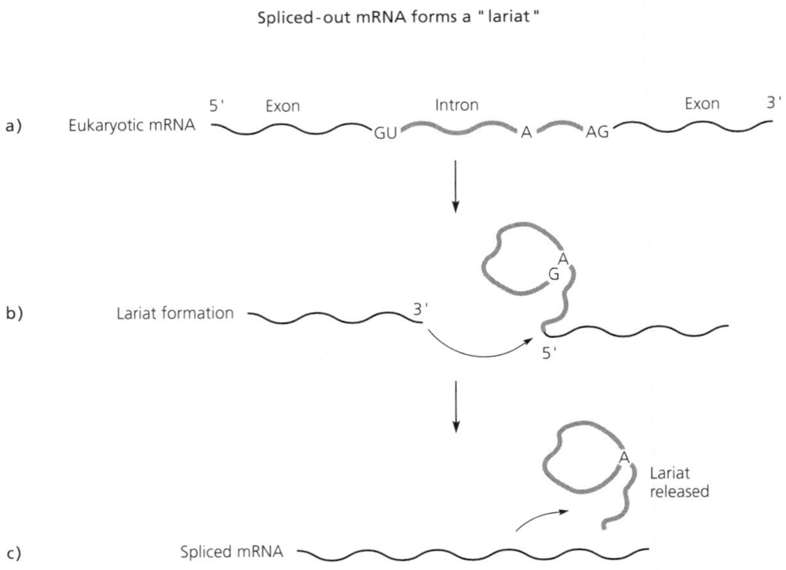

a) Eukaryotic mRNA

b) Lariat formation

c) Spliced mRNA

FIGURE 6–28. Many eukaryotic mRNAs have intervening sequences ("introns") that must to be removed so that the informational parts of the message ("exons") are next to each other and can be expressed as a protein (a). During the splicing process, the intron RNA forms a loop that looks a bit like a lariat. This connection is formed between an A base near the end of the intron and the G base at the beginning of the intron, while maintaining the normal linkages of the RNA chain (b). After the lariat forms, the 3' end of the first exon joins with the 5' end of the next, and the lariat is released (c).

which they then added to their extract system. Once the cell-free systems were developed, the RNA events in the splicing pathway were quickly worked out by Green, Maniatis, and associates at Harvard, and by Sharp and coworkers at MIT.

The splicing extract produced the expected "spliced" RNA that joined two discontinuous RNA segments together (Figure 6–28c). In addition, two "funny" RNAs carrying sequences that would be spliced out ("intervening sequences" or "introns") were observed (6–28b, c). By several criteria, these RNAs behaved as if they were circular on one end resembling a "lariat"; the odd RNAs also carried an unusual chemical linkage that was resistant to RNases. The properties of the RNAs matched those of a "branched" RNA structure detected previously in cells by John Wallace and Mary Edmonds. The "lariat" is formed because a particular A nucleotide in the intron makes a second chemical linkage to another nucleotide, while maintaining its normal linkage within the RNA chain. A variety of evidence argued that the "lariat" structure shown in the diagram was indeed an intermediate in the splicing pathway. Importantly, it is the obligatory G at the 5′ end of the intron that is joined to the internal A, leaving the cleaved mRNA end held by the splicing apparatus (Figure 6–28b). The two mRNA segments are then joined in a second step in which the intron "lariat" is discarded (Figure 6–28c).

The RNA sequence requirements for splicing are surprisingly flexible. The GU and AG sequences at the ends of the intron are almost always conserved, and other sequence properties are important but vary within constraints. The situation is somewhat akin to recognition between 16S RNA and the Shine-Dalgarno sequence (see Figure 6–18); so long as some base-pairing can occur within a certain spatial "window," the reaction proceeds. An A (called the "branch point") located near the AG end is always the nucleotide that forms an additional linkage with the 5′ G. For yeast, the branch point is defined explicitly by a seven-base sequence, but in mammalian systems a very degenerate form of this sequence is often present at lariat sites. The animal cell pathway seems to be designed to function with virtually any intron sequence, so long as the required ends are present.

The price of RNA sequence flexibility in splicing appears to be a complex mechanism. The splicing apparatus is an RNA-protein structure nearly as large as a ribosome, containing five different RNAs and more than 50 proteins. This very large spliceosome is composed of smaller RNA-protein structures, termed snRNPs (pronounced "snurps") for small, nuclear ribonucleoprotein particles. The association of snRNPs in the larger spliceosome was demonstrated by centrifugation experiments carried out by John Abelson, Phil Sharp, and Walter Keller with their associates. This rapidly sedimenting RNA-protein structure contains the precursor RNA, splicing intermediates, snRNPs, and additional proteins. Clearly, the spliceosome carries out the main event for the splicing reaction. The spliceosome in cells is not a stable entity like a ribosome—it is reassembled from snRNPs for each splicing event. In any case, the pathway to splicing appears at present to involve a level of complexity analogous to protein synthesis.

The discovery and initial understanding of snRNPs derived from experiments by Joan Steitz and associates. Steitz started out to study eukaryotic RNA processing at its earliest event—the formation of an RNA-protein structure involving hnRNA. The idea was to isolate RNA regions recognized by the processing proteins through an RNase-protection experiment (similar to her ribosome protection experiment; see Figure 6–17). To improve her chances of success, she wanted to use antibodies to collect the proteins and associated RNA. The project became feasible when Steitz learned that some people with autoimmune diseases produced antibodies directed toward a nuclear "RNA-protein structure." Michael Lerner and Steitz found that the antibodies generally worked as they were supposed to, except that the targets were the snRNPs rather than the hnRNA structures she had started to study. The snRNPs turned out to be exceedingly interesting, and the antibodies were extraordinary tools for understanding them, providing a way to identify and isolate individual snRNPs.

There are five abundant spliceosomal snRNPs in the cell nucleus, each with its own small RNA, designated U1, U2, U4, U5, U6. While these snRNPs share a subset of proteins in common, a few special proteins are unique to each particle (a feature that permitted the antibodies to distinguish between them). The appearance of these RNA-protein particles with a "ribosome-like" construction led Steitz to suppose that at least some of these particles participated in splicing, with the RNA providing a "Shine-Dalgarno"-type sequence to hold the splice-sites in the RNA. In another parallel to the ribosome, Christine Guthrie used a phylogenetic comparison between yeast and vertebrate snRNAs to show that certain sequences and structures of the snRNAs were extremely well conserved, strongly suggesting that RNA-RNA pairing was important for the function of spliceosomal RNAs. RNA-RNA pairing has since been confirmed using a genetic approach to create compensatory base-pair changes, mimicking the phylogenetic analyses of Woese and Noller for rRNA. This approach has been particularly effective for yeast, where Guthrie and colleagues have shown base-pairing between U2 snRNA and the branch site, and with U1 and U6 to the 5′ splice site. These RNA:RNA interactions have also been proven by photochemical crosslinking experiments by Steitz and others. In a surprisingly satisfying conclusion, the genetic and biochemical data combine to support a specific model for an active site of the spliceosome comprising a network of snRNA:mRNA and snRNA:snRNA base-pairing interactions. Whether the breaking and joining of the RNA is mediated by snRNP RNA, as seems likely, or is aided by proteins, remains to be established.

An interesting historical footnote is that eukaryotic tRNAs are also spliced but use an entirely different mechanism. Work largely from the Abelson laboratory demonstrated that three proteins are sufficient for this reaction. The mechanism of tRNA splicing was established several years before that of pre-mRNA splicing. During the intervening period, there were thoughts that the tRNA mechanism would be universal. Instead, pre-mRNA splicing proved to be enormously complex.

The ability of proteins to accelerate specific chemical reactions provides for the replication and expression of genes and for the metabolic activities of living creatures. Before 1981, biology was thought to have an orderly division of labor: nucleic acids stored and transferred genetic information to proteins; proteins carried out biological catalysis. Then, an RNA that could splice itself appeared in Tom Cech's lab like the unicorn in the garden in James Thurber's famous fable. The unicorn found a believer, and we now know that RNA can also accelerate specific chemical reactions. RNA catalysis may be a fossil from primordial life, but it is definitely not a mythical beast.

Cech came to catalytic RNA by a somewhat circuitous route, but the unicorn could not have chosen a better garden for communicating its existence. Cech had just the right background and personality for the task. With superb intuition, boundless enthusiasm, and a passion for chemistry, he very quickly took a fortunate observation into a textbook story for biology, molecular biology, and chemistry. In his extraordinary combination of biological and chemical insight and capacity for verbal communication, Cech might be termed the "Francis Crick" of his scientific generation. When RNA intruded in 1978, Cech was a beginning professor at the University of Colorado at Boulder, intending to study chromosome structure and transcription. He had worked on the organization of DNA sequences within chromosomes as a graduate student in chemistry at Berkeley, followed by a postdoctoral period at MIT spent in learning ways to visualize chromosome structure. To provide a focused study, Cech selected the ribosomal RNA genes of a protozoan, *Tetrahymena thermophila*, because it provided a technical advantage. This organism had many identical copies of the genes for its ribosomal RNA (called rDNA).

To locate the transcription unit responsible for rRNA, Cech used the "R-looping" method of electron microscopy (see Figure 6–26), adding ribosomal RNA (rRNA) to rDNA to identify the R-looped structure containing an RNA-DNA duplex as well as a looped-out single strand of DNA. This procedure identified the transcription unit. More important, it showed that the rDNA had an intron—a segment removed from mature rRNA by splicing (Figure 6–29). In 1979, splicing was a brand new phenomenon, and people were still groping for an optimal experimental system for biochemical studies. The ribosomal genes looked promising—they

Tom Cech

R-loops seen with electron microscopy showed an intron

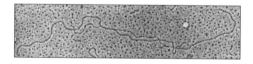

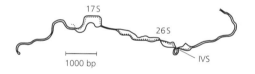

FIGURE 6–29. When Cech hybridized 17S and 26S ribosomal RNA from *Tetrahymena* with ribosomal DNA, he saw the two expected R-loops caused by 17S and 26S RNA hybridizing to the DNA (each R-loop consists of an RNA-DNA hybrid that displaces one strand of the duplex DNA molecule). He also saw a small loop structure that interrupted the R-loop between 26S RNA and DNA. This looped out stretch of DNA within the RNA-DNA hybrid was an intron, which is spliced out in the final RNA product. Determining how this splicing occurred allowed Cech to discover self-splicing. In the interpretative drawing shown below the micrograph, a solid line is single-strand DNA, a dashed line is RNA, and the intron is labeled IVS (intervening sequence). (Electron micrograph courtesy of T. Cech.)

provided an abundant RNA with a single intron. The approach looked even better when Cech and colleagues found that extracts of cell nuclei produced full-length precursor RNA and then spliced it. Everything seemed to be optimal for trying to isolate the splicing proteins; the precursor rRNA could be purified, added to the extracts, and the splicing components purified by fractionating the extract. One problem with this approach quickly emerged—RNA splicing occurred in control experiments when the extract was left out of the reaction. The RNA was splicing itself!

Cech spent a year refusing to believe in the unicorn and trying to find alternative explanations. The splicing protein might have clung to the precursor RNA during the isolation of the RNA. More and more vigorous purification of the RNA did not alter the result even though the hypothetical splicing protein should have been vastly reduced in quantity by the purification procedures. Perhaps there was a protein that held together the spliced-out intron and the mature rRNA. The test-tube reaction might simply remove the hypothetical linker protein. But that seemed unlikely because the self-splicing reaction absolutely required GTP, and a G was actually transferred to the end of the intron. So, something more than releasing a protein linker must be involved. In 1981, Cech and colleagues published their experimental results and suggested that the precursor RNA itself carried the catalytic activity. Considerable outside skepticism remained for a time from holdouts of the hidden protein school. But, in 1982, Cech and colleagues made the precursor rRNA from an rDNA gene cloned in *E. coli* using pure RNA polymerase in the absence of any cellular proteins. The RNA still insisted on splicing itself.

The Cech experiment demonstrated that an RNA molecule could exhibit the hallowed characteristics of a protein enzyme—exceptional specificity and rate acceleration. The precursor *Tetrahymena* rRNA is a "ribozyme." In the purest sense, the RNA in the self-splicing reaction is not a catalyst because the original RNA is not regenerated to act again. This point is more a technical detail of the biological system than a fundamental distinction. In 1986, Cech and colleagues engineered the RNA to work as a true catalyst.

The *Tetrahymena* self-splicing reaction is an example of a large class of similar reactions called "Group I" splicing. To date, more than 500 examples of this splicing pathway have been identified. The folded, tertiary structure of the intron RNA is critical for the Group I pathway, as might be expected for a reaction capable of cat-

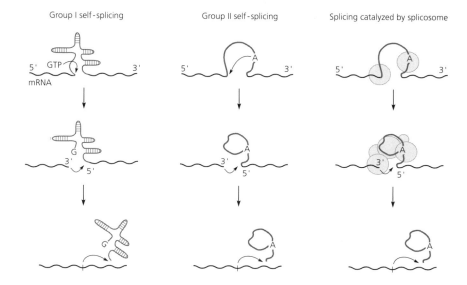

Group I self-splicing   Group II self-splicing   Splicing catalyzed by splicosome

FIGURE 6–30.   The self-splicing reaction, catalyzed by the RNA chains themselves, can occur by two distinct pathways, called Group I and Group II splicing. Both require that the intron RNA be folded into a complex tertiary structure that is able to act as the enzyme in the reaction. In the Group I reaction, a GTP nucleotide binds to the folded structure and initiates the splicing reaction. For Group II splicing, an internal A is involved in lariat formation, which leads to removal of the intron. Splicing events catalyzed by the spliceosome are chemically similar to the Group II self-splicing reaction. Darkened circles indicate the involvement of proteins in the spliceosomal reaction.

alytic activity. The mechanism of the *Tetrahymena* self-splicing reaction differs from the spliceosome pathway just described for mRNA. For *Tetrahymena*, the biochemical intermediate involves the addition of a G nucleotide to one end of the intron, rather than the lariat produced by the spliceosome. Interestingly, there is a second type of self-splicing, called the Group II pathway, that proceeds with chemistry like that of the spliceosome, including generation of a lariat. The structure of Group II introns, while distinct from that of Group I introns, is also thought to be critical for the self-splicing reaction (Figure 6–30).

The realm of RNA processing was keeping another big surprise about catalytic RNA under wraps. As mentioned earlier, processing of tRNA molecules in bacteria depends on an RNA-cutting activity termed RNase P. Starting in 1971, Sid Altman set out to purify and characterize this enzyme. The isolation of RNase P turned out to be an exceedingly difficult task. An RNA came along with the protein through many different purification steps. Initially, this was considered to be a major nuisance deriving from a general affinity of the protein for RNA. Worse yet, treatments that removed the "contaminating" RNA killed the cutting activity of the protein for tRNA. Perhaps the RNA was some sort of necessary "cofactor" for

Sidney Altman

Norm Pace

the activity of the protein. The RNA began to look important when Altman and his colleagues demonstrated that it was a single species, 400 bases in length, which they termed M1 RNA. In 1980, the Altman group obtained definitive proof for a functional role of the RNA. The isolated RNA or protein did not work for tRNA processing, whereas mixing the two components restored RNA-cutting activity.

The final stage in the exaltation of the RNA component of RNase P occurred in 1983—converting contaminating crud to the catalytic component after a decade. Following Cech's discovery of self-splicing RNA, Altman looked for RNA cutting activity by RNase P RNA using his normal conditions for cutting. He had no success. The big news came by surprise from a collaborative project between Altman and Norm Pace, a major previous contributor to the RNA processing field with his studies on the bacterium *Bacillus subtilis*. Pace and his colleagues had isolated RNase P from *B. subtilis* and discovered that the RNA component exhibited a surprisingly different base sequence from the RNA component of *E. coli*. The two groups wondered whether the RNAs were interchangeable between the *E. coli* and *B. subtilis* proteins. The control experiment was to use solely the RNAs in the processing reaction. The Pace group had determined that the *B. subtilis* RNase P produced its most effective cutting activity under rather unusual ("nonphysiological") conditions with very high magnesium. Both groups found that M1 RNA alone (whether from *E. coli* or *B. subtilis*) could process the tRNA under these extreme conditions. Although the RNase P RNA will work only with its protein component under more typical conditions, the Altman-Pace experiment clearly demonstrated that the RNA is the biological catalyst, with the protein performing some critical but ancillary function. RNase P RNA is a true catalyst, acting on another RNA molecule without undergoing a chemical transformation itself.

How frequent is RNA-mediated biological catalysis? So far, peptide bond formation is the major example *in vivo*. But there are some additional essential processes whose mechanism of catalysis is likely to rely at least in part on RNA: RNA splicing by the spliceosome, and finishing the ends of chromosomal DNA by the telomerase enzyme. A popular notion is that RNA catalysis represents a molecular fossil of an ancient "RNA world" in which RNA both carried the genetic information and catalyzed its self-perpetuation. In fact, RNA enzymes, or ribozymes as they are now known, can catalyze reactions as diverse as polymerizing nucleotides, ligating DNA, cleaving DNA phosphodiester bonds, and synthesizing peptides. The process of information transfer from DNA to protein has proven to involve a complex and fascinating array of biochemical events, and there are no doubt more surprises to come.

FURTHER READING

Cech, T. (1986) RNA as an enzyme. Sci. Am. 255, 64–75.

Chambon, P. (1981) Split genes. Sci. Am. 244, 60–71.

Gilbert, W. (1978) Why genes in pieces? Nature 271, 501.

Moore, P. B. (1988) The ribosome returns. Nature 331, 223–27.

Rich, A, and S. H. Kim (1978) The three-dimensional structure of transfer RNA. Sci. Am. 238, 52–62.

Steitz, J. A. (1988) Snurps. Sci. Am. 258, 56–63.

Watson, J. D. (1963) The involvement of RNA in the synthesis of proteins. Science 140, 17–26.

Witkowski, J. A. (1988) The discovery of "split" genes: A scientific revolution. Trends Biochem. Sci. 13, 110–13.

Zamecnik, P. (1984) The machinery of protein synthesis. Trends Biochem. Sci. 9, 464–66.

7

DNA ON ITS OWN:
GENETIC RECOMBINATION

DNA stores the genetic information. The accurate transmission of this information into new cells is essential for the survival of a multicellular organism and for the perpetuation of biological species. As we have seen, the organism goes to a great deal of trouble to have high fidelity of replication (see Chapter 4). However, the genes of an organism cannot be inherited completely unperturbed by succeeding generations. The survival of a species in a changing environment depends on the existence of genetic adaptability within the population; this genetic variation provides for the continued presence of new proteins with the potential to respond to new challenges. Genetic diversity in a biological population arises from mutation, which produces new genotypes, and from recombination, which rearranges preexisting genes.

As a rough guide, genetic recombination comes in three major forms: general or homologous recombination, site-specific recombination, and DNA transposition. We will consider all of them in this chapter. General recombination refers to the mode of genetic exchange studied by classical genetics. Here, new combinations of genes are created when two chromosomes exchange information by recombination between "homologous" regions of each chromosome (regions with identical or very similar DNA sequence). General recombination occurs with roughly equal probability along the homologous DNA segments. Site-specific recombination occurs only between unique, specific sequences on both recombining

DNAs and joins regions of DNA with little or no homology. The first and best-studied example of site-specific recombination is the integration of phage λ DNA into the bacterial genome by recombination between unique sites in the phage and bacterial DNA. DNA transposition refers to "mobile" segments of DNA that move from one place to another in the genome. In this reaction, the ends of the mobile element are joined to a random site on the DNA. This reaction has been most thoroughly studied for the bacteriophage Mu. The finding that the human AIDS virus integrates into DNA by an analogous reaction mechanism has made this an area of intense interest. In addition to these well-studied examples of recombination, cells also undergo what I call "extraordinary" recombination. This is a catch-all term for rare genetic exchanges between regions of DNA lacking extensive homology or a unique site. Extraordinary recombination can delete or duplicate genetic segments. Illegitimate recombination is an alternative term for extraordinary but seems to imply an unnecessary moral judgment. Less is known about the mechanism in these events.

GENERAL RECOMBINATION AND THE HOLLIDAY MODEL

Our understanding of general recombination, the process that promotes exchanges between homologous segments of DNA, started with genetic insights obtained by studying the behavior of DNA in cells and progressed to biochemical analyses with purified proteins and DNA. From the earliest studies on the genetics of fruit flies, it was clear that homologous chromosomes could exchange regions with each other. However, our initial molecular understanding of recombination came from work with phage or bacterial viruses, augmented by a giant leap of insight from the genetics of fungi.

The critical properties of recombination in phage necessary for developing a molecular understanding of this event came from experiments in the 1950's with phage T2, mostly by Al Hershey, Martha Chase, and Gus Doermann. When two phage that differ in their genetic makeup infect the same cell, genetic markers between the two phage will exchange places, allowing recombination events to be observed (Figure 7–1a). This process is akin to classical genetic exchange between chromosomes. Two additional properties of recombination in phage rapidly became apparent. First, phage that had undergone recombination often had a "heterozygous" region in their DNA, meaning that information from each parental genotype

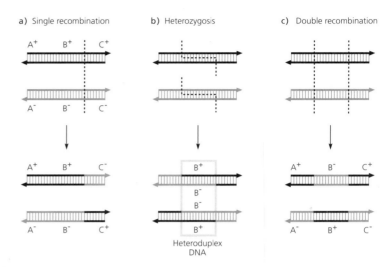

a) Single recombination   b) Heterozygosis   c) Double recombination

Heteroduplex
DNA

FIGURE 7–1.   When two genetically different phages infect a bacterial cell, recombination between phage chromosomes is observed (a). However, recombination exhibited two unexpected features. A region of the recombinant chromosome often had "heteroduplex DNA," with one DNA strand of the DNA duplex originating from each parent, a phenomenon called heterozygosis (b). In addition, phage that had experienced two recombination events close together were about as frequent as the single recombinants. This higher than expected frequency of clustered recombination events is called "negative interference" (c).

was represented in this region. After considerable confusion as to the nature of these heterozygous regions, the major class was correctly inferred to derive from DNA in which one strand of the duplex came from each parent. The DNA was said to be "heteroduplex" in this region (Figure 7–1b). Second, phage often seemed to have two genetic exchanges very close to each other. These "double exchanges" occurred almost as frequently as the single exchange, which was unexpected if each recombination event occurred independently. This phenomenon was called "negative interference," to signify that having one exchange in a region seemed to encourage rather than inhibit a second exchange in the same region (Figure 7–1c). These complex features could not be blamed on some bizarre feature of phage recombination because they turned out to have close analogues in fungal recombination. Clearly, a molecular picture of genetic recombination must explain not only the exchange of parental DNA but also the associated phenomena of heterozygosis and negative interference.

What might be happening to the DNA during recombination? Initially, many possible mechanisms were proposed, but none were really satisfactory. The first critical molecular experiments were performed in phage λ and showed that recombination in λ occurred by breaking the DNA of each parent and rejoining it to the DNA of the other parent. These experiments were possible because Jean

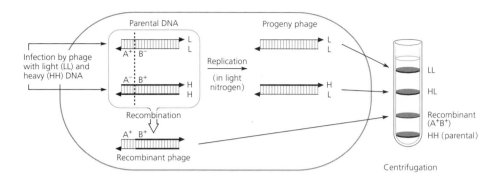

FIGURE 7–2. Meselson and Weigle were able to infer the mechanism of recombination by using cesium chloride density gradient centrifugation to determine the density of phage particles whose DNA had undergone recombination. They infected bacteria with two different phage: one having normal DNA (containing light ["LL"] nitrogen), and a second, genetically distinct phage having heavy DNA (containing heavy ["HH"] nitrogen). They separated the output phage by density gradient centrifugation and assayed the genetic markers of the phage present at different positions in the tube. As expected from the breakage and rejoining mechanism for recombination, the density of the selected recombinant (A$^+$B$^+$) was just slightly less than that of the HH parental phage and clearly distinguishable from that of replicated phage with HL DNA.

Weigle had found that density gradient centrifugation could be applied to phage as well as to DNA (see Figure 4–2). Amazingly, even after floating in high concentrations of cesium chloride for many hours during centrifugation, the phage could still infect bacteria. In 1961, Weigle and Matt Meselson, the ace of density gradient centrifugation, realized that they could infect bacteria with phage that differed not only in their genetic markers but also in the density of their DNA. Phage with different DNA density will be located in different regions of the density gradient after centrifugation. They could then ask whether the phage that had recombined their genetic markers had also changed their density. If so, then the DNA of the recombinants comes from both parents as required by the breakage and rejoining model of recombination.

Weigle and Meselson infected bacteria with one parental phage with heavy (HH) DNA (because it had been grown in medium with "heavy" [^{15}N] nitrogen) and another parental phage with light (LL) DNA (grown in medium with normal [^{14}N] nitrogen). A breakage and rejoining model predicts that recombination between the DNA from a heavy phage and the DNA from a light phage should yield a phage with an intermediate density because the DNA derives from both parents. The exact position of the recombinant phage band will depend upon where the breakage and rejoining occurred (Figure 7–2). The actual experiment was more complicated because, in addition to recombining, the phage DNA was also undergoing normal replication, using the ^{14}N in the medium to make new strands of DNA, thereby converting the heavy (HH) DNA into half-heavy (HL) DNA. In the example shown, the recombination

Robin Holliday

event examined was near one end of the DNA, so that the position of phage with a recombinant DNA molecule was significantly different from phage with newly replicated (HL) DNA. As predicted by the breakage and rejoining model, one class of recombinant phage was located at a position close to the HH parental phage. The Meselson–Weigle experiment argued strongly for a breakage-rejoining mechanism and ruled out some extreme ideas about genetic recombination—models in which genetic exchange occurred only during DNA replication by switching templates ("copy-choice"). However, the molecular events involved in breakage and reunion remained obscure.

The key insight into a molecular mechanism came in 1964 from the fungal geneticist Robin Holliday, following up on an earlier idea of Harold Whitehouse, who had suggested that gene conversion (the fungal analog of heterozygosis in phage) arose from the formation of heteroduplex DNA that occurred when broken parental DNA molecules rejoined in a new combination. Holliday noted that an appropriate construction of the recombination "joint" would explain both heterozygosis and negative interference. Following alignment of the two DNA molecules, Holliday presumed that recombination initiated (in some mysterious way) by the breakage of one DNA strand from each parent (Figure 7–3a). Each broken strand would then exchange with the other DNA molecule to form a "branch point" (Figure 7–3b). Because the recombining partners are homologous (have the same or nearly identical DNA sequence), the swapped strands can pair with the complementary parental strand of the other DNA molecule, facilitating further transfer of the exchanged strand. The result is creation of a region of "heteroduplex DNA" and migration of the branch point (Figure 7–3c). When the branch point is cut by a resolving nuclease, recombinants are produced. Holliday noted that the branch point can be cut in two different directions (Figure 7–3d). The markers at the ends of the chromosome (A, C) recombine when the branch point is cut in one direction, but not when it is cut in the other direction (Figure 7–3e). A recombination event in the heteroduplex region (B in this example) will be scored genetically as a single recombination event if the end markers recombine and as a double recombination event if they do not (Figure 7–3e). The two unusual features of recombination—a heteroduplex DNA region near the recombination event, and a high yield (about 50%) of double recombinants (negative interference)—are neatly explained by the Holliday model.

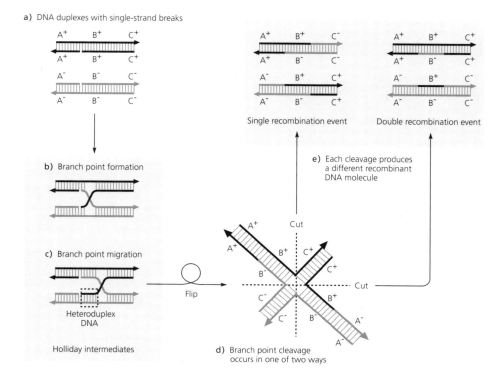

a) DNA duplexes with single-strand breaks

b) Branch point formation

c) Branch point migration

Heteroduplex
DNA

Holliday intermediates

d) Branch point cleavage
occurs in one of two ways

Flip

Cut

Cut

e) Each cleavage produces
a different recombinant
DNA molecule

Single recombination event

Double recombination event

FIGURE 7–3. Holliday proposed a mechanism to describe how DNA breakage and exchange could occur. If DNA duplexes containing the same genes ("A," "B," "C") are aligned with each other and acquire single-strand breaks (a), then strand exchange between the DNA duplexes can occur, forming a "branch point" (b). These crossover structures are termed "Holliday intermediates." Because the two DNA duplexes are homologous, the branch point can "migrate," extending the regions of heteroduplex DNA (c). After branch point migration, the Holliday intermediate is resolved when a nuclease cuts the branch point (d). This cleavage can occur in two different ways, yielding either single or double recombinants (e). This model successfully explained why recombinants often had a heteroduplex region and why double recombinants were as frequent as single recombinants.

The Holliday model is one of the great ideas of modern biology. The idea correctly explained the molecular basis of general genetic recombination. The Holliday model has required some modifications to explain certain features of recombination but has previewed with remarkable accuracy our current molecular picture of how recombination works. The new information has been mainly the discovery of how the mysterious initiating event is achieved and the identification and characterization of the fascinating proteins that execute the Holliday program.

HOLLIDAY INTERMEDIATES IN CELLS

The Holliday model makes explicit predictions about the nature of the recombination intermediates and the primary products of recombination. The intermediates should have single-strand branch points; the products should have extensive regions of heteroduplex DNA. Demonstrating the existence of long heteroduplex regions

FIGURE 7–4.   When two φX174 circular duplex molecules
undergo a single exchange, a "Figure 8" molecule is
produced. (Electron micrograph courtesy of D. Dressler and
H. Potter. Reproduced with permission from the *Annual
Review of Biochemistry* 51. © 1982 by Annual Reviews,
www.annualreviews.org.)

John Clark

was relatively straightforward; Meselson, Maury Fox, and their coworkers inferred the existence of such regions from experiments on the distribution of genetic markers after recombination between λ phage.

Getting at the postulated DNA intermediate in recombination was a tougher task. In 1975, Bob Warner, Irwin Tessman, and their colleagues extracted Holliday-type DNA molecules from cells infected with φX174, which has a duplex, circular "replicating form" (see Figure 4–17). If the two duplex circles had exchanged single strands, a "Figure 8" molecule would be formed. (Imagine joining the ends of the DNA molecules in the diagram of the Holliday model in Figure 7–3.) DNA molecules with this configuration were seen in the electron microscope (Figure 7–4), consistent with the idea of a single-strand branch point.

Further evidence for the Holliday model came from cutting the "Figure 8" molecules with a nuclease that recognized a single site in each duplex DNA. A structure that looks like the Greek letter "chi" was created. When these molecules were examined in the electron microscope, the position of their crossovers differed, showing that branch migration had occurred. Since recombination by φX174 depended on bacterial enzymes, the proteins responsible for the sequential pathway of recombination postulated by Holliday lurked somewhere in the *E. coli* cell. The major questions now became biochemical. As we have seen in our discussion of replication (Chapter 4), identification of the specific genes carrying out these processes is necessary to correlate biological function with the activities of the proteins.

GENES AND ENZYMES IN BACTERIAL RECOMBINATION

In 1965, John Clark began the genetic analysis that ultimately defined the genes and pathways for recombination in *E. coli*. Recall that when *E. coli* cells mate, Hfr (male) bacteria transfer their chromosome to F⁻ (female) bacteria; parts of the male chromosome are then incorporated into the F⁻ chromosome by recombination (see Figure 3–5). To find recombination defective *E. coli*, Clark looked for F⁻ mutants that were unable to carry out genetic recombination after DNA transfer by the Hfr. The first mutants identified were in a gene that Clark called *recA*. Its recombination defect prompted a multi-year quest for the role of the RecA protein in general genetic recombination. RecA turned out to be required for pairing homologous DNA duplexes and for strand transfer. In addition to a failure

Gene	Protein	Mutant (inactive) phenotype	Pathway	Function
recA	RecA	No recombination	RecBC	Strand transfer; formation of Holliday intermediate
recB *recC* *recD*	RecBCD	1% recombination	RecBC	Generate single strand (with bound RecA) needed to form Holliday intermediate
recF and others	RecF	No recombination, if RecBC is inactive	RecF	Affects RecA activity
ruvA *ruvB*	RuvAB	Mild recombination defect	RecBC	Branch migration
ruvC	RuvC	Mild recombinaton defect	RecBC	Resolves Holliday intermediate

of recombination, the *recA⁻* mutants had two other properties. When exposed to UV irradiation, they died much more rapidly than wild type cells, and they did not permit λ prophage to be induced to lytic growth. These two additional properties of RecA arise because they are involved in two pathways of DNA repair, recombinational repair and an inducible pathway of DNA repair and mutagenesis termed SOS. Both pathways are described in Chapter 8.

Clark, Paul Howard-Flanders, and their colleagues also defined two other genes involved in genetic recombination, *recB* and *recC*. Much later (in 1986), Gerry Smith identified another gene, *recD*, which works with *recB* and *recC*. Together, these three genes encode the protein subunits of the RecBCD enzyme. RecBCD turns out to be a powerful helicase-nuclease that prepares the DNA for RecA, generating the single strand necessary for loading RecA onto DNA. Interestingly, *recB⁻* and *recC⁻* mutants were substantially less defective in recombination than *recA⁻* mutants. These and other observations prompted the idea that *E. coli* had at least two recombination pathways; the RecA protein was required for both, whereas RecB and RecC proteins were together participants in only one of the two pathways (the RecBC pathway). Other work by Clark and colleagues identified mutants that eliminated the residual recombination by *recB⁻* or *recC⁻*; these experiments defined a second, normally minor recombination pathway termed the RecF pathway. This pathway uses nucleases other than RecBCD to prepare the DNA for RecA, explaining why *recBC* strains were less defective in recombination than *recA* strains. The recombination genes currently identified in *E. coli* are summarized in Figure 7–5.

Robert Lehman

An understanding of the role of the RecA protein was clearly the key to a biochemical picture of how recombination pathways work. Robert Lehman undertook this task. Lehman is as close as one comes to a "salt of the earth" biochemist. He has emphasized scientific fundamentals while making landmark discoveries in a career encompassing nearly every aspect of DNA metabolism. His unfailing good humor and integrity have won him respect and admiration. Although he does not seek the limelight, few can match him for accomplishment.

Purification of RecA presented some problems. Since the function of the protein was unknown, it was impossible to locate RecA during purification by assaying its activity. Instead, Lehman and Kevin McEntee used an approach that has become increasingly popular as gene cloning and overexpression techniques have been developed—following the protein as a "band on a gel." As a graduate student with Wolf Epstein, McEntee had shown where RecA migrates on gels (by selectively labeling RecA carried on a phage with radioactive amino acids; the strategy used by Mark Ptashne to purify the λcI repressor—see Figure 5–11). McEntee and Lehman could then purify RecA protein by standard techniques of protein fractionation, assaying for a "band on a gel" rather than for the activity of the protein. It turned out that RecA was sufficiently abundant to be detected without specific radioactive labeling. At each stage of purification, the proteins in a fraction were separated by gel electrophoresis, stained with a dye, and visualized as "bands on a gel" to see whether RecA was being selectively isolated with respect to other proteins. The good aspect of this approach is that a protein can be isolated without knowing what it does (no biochemical activity is necessary), so long as it is sufficiently abundant to be seen on the gel. The problem is then to figure out what the protein might do; in this case what RecA does for recombination. Purification without an assay can be risky, as we have no way of knowing whether the protein has been damaged during the isolation protocol and lost its biochemical activity.

The first biochemical activity identified for RecA was not recombination-related. Starting with a crude extract, Jeff and Christine Roberts with Nancy Craig found that RecA promoted cutting the λcI repressor protein into two pieces, thereby destroying its activity and inducing the resident λ prophage. This activity of RecA required both DNA and ATP. Independently, Tamoko Ogawa in Japan demonstrated that RecA was a DNA-dependent ATPase. These experiments suggested that ATP might be required for the

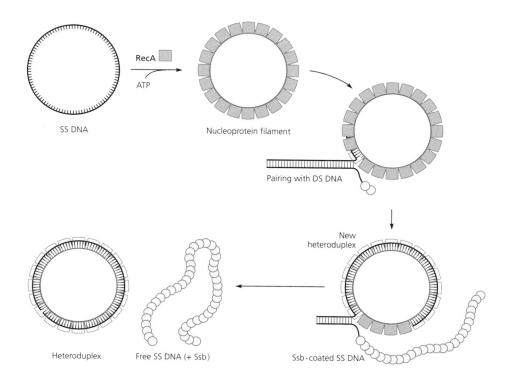

SS DNA

Nucleoprotein filament

Pairing with DS DNA

New heteroduplex

Heteroduplex

Free SS DNA (+ Ssb)

Ssb-coated SS DNA

various activities of RecA. Indeed, by using ATP in a variety of assays that simulated genetic recombination, Lehman, Howard-Flanders, and Charles Radding and their colleagues began to tease apart the role of RecA in recombination. The ability of RecA to cleave ATP in the presence of DNA also provided a simple biochemical assay for purification.

RecA protein was found to associate very tightly with single-strand (SS) DNA in the presence of ATP (ϕX174 DNA was the favorite substrate). This SS DNA-binding function is probably the very first step in recombination in *E. coli*. The product was a RecA-DNA nucleoprotein filament with a precise helical structure. When the RecA-coated, ϕX174 SS DNA was mixed with linear double-strand (DS) ϕX174 DNA, the "donor" SS DNA became stably interwound with the "recipient" DS DNA (Figure 7–6). It is exciting to note that this DNA-pairing reaction is analogous to the one expected for initiating recombination according to the Holliday model (provided that we adjust the model for a situation where one of the DNA molecules is single stranded).

FIGURE 7–6. When RecA binds to single-stranded (SS) DNA in the presence of ATP, it forms a nucleoprotein filament with a precise structure. This RecA-SS DNA filament can invade duplex DNA and form a heteroduplex by displacing one of the original strands of the DNA. Thus, RecA can carry out the initiation and branch migration reactions predicted in the Holliday model of recombination. SSB aids RecA in formation of the filament and displacement of the duplex strand by binding to SS DNA and keeping the DNA stretched out.

With more refined experimental conditions, Radding, Lehman, and associates observed a second-stage reaction with RecA in which the donor SS DNA was transferred into duplex DNA, mating with its complementary strand to form a circular duplex and displacing the original complementary strand. RecA could clearly execute a branch migration reaction starting with the single-strand "paired" intermediate. The branch migration occurred efficiently only in the presence of the single-strand DNA-binding protein used for DNA replication (SSB). SSB helps to form a reactive RecA filament and also to pull off the displaced SS DNA during branch migration (see Figure 7–6).

We now know that additional proteins are involved in the later stages of general recombination. Steve West has recently shown that a heterodimeric protein, the product of the *ruvA* and *ruvB* genes, can promote branch migration more efficiently than RecA. RuvB, a helicase-like motor protein that promotes rapid branch migration, is loaded onto Holliday junctions by a junction-binding protein called RuvA. The current picture is that RecA is required for initial formation of the Holliday intermediate. The RuvA•B complex then promotes branch migration to enlarge the initial heteroduplex generated by RecA. RuvC is a specific nuclease that resolves the Holliday Intermediate into the final products of recombination (see Figure 7–5).

The biochemical studies of RecA revealed the central role of this fascinating protein in genetic recombination. A RecA-coated single-strand of DNA could find homology in duplex DNA and begin the branch migration reaction. Two key questions remained. Where does the donor single-strand come from? How does the RecA filament locate homology? As we will see next, making the single-strand donor is the job of RecBCD. The quest for homology in a real-world recombination reaction is a spectacular (and unsolved) DNA transaction. Thousands of base-pairs of the duplex must be scanned quickly and accurately. It is likely that the major function of the RecA filament is to solve this problem. The search for homology probably involves unstacking the duplex base-pairs in the presence of the single-strand DNA to form a localized and mobile three-strand DNA intermediate within the RecA filament.

THE RecBCD PROTEIN: UNWINDING AND NICKING

The biochemical experiments just described established that RecA-mediated recombination required a single-strand donor DNA. But where did it come from? In the RecBC pathway, the source of the single-strand donor DNA is the RecBCD enzyme. This enzyme was initially isolated by Peter Goldmark and Stuart Linn because of its nuclease activity on duplex DNA. RecBCD is an exonuclease because it can degrade from the ends of double-strand or single-strand DNA starting either from a 5′ end or a 3′ end, but it is also an endonuclease because it cuts a single-stranded circle. Based on all of these cutting activities, RecBCD could fit into anyone's model of recombination. Later, Linn, Gerry Smith, and associates found yet another biochemical activity of RecBCD. This enzyme can unwind duplex DNA and is therefore a DNA helicase.

In the meantime, Frank Stahl and colleagues found a fascinating new property of the RecBC pathway—recombination depended almost entirely on specific sites in the DNA, which Stahl termed Chi. Largely because of the elegant work of Steve Kowalczykowski, we now know that Chi sites change the activity of the RecBCD helicase/nuclease. Presented with linear duplex DNA, RecBCD travels down the DNA executing its helicase activity and degrading one strand from the 3′ end. The other strand is coated with the single-strand binding protein, SSB. When RecBCD reaches a Chi site, it pauses and changes its state. RecBCD now loads RecA onto the strand it was previously degrading and starts degrading the other strand in the opposite direction (5′ to 3′). This leaves RecA bound to the 3′ end of a single strand of DNA, ready to start the strand invasion process that leads to recombination (Figure 7–7). The *E. coli* chromosome has about 1,000 Chi sites. So RecBCD is a traveling source of donor single-strands for recombination as it makes its way around the chromosome.

GENERAL RECOMBINATIONAL PATHWAYS: THE FUTURE

Currently, we understand only the RecBC pathway of general recombination in any detail; other recombinational strategies are certain to exist. However, a vision of one pathway provides a framework for thinking about others. Certainly the functions provided by RecBC—generation of single-strand ends and those provided by RecA—the search for homology, stable pairing, and branch migration are likely to be universal requirements for any pathway of recombination. Moreover, it has now become clear that this recombinational pathway has an additional function besides that of exchanging markers. The very same strategy of forming and resolving a Holliday intermediate is used by the organism to repair damage to DNA. Recombinational repair is one of the major ways

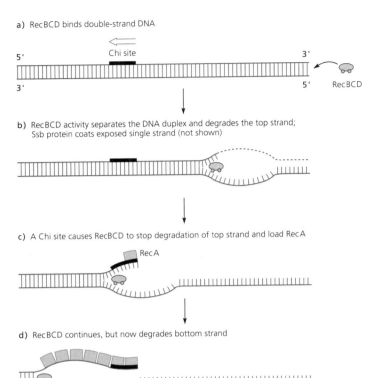

FIGURE 7–7. The RecBCD protein binds to linear, double-strand DNA (a) and travels down the DNA using its helicase activity to force the strands of the DNA duplex apart, forming a single-strand "bubble" and degrading one strand of the DNA from its 3′ end (b). When the protein complex encounters a special recognition site, termed "Chi," RecBCD pauses, stops degrading DNA from the 3′ end, and loads RecA onto that end (c). RecBCD now degrades the other strand of DNA in the 5′ to 3′ direction, creating RecA bound to a protruding 3′ end of a single DNA strand, ready to perform strand invasion (d).

to remove single-strand lesions from DNA and the only way to fix double-strand breaks in the DNA molecule (see Figure 8–14). In fact, the role of the RecA protein in repairing DNA is probably just as important to the organism as its role in mediating recombination.

RecA-like proteins are now established as central recombination proteins in prokaryotes, and the same is likely to be true for eukaryotes. The eukaryotic equivalent of RecA, called Rad51 protein, was recently isolated by several groups. Patrick Sung found conditions where it worked well *in vitro*, allowing him to prove that it carries out many functions similar to RecA. Yeast cells with mutations in the *RAD51* gene are abnormally sensitive to X-rays, indicating that Rad51 protein is needed for repair of double-strand breaks in DNA. Repair is believed to occur by a recombinational mechanism involving formation of one or more Holliday intermediates. Eukaryotes may even have a proliferation of RecA-like genes with special functions, since specialized RecA-like proteins dedicated to genetic exchange during meiosis have been described. In 1998, a RecA-like protein was also found in the third kingdom of organisms—the Archaea, indicating that RecA will likely be universally required for recombination and recombinational repair in all organisms.

The 1997 finding that the BRCA1 and BRCA2 proteins, encoded by the breast cancer susceptibility genes, interact tightly and specifically with the human version of Rad 51 brings home the importance of this seemingly academic process. This result implies that a deficit in the recombinational repair of double strand breaks results in an inherited predisposition to breast cancer.

SITE-SPECIFIC RECOMBINATION AND THE CAMPBELL MODEL

Our current rather good understanding of site-specific recombination owes its intellectual heritage to the Campbell model of prophage integration, originally proposed in 1962. By introducing the correct concept, this simple and elegant idea guided the design of the experiments that validated, refined, and generalized the model to other types of localized genetic exchange. Introduced in Chapter 5 (see Figure 5–6), the Campbell model is presented below in a more contemporary framework.

The linear λ DNA, injected from the phage into the bacterium, forms an intracellular circle by union of the short single-strand segments at the ends of λ, called the "cohesive ends." This brings the R and A genes next to each other (Figure 7–8, top). This circular form of λ is the initial substrate for transcription, replication and integration. When expression of the lytic genes is turned off by the λ repressor protein during the lysogenic response, the circular λ DNA molecule integrates into the larger circular bacterial chromosome. In this reaction, the phage Int protein, along with the host IHF protein, promotes recombination between the phage attachment site (POP′), and the bacterial attachment site (BOB′), generating the hybrid prophage attachment sites, BOP′ and POB', that flank the inserted phage DNA (Figure 7–8, middle). Prophage DNA then replicates passively as part of the *E. coli* chromosome (Figure 7–8, bottom).

Even in the absence of molecular data, Campbell's model had immense appeal because the idea explained so many features of the lysogenic lifestyle. The subsequent finding that, as predicted, intracellular λ DNA was indeed a circle, coupled with Marty Gellert's

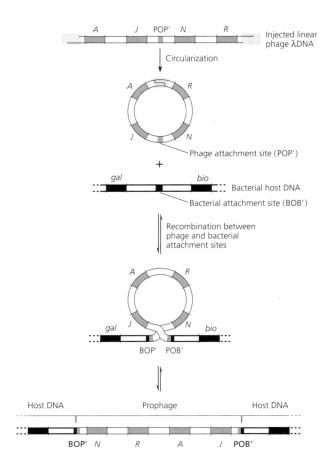

FIGURE 7–8. When the linear λ chromosome is injected into the bacterial cell, it circularizes and then integrates into the bacterial chromosome by recombination between the phage attachment site (POP') and the bacterial attachment site (BOB'). Recombination between the bacterial and phage attachment sites creates the hybrid BOP' and POB' prophage attachment sites that bracket the inserted prophage DNA. Note that the genetic markers flanking POP are separated after the recombination, creating a prophage whose order is permuted from that of the linear λ chromosome.

discovery of the enzyme that joined the ends together, convinced most people of the validity of the model. Naomi Franklin, Bill Dove, and Charles Yanofsky added a powerful argument for insertion into bacterial DNA by demonstrating that bacterial and prophage genes were contiguous in the chromosome. One critical aspect left unspecified by the model was the mechanism for the recombination event. Campbell had thought that the general recombination system carried out the recombination event, using homology between phage and host to accomplish the integration.

Was general recombination sufficient to explain the prophage insertion reaction, or did phage λ have a special system to insert itself into the chromosome, a "site-specific recombinase"? This question attracted a number of incipient lambdologists in the mid–1960's, including me. An obvious problem with the homology model was that prophage insertion was more efficient than general recombination between two phage DNAs. I think that the explicit notion of a site-specific recombination system was initially advanced by Ethan Signer, who was intrigued by the ability of each of the many different temperate phages to pick out its unique chromosomal site with remarkable precision and efficiency. If there were a special λ-directed recombination mechanism, then there must be one or more phage-encoded proteins that do the job. To test this fascinating idea, Jim Zissler and Allan Campbell, Max Gottesman and Michael Yarmolinsky, and Roy Gingery and I set out independently to look for integration-defective (*int⁻*) λ mutants.

Ethan Signer

The trick to the successful isolation of *int⁻* mutants is worth describing because it illustrates how the phage solves a major biological problem by integrating into the host chromosome. After integration, the prophage is happily maintained in cells by replication of the host chromosome. In contrast, a repressed, nonintegrated phage DNA finds itself in trouble—it can neither produce new phage by initiating the lytic response, nor be maintained in the cell as part of the host chromosome. Consequently, when the bacteria divide, the repressed phage DNA stays in only one of the two daughter cells. A phage defective in the postulated integration system should make repressed nonintegrated phage. Such phage would be lost from the growing culture.

Grete Kellenberger and Campbell found a large deletion mutant of λ DNA (called *b2*) that failed to carry out prophage insertion, most likely because it had lost the phage attachment site. The properties of this phage served as a guide for identifying mutants in the putative *int* gene. Both wild type (wt) λ and λ*b2* phage form turbid plaques on a bacterial lawn, indicating that bacterial cells are present in the middle of the plaque (see Figure 3–4). But the cells inside the plaque differ in the two cases. Cells in the center of a wt λ plaque contain an integrated prophage that produces cI repressor and protects them from killing by λ. But cells in the center of a λ*b2* plaque survived only because they transiently made λ repressor during the time they were infected with λ*b2*. These cells lack a prophage and are therefore killed when reexposed to λ. The quest for mutants defective in the putative Int protein relied upon this test. Wild type λ phage were treated with a chemical agent that introduced random single-base mutations, allowed to make plaques. The cells from the center of the plaque were then tested for their ability to survive on plates with λ. Using this test, all of us found the *int⁻* mutations. These mutations were likely to define the gene for a protein required for integration into bacterial DNA because their integration defect was restored when the wild type *int* gene was present.

The existence of a protein required for integration was exciting but did not prove a localized recombination mechanism. The way to test this point was to look at recombination between two phage DNAs in a standard phage cross. Whereas general recombination should occur throughout the phage genome, site-specific recombination promoted by the Int protein should occur only at the phage attachment site (POP′). I set out to test this idea, as did Jon Weil and Ethan Signer.

An obvious initial problem had to be overcome. The few known phage genes were so far apart that even a general recombination system would give a substantial number of recombinants between them. Our only hope of seeing the site-specific recombination system was to get rid of general recombination. In fact, we knew that there were two general recombination systems, a bacterial system and a separate λ system. We eliminated the bacterial recombination pathway by doing the phage cross in cells carrying a *recA⁻* mutation, which removed the RecA protein necessary for that system. To get rid of the λ recombination system, Signer and I sought and found recombination-defective phage mutants. We discovered that we had isolated the same mutants and decided to call them *red*. (Some people thought the name was for the political leanings of the discoverers, but of course the reference was to *recombination-defective*.) These mutants defined the genes for the exonuclease and β-proteins of λ, which contribute to a RecA-independent phage-specific pathway of general recombination.

Finally we were ready for the big experiment. We infected a *recA⁻* host with *red⁻* λ phage carrying markers that allowed us to score recombination in three different intervals of λ. There was 3% recombination between the markers that bracketed the interval containing the phage attachment site (the *J* to *N* interval) and no recombination in the two intervals lacking the phage attachment site (the *A* to *J* interval and the *N* to *R* interval) (Figure 7–9). This is the result expected if recombination were being carried out by a site-specific recombination system. Most important, when the same experiment was performed with λ phage that contained a mutation in *int⁻* as

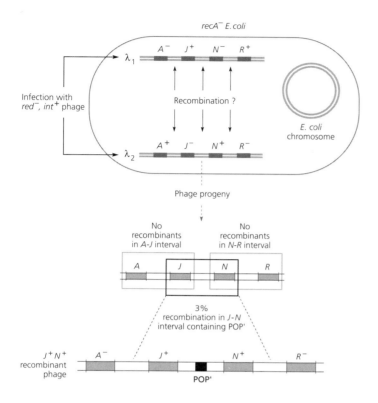

A site-specific recombination system mediates
recombination between attachment sites

FIGURE 7–9. To determine whether recombination at the attachment site was carried out by a specialized recombination system or by one of the general recombination systems (the RecA or Red system), a *recA⁻* bacterial host was infected with two different genetically marked *red⁻* phage. The progeny phage were scored for recombination in three different intervals of the λ chromosome. Recombinants were found in only one interval, between the J and N markers, which contained the attachment site. Because general recombination was disabled, any recombinants observed must have come from a site specific recombination system.

well as *red⁻*, recombination between J and N vanished. There was a new type of recombination, site-specific recombination, carried out by the Int protein at the attachment site! Our two groups published the definitive experiments in back-to-back papers. We each would have liked to be unique, but we could enjoy the cooperative spirit that permeated the field at that time.

PROPHAGE EXCISION AND EXCISIVE RECOMBINATION

Campbell had supposed that excision of λ DNA during prophage induction occurred by a reversal of the integration reaction. Indeed, once the *int⁻* λ mutants were inserted with the help of wild type phage, they were unable to excise. However, I doubted that simple reversal was the complete story because the two reactions occurred

FIGURE 7–10. Phage integrates into bacterial host DNA by recombination between the specific cross-over points POP' (in the phage DNA) and BOB' (in the bacterial DNA). This process needs the phage Int protein and the bacterial IHF protein. The integrated phage (or "prophage") can excise itself from the bacterial DNA by recombination between the cross-over points BOP' and POB'. Excision requires the phage Xis protein in addition to the two proteins required for integration.

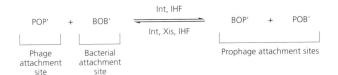

Max Gottesman

under different conditions. Integration was extremely efficient in the lysogenic response, and excision worked exceptionally well during prophage induction. There ought to be some asymmetric element that could provide this type of regulation. Gabriel Guarneros and I looked for and found phage λ mutants that integrated normally but could not excise. Guarneros located the mutants to a gene next to *int* on the λ chromosome and called the gene *xis* (for excision).

Could we show the asymmetry of integration and excision in a convincing way? The strategy that we utilized depended upon the emerging understanding of the nature of the phage and bacterial attachment sites. Experiments by Marc Shulman and Max Gottesman argued that a small region of the bacterial and phage attachment sites (called O) was identical. However, a variety of experiments, beginning with an inquiry into the integration problems of the λ *gal* phage by Franco Guerrini, argued that the sequences flanking O were different in the phage and bacterial sites. Since the prophage attachment sites are created by recombination between the bacterial and phage sites, they must differ in sequence from each of them. Therefore, integration and excision must use different attachment sites. Only the excision reaction, which recombines the prophage attachment sites, should require Xis. To test that prediction, we repeated the experiment that established the Int requirement for site-specific recombination (see Figure 7–9), except that now the phage carried prophage attachment sites. The answer was clear: recombination between the two prophage attachment sites required both Int and Xis, whereas recombination between the phage and bacterial sites required only Int. Although the integration-excision reaction proceeds in both directions, catalysis is asymmetric—Int for integrating the prophage, and Int and Xis for excising the prophage (Figure 7–10).

At this point, the key problem was to define the biochemistry of the reaction: the proteins, their substrates, and the reaction mechanism. Work on the biochemistry of site-specific recombination has involved a large number of scientists. Three groups in particular addressed various aspects of the problem, with largely complementary contributions. Howard Nash figured out how to do site-specific recombination in a test tube and worked out the basic reaction mechanism; Art Landy defined the DNA sequences and the protein-DNA interactions required for site-specific recombination; and I mainly contributed the concept that nucleoprotein structures localize and regulate the reaction.

Howard Nash

In 1975, Nash developed a recombination reaction for integrative recombination that worked in crude extracts. Initially a medical doctor without formal training as a biochemist, he had the requisite intellectual boldness and experimental insight to succeed with an approach that seemed almost impossible at that time—to produce a complete recombinant DNA molecule in a test tube. His clever idea was to make a parental λ DNA molecule that carried both the phage (POP′) and bacterial (BOB′) attachment sites and look for recombination between them. Two recombinants should result, each now having a single attachment site: a very small molecule consisting essentially of the biotin gene and a larger molecule containing the λ DNA (Figure 7–11, top). But how to detect either of the recombinant molecules, which were likely to be produced only rarely, at least in the beginning experiments? Because the recombinant λ DNA was 13% smaller than the parental λ molecule, Nash was able to selectively retrieve the recombinant. It turns out that λ phage with a higher than normal DNA content are killed very efficiently by chelators, such as EDTA, whereas λ phage with a smaller than normal DNA content are spared by this treatment. This gave Nash a way to get rid of phage carrying large parental DNA, so that he could find the rare recombinant phage with smaller DNA molecules.

In the actual experiment, circularized λ DNA was added to a crude extract made from cells that expressed the appropriate λ proteins. Recombination was allowed to proceed in the test tube. The resultant λ DNA molecules were separated from the extract, transfected into host cells and allowed to make phage (Figure 7–11, middle). These phage were treated with EDTA, to selectively kill

FIGURE 7–11.   To study the site-specific recombination reaction that led to phage integration, Nash developed an assay that would allow him to detect very rare recombinants. This assay depended on the fact that phage with larger than normal DNA molecules are very susceptible to killing by EDTA. As a substrate for site-specific recombination, Nash used a λ DNA molecule larger than wild type λ with the phage (POP') and bacterial (BOB') attachment sites flanking the bacterial biotin gene. If recombination occurred in the test-tube reaction carried out in a crude cell extract from bacteria infected with λ, two new DNA molecules would be produced; a circular DNA unable to replicate and a phage DNA molecule significantly smaller than the parental phage DNA. This mixture of DNA molecules was transfected into bacteria where the two phage chromosomes replicated and produced progeny. This mixture of phage was then treated with EDTA and reintroduced into bacteria. Because the large parental phage but not the smaller recombinant phage was destroyed by EDTA, only the recombinant phage grew. The surviving phage were recombinants because they had lost the biotin gene located between the recombination sites in the parental DNA.

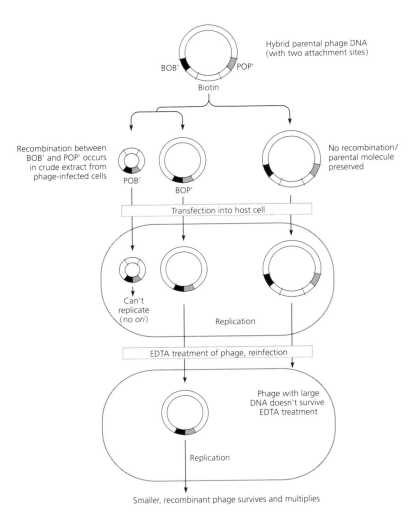

phage with large parental DNA. The surviving phage, presumed to carry recombinant DNA molecules, were then used to infect a host cell. The progeny phage were checked to make sure they had actually undergone the recombination (by determining whether they had lost the bacterial gene for biotin that was located in the DNA that should have been excised) (Figure 7–11, bottom). Although cumbersome, this assay was very sensitive—it could detect the production of extremely small numbers of recombinant λ phage that would not have been seen in any other assay. Nash was able to use

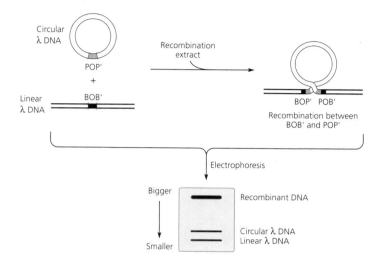

A simple assay for site-specific recombination
used circular and linear λ DNA molecules

Circular λ DNA

POP'

+

Linear λ DNA

BOB'

Recombination extract

BOP' POB'

Recombination between BOB' and POP'

Electrophoresis

Bigger

Recombinant DNA

Circular λ DNA
Linear λ DNA

Smaller

FIGURE 7–12. The product of recombination between a circular and a linear λ DNA is a linear molecule twice as long as the initial linear molecule. This recombinant product can easily be distinguished from both starting molecules by separating the mixture using gel electrophoresis. This and similar assays were used to determine all of the requirements for site-specific recombination.

the assay to demonstrate that recombination was occurring and then to refine the system to get a more efficient reaction.

When the efficiency of the reaction was high enough, Kiyoshi Mizuuchi and Nash switched to identifying the recombinant molecules directly. Eventually, Kiyoshi Mizuuchi and Michiyo Mizuuchi introduced a simple intermolecular assay in which a linear and a circular λ DNA molecule recombined into a single molecule (Figure 7–12). The ability to analyze the complete recombination reaction allowed the direct study of both the protein and DNA requirements of the site-specific recombination reaction, a capacity that yielded some remarkable insights.

Kiyoshi Mizuuchi, Gellert, and Nash found that the substrate DNA must be a circular molecule. In pursuing this point, they discovered an enzyme, DNA gyrase, which converts the cirucular DNA into a negatively supercoiled form. They found that the integrative recombination reaction not only used negatively supercoiled DNA but actually required it. Both the bacterial chromosome and the small circular DNAs that inhabit bacterial cells (plasmids) are normally negatively supercoiled. *In vivo*, DNA gyrase not only introduces negative supercoils into covalently closed intracellular DNA but also removes the "positive" supertwists that are introduced into DNA by the duplex unwinding reaction required for DNA replication (see Figure 4–19).

The role of Int was also clarified. We knew that Int bound to the substrate DNA, but was it responsible for executing the breaking and joining reaction? Yoshiko Kikuchi and Nash found that Int carried out a DNA topoisomerase reaction; that is, it removed negative supercoils from DNA. Since topoisomerases work by a DNA-nicking and closing reaction, Int was probably the nicking and closing enzyme in recombination as well. The activity discovered by Kikuchi and Nash was presumably a "side-reaction" in the normal breaking and joining pathway.

In the course of using the recombination assay to purify Int protein, Kikuchi and Nash also discovered the need for a host protein, which they termed IHF for "integrative host factor." Some years later, Robert Weisberg and David Friedman isolated host mutants defective in the integration and excision of λ. These host mutants defined the two genes that code for the two subunits for the host component of λ site-specific recombination, the IHF protein.

Art Landy

DNA-PROTEIN INTERACTIONS IN SITE-SPECIFIC RECOMBINATION

An understanding of site-specific recombination clearly required a knowledge of the DNA sites necessary for the reaction. In 1977, Art Landy and Wilma Ross completed the first giant step in this project by determining the DNA sequences involved. As guessed by genetics, the P, P′, B, and B′ sequences were all distinct, but there was a common "core" sequence in all attachment sites, 15 bp in length (the "O" site). The next goal was a functional analysis: how much DNA defined an attachment site; and what were the protein-DNA contacts? The first question was addressed by using the Mizuuchis' intermolecular recombination assay. Each individual substrate site was trimmed by genetic engineering techniques and then tested for function in this assay. The Mizuuchis, Landy, and colleagues found that the phage attachment site was large, requiring some 240 bp of DNA; by contrast the bacterial attachment site was small, only about 20 bp, just slightly more than the length of the common "core" sequence.

The Mizuuchis also determined the site of DNA recombination, using an elegantly conceived experiment, based on the observation that when a radioactive ^{32}P molecule in a DNA chain decays, the DNA chain breaks. They synthesized an attachment site with ^{32}P nucleotides at random positions and then recombined the labeled site with an unlabeled attachment site. When the recombinants were stored, the DNA chain was broken at (and only at) those positions containing ^{32}P, allowing them to determine the portion of the recombinant that came from the labeled partner. By repeating this experiment with label in each strand of both partners, they showed that the points of strand exchange were unique (rather than distributed over the core sequence) and offset from each other by 7 bp in the two strands (making a "staggered" rather than a "blunt" joint).

Ross and Landy next mapped out the detailed interactions of Int with DNA. Their remarkable results showed that Int had seven binding sites within the 240 bp phage attachment site and two binding sites in the bacterial attachment site, both of which overlapped the core region. Nancy Craig and Nash found that IHF was also a DNA-binding protein and had three binding sites within the phage attachment site. In Landy's words, "the phage attachment site was a busy region of DNA."

The DNA-protein binding studies revealed a fascinating complexity to site-specific recombination. Why was the phage attachment site ten times longer than the bacterial site, with five extra Int binding sites? My initial idea was that the extra Int proteins might somehow unwind the core DNA to facilitate recombination. I thought that electron microscopy might be useful in exploring this possibility as it might allow us to visualize both the proteins and the path of the DNA. I approached Marc Better, a skilled electron microscopist who, along with his thesis adviser Dave Freifelder, was visiting my lab. Better agreed to give the project a try. He was able to see a rather large DNA–protein complex on supercoiled plasmid DNA containing the phage attachment site. To localize the interaction, he formed the Int complex and cut the plasmid DNA with a restriction enzyme, producing a DNA segment 540 bp in length. That experiment yielded a most unexpected result—the lengthy stretch of DNA constituting the prophage attachment site was condensed within the protein complex. We could see that Int was not arrayed along the DNA in linear fashion but formed a nucleoprotein structure in which the phage attachment site DNA had to be folded or wound. We supposed that the reactive component in integrative recombination was the nucleoprotein complex or "Intasome" depicted in Figure 7–13.

Inspired by this result, Better and I decided to look at the role of Xis in excisive recombination. He had three weeks in Berkeley before he would leave for a postdoctoral position—an impossibly short time in which to do anything. We wrote all possible experiments on

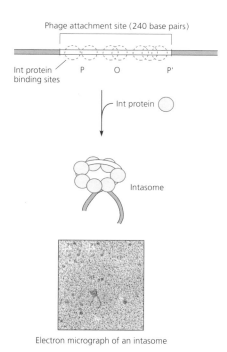

Int protein binds to the phage attachment site to form a complex called the intasome

Electron micrograph of an intasome

FIGURE 7–13. The phage POP' attachment site is 240 bp long and contains seven DNA-binding sites for the phage protein Int. When these binding sites are occupied by Int, this region of DNA is condensed into a compact structure termed the "intasome." An electron micrograph of the phage attachment site in the presence of Int protein depicts an actual intasome. (Electron micrograph courtesy of M. Better.)

the blackboard. Better tried them and, for the only time in my life, every projected experiment worked. Int formed a small nucleoprotein structure with the left prophage attachment site (BOP′) but not with the right prophage attachment site (POB′). With added Xis protein, a complex at the right prophage attachment site was observed. Thus, we could conclude that Xis allowed formation of a reactive nucleoprotein structure on the right prophage attachment site POB′ DNA.

Why bother with the complicated nucleoprotein structures? I could think of two obvious advantages for such an approach to the selectivity and regulation of this complex DNA transaction. First, the use of multiple binding interactions allowed a more precise localization of a DNA target than a single protein-DNA interaction. And reactions such as site-specific recombination require exceptional precision because a misintegrated λ prophage cannot excise. Second, the DNA-wound nucleoprotein structures could allow directional control over site-specific recombination by defining the appropriate juxtaposition of DNA sites.

Scientific discovery is a popular but truly obscure term, because it describes several different types of scientific endeavor. On one level we seek to refine an accepted concept to a greater stage of understanding ("concept processing"). This effort acquires new information, and so we are engaged in discovery; we may even discover that the accepted concept is wrong. Concept-processing is what most of us do most of the time. On a second level, we sometimes manage to do an experiment in which we are deciding whether an emerging concept is right or wrong ("concept testing"). This stage of discovery is tremendously exciting because we hope that we will need to look at our segment of the scientific universe in a different way after our experiment. The genetic test of site-specific recombination described above (Figure 7–9) is an example of concept-testing. At a third level, we might find ourselves suddenly in possession of a brand new concept ("concept formation"). This last stage provides a super rush of excitement because we are convinced that we have discovered something completely new and unexpected—an ultimate stage of novelty (we may be wrong). The specialized nucleoprotein structure was a "discovery of the third kind." DNA-bound proteins did not have to act in a linear array but could interact over long distances, bending or winding the intervening DNA. Perhaps all multiple DNA-binding interactions that worked from distant sites were examples of specialized nucleoprotein structures.

In spite of the preceding buildup, I was nervous about whether or not the specialized nucleoprotein concept was correct. I was relieved to learn that, based upon completely different experiments, Nash had also been thinking about the *att* sites in terms of such a structure. Nash had discovered that there was always a "left-over" supercoil after integrative recombination, suggesting that DNA looped around the recombining Int proteins to form a nucleoprotein structure. Landy and Nash added evidence for long-distance "cooperativity" in DNA-binding. Work with other site-specific recombination reactions and with the transposition reactions described in the next section has supported the general validity of the concept. In previous chapters, I have noted that specialized nucleoprotein structures are also used in the initiation of DNA replication and to promote distant interactions in transcriptional regulation.

So Int and IHF were responsible for a reactive nucleoprotein complex that governed pairing and recombination, and Int was the breaking and joining enzyme. How did Int execute the seven-base staggered cleavage in the homologous core region? Landy and Nash and their colleagues have demonstrated that the reaction is sequential. A break and join on one side of the homologous region is followed by a second break and join on the other side that completes the exchange of DNA strands. The seven-base homologous segment does not specify the initial DNA pairing but only provides a platform for testing homology. This "homology check" may be another level of control to reduce misintegration at an improper site.

OTHER SITE-SPECIFIC RECOMBINASES

Site-specific recombination by phage λ provided the initial discovery and molecular understanding of a DNA transaction that is widely used in nature for a variety of biological purposes. "Inversion" and deletion reactions provide a second example of this type of reaction. When the two sites for the recombination reaction are in the opposite (inverted) orientation, site-specific recombination between them results in inversion of the segment (Figure 7–14). When the two sites for the recombination reaction are in the same orientation, site-specific recombination deletes the DNA segment between them. The most spectacular example of an inversion event is the Hin system of *Salmonella*, studied by Mel Simon and his colleagues. In this system, inversion of a DNA segment changes the particular flagellar genes hooked up to a promoter. After inversion, a

FIGURE 7–14. When the recombination sites recognized by a site-specific recombination system are in opposite orientation on the same DNA molecule ("inverted repeats"), the DNA segment in-between them is inverted by the recombination reaction. This event is used to regulate gene expression.

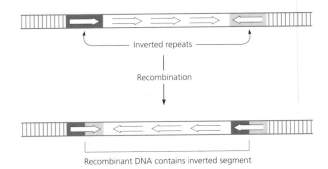

Inverted repeats

Recombination

Recombinant DNA contains inverted segment

different flagellar gene is expressed, so a different flagellin protein is used to make flagella, the swimming appendages of the bacteria. Flagella are a major target of our immune system. The use of alternative flagellar proteins allows the bacteria to dodge the antibody protection system of an animal. The animal tries to disable the bacterium by making an antibody that binds to the flagella; when the flagellar target protein is changed, the antibody no longer works. Work by Simon on Hin and by Nick Cozzarelli on a related system called Gin has demonstrated that, in reactions of this type, a specialized nucleoprotein structure and DNA supercoiling are used to direct the appropriate specificity.

Site-specific recombination is not limited to these two examples. Bacteriophage P1 has the site-specific Cre recombinase and a yeast plasmid, called the 2 micron plasmid, encodes a site-specific recombinase called Flp. Flp and Cre belong to the same enzyme family as the lambda recombinase.

Site-specific recombination combines the functions of DNA cutting (using a site-specific endonuclease) and DNA rejoining (ligase) into a single enzyme or enzyme complex. Practical applications are inevitable. The basic reaction can be used in the laboratory to modify the genome of almost any organism. The Cre and Flp recombinases are generally used for this purpose because in each case only a single protein is needed to promote recombination, and both systems will readily promote either chromosomal insertions or deletions. Cre- or Flp-promoted site-specific recombination can insert any gene into a particular site in any chromosome (after the chromosome has been modified to possess a recombination target site at that location). Even more important, site-specific recombinases are now being used to study gene function. Recombination target sites in the same orientation are placed on either side of a gene with an unknown function in a genome that has been modified so that it contains the site-specific recombinase under the control of a regulatable promoter. When the recombinase is induced, either at a particular point in development or in a small number of cells, the sites flanking the gene are recombined, resulting in deletion of the gene. The fate of the cells lacking the gene can then be monitored. The results have often been dramatic. With these developments, recombinases have followed the scientific trail blazed by DNA polymerase, DNA ligase, nucleases, and other enzymes that act on DNA. As was the case for these enzymes, the site-specific recombinases have not only provided an improved understanding of DNA transactions, but also a new approach to the exploration of molecular and cell biology.

TRANSPOSITION AND REPLICATIVE SITE-SPECIFIC RECOMBINATION

Transposition is used extensively to move DNA from one place to another. Many antibiotic resistance genes are located in transposons and move between different bacteria and even different species by transposition. Bacterial and animal viruses have borrowed transposition as a means for integration and replication. Notably, retroviruses, including HIV, the virus that causes AIDS, integrate by a transposition mechanism.

Barbara McClintock discovered transposable genetic elements in corn and correctly inferred their properties by the late 1940's, before genes were definitely known to be DNA. Despite McClintock's amazing achievement, an understanding of the molecular properties of these mobile segments of DNA awaited their rediscovery and detailed study in bacteria and phage. A transposable genetic element (or transposon) in its simplest incarnation is a DNA unit with a self-directed capacity to move from one region of a genome to another. Sometimes transposons also contain additional genes, most especially those coding for resistance to an antibiotic. Transposons move to a new location by a recombination event in which the ends of the transposon are joined to a site on the target DNA, which is generally chosen at random. So the recombination event is localized with respect to the transposon DNA, but not the recipient (a sort of hemi-site-specific recombination event). The enzyme responsible for transposition (transposase) is carried by the transposon itself. The

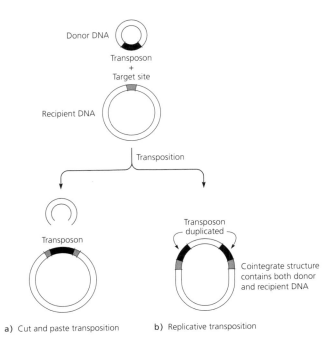

FIGURE 7–15. A transposon is a piece of DNA that can move from one place to another in the genome. This process is called "transposition" and can occur in two ways. In cut and paste transposition, the transposon "cuts" itself out of the "donor" DNA and then "pastes" itself into a new target site (a). In replicative transposition, the transposon is duplicated during the transposition process, creating a "cointegrate" structure that contains two transposons flanking the donor DNA (b). In either type of transposition, the target site is duplicated on either side of the inserted DNA.

hallmark of transposition is that the DNA in the target sites is changed by the introduction of a small duplication (usually five or nine base-pairs), which then flanks the inserted transposon. Because the DNA is changed during transposition, we know that this process does not proceed by the completely conservative pathway characteristic of the λ site-specific recombination event, which involves no DNA synthesis.

There are two pathways for transposition in bacteria. One way to achieve transposition is by a simple insertion reaction: the transposon is directly transferred to a new location and DNA synthesis is necessary only to repair a small gap. In this "cut and paste" transposition, transposase completely excises the transposon from its surrounding DNA, cuts the target DNA, and then joins the transposon to this new DNA site (Figure 7–15a). In an alternative pathway, the transposon itself is replicated during transposition, so that one copy remains linked to the old target site and the second is linked to the new target site. The product of this reaction is called a "cointegrate" because donor and recipient DNA molecules (usually circular in bacteria) are fused into a large circle (Figure 7–15b).

Our current understanding of how transposons move comes mostly from a fortunate parallel study of two mutational manifestations of transposition that were initially obscure and not obviously related: "insertion elements" and a phage that causes many mutations, called Mu. In addition, study of the mobile antibiotic resistance element Tn3 by David Sherratt and Stan Falkow provided the first evidence that a cointegrate structure was an intermediate in transposition.

THE IS INSERTION ELEMENTS

In 1965, Jim Shapiro made the initial observation of an insertion element. He was a graduate student working on the galactose operon of *E. coli* with Bill Hayes in Cambridge, England. Because Sankar Adhya, Chuck Hill, and I shared an interest in this operon, Shapiro made several visits to our lab in Madison, Wisconsin. He was an independent, enthusiastic, argumentative, and generally provocative character, and his sojourns to our place were always fun. We were all puzzled by a bizarre mutation, called *gal3*, originally isolated by Esther Lederberg. The *gal3* lesion was a "polar" mutation (see Figure 5–16), meaning that it turned off expression of all three genes in the galactose operon, preventing cells from growing on galactose. There were two strange things about *gal3*. First, it regained the ability to grow on galactose ("reverted" to a *gal*+ phenotype) more frequently than most mutations. Second, some of the revertants had properties that differed from the parental strain. wild type cells express the *gal* genes only when galactose is present (like the *lac* operon [see Figure 3–14], *gal* is inducible). Hill found that, rather than expressing the *gal* genes only when galactose was present, some of the "revertants" expressed the *gal* enzymes all of the time (constitutively). Shapiro announced one day that he understood the *gal3* mutation—the lesion was caused by insertion of a large piece of DNA into the *gal* operon. It might be easier to remove the inserted DNA than to make a base change to revert a typical mutation. Moreover, partial removal of the insertion might give odd phenotypes, such as constitutivity.

Shapiro realized that there was a way to test this idea. As we have seen, the position of phage band in a cesium chloride density gradient depends upon its density, which is determined by the relative ratio of DNA and protein in the phage (see Figure 7–2). If *gal3* were a large insertion into DNA, then λ*gal3* phage should band at a different place in the gradient than λ*gal*+ phage. He tested his idea by

Jim Shapiro

FIGURE 7–16. A peculiar *gal⁻* mutation was explained by the discovery that this mutant was caused by insertion of foreign DNA into the *gal* operon. The foreign DNA contained a transcription stop signal, preventing expression of downstream genes in the *gal* operon. The discovery of several such insertions led to the study of mobile elements and transposition.

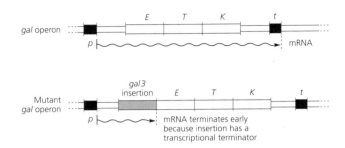

comparing the density of λ*gal3* with that of a λ*gal⁺* revertant phage. The experiment worked; the two phage banded at different places in the gradient, and Shapiro could calculate that the reverted λ*gal⁺* DNA was smaller by about 1,000 base pairs than λ*gal3*. A similar test of the insertion idea by Elke Jordan and Peter Starlinger gave the same answer. Mike Malamy found that a number of strongly polar mutations in the *lac* operon were insertions—all of about the same size. He suggested that the DNA insertion prevented expression of downstream genes because it introduced a terminator site for RNA synthesis. The notion of an insertion mutation explained many "strong polar" mutations but left unsolved how a large segment of DNA suddenly arrived in the middle of operons. The apparently common size of the insertion sequences was an intriguing clue (Figure 7–16).

Were the insertion sequences all the same? The appropriate technology for answering this question, heteroduplex DNA analysis, had just been developed by Norman Davidson, Waclaw Szybalski, and their colleagues. As noted earlier, duplex DNA could be separated and isolated as individual strands. If the complementary DNA strands were mixed, then base-pairing occurred, and a complete duplex reformed (see Figure 5–3). However, if one strand carried an insertion that was not present on the complementary strand, then a single-strand region representing the insertion sequence would remain when duplex DNA reannealed. This single-strand DNA could be visualized by electron microscopy (Figure 7–17).

Starlinger, Szybalski, and their colleagues used heteroduplex mapping to look at a number of known insertion mutations. The heteroduplex techniques allowed relatively precise measurements of size and sequence homology between different insertion sequences. The results were spectacularly interesting. All of the insertion muta-

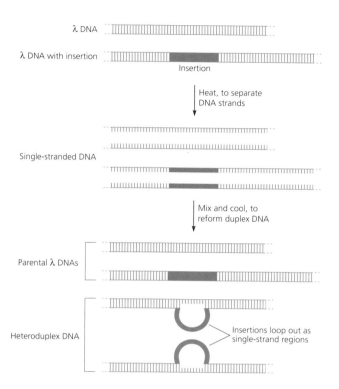

λ DNA

λ DNA with insertion

Insertion

Heat, to separate
DNA strands

Single-stranded DNA

Mix and cool, to
reform duplex DNA

Parental λ DNAs

Heteroduplex DNA

Insertions loop out as
single-strand regions

FIGURE 7–17.   When λ DNA with and without an insertion was separated into single strands and mixed, both parental and "heteroduplex" (having one strand from each parent) DNA was formed. When viewed under the electron microscope, the parental molecules looked like normal double-strand DNA, but the heteroduplex molecules contained a single-strand loop that corresponded to the position of the insertion sequence. Examining many different insertion mutations revealed that there were only three different types of insertions in *E. coli* DNA. These three different sequences were the three major insertion elements found in *E. coli*.

tions represented only three DNA sequences, which were designated IS1, IS2, and IS3. So these special DNA sequences were somehow endowed with the ability to move about. Of course, this immediately raised the question of the mechanism of mobility. Although further work on these sequences has answered this question, the first and most complete answer came from seemingly unrelated studies on phage Mu. We now turn to the fascinating story of how an understanding the Mu life cycle led to the discovery of the mechanism of transposition.

MU: A PHAGE THAT DOES NOT INTEGRATE LIKE LAMBDA

The Campbell model for how λ integrated into the host chromosome quickly became the predominant paradigm for a specialized genetic recombination reaction. In this model, a circular λ molecule integrated with a single recombination event at a unique position in the chromosome and then excised by the reverse of this process (see

Figure 7–8). This extraordinarily successful model not only explained why the order of genetic markers in linear phage DNA was permuted from that in the prophage but also provided a fertile intellectual climate for identifying the biochemistry of this fascinating reaction. In short order, geneticists and biochemists had determined that the λ Int and Xis proteins bring the bacterial and phage attachment sites together and then mediate a site specific recombination event between them.

Given the success of the Campbell model, it is not surprising to find that this paradigm guided the experiments examining how another phage, called Mu, integrated into the chromosome. However, starting from its initial characterization by Larry Taylor in the early 1960's, it was clear that Mu was different from λ; every subsequent test designed to demonstrate that Mu used the λ integration paradigm failed. A small group of creative scientists, most notably Ariane Toussaint, Piet van de Putte, Ahmad Bukhari, and Martha Howe took on the challenge of trying to understand how Mu integrated. Some of the early work was also contributed by Ellen Daniell and John Abelson. Eventually, it became evident that the Mu phage was a giant transposon, inserting into the chromosome and replicating in the cell by multiple transposition events. Clearly, Mu was specialized to carry out this reaction frequently and efficiently, making it the system of choice to dissect the mechanism of transposition.

The first inkling that Mu used a significantly different mechanism for integration than λ came from initial characterization of Mu lysogens. Bacteria into which Mu phage integrated had a high frequency of spontaneous mutations; interestingly, these mutations were always located very close to the site of Mu integration. The simplest explanation for this was that Mu could insert at many places in the host chromosome, often disrupting a gene and causing a mutation during the process of integration. In fact, it is this property that gave this phage the name "Mu," which is short for "mutator." Careful genetic analysis of the insertion sites showed how little Mu cared about where in the genome it integrated. Two different groups—Bukhari and David Zipser, and Daniell and Abelson—isolated a great many independent Mu insertions in the *lacZ* gene and showed that nearly all of them could recombine with each other, indicating that each insertion was at a different site in the gene. These experiments certainly called into question the idea that Mu used a bacterial attachment site, analogous to the one used by λ. If Mu used such a site, it must be a very redundant sequence indeed.

The problems with applying the Campbell model to Mu integration were more fundamental than the inability to find the bacterial attachment site. No evidence could be found for a basic premise of the model—that the phage circularized and then integrated by recombination. Piet van de Putte found that, unlike λ, the order of genetic markers in prophage and linear Mu DNA were the same. Thus, if there were a phage attachment site, then it must be located at or very near the "ends" of the phage DNA, following all of the known genetic markers in Mu. Analysis of the structure of the infecting phage genome with the electron microscope provided further disturbing information. Daniell, Abelson, and colleagues performed the same kind of heteroduplex analysis on Mu DNA that had previously revealed insertions in phage λ (see Figure 7–17). These investigations indicated that each molecule of Mu DNA had variable sequences of DNA at its ends. This DNA was subsequently shown to be random selections of host DNA. This made the idea that Mu circularized by means of complementary single-strand regions at each end of the phage (like the cohesive ends of λ) appear very unlikely. Indeed, a search for a circular Mu DNA was unsuccessful. Circular Mu replication intermediates were found, but in contrast to those characterized for phage λ, the Mu intermediates always contained different, variably sized host DNA sequences. Clearly the Campbell model did not work for Mu.

Analysis of the DNA structures generated during phage replication eventually lead to the insight that Mu was a huge transposable element. Two observations were especially critical. First, in 1977 Elisabeth Ljungquist and Bukhari compared the junctions between an individual Mu phage and its host DNA when it was in the prophage state to the junctions present after induction of lytic growth. Although replicating Mu formed new junctions with host DNA after induction, the original junction with host DNA established by the prophage was always also present. This experiment unequivocally showed that the Mu genome did not excise from host DNA prior to the onset of replication. Second, replicating Mu molecules visualized in the electron microscope were complex fusions between circular and linear segments of DNA that had branch points and Y-structures. These molecules looked much more like intermediates in DNA recombination than in replication. Together, these observations lead to the conclusion that phage replication and genetic recombination were deeply intertwined processes. Along with these insights about the physical structure of DNA during Mu replication, an increased appreciation of the strong association between genome rearrangements and phage replication emerged.

These genetic consequences of Mu growth strengthened the idea that Mu behaved similarly to transposons and IS elements. Even some details of the structure of Mu and IS element insertions were similar. In the late 1970's, it became clear that the junctions between the IS elements and bacterial DNA consisted of a few base-pairs of a duplicated sequence present in the form of a direct repeat. Mu insertions were flanked by these direct repeats as well.

The accumulating information about transposition spawned a series of molecular models suggesting how this process might occur. Nigel Grindley and David Sherratt were the first to point out that the short duplication at the ends of the IS element could result from a staggered cut introduced into the DNA at the target site. If the cuts in each strand of the target site are offset by a few base-pairs, a complementary single-strand region will be present at each side of Mu after it inserts into the target. Repair synthesis to restore double-strand DNA would generate the observed duplication. In 1979, Shapiro, who had been involved in the initial discovery of insertion sequences (see Figure 7–16), proposed a detailed and insightful model for Mu transposition and replication. The "Shapiro model" incorporated the idea of a staggered cut at the Mu target site and explained how Mu could be completely replicated by the transposition event. According to this model, Mu DNA is nicked at the 3′ end of each junction with the chromosome (the "donor site"). Both strands of the "target site" DNA are also cleaved, yielding a staggered cut with a 5′ overhang at each end. These 5′ ends of the target DNA were proposed to join to the 3′ ends of the Mu DNA, resulting in a DNA structure with two branches between the Mu and host sequences. Each branch could be recognized by the host DNA replication machinery; replication through the Mu portion of this "Shapiro intermediate" would yield a cointegrate structure, the predominant product of Mu replication (Figure 7–18). The Shapiro model demystified transposition in the same way that the Campbell model explained site-specific recombination.

IDENTIFYING THE MU COMPONENTS NEEDED FOR TRANSPOSITION

As was the case for elucidating the mechanism of chromosomal DNA replication, genetics and *in vivo* experiments were the keys to determining the requirements for Mu transposition. The necessary Mu components turned out to be very simple: the MuA and MuB

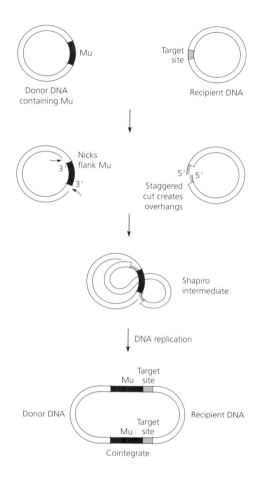

The Shapiro model explained the mechanism
of Mu transposition and replication

Donor DNA containing Mu — Mu

Recipient DNA — Target site

Nicks flank Mu — 3', 3'

Staggered cut creates overhangs — 5', 5'

Shapiro intermediate

DNA replication

Donor DNA — Mu, Target site — Recipient DNA

Mu, Target site

Cointegrate

FIGURE 7–18. Shapiro proposed a model that explained how the Mu phage was able to insert itself into recipient DNA, and, in doing so, replicate itself. The process begins with cutting ("nicking") the DNA strand at each 3' end of Mu DNA and making a staggered cut in the target site of the recipient DNA. The 3' ends of Mu are joined to the 5' ends in the target site; this joined structure is the Shapiro intermediate. DNA replication beginning at the branches in the intermediate give rise to a cointegrate structure containing both donor and recipient DNA, as well as two copies of Mu and of the target site.

proteins, and sequences of a few hundred base-pairs from each end of the Mu genome. By systematically isolating conditional lethal mutations in Mu, Michel Faelen and Ariane Toussaint, and Kathy O'Day and Martha Howe, showed that the Mu *A* and *B* genes were essential for both replication and phage growth. Both *A* and *B* are expressed early in infection, prior to integration. As the *A* gene product (MuA) is absolutely essential for both integration and replicative transposition, it was considered most likely to be the transposase. In contrast, although MuB was essential for replication, it stimulated integration only about 100-fold. It was clearly also critically involved in transposition but appeared to play an "enhancing" role in this process.

Establishing that the ends of the Mu genome were essential for transposition resulted from constructing artificial "mini-Mu" transposons and observing their behavior. These artificial constructs converted Mu into a more classical transposon. The two ends of the Mu DNA were cloned into a plasmid on either side of a gene imparting resistance for some antibiotic (like resistance to tetracycline). When the Mu A and B genes necessary for transposition were expressed in the cell, this "mini-Mu" could transpose. Transposition could easily be followed by looking for the ability of the "mini-Mu" to move tetracycline resistance from one piece of DNA to another piece of DNA. Following the movement of mini-Mu to an F^+ factor was particularly convenient. This transposition event could be monitored by mating the transposing F^+ bacteria to F^- bacteria (as explained in Figure 3–5) and then determining how many F^+ factors also transferred tetracycline resistance to the mated cells. The fraction of tetracycline resistant F^+ factors is a measure of transposition frequency. By systematically shortening the ends of mini-Mu DNA, the minimal sequences required for transposition were determined.

A UNIFYING MODEL FOR TRANSPOSITION

Most transposons move very infrequently, making it difficult to study the recombination process in the test tube. The realization that Mu replication was likely to be an extremely efficient example of transposition made it the system of choice for *in vitro* analysis. Pat Higgins and George Chaconas both sought to do exactly that. However, it was Kiyoshi Mizuuchi, fresh from his study of the mechanism of λ integration, who played the key role. Upon meeting Mizuuchi, one does not soon forget him. A small man, native of Japan, he has a long beard and long hair (held in place by a head band), and an extremely deep voice. This, and a significant accent, makes understanding him a challenge, but it is well worth the effort. Mizuuchi has been at the forefront of all major breakthroughs relating to the mechanism of transposition. Working with a very small group at the National Institute of Health, and often performing the critical experiments himself, Mizuuchi is famous for his biochemical vision and beautifully executed, elegant experimental designs.

Mizuuchi set out to recreate the Mu replication-transposition reaction in the test tube. Because Toussaint had established that host replication proteins were needed for Mu DNA replication, Mizuuchi's initial experiments used the very same extracts that sup-

Kiyoshi Mizuuchi

ported *E. coli* chromosomal DNA replication. The extracts used by Bob Fuller and Arthur Kornberg to study the replication of plasmids carrying *oriC*, the *E. coli* origin of replication, were especially useful. Mizuuchi mixed an extract that supplied the MuA and MuB proteins with this Fuller-Kornberg replication extract. He added the DNA substrates for transposition, mini-Mu on a plasmid and phage λ DNA, to this mixed extract. If transposition occurred in the test-tube reaction, the mini-Mu and its antibiotic resistance marker would be transferred to the λ DNA, creating a λ that conferred resistance to this antibiotic (for example, resistance to tetracycline). He assayed for this event by packaging the DNA into λ phage in a test-tube reaction, adding the phage to bacteria and growing the bacteria on agar plates that contained tetracycline. Only cells whose λ had incorporated the antibiotic resistance gene would grow. This sensitive *in vivo* assay proved unnecessary. Because Mu transposition *in vitro* was as vigorous as it was *in vivo*, transposition could easily be followed with a biochemical assay that relied upon detecting the products of transposition using gel electrophoresis (this was similar to the method used to detect λ integration; see Figure 7–12). The first transposition events captured *in vitro* came from this study of the replicative transposition of Mu. Bob Craigie, working with Mizuuchi, characterized the structures of the DNA molecules that were generated.

Although some insight into the molecular mechanism of transposition could be gleaned from analysis of the final products of Mu replicative transposition, the major breakthrough occurred when the overall reaction was divided into its two separate phases: recombination and DNA synthesis. One might imagine that this important feat was accomplished as a result of careful fractionation studies. In reality, this breakthrough depended on serendipity. Whereas DNA replication required many complex, multi-subunit enzymes, the only *E. coli* protein required for Mu recombination was the DNA bending protein Hu. A close relative of IHF, Hu is a small and very stable protein. As a result, it was easy to prepare cell extracts that vigorously supported the recombination steps but very difficult to prepare extracts that were highly proficient in replication. At one point, Craigie and Mizuuchi noticed that a "suboptimal" replication extract efficiently produced novel DNA products visible in their gel assay. The new bands turned out to be transposition intermediates, which now accumulated in large amounts because the replication process that removed them was blocked.

The structure of the true recombination intermediate now could be directly compared to the intermediates predicted by the various transposition models.

In a very exciting study, Craigie and Mizuuchi determined that the DNA intermediate formed during Mu transposition *in vitro* had exactly the structure predicted by the Shapiro model of transposition. Cleavage of Mu DNA at its junctions with host DNA exposed the 3′ ends of the Mu element; these ends were then joined to cut ends in the target DNA. The newly inserted portion of the branched DNA molecule is replicated, creating a cointegrate (Figure 7–19, right). Alternatively, the old flanking host DNA could be degraded, followed by the repair of the single-stranded DNA left by the staggered target site cleavage. In this case, a simple "cut and paste" insertion product is the result (Figure 7–19, left). The major feature distinguishing elements that replicate with each round of transposition and form a cointegrate structure (like Mu) from those that do "cut and paste" is whether only one or both strands at the element-host DNA junctions are cleaved. In replicative transposition, only one strand (the 3′ end) is cleaved at each junction; in the "cut and paste" pathway, both strands are cleaved. In both types of transposition, the cleaved 3′ end of the transposon is joined to the target DNA in the fashion first shown for phage Mu.

In vitro analysis of Mu transposition also established unequivocally that the MuA protein was the transposase for Mu. MuA recognizes specific sequences near the two ends of the Mu genome, binds these sequences, and pairs the two ends to form a synaptic complex. With only the assistance of divalent metal ions (such as magnesium), MuA carries out all of the DNA cleavage and DNA joining reactions necessary to generate the Shapiro transposition intermediate. As might be expected from the highly conserved nature of the chemical steps in transposition, MuA is a member of a large protein family of transposases and retroviral integrases; these proteins use homologous catalytic domains to carry out these reaction steps.

TRANSPOSITION: WHY HAVE IT?

Possessing only a transposase and some substrate DNA at its ends, the simplest transposons appear to contribute nothing to the creatures they inhabit. Some people think that transposons spread through the population simply because they have developed the mechanism to do so, an exemplar of the so-called "selfish DNA."

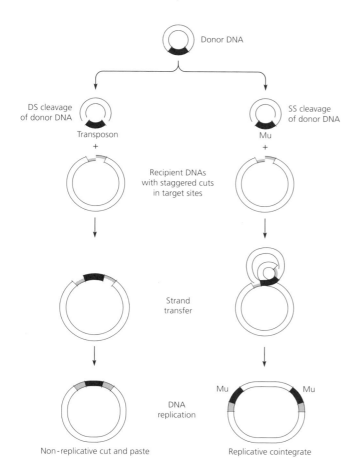

Phage Mu uses replicative, as opposed to simple cut and paste, transposition

Donor DNA

DS cleavage of donor DNA

Transposon
+

Recipient DNAs with staggered cuts in target sites

SS cleavage of donor DNA

Mu
+

Strand transfer

DNA replication

Non-replicative cut and paste

Mu   Mu

Replicative cointegrate

FIGURE 7–19.   While transposition can occur by either the cut and paste or replicative routes, phage Mu uses the latter. The two pathways are distinguished by the cleavage of the donor DNA: cut and paste cleaves both strands, but replicative transposition cleaves only one strand at each transposon end. Both types of transposition use a staggered cut in the target site of the recipient DNA, and, in both, the 3' ends of the inserted DNA (transposon or phage Mu) are joined to these cuts. Completion of the cut and paste transposition event requires synthesis of only a few bases, which duplicates the target site. In contrast, in replicative transposition, the entire transposon, as well as the target site, is replicated.

Although this is undoubtedly the case, transposons also change the organisms they inhabit for better or for worse. Transposons may have survived because of their value to organisms and species.

Because there are few barriers to transposition, genes carried by transposons can spread rapidly through a population. A prime example of this is the current medical crisis resulting from antibiotic-resistant bacteria. Many transposons have picked up genes that confer resistance to particular antibiotics. In fact, some transposons, called "multi-drug resistance elements," actually carry resistance genes to five or more antibiotics. Multi-drug resistance elements have traveled rapidly through bacterial populations and in some cases have made some bacteria resistant to all of the antibiotics currently used to treat a particular disease. Clearly, these transposons facilitate the survival of their bacterial hosts in a hostile environment. Their existence is responsible both for current efforts to limit antibiotic use and for the frantic efforts of drug companies to develop additional classes of antibacterial drugs.

Transposons are also a very important source of variability in organisms. Transposons can turn on or turn off expression of genes at the site of an insertion. In addition, when transposons move, they may cause genome rearrangements. Many of the bacterial mutations analyzed by early molecular biologists have turned out to be caused by transposons, indicating that insertion of transposons in genes is a very frequent cause of spontaneous mutations in bacteria. Even in humans, new occurrences of genetic diseases (such as hemophilia) in families without a history of this disease have recently been traced to new transposition events. The variability generated by transposons, although sometimes deleterious to a particular host, may allow the species to adapt to changing environments.

Finally, the variability caused by transposons has been harnessed to perform important functions. Humans and other animals owe their survival to their ability to generate an enormously diverse repertoire of antibody proteins. The immune system has the ability to produce 100 million different antibodies, each with a unique capability to recognize foreign compounds. Antibody diversity derives from a specialized recombination reaction that joins segments of antibody genes in different combinations. This recombination system is mechanistically very similar to transposition and is thought to have arisen from a transposon rather recently during vertebrate evolution.

FURTHER READING

Campbell, A. (1980) Some general questions about movable elements and their implications. Cold Spring Harbor Symp. Quant. Biol. 45, 1–9.

Cohen, S. N., and J. A. Shapiro (1980) Transposable genetic elements. Sci. Am. 242, 40–49.

Holliday, R. (1990) The history of the DNA heteroduplex. Bioessays 12, 133.

8 REGULATING THE REGULATORS:
DEVELOPMENTAL AND SALVATIONAL DECISIONS

For most nonbiologists, the process of animal development must be the premier example of biological regulatory mechanisms at work. In a visually dramatic, temporal tapestry, the single fertilized egg becomes an adult animal. Seeing this process unfold in our own children provides a spectacular and moving demonstration of the exquisite precision of a biological regulatory network of astonishing complexity. How does it all work? At present we lack a clear picture of this process; however, I believe that there is now enough insight to define some general principles. My view of the molecular basis of developmental biology is based on the belief that solutions to problems are used over and over again, from the simplest to the most complex organisms. From this vantage point, the principles of development in phage λ will give us ideas about how sequential gene expression and pathway choices in response to temporal and spatial cues are accomplished in all organisms.

Another complex biological problem is the coordinated cellular response of a set of genes to an external environmental challenge— I consider this to be a "salvational" response. For example, unusually high temperature provokes a "heat shock" response in the cells of all organisms; damage to DNA in bacteria induces a rescue response termed "SOS." As judged by work so far, remarkably similar molecular mechanisms are used to carry out developmental and salvational decisions. Multi-gene regulatory proteins ("master" regulators) control the large array of genes that execute salvational or developmental responses by regulating their transcription. In turn, the signaling system for the response itself acts to control the activity of these master regulators. Interestingly, the regulators themselves are almost always regulated post-transcriptionally. The amount of the regulator can be changed by altering its stability or how well its mRNA is translated. The activity of the regulator can be changed by modifying its amino acids (e.g., by adding phosphates) or by binding small-molecules or proteins to the regulator.

In this chapter, we will consider three complex biological responses of bacteria and phage that have been critical in our thinking about how developmental and salvational responses are regulated: the lysis-lysogeny decision by phage λ; the heat shock pathway; and the SOS response. The principles learned from each of these "model" systems are proving to be widely applicable to regulating similar responses in both prokaryotic and eukaryotic cells.

THE LYSIS-LYSOGENY DECISION BY PHAGE λ: cII AS THE MASTER REGULATOR

The life cycle of λ exhibits the general features of development found in more complex creatures, making it an attractive model system for the study of developmental regulation. After infection of a bacterial cell, the λ genome executes a choice between two temporal pathways: the lytic pathway (phage multiplication and cell lysis) and the lysogenic pathway (integration of a dormant prophage into the host chromosome). In a rudimentary sense, a bacterial cell infected with λ is analogous to a fertilized egg: it makes a molecular decision between available temporal pathways. Of course, the λ decision is two-dimensional, whereas the process of animal development is multi-dimensional, a succession of precisely defined choices between potentially available pathways.

How does λ choose between the lytic and lysogenic pathways? Dale Kaiser established the critical biological features of the λ lysogenic pathway in 1957. He identified the three genes governing lysogeny and provided a general understanding of their biological roles in this process (see Figure 3–9). One gene, called cI, was necessary for the cell to maintain lysogeny. The cI protein accomplishes this task by repressing transcription from the two major promoters expressed early in lytic development (see Figure 5–14). Two additional genes, called cII and cIII, were required only for establishing lysogeny; they must play a key role in the decision-making process.

Over the next 25 years, the molecular mechanisms operating this lytic-lysogenic switch were identified and studied.

By 1970, two rival ideas for the molecular role of the cII and cIII proteins had developed. Based on a rather complex genetic argument, Harvey Eisen suggested that cII and cIII were positive regulators that turned on synthesis of the cI protein. Based upon measuring transcription of various genes in *cII* and *cIII* mutant cells, I was claiming that cII and cIII worked as negative regulators to delay the expression of lytic genes expressed late during infection, allowing time for cI production and lysogeny to take over. In an unusual resolution to such disputes, both ideas turned out to be correct; cII and cIII positively regulate the *cI* gene and also negatively regulate the late lytic genes.

The experiments that established clearly how *cI* was regulated were direct measurements of cI protein production after infection, carried out by Lou Reichardt, a student with Dale Kaiser, and by Linda Green and me. The need for such an approach had been obvious for some time, but no assay was readily available to accomplish this task. Reichardt purified the cI protein, prepared specific antibodies, and then measured the amount of cI by its ability to form complexes with the antibody. We developed an assay for cI protein that was based on its ability to bind to operator DNA.

As for much of molecular biology, the cI measurements were a case in which it took years to develop the assay and only a few weeks to do the critical experiments. We were pleased that both assays gave the same answers. The first experiments showed that high cI production began only after a delay, indicating that its synthesis was regulated. The second set of experiments demonstrated that both *cII⁻* and *cIII⁻* mutants were deficient in producing cI. Since cI was not turned on in the mutants, cII and cIII probably positively regulated the *cI* gene (Figure 8–1). The site of action of these regulators was inferred from a regulatory mutant of λ (termed *cy⁻*) defined by René Thomas. The *cy⁻* mutant could not turn on cI production even when cII and cIII were supplied from another phage, suggesting that this mutant was defective in the site at which cII and cIII acted rather than in the production of these proteins. Thus, the operator site from which cII and cIII turned on transcription of *cI* was defined. The inferred promoter site for this positive regulation was termed p_{RE} for establishment of repression(Figure 8–2a).

The inference that cII and cIII turned on *cI* at promoter p_{RE} made sense, but could not be the complete story of *cI* transcription.

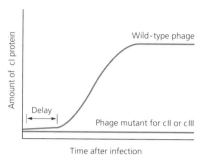

Production of cI protein is regulated

FIGURE 8–1. Following λ infection, the level of cI protein in the cell rises dramatically after a short delay. The delay period indicates that some event must occur to allow high cI synthesis. The inability of *cII⁻* and *cIII⁻* mutants to achieve a high rate of cI synthesis suggested that cII and cIII were positive regulators of cI synthesis.

FIGURE 8–2. Immediately after infection, the λ cII protein activates transcription from the p_{RE} promoter, allowing high level expression of the *cI* gene (*a*). In the lysogen, cI protein turns off almost all of the genes on the λ chromosome, including *cII*. Without cII protein to act at the p_{RE} promoter, cI protein does the job itself and activates low-level transcription of its own gene from the p_{RM} promoter (*b*).

a) After infection, cI is transcribed from p_{RE}

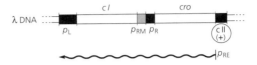

b) In the lysogen, cI is transcribed from p_{RM}

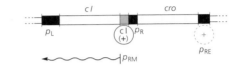

The job of the cI protein is to maintain repression for many, many generations of lysogenic cells. Lysogenic cells contain neither cII nor cIII because cI represses their transcription. So, *cI* must be transcribed even in the absence of cII and cIII. In principle, this "maintenance" transcription might have come from a low rate of transcription of p_{RE}. However, genetic experiments by Eisen and colleagues and the measurements by Reichardt of cI protein argued strongly that the maintenance synthesis of cI depended on a different promoter than p_{RE}; one that was positively regulated by the cI protein itself. Eventually Gary Gussin precisely defined the maintenance promoter, p_{RM}, by isolating point mutations that disrupted its activity (Figure 8–2b).

Because we were carrying out complementary approaches to the same problem, I suggested to Reichardt and Kaiser that we publish our papers together. They were happy to do so, and the results were a good deal for everyone. Our experiments were finished earlier, but Reichardt's were more quantitative because he had a more sensitive measurement of cI. We shared a definitive presentation, and other scientists learned the complete story all at once.

It gradually emerged from incidental observations that cII and cIII also positively regulate production of the Int protein, which is responsible for integrating λ into the bacterial genome. Work from Allan Campbell's lab showed that, in addition to being transcribed from p_L, int was also transcribed from a separate promoter, p_I. This provided a mechanism for specific regulation of Int expression. Marlene Belfort and Amos Oppenheim then found that efficient Int

FIGURE 8–3. Immediately after infection, the λ cII protein activates transcription from the p_I promoter, allowing the high level expression of Int protein required to carry out integration of λ DNA into the bacterial chromosome.

production depended on cII, and Steve Chung and I showed that cII and cIII stimulated Int expression without stimulating expression of the adjacent excision-specific Xis gene. Together, these experiments placed p_I between the *int* and *xis* genes (Figure 8–3). Efficient synthesis of Int coupled with low expression of Xis drives effective integrative recombination. The two key events in establishing lysogeny are repression of lytic growth and prophage integration. The cII and cIII proteins provide coordinate positive regulation of both events.

But how did cII and cIII negatively regulate the lytic genes? In later experiments, Barbara Hoopes and Will McClure as well as Yen Sen Ho and Marty Rosenberg found that cII and cIII used a very cute mechanism to carry out this function. There is a cII dependent promoter, p_{aQ}, located within the Q gene. Activated transcription from p_{aQ} is in a leftward direction. This transcript can pair with the complementary transcript from p_R, which is in the rightward direction. This "anti-sense" RNA covers the translation initiation signals for Q, decreasing the amount of Q that can be made from the p_R transcript. When cII is high, a great deal of the p_{aQ} transcript is made, leading to an under-supply of Q, the positive regulator of late gene transcription, thereby reducing the transcription of late lytic genes. Thus, cII and cIII provide an efficient bifunctional transition into the repressed state: they promote a burst of cI repressor and Int necessary for the prophage state and inhibit expression of genes required for lytic development.

By the late 1970's, the focus of work on the lysis vs. lysogeny decision turned to the biochemical functions of cII and cIII. In the meantime, new evidence and rethinking of old data had fingered cII as the central regulator, with cIII supplying a critical but ancillary role. Most dramatically, Belfort and Dan Wulff found that the cIII protein was completely dispensable for establishment of repression in some bacterial mutants, called *hfl* (for high *frequency* of *lysogenization*). We will soon see that these mutants were vital for our understanding of how cII itself was regulated.

The immediate biochemical goal became isolating the cII protein to analyze its mechanism of action. The best guess was that cII was a DNA-binding transcription activator. Using genetic engineering, Hiroaki Shimatake and Marty Rosenberg put the cloned *cII* gene under the control of the powerful λp_L promoter carried in a plasmid vector. They were able to make sufficient cII protein to allow purification without a biochemical assay; the successful fractionation of cII was carried out by visualization of stained proteins after acrylamide gel electrophoresis (the "band in a gel" purification described in Chapter 7 for the RecA recombination protein). Their engineering and purification efforts were rewarded—cII activated transcription by RNA polymerase from p_{RE} and p_I. This work was an early spectacular success of cloning technology in "overproducing" and isolating a protein. Within a short time, nearly all protein purification began with a plasmid overproduction system.

The next question was how cII worked. DNA sequence analysis carried out by several groups identified a closely similar sequence present in the p_{RE} and p_I cII-activated promoters, which could serve as the binding site for cII. Using DNase I footprinting (see Figure 5–29), Rosenberg and colleagues demonstrated that cII indeed bound to this operator site. Interestingly, the –35 region sequences in these promoters are predicted to be poor for binding RNA polymerase. Since the cII binding sequence overlaps this region of the promoter, protein-protein interactions between cII and RNA polymerase are likely to compensate for ineffective promoter sequences, thereby increasing the strength of the p_{RE} and p_I promoters.

The determination of the properties of cII protein clarified the role of cII in the lysogenic pathway; cII acts as a master transcriptional regulator to control transcription of three distinct genetic regions of λ DNA: the *cI*, *int*, and Q genes. When cII activity in an infected cell is high, cI and Int expression are high and Q expression is low; consequently the lysogenic pathway will ensue. When cII activity is low, the opposite situation obtains and the lytic pathway will ensue.

REGULATING THE REGULATOR: CONTROL OF cII ACTIVITY

The lysis-lysogeny switch is controlled by the amount and activity of the master regulator cII. What were the inputs that regulated the amount and activity of cII itself, so that the appropriate decision is made by the cell? This brings us to the role of the phage cIII and host Hfl proteins.

Chris Epp's observation that cII was an unstable protein, appar-

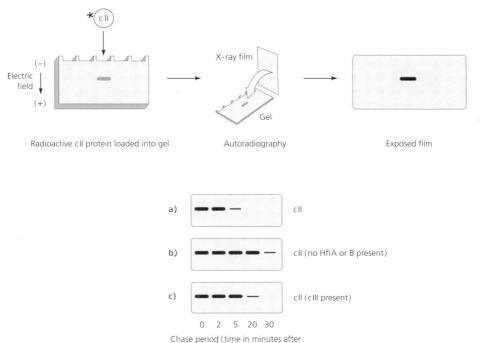

FIGURE 8–4. The stability of proteins can be determined by a pulse-chase protocol. A radioactive amino acid is added to bacterial cells for a short time (the pulse) to label proteins radioactively. A large amount of the nonradioactive amino acid added during the "chase" prevents additional labeling of proteins synthesized after the pulse, so that the stability of the previously synthesized molecules can be determined. At various times after labeling, bacterial proteins are extracted, separated on a polyacrylamide gel, and exposed to photographic film, which shows each protein as a blackened band, whose darkness is proportional to the amount of protein present. The cII protein band disappears during the chase period, showing that it is unstable in normal cells (a). In hfl mutant cells, the cII protein band disappears more slowly during the chase, indicating that cII is more stable. This suggests that the Hfl proteins function in cII degradation (b). When cIII is synthesized along with cII, the cII protein band disappears more slowly than when cII is synthesized alone, suggesting that the function of cIII is to stabilize cII (c).

ently subject to intracellular proteolytic degradation, provided an important clue to how cII was regulated. Might the stability of cII itself be subject to regulation? To examine cII stability, my student Andy Hoyt and I used a "pulse-chase" protocol. Following labeling of cellular proteins with a radioactive amino acid for a short time (the "pulse"), addition of a large amount of the nonradioactive amino acid stops labeling (the "chase"). The fate of a radioactive protein during the chase indicates whether or not it is stable. When a protein is stable, the amount of radioactive protein does not change during the chase. However, an unstable protein will be degraded during the chase, resulting in a decrease in the amount of the radioactive protein. To do these experiments, we expressed cII from a plasmid, performed the pulse-chase experiment, separated the proteins by gel electrophoresis, and visualized the radioactive proteins by autoradiography. We could follow cII protein because it was a unique band on the gel (Figure 8–4, top).

The experiments had an exciting ending—both *hfl* and *cIII* regulated cII stability. The cII protein was more stable in *hfl* mutant strains than in normal strains, indicating that the Hfl proteins regulate the lysis-lysogeny decision by specifying the turnover of cII protein (Figure 8–4; compare a and b). My student David Knight showed that cIII had the opposite effect on cII stability: when cIII was overexpressed along with cII, cII was more stable (Figure 8–4; compare a and c). To be sure that we were not being fooled by the plasmid system, Hoyt repeated these experiments using phage infection and got the same results. We now knew that a key element of the lysis-lysogeny decision was regulating the stability of the regulator; Hfl proteins enhance cII degradation to favor lysis, whereas cIII enhances cII stability to favor lysogeny.

The amount of cII is regulated in an additional way. Harvey Miller had found that mutations in the genes specifying IHF protein (required for integrating λ into the host genome) also affected the frequency of lysogenization. With Miller, we found that IHF increases synthesis of cII without affecting its stability. IHF is likely to affect *cII* translation, but the mechanism is not yet understood.

Although we are convinced that the Hfl proteins and IHF transduce environmental information into the lysis-lysogeny decision, we do not know the signaling mechanisms for these inputs. Hfl activity may be related to energy availability in an infected cell. In contrast, cIII is likely to monitor a completely different parameter. The action of cIII is most efficient when several phage infect a cell. So the role of cIII is most likely to swing the pendulum towards the lysogenic response when the phage population exceeds the number of available bacterial hosts—a birth control mechanism to allow the regrowth of the food source by temporarily preventing lytic infection. Clearly, several different inputs are required to correctly adjust the amount of cII so that λ can make an informed decision about whether to choose the lytic or lysogenic pathway (Figure 8–5). Given the complex control necessary for the simple λ switch to function well, we imagine that the major decision points in eukaryotic development will be controlled by a very large number of independent inputs.

MOLECULAR DECISION MAKING:
PRIMARY AND SECONDARY SWITCHES

From the foregoing discussion, we can see that the amount of cII mediates the molecular decision for lysis or lysogeny faced by phage λ. When cII is abundant, the cI repressor and Int are in large supply,

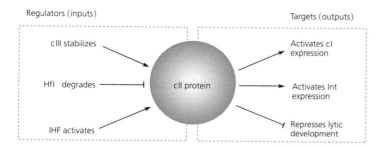

FIGURE 8–5.

lytic development will be inhibited, and the lysogenic pathway will be chosen; when cII is in short supply, the lytic pathway will prevail. Thus, cII can be considered to execute the primary switch between the two pathways of λ development. Although cII provides the initial partition between pathways, the choice of pathway is stabilized by two secondary regulatory switches that make the decision irreversible for the infected cell. One switch controls the activity of early promoters and the second controls the direction of site-specific recombination.

The first of these switches uses a competition between the cI and Cro repressors to regulate the closely spaced p_R and p_{RM} promoters. Three operator sites, o_{R1}, o_{R2}, and o_{R3}, which collectively constitute the right operator, control the activity of these promoters (Figure 8–6, top). Both cI and Cro bind to these sites but with distinctly different effects on the λ lifecycle: cI promotes and maintains lysogeny, whereas Cro is essential for the lytic response. What is the functional distinction between the two?

The critical differences between cI and Cro turned out to be twofold: their relative affinity for the three operator sites in the o_R region and the ability of cI, but not Cro, to stimulate transcription of the *cI* gene. These properties were determined mainly by a spectacular series of experiments by Sandy Johnson and Barbara Meyer working with Mark Ptashne. Johnson analyzed the binding of cI and Cro to their operator sites using the newly developed chemical protection and DNase I footprinting techniques. Meyer studied the effects of cI and Cro on transcription, using an isolated segment of λ DNA carrying the p_L and p_R promoters and the o_R operator region. Yoshi Takeda in my lab also noted the transcription switch between cI and Cro. The principal conclusions of this work are summarized in Figure 8–6.

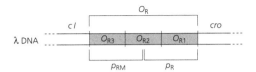

λ DNA

FIGURE 8–6.   The right operator (O_R) of λ contains three operator sites (O_{R1}, O_{R2}, O_{R3}) that control two promoters: p_{RM} (for *cI*), and p_R (for early lytic genes including *cro*). Both the cI and Cro λ repressors bind to all three operator sites but with different affinity. At relatively low concentrations, cI binds simultaneously to O_{R1} and O_{R2}, which activates transcription of *cI* from p_{RM} and shuts off transcription from p_R (*a*). At higher concentrations of cI protein, all three operator sites are filled, and transcription from both promoters is shut off (*b*). At relatively low concentrations, Cro binds only to O_{R3}, shutting off transcription of *cI* from p_{RM}, while allowing p_R transcription to continue (*c*). At higher levels of Cro, all three operator sites are filled, and transcription from both promoters is shut off (*d*).

a) Low cI concentration (cI transcribed, lytic genes repressed)

c) Low Cro concentration (cI repressed, lytic genes transcribed)

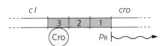

b) High cI concentration (cI and lytic genes repressed)

d) Medium to high Cro concentration (cI and lytic genes repressed)

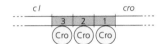

The cI protein binds most avidly to o_{R1} and o_{R2}, effectively occupying both sites simultaneously because two cI dimers associate on the DNA to give a "cooperative" binding interaction. This binding interaction shuts off the lytic response by repressing transcription from p_R, and simultaneously promotes lysogeny by stimulating cI production from the "maintenance" p_{RM} promoter (Figure 8–6a). Only when cI becomes excessively high in the cell does it occupy the o_{R3} site, preventing excessive transcription from p_{RM} (Figure 8–6b).

The Cro protein binds to the operator sites in reverse order from cI. Cro occupies o_{R3} most efficiently. This immediately blocks the lysogenic pathway by preventing cI production from p_{RM} while leaving transcription from p_R intact to promote the lytic pathway (Figure 8–6c). Only later, when much higher amounts of Cro have accumulated, does Cro occupy o_{R1} and o_{R2} to shut off transcription from p_R (Figure 8–6d). It turns out that this repression of the p_L and p_R transcription units is essential for the successful execution of the late stage of lytic development (for reasons that are still not very clear).

The ordered binding of cI and Cro to the right operator provides

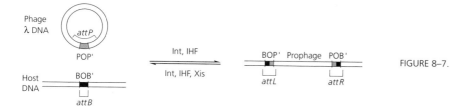

FIGURE 8–7.

a bidirectional switch that stabilizes the initial decision for lytic or lysogenic development. If cI is in relative excess, then the *cro* gene (which is transcribed from p_R) is kept off, and production of cI is self-regulated by cI protein. If Cro is in relative excess, then the *cI* gene is kept off, and Cro accumulates to the high concentrations needed for lytic regulation. So cI excess (produced by abundant cII) dictates a stable switch down the lysogenic pathway. Cro excess (a consequence of deficient cII) dictates a stable switch down the lytic pathway.

The second switch that stabilizes the choice of lytic or lysogenic pathway uses the relative ratio of Int to Xis to control the direction of site-specific recombination. Integrating the phage into the chromosome to establish the lysogenic pathway requires high amounts of Int relative to Xis, excising the prophage requires balanced Int and Xis synthesis, and lytic growth requires low amounts of Int to prevent integration. The regulatory problem is how to achieve the requisite amounts of Int and Xis for each developmental pathway (Figure 8–7).

We already knew how high Int concentrations were achieved. When cII is high, transcription from p_I produces Int far in excess over Xis, thereby favoring the integration reaction required for the lysogenic pathway. The problem was how to achieve the very low levels of Int required for lytic growth. Simply lowering cII concentration would not suffice, as Int should be made in high amounts from the p_L transcript once lytic growth is activated.

To prevent Int synthesis from p_L, λ uses a post-transcriptional control mechanism that Dan Schindler and I called "retroregulation" because it works from a regulatory site downstream from the gene that it regulates. Because λN protein prevents termination of the p_L transcript, it is very long, reading past many termination sites.

FIGURE 8–8.   The presence of λN protein allows the p_L transcript to pass through multiple transcriptional terminators so that it includes *sib*. The *sib* site promotes degradation of *int* mRNA, preventing expression of Int from p_L (*a*). In contrast, the p_I promoter is not anti-terminated by N. When the *int* gene is transcribed from p_I, the transcript terminates before reaching the *sib* regulatory site. Because *sib* is not present on the transcript, this Int mRNA molecule is not specifically degraded and Int protein is produced (*b*).

a) The presence of *sib* in the p_L transcript causes this mRNA to be partially degraded

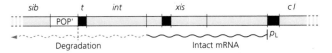

b) The p_I transcript does not include *sib*, and is not degraded

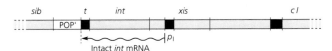

As a consequence, the transcript contains a regulatory site downstream of Int called *sib*, which specifically targets *int* RNA for degradation. So, Int is not made from the p_L transcript during lytic growth, and integration into the host chromosome is prevented (Figure 8–8a). In contrast, the p_I transcript, which is activated by cII, does not contain a binding site for the N antiterminator protein. Because this transcript terminates before the *sib* site, it is not degraded and the high levels of Int necessary for lysogeny are achieved (Figure 8–8b).

A cute feature of this regulatory mechanism follows from the fact that *sib* is located to the left of phage attachment site (POP′). Recombination at this site to insert λ into the chromosome separates *sib* from Int. Because *sib* is no longer near *int*, during prophage induction both Int and Xis will be made from the p_L transcript, providing the balanced production of Int and Xis necessary for excision from the bacterial chromosome.

In summary, phage λ executes a complex developmental pattern requiring the ordered expression of a large number of genes and a choice between alternative pathways. However, the general principles are rather simple. Each pathway is specified by the sequential synthesis and activity of a few regulatory proteins. The choice of pathway depends on the activity of a single master transcriptional regulator, which is in turn controlled by post-transcriptional signaling pathways. The decision for a given pathway is stabilized by subsequent secondary switching interactions. As far as we know, similar principles govern the pathway choices in all organisms.

In common with all other creatures, a bacterial species seeks to survive in a dangerous, hostile, and changeable environment. To accomplish this goal, the bacteria must be able to grow rapidly when plentiful food makes for good times, keep growing at some pace when times are poor, and survive when disaster strikes. As an evolutionary consequence of this lifestyle, our beloved *E. coli* devotes many of its genes to what I have termed salvational decisions—regulated responses to difficult environmental situations. Of the many examples of salvational decisions, I have selected two for discussion: survival at high temperature, and DNA repair and mutagenesis. These examples are notable for our level of understanding of molecular mechanisms, and are interesting for their diversity, general significance, and historical importance. Both of these crisis responses are currently under intensive study in prokaryotes and eukaryotes.

THE HEAT SHOCK RESPONSE AND MOLECULAR CHAPERONES

High temperature is a common environmental stress for all organisms and a potentially dangerous one because the proteins that make cells work are especially sensitive to damage from elevated temperature. All cells react to a heat shock with a coordinated, multigene regulatory response that produces elevated amounts of a class of proteins termed the heat shock proteins. Following the general theme for such responses, the genes for the heat shock proteins are regulated at the transcriptional level; however, the master transcriptional regulator of the response is, in turn, controlled by a variety of post-transcriptional mechanisms. Other environmental stresses besides heat also induce the heat shock proteins, so that the synthesis of these proteins appears to be a rather general response to cellular buffeting. For *E. coli*, ethyl alcohol is an efficient inducing agent; work in this research field might have attracted more notice if the term "intoxication response" had been used.

Many of the heat shock proteins function to refold or to facilitate the degradation of proteins damaged by the heat treatment—a molecular cleanup function. An exciting development was the realization that heat shock proteins exert a vital cellular function under normal conditions as well. The heat shock proteins associate with many newly synthesized proteins to promote their folding. They also facilitate the assembly of monomer units into active multimers and the disassembly of multimeric complexes. Because of their skills in directing polypeptide matings, these heat shock proteins have been termed "molecular chaperones."

HEAT SHOCK: FROM CELL BIOLOGY TO MOLECULAR BIOLOGY

The heat shock response was initially uncovered in 1962 in the fruit fly, Drosophila, a favorite eukaryotic organism for genetics and cell biology. The chromosomes in the salivary glands of Drosophila multiply many times and then form an ordered packed array in which chromosomal regions are visible in the microscope as banded structures. These banded structures undergo transient and developmentally ordered modifications, termed "puffing," which were considered likely to represent gene expression from that chromosomal region. Ferruccio Ritossa found that heat and certain chemical agents produced a series of characteristic chromosome puffs. The phenomenon was explored further and publicized considerably in the next few years by Michael Ashburner. As a new biologist, the first time I heard Ashburner talk about heat shock puffs, I had not the slightest idea what he was talking about, but I was convinced that the subject was important.

Heat shock became food for molecular biologists in 1973 when Alfred Tissières and Hershel Mitchell used gel electrophoresis to look at radioactively labeled Drosophila proteins under heat shock conditions. They learned that a small number of new proteins appeared, whereas production of most other proteins was repressed. This work engendered a burst of activity among molecular biologists because heat shock was clearly one of the few examples of inducible gene expression in an experimentally accessible eukaryote system. Experiments by Susan Lindquist in Matt Meselson's lab and by Allan Spradling, working with Sheldon Penman and Mary Lou Pardue, showed that heat shock RNA actually hybridized to the thermally induced puffs in the chromosome. So heat shock RNA is actually transcribed from the puffs that appear after heating; puffs are sites of active gene expression. Because of their abundant RNA production, heat shock genes were among the first of the eukaryotic genes to be cloned.

At this point, work on heat shock regulation in Drosophila slowed down because there was no way to analyze the regulatory circuits by genetics. Moreover, there was no clear insight into what the heat shock proteins did in the cell. Realizing this problem, Betty Craig, one of the first to clone a Drosophila heat shock gene,

Two-dimensional electrophoresis revealed that some proteins
are synthesized at higher rates after heat shock

a) Low temperature

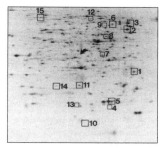

b) Heat shock

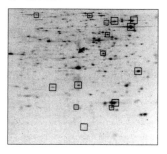

FIGURE 8–9. Bacterial cells were labeled for a short amount of time with a radioactive amino acid, either while growing at low temperature (28° Celsius) (*a*) or immediately after they were shifted to high temperature (42° C.) (*b*). Cellular proteins were subjected to electrophoresis in two different directions: the proteins were separated by size from top to bottom, and by electrical charge from left to right. Numbered boxes indicate the proteins whose synthesis rate increased when bacteria growing at 28° C. were shifted to 42° C. (Image of autoradiogram courtesy of F. Neidhardt from Ruth A. VanBogelen, Molly A. Acton, and Frederick C. Neidhardt, Induction of the heat shock regulon does not produce thermotolerance in *Escherichia coli*, *Genes and Development* 1 [1987].)

switched to an analysis of the role of heat shock proteins in yeast. At about the same time, regulation of gene expression by heat had been demonstrated in a host of other organisms, including *E. coli*.

REGULATION AND FUNCTION OF
HEAT SHOCK PROTEINS IN *E. COLI*

Beginning in 1978, Takashi Yura, Fred Neidhardt, and their associates carried out a series of important experiments that defined the heat shock regulatory response in *E. coli*. Since the native environment of *E. coli* is the human colon, the "normal" growth temperature of *E. coli* is 37°C. Neidhardt and Yura asked what happened to protein production when the bacterial population was suddenly shifted from 28°C to 42°C (high hot-tub temperature). Their results showed that *E. coli* knew about hot water—a class of proteins was induced.

To examine heat shock, the proteins in *E. coli* cells were briefly labeled with radioactive amino acids, either at low temperature or just after shift to high temperature, and then separated by gel electrophoresis. Neidhardt's analysis used a particularly sensitive, newly developed approach of two-dimensional gel electrophoresis. In this technique, worked out by Pat O'Farrell, proteins are separated by their content of charged amino acids in one dimension and by their size in a second dimension. As a result, a pattern of spots appears on the photographic film used to detect the radioactivity; each spot represents a protein (Figure 8–9). The darker the spot, the more highly expressed the protein. Because of the two-dimensional separation by two different properties of the proteins, a large fraction of *E. coli* proteins (almost 1,000 proteins) can be "seen" by such an analysis. After a shift from 28°C to 42°C, about 20 proteins gave darker spots, indicating that they were induced. Yura and colleagues provided strong evidence that the induced protein synthesis depended on induced transcription.

The heat shock response was clearly a major regulatory endeavor of *E. coli*. How did the regulation work? At this point, experiments originally directed toward other questions came to the rescue. If heat shock involved a positive regulator, mutants that inactivate this gene should be unable to produce heat shock proteins after shift to high temperature. Yura and Neidhardt independently uncovered the positive regulator by analyzing a mutation that caused death at high temperature, previously isolated by Stephen Cooper. Bacteria carrying this mutation failed to produce several proteins at elevated temperature. Upon closer analysis, Neidhardt and Ruth Van Bogelen as well as Yura and Tetsuo Yamamori found that the mutant strain failed to produce the entire class of heat shock proteins at high temperature. The regulatory gene was termed *htpR* by one group and *hin* by the other. So HtpR was probably a multi-gene regulator turning on transcription at the promoters for heat shock proteins.

In the meantime, through devoted study of phage λ, information began coming in about the function of some heat shock proteins. Costa Georgopoulos was λ's special scientific representative. Starting as a graduate student at MIT in 1970, Georgopoulos had studied mutants of *E. coli* that prevented the lytic growth of phage λ. He then identified mutations in phage genes that would overcome the host defect (presumably by a protein-protein interaction between an altered λ protein and the damaged host protein) (see Figure 4–26). His genetic abilities and imaginative insights provided two most valuable developments: identification of a number of important *E. coli* genes; and a major general approach to studying protein-protein interactions. Among the major classes of bacterial mutants preventing growth of λ were those defective in replicating λ (*dnaK*, *dnaJ* or *grpE*) and those defective in assembling its head structure (*groEL*, *groES*).

The connection between the host proteins that helped λ grow and heat shock proteins began to emerge in the early 1980's. Roger Hendrix noted that his purified GroEL protein appeared to be identical to the most abundant heat shock protein, a conclusion established in detail by Neidhardt. Georgopoulos, Hendrix, and asso-

Costa Georgopoulos

ciates determined that DnaK was the second most prominent heat shock protein. Further experiments established that DnaJ, GrpE, and GroES were also heat shock proteins. Then Jim Bardwell, working with Betty Craig, showed that DnaK was extremely similar to the major group of eukaryotic heat shock proteins, the Hsp70 class (70 refers to the molecular mass of the protein, 70,000). He followed up this work by showing that the major Hsp90 family of eukaryotic heat shock proteins also had a bacterial homologue.

The extraordinary degree of homology between the prokaryotic and eukaryotic heat shock proteins implies an exceptional conservation of some vital function over millions of years of evolutionary separation. But what was that function? The impetus for Craig's search for prokaryotic counterparts to the major eukaryotic heat shock proteins had been a search for this elusive function. But, *E. coli* proved disappointing in this regard. Mutations in *dnaK* had so many consequences that it was hard to figure out the primary function of the protein. In retrospect, this confusion was not surprising. The primary role of the DnaK chaperone is to help proteins fold; defective folding led to all sorts of apparently unrelated deficiencies in *dnaK* mutant cells.

Although the function of heat shock proteins in λ growth did not specify their activity in *E. coli*, the correlation established two important points. First, the heat shock proteins were probably vitally important for some normal *E. coli* activities. (Otherwise, why were nearly all host proteins important for λ growth found in the heat shock protein class?) Second, based on their roles in λ development, one general function could be ascribed to heat shock proteins: these *E. coli* heat shock proteins mediate protein assembly (GroEL/ GroES) and disassembly reactions (DnaJ/DnaK/GrpE) of multiprotein complexes.

REGULATION OF THE HEAT SHOCK RESPONSE: NEW PROMOTER RECOGNITION BY RNA POLYMERASE

The identification of the *htpR* gene had pinpointed a regulator of the heat shock genes. But, how was the coordinate induction of heat shock genes achieved? Our understanding of the biochemistry of bacterial heat shock regulation has been mainly dependent on the efforts of Carol Gross and her colleagues. Gross began her independent academic career at Wisconsin in 1981 and almost immediately catapulted into a spectacular (and initially unintended) series of experiments on heat shock. She combined an exceptional ability for physiological insights with rigorous biochemistry. Gross' initial goal as a new faculty member was to follow up her postdoctoral work on regulation of the sigma subunit of RNA polymerase (now called σ^{70} for its molecular mass of 70,000). It is this subunit that guides RNA polymerase to its promoter sites on the DNA. With Hope Liebke, Gross set out to study a temperature-sensitive mutant in the gene coding for σ^{70}. Liebke and Gross found that one class of proteins was still produced despite heat inactivation of σ^{70}—the heat shock proteins. In order to understand how the heat shock genes could still be transcribed when σ^{70} was inactive, Gross and Alan Grossman set out to purify HtpR, the putative high-temperature regulator. By looking for RNA synthesis from a heat shock gene in a crude extract of *E. coli* cells that contained a plasmid overproducing HtpR, Grossman succeeded in purifying the protein responsible for the heat shock activity. He then demonstrated that the heat shock activator was the product of the *htpR* gene.

How did the heat shock activator work? *E. coli* was thought to have only one form of RNA polymerase, with one subunit, sigma, that allowed it to initiate transcription specifically at promoter sites (see Chapter 5). Other known positive regulators, such as cII, worked by enhancing the activity of RNA polymerase. However, since the HtpR protein evidently functioned in cells in which the standard sigma subunit of RNA polymerase failed to work, Grossman and Gross suspected that the protein might be a new sigma subunit, directing RNA polymerase to heat shock promoters rather than standard promoters. This inference nicely dovetailed with data on amino acid homology from the DNA sequence of the *htpR* gene, work done originally by Bob Landick in the course of sequencing a neighboring genetic region. The amino acid sequence of HtpR protein was highly similar to that of σ^{70}, except that HtpR was much smaller, with a molecular mass of 32,000. With additional experiments, the HtpR protein was defined as σ^{32}, and the gene was renamed *rpoH*. Work from the Gross group, in collaboration with Craig and Hendrix, defined a special promoter sequence for heat shock genes, providing a clear picture of transcriptional regulation of heat shock genes. The master regulator is a special sigma factor, directing the core transcription apparatus to the promoters of the heat shock genes so that their expression could be coordinately regulated (Figure 8–10).

The existence of alternative sigmas was first demonstrated by

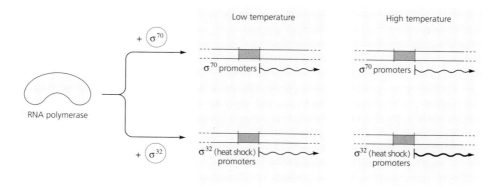

Rich Losick, who found that they controlled the development of *Bacillus subtilis* into a spore form able to withstand harsh environmental conditions. Using an alternate sigma factor as a master regulator is one way in which a large number of operons can be coordinately regulated. We now know that almost all bacteria have more than one sigma factor to help cope with the complexities of life under different environmental conditions.

FIGURE 8–10.   Bacterial RNA polymerase requires a sigma (σ) subunit to recognize specific promoters. σ^{70} specifies transcription of most of the genes that are active during normal growth in *E. coli*. However, the promoters of genes responsive to heat shock are recognized by RNA polymerase with a different sigma factor, called σ^{32}—the heat shock sigma. After shift to high temperature, heat shock promoters are more highly transcribed.

REGULATION OF THE REGULATOR: POST-TRANSCRIPTIONAL CONTROL OF σ^{32}

Although finding σ^{32} and the heat shock promoters was big news, this only pushed the regulatory problem one step backward: what controlled the activity of σ^{32} RNA polymerase? The early work of Yura and Neidhardt had defined two components of the heat shock response: a rapid induction of heat shock proteins, followed by repression. Gross and David Straus measured the amount of σ^{32} present in cells during the heat shock response. They found that synthesis of heat shock proteins correlated precisely with the level of σ^{32}. Therefore the amount of σ^{32} determined both the induction and repression phases of the heat shock response: σ^{32} itself was the master transcriptional regulator (Figure 8–11).

The regulatory question then shifted to finding the source of the rapid changes in the amount of intracellular σ^{32}. Straus and Gross found that the master regulator for heat shock was controlled by regulating its synthesis and stability, just as was true for λcII. Immediately after temperature upshift, both the stability and the

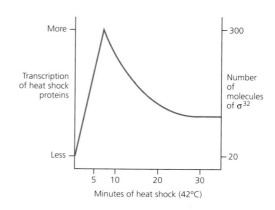

FIGURE 8–11.   When bacteria experience high temperature, transcription of the heat shock genes is increased to help the cell survive the stress. Gross discovered that the number of molecules of σ^{32} present in the cell rose in exactly the same way as the transcription of the heat shock genes did. This finding revealed that the production of heat shock proteins is directly tied to the amount of σ^{32} that is present in the cell.

translation of σ^{32} mRNA abruptly increased. Together, these accounted for the rapid increase in the amount of σ^{32}. At later times, σ^{32} became unstable again, even more unstable than at low temperature, accounting for the repression phase.

There is even some interplay between the regulatory circuit controlling λ development and the one controlling the heat shock response. Hubert Bahl and I found that λcIII provoked the heat shock response. This occurs because cIII stabilized σ^{32}, just as it stabilized cII. Very recent work from the groups of Richard D'Ari, Bernd Bukau and Teru Ogura has pushed these two regulatory systems even closer. Both master regulators, σ^{32} and cII, are degraded primarily by the same protease, HflB.

What causes the changes in σ^{32} synthesis and stability? Georgopoulos and Kit Tilly provided one important regulatory insight with the observation that *dnaK* mutants have a high basal level of heat shock protein synthesis and are defective in repression after temperature upshift. Straus and Gross subsequently found that these mutant cells are defective in degrading σ^{32}. They also found another mode of regulation. When there is too much σ^{32} in the cell (as occurs when cells are abruptly shifted to low temperature), σ^{32} is inactivated. The heat shock protein, DnaK, along with its associated proteins DnaJ and GrpE, is also essential for inactivation of σ^{32}. They proposed that these three heat shock proteins modulate the heat shock circuit through their ability to control the synthesis, turnover, and activity of σ^{32}. Recently, Bukau, Georgopoulos, and their collaborators showed that DnaK and DnaJ bind to σ^{32}. How the binding of these proteins to σ^{32} affects its function remains to be determined.

THE HEAT SHOCK RESPONSE IN EUKARYOTIC CELLS

The heat shock response in eukaryotic cells provided a most successful beginning for another independent research career, that of Hugh Pelham of the MRC Lab in Cambridge, England, in 1981. Like others before him, Pelham was intrigued with the heat shock genes because they defined a eukaryotic regulatory system. At that time, heat shock regulation was hard to study in depth in Drosophila, but Pelham reasoned that a universal response should have common regulation in eukaryotic cells. And, swapping genes among organisms was now possible for animal cells using genetic engineering. Pelham placed the DNA coding for a Drosophila heat shock gene on a plasmid vector and introduced it into animal cells. The genes were inducible by high temperature, indicating that they were regulated correctly, even in the foreign host. He then analyzed the regulatory region upstream of the gene and identified a short region of DNA that was necessary for thermal induction. Pelham was able to compare his defined sequence to the sequences upstream of other heat shock genes and pick out a likely regulatory element (termed HSE, for heat shock element). He and Mariann Bienz then carried out a beautifully definitive experiment to show that HSE was indeed the regulatory sequence. They introduced a chemically synthesized HSE upstream from a gene not normally subject to heat shock control and showed that it was now subject to heat shock regulation.

In 1984, Carl Wu and Carl Parker identified a protein that bound to the HSE site. The Drosophila heat shock transcription factor (HSF) was finally purified in 1987 by Wu. At the same time, Peter Sorger, working with Pelham, purified HSF from yeast. As was the case for prokaryotes, heat shock genes in eukaryotic cells are controlled by a dedicated transcription factor. HSF binds to the heat shock element in the promoter and then activates transcription. As expected for a salvational response, HSF activity is clearly controlled post-transcriptionally. HSF is present in the same amounts both before and after heat shock, and the mechanism by which it is activated is not yet clear. As in prokaryotic cells, part of the control mechanism may be a feedback regulation model in which the amount of Hsp70 (possibly along with other heat shock proteins) is a critical determinant for specifying the expression of heat shock genes.

From an early period in which heat shock proteins had no known function, we have now progressed to a state in which some people think that heat shock proteins do everything (at least everything involved in protein folding and assembly). Much of the switch in perception comes from Betty Craig's work. She showed that the lowly yeast has at least ten Hsp70 genes and that some of them function in each compartment of the eukaryotic cell. Mutations in many of these genes lead to defects in protein folding and in the translocation of proteins across membranes. Moreover, the functions of these genes are required during normal cell growth as well as after heat shock. Work from the Pelham laboratory (and others) showed that BiP, an Hsp70 homologue in higher eukaryotic cells, participates in assembly of the polypeptide chains of antibodies. Work from the Lindquist laboratory showed that another class of heat shock proteins, represented by Hsp104 in yeast, is required for the disaggregation of certain large protein complexes after heat shock. Hsp60 (a homologue of GroEL) participates in transport and folding of proteins in subcellular organelles—the chloroplasts of plants and mitochondria of yeast. Jim Rothman and Bill Welch have suggested that the molecular chaperones participate in general in the folding of newly synthesized polypeptides; this spectacular postulated role has been demonstrated by studying prokaryotic cells lacking selected classes of heat shock proteins, and it is likely to be true in eukaryotic cells as well.

REGULATED DNA REPAIR AND MUTAGENESIS

The result of excessive high energy radiation for higher organisms is well known: death, and elevated cancer rates among survivors. These consequences arise primarily because X-ray and other energetic radiation damage DNA. The resultant DNA damage interferes with DNA replication and introduces mutations (facilitating the production of malignant cells). Because radiation and other agents for DNA damage are normal environmental hazards, all organisms depend on DNA repair mechanisms for their survival. These repair mechanisms are sufficient to allow cells and organisms to cope with the damage under most conditions. For *E. coli* at least, DNA repair and increased mutation rate ("mutagenesis") involve an induced multi-gene response termed SOS. The nature of DNA repair mechanisms and their regulation constitute a fascinating intertwined story, which I will sketch in a regretfully abbreviated version.

Among many contributors to this field of research, Evelyn Witkin stands out especially because her 40 years of devotion provided the intellectual boldness, enthusiasm, and continuity that kept the field moving. Arthur Kornberg developed and dominated DNA replication. Witkin nurtured DNA repair and mutagenesis, combining original insights with enthusiastic cheerleading for the field as a whole. I once termed Witkin the "fairy godmother" of this research area (to her considerable embarrassment). Whenever my scientific or personal life seemed to be in a state of total disarray, by some miracle Witkin would be on the phone telling me that everything was wonderful and exciting and that there was all of this great work to be

Evelyn Witkin

done. Many others in this field also benefited from her personal concern and scientific inspiration.

The profound effect of X-rays on the genetic material was initially demonstrated in the 1920's. Herman Muller used X-rays to produce mutations in fruit flies, and Lewis Stadler did the same in barley. (Remarkably, X-rays of feet were still used casually to fit shoes for humans when I was a child in the 1940's.) For microorganisms, UV-irradiation was the major agent used to induce radiation damage, primarily because the origin of the lethal and mutagenic effects of UV were understood—they clearly resulted from absorption of photons by the DNA.

Initially the consequences of radiation damage were thought to be instantaneous and irreversible. However, two discoveries in 1949 at the Cold Spring Harbor Laboratory showed that the lethal effects of UV light on bacteria could be partially reversed. Albert Kelner found that much of the otherwise lethal UV damage could be reversed by visible light ("photoreactivation"); Richard Roberts and Elaine Aldous discovered that incubation of irradiated bacteria under nongrowing conditions produced a similar salvation ("liquid holding recovery"). Although these observations preceded the general acceptance of DNA as the genetic material, they provided the first inklings that genetic damage could be repaired. Working as a postdoctoral scientist at Cold Spring Harbor, Witkin bolstered the notion of repair by the observation that a class of UV-induced mutations was prevented if growth in the dark was slowed (presumably allowing time for repair to occur before the mutation was "fixed," or made permanent, by DNA replication). These and other observations led her to speculate on the existence of a "dark repair" mechanism complementary to photorepair by visible light.

THE UVR PATHWAY OF EXCISION REPAIR

The 1960's saw remarkable progress in understanding the DNA damage induced by UV and its repair. The first catalytic event was the identification of the major lesion in DNA induced by UV: a stable chemical linkage between two adjacent pyrimidine bases ("pyrimidine dimer"). Initially noted in frozen solutions of thymine, the pyrimidine dimer was found in isolated DNA by Adolph Wacker and in the *E. coli* genome by Dick Setlow and his colleagues. The second crucial conceptual advance was the intro-

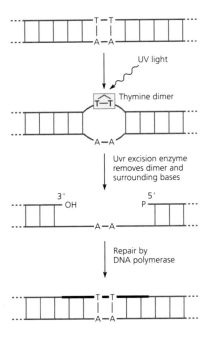

Thymine dimers can be
removed by excision repair

UV light

Thymine dimer

Uvr excision enzyme
removes dimer and
surrounding bases

3'

OH

5'

P

Repair by
DNA polymerase

FIGURE 8–12. UV light can cause a stable chemical linkage to form between adjacent thymine bases in DNA. These "thymine dimers" can block DNA replication, so repairing them is a priority of the cell. Repair is carried out by the UVR "excision-repair" pathway: the dimer is removed by the Uvr enzyme, and the gap is filled in by DNA Polymerase and closed by DNA ligase.

duction of genetic analysis to define clearly the repair and muta-genic pathways and to allow the eventual association of a pathway with an enzyme. Ruth Hill pioneered this effort by isolating bacterial mutants that were highly sensitive to UV.

These two approaches achieved their initial triumph in 1964 when Setlow and William Carrier found that thymine dimers were cut out of DNA during repair and that dimer excision failed to occur in some of Hill's mutants. But these mutants could not be examined genetically. An important advance was the isolation of mutants unable to repair UV damage in the standard strain of *E. coli* used for genetic analysis (K-12) by Dick Boyce and Paul Howard-Flanders. These mutants were isolated by a clever trick. When a lytic bacteriophage (like phage T7) is irradiated with UV and then used to infect the cell, *E. coli* repairs much of the damage to the phage. As a consequence, the phage grow and lyse the bacterial cells. Boyce and Howard-Flanders reasoned that the surviving *E. coli* cells might have mutations in the repair genes. They used this selection to isolate the *uvrA*, *B*, and *C* genes. The demonstration that mutants in defined genes behaved similarly to Hill's mutants opened the door to biochemical analysis. One major component of DNA repair was now clearly defined—by excising T-T dimers, the UVR pathway of excision repair saves about 99% of UV-irradiated *E. coli* from death. After the dimers were removed, a special type of repair DNA synthesis presumably filled the gap. This mode of DNA replication was demonstrated by Phil Hanawalt and his associates. The repair polymerase is DNA polymerase I, the first polymerase discovered by Kornberg (Figure 8–12).

RecA-MEDIATED RECOMBINATIONAL REPAIR

Although excision and photorepair are crucial pathways for survival from DNA damage, evidence quickly accumulated for additional pathways for repair, which were dependent on the RecA protein. John Clark found that his *recA⁻* mutants were as sensitive to the lethal effects of UV light as the *uvr⁻* strains. Howard-Flanders then showed that a *recA⁻ uvr⁻* double mutant was extraordinarily sensitive to UV-killing, much more sensitive than either of the single mutants. This was unexpected if the *recA* and *uvr* mutations inactivated the same pathway. For example, a strain with mutations in two different *uvr⁻* genes was no more sensitive to UV irradiation than either single mutant strain. Because the effect of the *uvr⁻* and *recA⁻* mutations was additive, Howard-Flanders concluded that Uvr and RecA repair occurred by different pathways. Whereas wild-type cells can survive 1,000 or so UV lesions, it is likely that a single pyrimidine dimer is lethal in the *recA⁻ uvr⁻* strain. These profound conclusions were derived from a very simple experiment. Bacteria were irradiated with UV light at different doses and then spread on agar dishes to count survivors (Figure 8–13).

Because RecA is required for genetic recombination (see Figure 7–5), the effect of *recA⁻* mutations on survival was initially thought to derive solely from recombinational repair of the DNA damage caused by UV. In support of this idea, Dean Rupp and Howard-Flanders showed that pyrimidine dimers served as blocks to DNA replication, leaving behind gaps in the replicated DNA that could then be repaired by recombining with successfully replicated complementary strands (Figure 8–14).

We now know that recombinational repair is a major function of the RecA protein. Even without UV irradiation, exposure of cells to the free radicals made from the reactive oxygen species in our environment results in tremendous DNA damage. It is estimated that up to half of all replicating DNA forks are stopped at some point by such damage, resulting in either single-strand or double-strand breaks. Most of these arrested forks are restored by recombinational repair; indeed, recombination is the only way to repair double-strand breaks. Evidence for the key role of RecA in repairing oxidative damage to DNA comes from the finding that, under standard laboratory conditions, *recA⁻* cells grow slowly and only about half of cells in a culture can form colonies. When *recA⁻* cells are grown in the absence of oxygen, their growth rate improves and almost all of the cells form colonies. After the damage at stalled replication forks is repaired by RecA-mediated recombination, replication must be restarted at these forks. Interestingly, the proteins utilized by phage ϕX174 to create the priming site for DNA synthesis (see Figure 4–17) are used to restart replication following recombinational DNA repair.

It turns out that RecA has a second major role to play in combating DNA damage. Remarkably, this role depends on yet another biochemical activity of the RecA protein. This story is described in the next section.

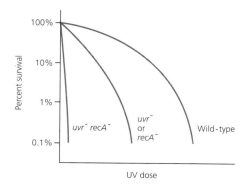

uvr and *recA* mutants have a synergystic effect upon the cell's ability to survive UV light

FIGURE 8–13.   Although most wild-type cells can survive irradiation by UV light, bacteria with mutations in either the *uvr* or *recA* genes do not fare as well. A mutation in either of these genes increases sensitivity to UV irradiation because cells have reduced ability to repair DNA. When a bacterial strain carries both mutated genes, then it cannot tolerate even a slight exposure to UV light. Because the double mutant strain is more sensitive than either single mutant, each mutation must inactivate a different pathway that repairs UV damage to DNA.

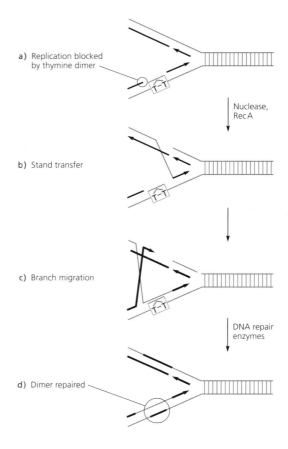

a) Replication blocked by thymine dimer

Nuclease, RecA

b) Stand transfer

c) Branch migration

DNA repair enzymes

d) Dimer repaired

FIGURE 8–14. The DNA replication apparatus cannot replicate past a thymine dimer (a). In order to repair this damage, a parental DNA strand is nicked by a nuclease, and the strand is transferred by RecA to the site of damage (b). Branch migration (also catalyzed by RecA) covers the area of damage and brings the interrupted newly synthesized strand up to take the place of the missing parental strand (c). Various DNA repair enzymes, along with DNA polymerase and ligase, excise the thymine dimer and repair the gap (d).

THE SOS RESPONSE TO DNA DAMAGE

The response of the cell to DNA damage turns out to be vastly more complicated than suspected, and RecA is the central player in this multifaceted response. The critical insights into the nature of this process came from studies of three of the "side effects" induced by UV damage: cell filamentation, prophage induction, and UV-induced mutagenesis.

Damaged bacteria do not continue to divide into new cells but form elongated cells or "filaments." Eventually, normal cell division resumes in surviving bacteria. The temporary blockage of cell division allows time for the recovery of DNA replication, so that successful genome duplication will precede cell division and the normal complement of DNA in the cell will be preserved. A second

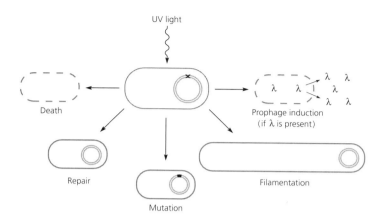

FIGURE 8–15.

consequence of DNA damage is that λ prophage is induced to lytic growth, allowing the phage to divorce itself from host cells potentially doomed by DNA damage. Finally, DNA damage enhances mutation, possibly as a last gasp effort of the cells to survive. The source of these mutations was an enigma for quite a while. Studies on the *uvr* mutants by Hill, Witkin, and Bryn Bridges had shown that excision repair was error-free. Witkin and Bridges proposed a mutation-inducing "error-prone" type of repair replication past the lesion site, and this later turned out to be the case. Interestingly, Witkin found that *recA* mutants fail to exhibit UV-induced mutagenesis, indicating that RecA is involved in this process. So there were four related cellular phenomena that required understanding: DNA repair, mutagenesis, filamentation, and prophage induction (Figure 8–15).

Understanding the physiology and ultimately much of the biochemistry of these puzzling observations came from studies that showed all of these phenomena to be manifestations of the same underlying regulatory network. A key insight connecting these phenomena came from Witkin's discovery in 1967 that the same physiological conditions that induced prophage also caused filamentation. Since prophage induction resulted from inactivation of the λcI repressor protein, Witkin suggested that filamentation might result from inactivating a repressor of a cell division inhibitor. Immediately, there was independent evidence to confirm her hypothesis that these two processes were regulated together. François

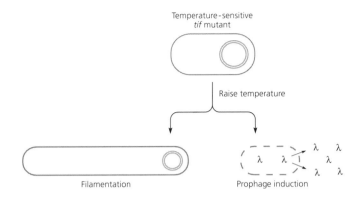

FIGURE 8–16.   Upon exposure to high temperature, the temperature-sensitive mutation *tif* was shown to both cause filamentous growth and induce lytic growth (if a λ prophage was present). This demonstrated that these two different outcomes were somehow linked.

Miro Radman

Jacob had isolated a bacterial mutant that induced λ at high temperature, and David Goldthwait showed that filament formation was also induced at high temperature in the mutant (which he named *tif*, for temperature induced filamentation) (Figure 8–16).

In 1971, Miro Radman further generalized this idea, first in a paper he co-authored with Martin Defais and others, and then in a separate "working paper." Radman, then a postdoctoral fellow at Harvard, had been working on a phenomenon called "Weigle mutagenesis," which refers to the fact that UV irradiation of λ is mutagenic to the phage only when they are grown in host cells that were themselves irradiated with UV. He and his coworkers had found that, when cells are mutant for either the *recA* or *lexA* genes, Weigle mutagenesis is prevented. Radman was struck by the realization that mutations in these two genes also prevent prophage induction. He suggested the critical concept that the *recA* and *lexA* genes regulated both prophage induction and Weigle mutagenesis. At the same time, he hypothesized that these genes might also control UV-induced mutagenesis of bacteria, since Witkin had earlier shown this process to be *recA* and *lexA* dependent. Radman termed this mutagenic rescue activity "SOS replication," in honor of its salvational purpose. Radman is a born storyteller, a man who can spin out a tale that encompasses all of biology from a single experiment. However, even he did not yet realize the profound biological importance of the SOS idea.

Radman's proposal soon received support from further studies on the *tif* mutation. In addition to inducing prophage and filamentation at high temperature, Marc Castellazi, Jacqueline George, and Gerard

Buttin now found that *tif* cells exhibit Weigle mutagenesis. Even without irradiation, at high temperature *tif* cells were able to promote mutations of irradiated λ phage. Then, Witkin found that *tif* induces UV mutagenesis at high temperature. It seemed that many of the functions induced by UV were expressed in *tif* cells at high temperature. The *tif* mutation was suspiciously close to *recA* on the *E. coli* genome, and eventually Lorraine Gudas and David Mount and, independently, Kevin McEntee showed that *tif* was in the *recA* gene, producing an altered form of RecA.

Radman's ideas were not immediately embraced by others. However, after some initial reservations, Witkin became an enthusiastic supporter. In papers published in 1973 and 1974, Witkin and Radman expanded the list of putative UV-inducible functions whose regulated expression depended upon *recA* and *lexA*. These functions were designated collectively "the SOS response." The concept of a unitary response to UV synthesized complex physiology into a simple and elegant molecular model, always a joyous event for a molecular biologist. But included in the idea is also the general notion of a salvational response—the coordinate control of a variety of biochemically disparate events that together respond to a special lifestyle crisis (Figure 8–17).

If the SOS response resulted from inactivating a cellular repressor, just as λ prophage induction resulted from inactivating the cI repressor, then which gene coded for the repressor? This gene could not be *recA*, because the genetics were wrong. Inactivating the mythical repressor gene should make SOS constitutive—expressed all of the time. But, inactivating RecA had the opposite phenotype: *recA⁻* mutants were unable to turn on the SOS response at all. Only the special "activated" form of RecA, present in the *tif* mutant strain, was constitutive for SOS.

Maybe LexA was the putative repressor. If so, the existing *lexA⁻* mutations also had the wrong phenotype. They, too, blocked the response, rather than making it constitutive. In the *lac* system, some rare mutations in the *lac* repressor gene had a dominant, noninducible phenotype. Such mutant repressors had lost the ability to bind the inducer. Perhaps the existing *lexA* mutants were of the same type. This turned out to be the case. Mount did the experiment and found that the *lexA* mutations defective in SOS functions were dominant to the *lexA⁺* gene in diploid cells. But, why had recessive mutations in *lexA* that would make cells constitutive for the SOS response not been identified? Mount reasoned that such

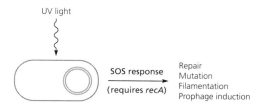

All cellular responses to UV were united into a single pathway by the SOS response

UV light

SOS response
(requires *recA*)

Repair
Mutation
Filamentation
Prophage induction

FIGURE 8–17. Miro Radman suggested that all of the various effects of UV irradiation were the result of a single, multi-gene pathway that was responding to stress in the cell's environment. Later work confirmed this hypothesis and showed that the *recA* gene was a key player in SOS.

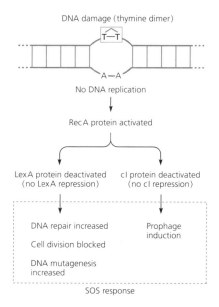

The SOS response is turned on when
RecA inactivates LexA and cI

DNA damage (thymine dimer)

No DNA replication

RecA protein activated

LexA protein deactivated
(no LexA repression)

cI protein deactivated
(no cI repression)

DNA repair increased

Cell division blocked

DNA mutagenesis
increased

Prophage
induction

SOS response

FIGURE 8–18. When the RecA protein is activated in
response to signals that induce the SOS pathway (such as
DNA damage), it disables LexA, the repressor of the SOS
response. This allows the functions of the SOS pathway to be
expressed. RecA also deactivates the λcI repressor, which in
turn leads to prophage induction.

mutations might be lethal because of excessive filamentation.
Bolstered by this insight, Mount looked for *lexA⁻* mutants in a strain
where filamentation is blocked by mutation. He proceeded to isolate the anticipated *lexA* mutants that were constitutive for SOS.

The SOS regulatory pathway was now defined. LexA is a negative regulator, presumably the SOS repressor, and prevents transcription of the genes in the SOS pathway. RecA is a positive regulator
somehow involved in turning on the response. The emerging story
can be followed in Figure 8–18.

THE BIOCHEMISTRY OF SOS REGULATION:
INDUCTION BY REPRESSOR CLEAVAGE

The initial biochemical insight into the mechanism of SOS induction came from studies by Jeff and Christine Roberts addressing the
fate of the λcI repressor in SOS induced cells. In 1975, they made
the startling observation that prophage induction led to the cleavage
of λcI repressor into two pieces. Although the cleavage might have
been a secondary consequence of inactivating cI, this work introduced the intriguing possibility that proteolysis of cI might itself be
the induction mechanism. Within a few years, the biochemistry of
the proteolytic event was worked out. Nancy Craig and Jeff Roberts
found that the RecA protein mediated the cleavage of cI. This reaction was dependent on two "cofactors": a polynucleotide such as
single-strand DNA and ATP or a related compound (Figure 8–19).
In parallel experiments on recombination described earlier,
Lehman, Radding, and their associates found that RecA bound to
single-strand DNA in the presence of ATP would catalyze a DNA-
pairing and strand transfer reaction with a double-strand DNA substrate. So RecA could be activated by the same molecular cofactors
to perform two completely different reactions: a DNA strand-
transfer reaction and repressor cleavage reaction.

Cleavage of cI was not, of course, the reaction for which the bacterial SOS response was designed. No doubt, phage λ was stealing
the cellular signaling system to escape from the damaged cell.
However, cleavage of LexA, the putative cellular repressor of SOS
functions, could now be inferred as the key event inducing SOS-
regulated genes. This pathway was soon demonstrated in considerable detail by John Little and his associates, first in cells and then
with purified LexA. Little also uncovered a fascinating property of
the cleavage reactions—LexA and cI could actually cleave them-

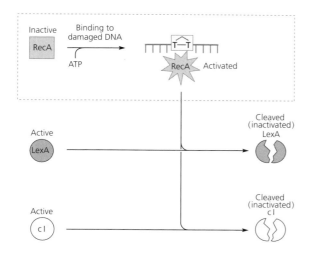

Activated RecA stimulates the cleavage
of the cl and LexA repressors

FIGURE 8–19. When activated by association with damaged DNA, RecA protein stimulates the self-cleavage reactions of the cl and LexA repressors. RecA does not actually act as a protease in these reactions; rather, it is a catalytic mediator of proteolysis.

selves under the appropriate (nonphysiological) conditions. So RecA is not actually a protease but a catalytic mediator of a "self-destruct" domain built into these particular repressors (see Figure 8–19). To complete the biochemical analysis of the SOS regulatory system, both Little and Roger Brent demonstrated that pure LexA can bind to the promoter regions of SOS regulated genes and prevent their transcription by RNA polymerase.

In the meantime, the scope of the SOS response was enlarged by identification of new SOS-inducible genes, using genetic fusion techniques. Malcolm Casadaban and Jon Beckwith had developed to a fine art the technique of "genetic fusion," in which expression of one gene is measured by assessing expression of another easy-to-assay gene such as the *lacZ* gene for β-galactosidase (often called a "reporter" gene). In this technique, the reporter gene is placed downstream of the gene whose expression you want to measure, so that both genes will be part of the same mRNA transcript. When expression of your gene is induced by some condition, expression of the reporter gene will also be induced. Casadaban developed a particularly clever technique for making fusions, which utilized a *lacZ* gene carried by phage Mu, the transposition phage able to insert anywhere (Mu is described in Chapter 7). Cynthia Kenyon and Graham Walker used this phage to identify SOS-inducible genes by identifying Mu insertions whose *lacZ* gene was induced by DNA

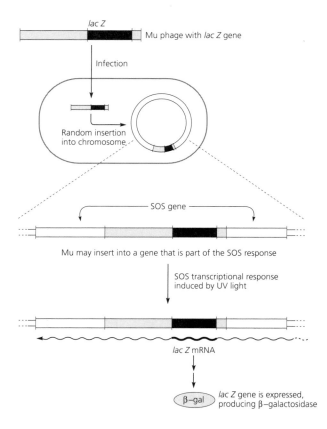

FIGURE 8–20.   By using a Mu transposon carrying the *lacZ* gene, Walker and Kenyon discovered which genes were involved in the SOS response. Transposons integrate into the chromosome at random. So, if the hybrid Mu transposon were to insert itself into a gene involved in the SOS response, then β-galactosidase (the product of the *lacZ* gene) would be produced when the gene was transcribed in response to SOS induction.

damage (Figure 8–20). Kenyon and Walker identified some ten distinct transcription units by this procedure; others have since been identified by direct measurement. These include the genes that carry out the classical physiological SOS responses: *uvrA*, *uvrB*, and *recA* responsible for UV-induced repair; *sulA* and *sulB* responsible for UV-induced filamentation; and *umuC* and *umuD* responsible for UV-induced mutagenesis. In addition, a number of genes of unknown function have been identified in different chromosomal locations.

The SOS response clearly involved inducing the synthesis of a large number of genes and many different promoters. The LexA protein is the master regulator, binding to multiple operator sites to turn off the multiple SOS promoters. In turn, the activity of LexA is controlled by the proteolysis reaction catalyzed by RecA.

What is the initial signal that starts the SOS response in motion? This aspect has not yet been completely worked out. One signal is likely to be single-strand DNA. Single-strand DNA is likely to be generated when DNA replication is blocked (see Figure 8–18), and activation of RecA for proteolysis can occur in the test tube with single-strand DNA and ATP. Another signal is likely to be damaged double-strand DNA. Chi Lu and I found that, in addition to its association with single-strand DNA, RecA also associates with double-strand DNA carrying a dipyrimidine lesion, presumably because the distortion of the B-DNA configuration at the lesion looks somewhat like single-strand DNA to RecA. Presumably, RecA bound to this lesion can also be activated for proteolysis. Based on other evidence that there is more than one route to the inducing signal, both pathways are likely to be involved in inducing the response. In addition, the SOS response might be induced by altering the mononucleotide content of cells, perhaps by generating a special mononucleotide cofactor. Clearly, the cell wishes to use as many avenues as possible to identify this potentially lethal condition.

INDUCED MUTATION IN THE SOS RESPONSE

One fascinating biological aspect of SOS is induced mutation, the property that has involved me in the field. In this process, single-base mutations are introduced by replication errors. However, SOS-mediated point mutations do not arise as a passive response to the presence of lesions. Instead, they occur as a consequence of a highly regulated pathway requiring two SOS-induced proteins, UmuC and UmuD (named for UV mutagenesis by Takesi Kato and Yukiko Shinoura, who first identified the genes). Walker and his colleagues showed the *umuC* and *umuD* genes are induced by the SOS response. Then, three labs found that SOS mutagenesis requires the RecA-mediated cleavage of the UmuD protein to a smaller form UmuD′. Hideo Shinagawa and his associates showed that UmuD was cleaved to UmuD′ in cells; we demonstrated the RecA-mediated cleavage reaction with pure proteins; and Walker and coworkers found that an engineered gene producing only UmuD′ was sufficient for mutagenesis. We subsequently showed that UmuC associates tightly with UmuD′, so that the UmuC/UmuD′ complex is probably the active agent for mutagenesis. In addition to initiating the SOS response by facilitating the cleavage of LexA, RecA is also involved in generating the form of UmuD essential for induced

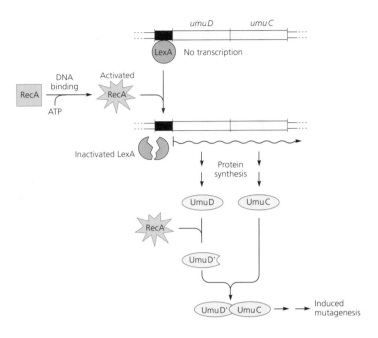

FIGURE 8–21. Two genes repressed by LexA, *umuD* and *umuC*, are involved in inducing mutations in response to the SOS response. When RecA protein inactivates LexA, these genes are expressed. However, before the UmuD and UmuC proteins become active in mutagenesis, UmuD is cleaved (in a reaction mediated by RecA) into a smaller protein, UmuD'. UmuD' binds to UmuC, and it is this complex that is necessary for UV-induced mutations to occur.

mutagenesis (Figure 8–21). Why is RecA involved at two levels of the SOS response? UmuD is less sensitive to RecA cleavage than LexA, suggesting that the induced mutation response is triggered only at higher levels of DNA damage. As might be expected from the conservation of salvational responses, UV-mutagenesis in eukaryotes also depends upon UV induction of specific proteins to carry out this task.

The biochemistry of the induced mutation response is well on its way. Our work indicated that, in addition to UmuC/UmuD', RecA itself is also involved in replicating past the lesion. Polymerase III holoenzyme is stopped at the DNA lesion in the absence of SOS intervention. We supposed that RecA, UmuC/UmuD', and pol III interact at the site of the lesion to facilitate DNA replication across the lesion, with the introduction of incorrect bases opposite the damage-distorted template bases, and demonstrated such replication bypass *in vitro*. Surprisingly, very recent results indicate that UmuC/UmuD' itself has the polymerase activity in this reaction.

Most mutations are bad for cells, because damaged proteins are produced. If error-free repair mechanisms exist, why does *E. coli* embark on such a highly regulated mutagenic pathway? Since this pathway is induced only when DNA damage is extremely high, the UmuCD mutagenic pathway is likely to be a "last resort" effort to bypass potentially lethal DNA damage. In addition, induced mutagenesis is also likely to be a population survival mechanism for endangered cells. Although most mutations are deleterious (or neutral), an occasional mutation will produce a bacterial cell better able to cope with the new hostile environment. An increase in mutation rate at this critical juncture would increase the likelihood that a fortunate variant would be produced that could give rise to a new, better adapted bacterial population. This may be the major function of the induced mutagenic pathway.

FURTHER READING

Hall, S. S. (1987) *Invisible Frontiers.* New York: Atlantic Monthly Press.

Ptashne, M. (1996) *A Genetic Switch: Gene Control and Phage* λ. Cambridge, Mass., and Oxford: Cell Press and Blackwell Scientific Publications.

9

MAKING DNA FROM RNA:
THE STRANGE LIFE OF THE RETROVIRUS

By the early 1970's, the pathway for transfer of biological information from gene to protein had been clearly defined: DNA→RNA→Protein. To sanctify this description of information flow, Francis Crick coined a term with an appropriately ecclesiastical ring, "Central Dogma." At first there was nothing to challenge the idea that information always flowed from DNA to RNA. A few viruses had genes consisting of RNA, but these could easily be explained as simple, primitive species that had survived solely by making and using an RNA message; certainly they did not impinge on the "Central Dogma." However, as we have seen already, Nature is often more versatile than the decrees of molecular biologists, and an anti-establishment type of RNA virus was lurking in wait for an appropriate disciple to announce its presence.

The "retrovirus" (backward virus) is an RNA virus that copies the information in its RNA into DNA upon entering the cell, without any consideration of the direction of information flow in the Central Dogma. This viral DNA is then integrated into the host genome and finally copied into RNA to make new viruses. Because they integrate into the host chromosome, retroviruses can acquire adjacent host genes, just as bacteriophage λ acquired the nearby *gal* gene. In fact, retroviruses were first studied because they can acquire host genes that then cause cancer, making these retroviruses "tumor viruses." The study of these viruses has provided a unique window into the genesis of cancer. Since the discovery that HIV, the causative virus of AIDS, is a human retrovirus, there has been a pressing need for greatly increased understanding of the retroviral life cycle in general, and that of HIV in particular.

QUANTITATIVE ANIMAL VIROLOGY: THE PLAQUE AND FOCUS ASSAYS

Max Delbrück made two enormous contributions to early molecular biology: a commitment to bacterial viruses (phage) as the simplest biological entity that directed its own destiny; and the notion that biological phenomena can and should be defined in quantitative terms. Although Delbrück sometimes thought that complex biological phenomena could be defined more quantitatively than was warranted by current understanding, his point of view was critically important in transforming phage research from a descriptive to a quantitative science. Delbrück's influence engendered a similar transformation in animal virology.

Cal Tech was given an endowment for studying animal viruses, but the institution lacked a research program in this field. In 1950, Delbrück proposed to Renato Dulbecco, a member of the phage group, that he should take up animal virology to fill the research gap. Dulbecco, who had earlier worked with the culture of animal cells, agreed to the idea and set out to learn what was known about the subject. He promptly concluded that animal virology was severely limited by the lack of a quantitative approach.

The quantitative assay for phage employed a "lawn" of bacteria embedded in an agar layer; as the bacteria grew, each phage produced a hole or "plaque" in the bacterial layer—a circle of lysed cells generated by the process of successive infection (see Figure 3–4). The plaque assay had provided the critical entrée into the physiology and genetics of phage growth in bacteria. Dulbecco aimed to develop a similar assay in which the growth of animal viruses would leave a hole in a layer of animal cells. Because techniques to culture animal cells were then in their infancy, this was a difficult task. Dulbecco chose to adapt a technique developed by Wilton Earle that permitted chicken cells to grow on the bottom of glass dishes under a layer of agar. Delbrück added some horse encephalitis virus to the growing layer of cells and then covered this with a layer of agar; he found circular holes in proportion to the number of viruses added (Figure 9–1). Quantitative animal virology had been born! In the case of animal viruses, the plaque assay depends on the ability of

The phage plaque assay was adapted for animal viruses

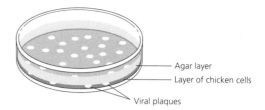

Agar layer
Layer of chicken cells
Viral plaques

FIGURE 9–1.   Dulbecco used the principle of the bacteriophage plaque assay to count animal viruses and observe their growth. He infected chicken cells that were growing in a dense layer on the bottom of a glass dish with horse encephalitis virus and then covered the cells with a layer of agar to prevent the virus from moving around (diffusing). When the virus infected a cell, it produced more virus, killed the cell, and moved into the surrounding cells. Over a period of time, this cycle of infection produced holes, or "plaques," in the lawn of living cells.

Renato Dulbecco

the virus to kill cells, rather than on a phage-type lysis reaction, but the resulting information is the same.

News of Dulbecco's success in developing a plaque assay for animal viruses attracted Harry Rubin to Cal Tech. Rubin wanted to study a special type of animal virus, one that caused cancer. He had come to this extremely descriptive field intent on developing a quantitative approach to the study of cancer-causing "tumor viruses." Trained initially as a veterinarian, Rubin had worked for several years on viral epidemiology with the U.S. Public Health Service before deciding to turn to basic research. He came to Wendell Stanley's Virus Lab at Berkeley with a postdoctoral fellowship but no background for his new career beyond having taken a course on phage and animal viruses. Although Stanley's scientific interests were focused on the tobacco mosaic plant virus, Rubin wanted to work on animal viruses. A fiercely independent, innovative, and determined person, he eventually decided to work on tumor-causing viruses based on hearing about the topic in some class lectures. Rubin had started to work on his own at Berkeley with Shope papilloma virus, which caused tumors in rabbits. However, when he moved to Cal Tech, he decided to switch viruses as well as locations. Rubin began work with Rous sarcoma virus because this chicken virus was known to grow and produce tumors rapidly when injected into birds.

The Rous sarcoma virus (RSV for short), discovered by Peyton Rous in 1911, is the historical grandfather of all RNA tumor viruses. Although retroviruses related to RSV are common in chickens, those causing tumors are extremely rare. So RSV is not a serious health hazard for chickens. Nevertheless, the Rous virus has become one of the most famous viruses in scientific history. As Peter Duesberg has described it, "The Rous virus has led to five Nobel prizes, 15 members of the National Academy of Sciences, several thousand research papers, and the death of 10 chickens." However, only a few people in the world were interested in RSV in 1953.

Rubin's initial goal was to develop a quantitative assay for Rous sarcoma virus along the lines of Dulbecco's work with horse encephalitis virus. He faced a severe problem in this endeavor. RSV had been identified by its ability to form tumors rather than its ability to kill cells. Perhaps not surprisingly, the plaque assay did not work—RSV did not kill the infected cells. After several months of looking unsuccessfully through the microscope at infected cells to see if he could discern some mystery event signifying transformation to tumor-pro-

Harry Rubin

ducing cells, Rubin decided that he needed something positive to show for his postdoctoral work. He switched his efforts to more laborious ways to measure RSV infection and also began to study a different virus, one more amenable to the cell-killing plaque assay.

In 1956, Rubin saw a paper by Robert Manaker and Vincent Groupé which reported that, when chicken cells were grown with RSV, their appearance was altered. Clearly, the virus could induce a "morphological transformation" when growing in cultured cells. Rubin realized that this transformation should allow development of a quantitative assay for RSV, but he was so heavily involved in other experiments that he decided to recruit a graduate student to the project. He convinced Howard Temin, a beginning student studying an embryology problem, to switch to the Rous sarcoma virus—a highly successful decision for both parties. After some struggles, Temin and Rubin worked out the "focus assay," which revealed an initial virus infection in a layer of cultured cells through the growth of distinct colonies of morphologically altered cells, termed "foci." The virus-altered cells in each focus represented the cell culture equivalent of malignant transformation in the chicken (Figure 9–2).

Using this new assay to measure virus production, Temin and Rubin confirmed an earlier inference that infected cells release RSV continuously while themselves continuing to multiply in an apparently unperturbed fashion. This result suggested that Rous sarcoma virus might need a different conceptual framework than the one used to think about the cell-killing viruses under study at Cal Tech. In support of this idea, experiments on the radiation sensitivity of RSV growth indicated that reproduction of RSV was more dependent upon the host cell than was reproduction of the cell-killing viruses. Ever since hearing François Jacob speak about the λ prophage in 1953, Rubin had been fascinated with the idea of a "provirus" model for tumor viruses. Perhaps Rous sarcoma virus, just like the λ prophage, knew how to integrate into the host chromosome. Although gene expression of Rous sarcoma virus would not be repressed in the hypothesized proviral state, somehow RSV would be able to co-exist with the host in this state. In 1959, Temin carried out experiments on the pattern of viral inheritance in cells that encouraged his belief in the provirus mode.

Rubin moved back to Berkeley in 1958, where he continued to study the growth cycle of Rous sarcoma virus. In 1960, to everyone's surprise, Lionel Crawford learned that RSV packaged RNA, not DNA, into its virus particles. This discovery ended Rubin's—

Infection by tumor virus can be detected by a change in the appearance of cells

Normal cells growing in an ordered array

RSV infection produces a transformed focus

FIGURE 9–2.   Rubin and Temin measured infectivity of the Rous sarcoma virus (RSV) using an assay based on the nature of a tumor virus: infection leads to cells that grow in uneven and uncharacteristic ways. Whereas a normal population of cells would grow in a somewhat ordered array, infected cells grow in a less ordered fashion, piling on top of one another to form a "focus" that is easy to detect with a microscope.

Howard Temin

and almost everyone else's—entrancement with the idea of an integrated provirus; clearly an RNA virus could not integrate its nucleic acid into host chromosomal DNA. Only Temin remained a believer. He moved to Wisconsin in 1960 and spent a decade of research before convincing most of the world of the reality of the provirus.

MAKING DNA FROM RNA

Howard Temin and I both started on the faculty at Wisconsin in 1960, though in different departments. Of the several hundred biologically oriented scientists at Wisconsin, only about four of us were molecular biologists. We formed a research discussion group to help ourselves both with science and with the fact that we were embedded in very classical and not always very sympathetic departmental surroundings. When I first met Howard Temin, I mainly noted his supreme self-confidence—reminding me of the prototypical kid in the class who has all the answers. Then I realized that he had a number of special qualities besides self-confidence: vast enthusiasm, exceptional biological insight, abundant scientific curiosity, and a passionate love for science. He also had a remarkable naiveté about the biochemical behavior of proteins and nucleic acids that probably helped sustain his faith in the proviral model, though perhaps delaying his eventual success in proving its reality.

If the Rous sarcoma virus was to have a proviral existence in the host genome, the viral RNA must somehow become duplex DNA, in wanton disregard of the Central Dogma. Alternatively, there might be an "RNA provirus" of an ill-defined nature. During the early and mid-1960's, Temin carried out a series of experiments with compounds that inhibited nucleic acid synthesis, designed to explore the pathway by which viral RNA was duplicated and determine whether DNA was part of the pathway. Temin found that virus production was blocked by an inhibitor that prevented the synthesis of RNA from cellular DNA—an argument for a DNA intermediate on the path to viral RNA synthesis. Virus growth was also prevented by general inhibitors of DNA replication. Temin proposed the following pathway for viral replication: RNA→DNA→RNA.

Temin found the "early experiments completely convincing," but most scientists were more skeptical. Experiments with metabolic inhibitors are tricky because sorting out the direct target of the inhibitor is often difficult. For example, the *lac* operon repressor had originally been identified as an RNA molecule by inhibitor experiments. Having been burned by such experiments myself, my own rule of thumb was "Inhibitor experiments are most helpful when you know the answer already." By 1969, most molecular biologists in the virus field were convinced that RSV was indeed different from other RNA viruses but not that the RNA replicated through a DNA intermediate.

During the late 1960's, Temin tried unsuccessfully to find the postulated "reverse transcriptase" enzyme that copied RNA into DNA in extracts of infected cells. In the meantime, another possible source for reverse transcriptase emerged from work on a DNA virus called vaccinia virus. When bacterial viruses enter cells, they inject their nucleic acids into the cells, leaving their protein coat outside (see Figure 3–8). However, the nucleic acid of animal viruses is generally "uncoated" from its capsid within the infected cell, and so an enzyme can come along with the nucleic acid. In fact, transcription studies indicated that, for several viruses, the viral RNA polymerase was in the virus particle itself. Perhaps the same was true for the mythical reverse transcriptase. This idea would be consistent with new inhibitor experiments by Satoshi Mizutani and Temin in 1969, which suggested that RSV could replicate its genome even if the host cell was prevented from making any new proteins. The crucial experiment turned out to be extremely easy. The virus coat was disrupted by treatment with detergent to expose the viral RNA, deoxynucleoside triphosphates were added as substrates for DNA synthesis, and the long-awaited RNA-directed DNA synthesis appeared (Figure 9–3). Reverse transcription really occurred!

Lightning suddenly struck twice in the retrovirus field. At this same time, David Baltimore at MIT found reverse transcriptase activity in mouse leukemia virus particles. Baltimore usually worked with other viruses but had decided to include a "What if Temin is right" experiment in addition to his other efforts. His scientific curiosity was generously rewarded. The demonstration of reverse transcriptase activity in two separate viruses clinched the case for this new type of DNA polymerase that copied RNA into DNA.

THE PATH FROM VIRUS TO PROVIRUS AND BACK AGAIN

The identification of reverse transcriptase in 1970 provided a very strong case for the validity of the DNA provirus model, although this finding did not directly verify the idea. With reverse transcriptase in hand, the nature of the fascinating molecular pathway from RNA to

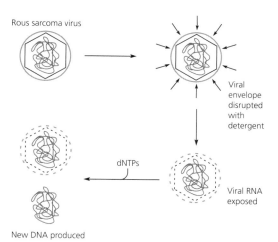

Rous sarcoma virus has a built-in reverse transcriptase

Rous sarcoma virus

Viral envelope disrupted with detergent

dNTPs

Viral RNA exposed

New DNA produced

FIGURE 9–3.   Temin wanted to see if the Rous sarcoma virus carried a reverse transcriptase enzyme (which would allow it to make a DNA copy of its RNA genome). He dissolved the virus's outer envelope and damaged the capsid structure so that the deoxynucleotides (dNTPs), which are substrates for DNA synthesis, could enter. He saw the formation of new DNA—the virus needed only raw materials to create a DNA copy of itself from the viral RNA genome in the capsid.

David Baltimore

FIGURE 9–4. RSV has a single-strand RNA genome of about 10,000 nucleotides. This is long enough to contain four genes: *gag* and *env* (which code for proteins in the viral capsid and envelope), *pol* (which codes for the viral reverse-transcriptase polymerase and integrase), and a gene called *src*, which is carried at the right end of RSV RNA. The *src* gene can disrupt normal cell growth and cause tumor growth (cancer).

DNA and back again was now open for investigation. Moreover, the probable existence of a genomic copy of the virus opened up new ways of thinking about the relationship between viral cancer genes and cellular counterparts. Finally, the association of a new type of replication reaction with a cancer virus attracted great interest. If viruses were the general agents of cancer, then an understanding of the reverse transcription process might be key to prevention of the disease. As a result of new ideas and the possibility of new approaches to understanding cancer, an explosive expansion of interest in RNA tumor viruses followed.

The first step in understanding the pathway from an RNA virus to a DNA provirus was to define the viral RNA genome. The initial studies in this area were carried out in the late 1960's and early 1970's, mostly by associates of Rubin. Two crucial early discoveries showed that the ability of the virus to multiply could be separated from its ability to "transform" normal cells to tumor-like cells. Steve Martin isolated a temperature-sensitive mutant of RSV that grew normally at high temperature but was unable to transform cells at that temperature. Independently, Peter Vogt and Hidesaburo Hanafusa isolated a deletion mutant of RSV that grew normally as a virus but failed to transform cells. So RSV could function as a virus without the gene specifying tumor formation (later termed *src* for *sarcoma*). Just as the *gal* operon was added to the λ phage to make the λ*gal* phage, so *src* might be added to the basic RSV virus. This analogy to λ was strengthened by the realization that avian leukosis virus, closely related to Rous, lacked a *src*-type gene. So, the RNA tumor viruses are simply a small subgroup of a particular kind of virus. Eventually, the RNA tumor viruses and their relatives were renamed "retroviruses" clarifying that fact.

Duesberg at Berkeley, Vogt at USC, and their colleagues eventually worked through a number of complexities to define the essential viral genome as an RNA molecule of about 10,000 nucleotides (present in two copies per virus particle). This length of RNA turned out to be about that needed to encode the viral proteins found in the virus particle. We now know that these proteins are first

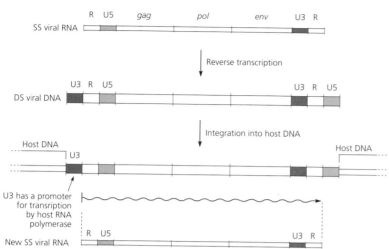

FIGURE 9–5. Genomes of retroviruses like RSV undergo several transformations as they replicate within host cells. Their single-strand RNA genome is copied into double-strand DNA by reverse transcriptase and integrated into host cell DNA. This viral DNA copy is then transcribed into new viral RNA. During replication, reverse transcriptase adds an extra viral "U3" sequence to the left end and an extra viral "U5" sequence to the right end of the viral DNA genome to make the ends of the viral genome. These ends, called long terminal repeats (or "LTRs"), play important roles in replication, integration, and transcription of the viral genome.

made as precursor polyproteins and then cleaved to form the mature proteins: the Gag and Env protein components of the viral capsid and envelope, and the reverse transcriptase polymerase and integrase enzymes. The gene order was determined by a combination of RNA sequence studies and genetic comparisons between related viruses. Acutely oncogenic, cancer-causing retroviruses have an extra gene added to their genome, almost always, as with λ*gal*, at the expense of some viral genetic material. As it turns out, Rous sarcoma virus is the single exception in which all of the viral genes are intact (Figure 9–4).

The next step in understanding the viral life cycle was to determine the structure of the provirus. Initial experiments proceeded in two general directions: determining the properties of the viral DNA in infected cells; and determining the route by which reverse transcriptase copied viral RNA into DNA. These two approaches coalesced to reveal a fascinating solution to a complex biological problem. The viral DNA must not only have a complete genetic copy of the RNA (a difficult problem already) but must also supply a promoter positioned so that the host RNA polymerase could make a complete copy of viral RNA from the integrated proviral DNA. In brief, the solution turned out to be the production of a DNA copy of the virus with a duplicated portion of the viral RNA at each end (Figure 9–5, top, middle). The region duplicated at the left end carries a strong promoter sequence. The RNA transcript initiating from this promoter yields a complete viral RNA that can be used either as an mRNA or as the single-strand viral RNA that is packaged into new viruses (Figure 9–5, bottom). The portions of the viral RNA that are duplicated at each end of the proviral DNA are termed LTR for "long terminal repeat" (to distinguish them from the R segments, which are the short terminal repeats found at the ends of viral RNA). In addition to providing the promoter for making viral RNA, the ends of the LTRs encode segments required for integration.

The studies that revealed the DNA intermediates in the proviral pathway were derived mainly from a long-term collaborative effort between Harold Varmus and Mike Bishop at the University of California–San Francisco. Based largely on the innovative and precise applications of DNA-DNA hybridization (this technique is described in Figure 5–3), their work not only revealed the structure of the provirus but also clarified the greater mystery surrounding the source of the cancer genes in the retroviruses.

Mike Bishop developed an interest in virology as a medical student, which led him to do postdoctoral work on polio virus with Leon Levintow at NIH. Levintow moved to UCSF and, in 1968, Bishop followed with an appointment as a new faculty member. Bishop acquired his interest in Rous sarcoma virus from Warren Levinson, a former associate of Rubin and a fellow faculty member. He began to work on RSV shortly before the Temin-Baltimore discovery of reverse transcriptase.

Harold Varmus was another physician-turned-molecular biologist, who came to UCSF in 1970 to join Bishop for postdoctoral work and then remained there as a faculty member. Varmus had been attracted to molecular biology by a fellowship at NIH, where he had worked with Ira Pastan measuring transcription of the *lac* operon using DNA-RNA hybridization techniques (see Figure 5–3).

The Bishop-Varmus collaboration brought a most successful mix of poet and scholar personalities to the retroviral field. Bishop contributed especially his virological understanding, biological insight, and passion for solving big problems. Varmus supplied a quantitative and analytical wizardry that converted biological concepts into molecular reality.

Verifying the existence of the expected proviral DNA was the first crucial point that needed to be established after the discovery of

Mike Bishop

the reverse transcriptase enzyme that made viral RNA into DNA. Varmus and Bishop tackled this problem by using a DNA-DNA hybridization assay to detect proviral DNA. They made a radioactive viral DNA probe from RSV viral RNA with reverse transcriptase and then asked whether it could hybridize to DNA from infected chicken cells. In this experiment, the cellular DNA is denatured so that it is single stranded and can reassociate with the radioactive viral DNA. If viral DNA was present in the cell, it should be detected because it would hybridize to the radioactive probe. Although hybridization was detected, the experiment turned out to be difficult to interpret because DNA from uninfected chicken cells also hybridized. In the meantime, Miroslav Hill and Jana Hillova in Europe verified the existence of viral DNA in a different way. They extracted DNA from RSV-infected cells, added it to uninfected cells, and demonstrated that new viruses are produced. So, proviral DNA must exist. Moreover, RNA polymerase from uninfected cells must be able to transcribe this DNA into an RNA virus copy.

Harold Varmus

Meanwhile, Varmus and Bishop discovered that RSV-related DNA sequences were not present in the cellular DNA from uninfected cells of other birds such as ducks or quails. By switching to duck or quail cells infected with RSV for their hybridization studies, they would be able to readily identify RSV DNA. Using a combination of fractionation of cell extracts and hybridization studies, various forms of RSV DNA were identified: linear, circular, and integrated into the host genome. So the proviral DNA clearly became integrated, but how was the linear DNA created, and how did the integration event occur?

As a result of the experiments on replication of RSV in the test tube (described below), the ends of proviral DNA attracted great interest. Each end of the viral RNA had been found to have an identical 20-base sequence (the R sequences in Figure 9–5). How was this sequence related to the ends of viral DNA? Varmus, Bishop, and colleagues set out to analyze the structure of the free and integrated viral DNA, with a focus on their terminal sequences. By the mid-1970's, the availability of restriction enzymes (described in Chapter 10) allowed the viral DNA to be separated into individual restriction fragments, greatly increasing the sensitivity with which DNA sequence arrangements could be investigated. Examining the pattern of restriction enzyme cuts at each end of the virus revealed a remarkable terminal structure for the viral DNA. Instead of the short 20 bp repeat sequences present at each end of viral RNA, these experiments identified a long 300 bp terminal repeat sequence present at each end of proviral DNA (the LTR sequences in Figure 9–5). John Taylor and colleagues at the Fox Chase Cancer Center in Philadelphia derived a similar conclusion from other work. The same terminal structure was present in the integrated provirus and in the free viral DNA. What was the genesis of this extra DNA at the ends of the proviral DNA? The answer turned out to be that reverse transcriptase uses a jumping primer to make the double-strand proviral DNA from the single-strand viral RNA genome.

THE WILD RIDE OF REVERSE TRANSCRIPTASE

The initial insights into the complexities of making the provirus came from test-tube experiments on DNA synthesis by reverse transcriptase. Like all DNA polymerases, reverse transcriptase requires a primer to start replication. Early work by Bishop and collaborators established that the primer was a transfer RNA. Experiments by Taylor, John Coffin, Bill Haseltine, and others determined that the primer binding site, "P" was in a remarkable location, close to the left (5′) end of the RNA. So the initial DNA copy of the RNA was a short chain headed in what seemed to be the wrong direction (Figure 9–6a). The realization that both ends of viral RNA had identical 20 nucleotide "R" sequences provided a way for this "wrong direction" DNA fragment to be used as a primer for synthesis of the remainder of the viral DNA (Figure 9–6b). But first, the short initiator DNA had to be released from its original template, which required removing the RNA that had just been copied. This job is carried out by a second enzymatic activity of reverse transcriptase, an RNA-degrading activity termed RNaseH. Conveniently, RNaseH degrades only RNA that is in an RNA-DNA hybrid (hence the H designation, for hybrid). Upon digestion of the original template RNA, the short initiator DNA is free to anneal with the R sequence in the 3′, "right end" of the viral RNA and from there continue copying. This stage of replication yields a complete but permuted DNA copy of the viral RNA, with a tRNA molecule attached at its 5′ end. The right LTR had been acquired, and the next question to be clarified was the mechanism for producing the left LTR (Figure 9–6c).

The best guess for a mechanism for making the left LTR was to copy the right LTR, and this turned out to be correct. The RNA

Viral DNA is made from a jumping primer

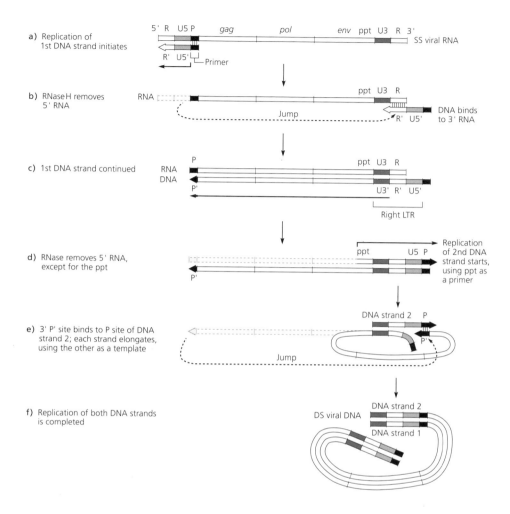

a) Replication of 1st DNA strand initiates

5' R U5 P *gag* *pol* *env* ppt U3 R 3'

SS viral RNA

R' U5' └─Primer

b) RNase H removes 5' RNA

RNA

ppt U3 R

Jump

DNA binds to 3' RNA

R' U5'

c) 1st DNA strand continued

RNA

DNA

P

ppt U3 R

P'

U3' R' U5'

Right LTR

d) RNase removes 5' RNA, except for the ppt

Replication of 2nd DNA strand starts, using ppt as a primer

ppt

U5 P

P'

e) 3' P' site binds to P site of DNA strand 2; each strand elongates, using the other as a template

DNA strand 2 P

P'

Jump

f) Replication of both DNA strands is completed

DNA strand 2

DS viral DNA

DNA strand 1

FIGURE 9–6.   RSV synthesizes a double-strand DNA copy from its single-strand RNA chromosome. Replication of the first DNA strand starts from a tRNA primer that binds to a primer binding site (P) near the 5' end of the viral RNA (*a*). After a short DNA has been synthesized, the 5' end of the viral RNA is removed by the RNase H activity of reverse transcriptase, freeing the short DNA to "jump" to the 3' end of the viral RNA, using base-pairing between R' and R (*b*). Replication of the first DNA strand continues from the right LTR, producing a DNA copy of the rest of the RNA genome (*c*). RNase H then removes most of the viral RNA, except for the polypurine tract ("Ppt"), which then serves as a primer for initiating synthesis of the second DNA strand (*d*). The creation of a P site on the second DNA strand allows the P' site from the first strand to "jump" and pair with the new P site (*e*). In this configuration, each DNA strand can be used as a template to complete the other strand (*f*).

primer for this reaction is the polypurine tract (ppt) of RNA close to the left end of the viral RNA used to make the complete DNA copy. It turned out that this ppt segment is resistant to RNaseH, probably because of its polypurine composition. So, when RNaseH munched up the long RNA bound to the completed first strand of the DNA, it left the ppt segment intact (Figure 9–6d). The ppt site then functions as the primer for synthesis of the second strand of viral DNA, using the first strand as a template. Its extension produces a copy of the right LTR (Figure 9–6d). The Varmus-Bishop group as well as Baltimore and colleagues were primarily responsible for elucidating this part of the pathway.

The mechanism suggested for finishing the left LTR of the first DNA strand was based on a consideration of the potential base-pairing opportunities of the newly synthesized right LTR of the second DNA strand. The P segment at the very right end of this LTR could base-pair with its P′ complement at the left end of the other DNA strand, resulting in both ends of the first DNA strand being base-paired with the LTR segment of the second DNA strand (Figure 9–6e). This pairing generates new primers for replication at the P and P′ sites. Replicative extension of P′ would give the first DNA strand its left LTR; similarly, extension of P would finish the second DNA strand. A blunt-ended, linear duplex DNA is the product of this last stage of DNA replication (Figure 9–6f). The essential features of this model were demonstrated by the experimental work of Varmus, Bishop, Baltimore, Taylor, and Ann Skalka at the Roche Institute of Molecular Biology, who came to be interested in retroviruses from her studies with λ phage.

THE RETROVIRUS AND ITS CANCER GENE

For most scientists in the field, learning something about cancer was the primary motivation for investigating the life-style of the Rous sarcoma virus and other retroviruses. At the beginning, there were two classes of ideas for how the virus promoted formation of tumors: infected cells might experience a general malaise that changed ("transformed") their cellular behavior to that of a tumor cell, or a specific viral gene product might disrupt cellular mechanisms controlling growth. As early as 1959, Temin had argued for the notion of a cancer gene (later termed an "oncogene"), based upon his early experiments showing that some RSV mutants had altered patterns of transformation. Of course, the notion of a specific cancer gene (*src* in Rous) was nearly inescapable once the RSV temperature-

Three models explain how viruses could cause cancer

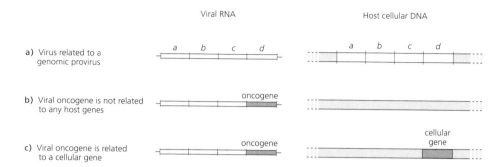

sensitive and deletion mutants that grew normally but failed to transform were isolated in the early 1970's. The experiments of Duesberg, Vogt, and colleagues mapped the *src* gene to one end of the viral RNA (see Figure 9–4). Where did the *src* gene come from, and how does it cause cancer? We now can answer the first question but not the second.

Given that *src* behaved as a specific oncogene, there were three quite different possibilities to explain its capacity to cause cancer. First, *src* might activate a provirus already residing in the chicken genome (and possibly in the genome of other organisms) (Figure 9–7a). This is a specific formulation of the more general virogene-oncogene hypothesis for cancer proposed by George Todaro and Robert Huebner in 1969. They suggested that, as a consequence of an ancient infection deep in evolutionary history, animal genomes carried proviral DNAs, similar to the λ prophage. Activation of these quiescent viruses by viral infection or exposure to environmental agents such as carcinogenic chemicals and radiation would initiate cancer. Second, *src* might function specifically during viral replication to cause transformation of host cells, without being related (or only very distantly related) to any cellular gene (Figure 9–7b). Third, *src* might be a close relative of a cellular gene whose normal function was to regulate cell growth. However, when present in the tumor virus, the *src* gene could lead to cancer either because viral growth resulted in an overabundance of the Src protein or because the virus carried a mutant form of this protein (Figure 9–7c).

The experiments that located the *src* gene in RSV RNA also indicated a road to discerning its evolutionary history. Bishop, Varmus, and their colleagues realized that they could make pure *src* gene

FIGURE 9–7. Viruses could cause cancer in three different ways. First, the virus might be related to a dormant provirus in the host cell genome; viral infection might activate a dormant provirus in the host cell genome, leading to cancer (a). Second, the virus could have an oncogene, unique to the virus, whose expression in the host cell could lead to cancer (b). Finally, the virus could have an oncogene related to a cellular gene controlling growth (or the viral gene might actually *be* a cellular gene that has been hijacked by the virus), and expression of this gene from the virus could alter cell growth and lead to cancer (c).

DNA (free from other viral DNA) in test-tube experiments and then use this DNA to ask if various organisms had a cellular gene similar to *src*. They used reverse transcriptase to copy RSV RNA into RSV DNA, but by damaging either the DNA or RNA they made small pieces of DNA rather than a single long DNA molecule. Some of these small pieces should contain only *src* gene DNA. *src* DNA could be separated from other viral DNA by hybridization to the deletion mutant of RSV that lacked the *src* gene. This step will remove DNA pieces from the entire retroviral genome except for *src*; the nonhybridizing DNA that remains is solely *src* DNA (Figure 9–8a). The purified *src* DNA was then used to ask whether genomic DNA of various creatures had a gene similar to *src*. Cellular DNA was denatured to single strands and reassociated with the radioactive *src* DNA. If cellular genes similar to *src* existed, they should be detected because they would hybridize to the radioactive probe (Figure 9–8b). At the same time, genomic DNA of various creatures was tested for genes similar to the essential viral genes (*gag*, *pol*, and *env*) by using DNA pieces from the deletion mutant of RSV that lacks the *src* gene as a probe, to see if they contained proviral DNA (Figure 9–8c).

The resultant hybridization study by Dominique Stehelin, Varmus, Bishop, and Vogt yielded a spectacular and clear conclusion. Both *src* DNA and the *gag*, *pol*, *env* viral DNA associated efficiently with cellular DNA from chickens (Figure 9–8b, c, left). So chickens did apparently carry something like an inactive provirus. But for all other avian species—duck, quail, and even emu (separated from chickens by 100 million years of evolution)—only *src* DNA hybridized efficiently; other viral DNA probes did not hybridize (Figure 9–8 b, c, right). Based on the absence of other viral DNA in these genomes, the *src* gene looked like a cellular gene that had been highly conserved in evolution. Later work reinforced this conclusion by demonstrating homology between the *src* DNA and the cellular DNA of mammals. The Rous sarcoma virus strongly resembled a λ*gal* phage in that it had incorporated a cellular gene; for RSV, a misplaced cellular gene led to a dramatic result—a typical retrovirus was converted into a cancer-causing agent.

The logic of the *src* approach was extended to ask if other tumor viruses had coopted cellular genes. Only for *src* did a rigorous deletion approach exist: two retroviruses that were identical except for the presence or absence of the *src* gene. However, other tumor viruses were matched to closely related nontransforming retroviruses to isolate nonhybridizing DNA segments that defined

Most genomic DNA has the *src* oncogene but not other viral genes

a) Isolate pure *src* oncogene DNA

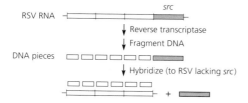

b) Probe chicken and duck DNA with radioactive (★) *src* oncogene DNA

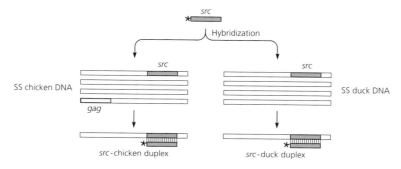

c) Probe chicken and duck DNA with radioactive (★) *gag* DNA

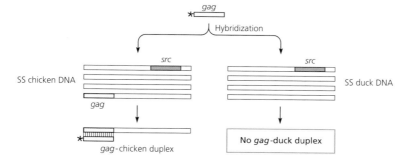

FIGURE 9–8. A hybridization experiment was used to find out if genomic DNA from different organisms contained either the *src* oncogene or the postulated provirus. A *src* DNA probe was made by replicating RSV RNA with reverse transcriptase, fragmenting the single-stranded DNA, and hybridizing the DNA fragments to the RSV RNA mutant lacking *src*. The nonhybridized DNA fragment remaining contained the *src* gene (a). A radioactive (*) *src* fragment hybridized to single-strand genomic DNA from chicken and duck, indicating that genes homologous to *src* were present in both animals (b). Chicken DNA also hybridized to a radioactive probe containing the viral *gag* gene, indicating that chicken had a provirus. However, duck DNA did not hybridize to *gag*, so it did not have a related provirus. Therefore, *src* must be hybridizing to a related cellular gene (c). Results similar to those with duck were found for many other avian species and for mammals. This experiment established that the viral *src* gene was related to a cellular gene found in many animals.

presumptive cancer genes. In turn, these "oncogene sequences" were tested for homology to cellular DNA. From these studies, a number of different cellular genes were identified that exhibited strong homology to the retroviral oncogenes. Thus, the concept that these viruses can carry cellular genes was rapidly extended to all known retroviruses. Work by Coffin, Temin, and colleagues made the analogy of λ*gal* and oncogene-containing retroviruses even closer. The only difference was how the virus captured the cellular gene. Whereas λ*gal* captures bacterial *gal* by abnormal excision (see Figure 5–6), retroviruses capture oncogenes from adjacent cellular proto-oncogenes by transcribing into cellular genes and then making an aberrant virus by recombination. Further study by Susan Astrin at the Fox Chase Cancer Center, Bill Hayward at Rockefeller University, and Varmus and his colleagues revealed how retroviruses without an oncogene could also cause cancer, although with much lower frequency. In these cases, cancer resulted from integrating the retroviruses within, or even in the neighborhood of a proto-oncogene, causing its unregulated expression.

At this point, the study of retroviruses and their mechanisms of oncogenesis had led to an exciting set of conclusions but a sobering view of cancer. As most scientists had believed earlier, the initiation of cancer now seemed likely to result from mutation in a variety of critical cellular regulatory genes. Many of these genes could now be identified by the study of the oncogenic retroviruses. However, it became clear that neither the retroviruses that exist today nor ancient proviruses were likely to be general causative agents for cancer. So the incipient hope for an antiviral "magic bullet" against cancer had been dashed. Retroviruses carrying oncogenes were now seen to be rare variants, selected from a general retroviral population by virologists (just as the rare variant λ*gal* had been selected from an integrated λ phage by lambdologists); these viruses would be extremely useful in understanding the origins of cancer but were not general agents of cancer in natural animal populations.

RETROVIRUSES AS TRANSPOSONS

The lifestyle of retroviruses, where integration into the host is an essential part of the cycle, is unique among animal viruses. The rare integration events of DNA viruses such as SV40 and polyoma are side reactions rather than a stage in the normal infection cycle. The general features of the retroviral integration reaction have now been characterized in infected cells and in test-tube experiments. The conclusions of this work have defined the insertion mechanism as a transposition event, relatively random with respect to host target sequence. Nonviral examples of "retrotransposons" have also been uncovered. As judged by DNA sequence arrangement, much of the highly repeated DNA in animal genomes was produced by ancient retrotransposition events.

In the early 1980's, DNA sequence analysis by Ann Skalka, George Vande Woude at NIH, and others established two essential properties of the retroviral integration reaction. First, the insertion event was completely precise with respect to the proviral ends; two bases were lost from each end of the LTR. Second, a small segment of host target DNA appeared as a duplicated segment adjacent to the retroviral ends (four to six bases were duplicated dependent on the particular retrovirus integrated). These two features of integration established that the reaction almost surely proceeded by a virus-directed pathway analogous to that used for moving bacterial transposons and integrating phage Mu (see Figure 7–16). By the mid-1980's, the retroviral integration protein had been identified, confirming the implied virus-specific reaction. Duane Grandgennett at Washington University, and Skalka together with Jon Leis at Case Western Reverve, had first noted and studied an endonuclease activity associated with preparations of reverse transcriptase; their work suggested a possible association between integrase and polymerase. Steve Goff at Columbia, Varmus, Temin, and collaborators then identified the *int* coding sequence at one end of the *pol* gene by mutational studies. The integration protein turned out to be produced by proteolytic cleavage of the larger Pol polyprotein precursor that contains both the polymerase and integrase proteins.

The biochemical analysis of the integration reaction began when Pat Brown and Bruce Bowerman in the Varmus/Bishop lab developed a test-tube reaction in 1987. Initially, the integration reaction was studied with a highly sensitive biological assay to measure recombination, but further experiments employed a more direct measurement with restriction enzymes. The source of integrase activity for the breaking and joining reaction was a large nucleoprotein complex found in infected cells, called the pre-integration complex, which contains the newly replicated viral DNA and integrase in a configuration ready to react with host DNA. Skalka and Leis, and Bob Craigie at NIH, have now developed a highly simplified biochemical system in which the purified integration protein executes

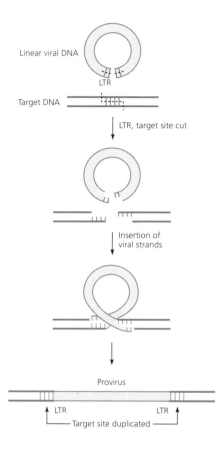

Retroviruses integrate into host
DNA much like transposons

Linear viral DNA

LTR

Target DNA

LTR, target site cut

Insertion of
viral strands

Provirus

LTR   LTR

Target site duplicated

FIGURE 9–9. The LTR regions of DNA from a retrovirus allow it to integrate into DNA in manner similar to that used by transposons. For clarity, cleavage of the host DNA is shown in a distinct reaction. In fact, the cleavage and joining reactions occur in a concerted fashion.

a breaking and joining reaction with short DNA segments that represent the LTR ends.

With the availability of a completely reconstituted system, the mechanism of the integration reaction could now be studied. Kiyoshi Mizuuchi and his colleagues at NIH entered the field to show that the chemistry of the retroviral integration reaction was essentially identical to that which they had worked out for phage Mu (Chapter 7). The nicked 3′ ends of viral DNA are joined to staggered phosphates in the host DNA backbone in a concerted cleavage and ligation reaction (a transesterification). Retroviruses were truly like DNA transposons (Figure 9–9).

THE HIV RETROVIRUS

In 1981, clusters of an unusual cancer (Kaposi's sarcoma) in unmarried young men in Los Angeles revealed the presence of a newly recognized disease that has since grown into a global pandemic, AIDS. This disease is menacing developing countries everywhere and devastating inner cities in the United States. By 1983, a French group headed by Luc Montagnier had isolated a strange new retrovirus from a pre-AIDS case, and in 1984 American scientists, headed by Robert Gallo at the NIH had established that this virus caused AIDS. Although both groups thought they had different viral isolates, recent work showed that contamination in France and the United States had resulted in both groups studying the same virus. This new virus, called human immunodeficiency virus, HIV-1, proved to be a retrovirus. HIV has since been studied using the techniques and ideas developed for other retroviruses. In fact, AZT—the first and, until recently, the most successful drug against HIV-1—was developed in the 1970's in a cancer therapy program.

HIV-1 is more complex than the previously studied oncogenic retroviruses. It belongs to a distinct group called the lentiviruses. In addition to the *gag*, *pol*, and *env* genes present in all retroviruses, lentiviruses contain additional genes (Figure 9–10). Among these, HIV has two genes called *tat* and *rev*, which defined new types of regulation. The *tat* and *rev* RNAs are among the earliest made after HIV infects a cell. As soon as Tat protein (short for *transactivating* protein) appears in the cell, it activates viral transcription. Tat binds to a sequence at the beginning of viral mRNA, called Tar, and helps this mRNA to be elongated efficiently. Rev protein controls the transport of viral mRNAs from the nucleus. As Rev accumulates in

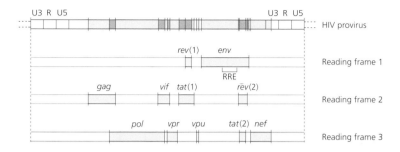

FIGURE 9–10. The genetic information in HIV RNA is contained in three separate, yet overlapping reading frames (see Figure 2–11 for description of reading frames). The HIV provirus uses all three of its reading frames to encode viral proteins. The schematic at the top shows the location of these coding segments; the individual proteins translated in each frame are shown below. Sometimes the same portion of RNA is translated in different frames for different proteins (for example the RNA at the end of *gag* and the beginning of *pol*). Portions of the virus that are translated in more than one frame are darkened in schematic at the top. Two proteins (Tat and Rev) are translated partly in one reading frame and partly in a different reading frame. For these two proteins, splicing of the mRNA joins the reading frames so that they are translated into a continuous polypeptide. These strategies allow a relatively small piece of DNA to encode many different proteins.

the cell, it binds to a specific RNA sequence, called the RRE (short for *Rev*-responsive element) present in HIV RNA. The Rev-RNA complexes are efficiently transported to the cytoplasm, where the mRNA is either translated into the viral structural proteins and enzymes, or incorporated into virus particles as genome RNA. The other proteins unique to HIV (Nef, Vif, Vpr, and Vpu) are called accessory proteins because they are nonessential for growth of the virus in cultured cells. However, they exert important effects on infected cells *in vivo* that contribute significantly to the pathogenesis of the virus.

HIV is an especially devastating virus because its major targets for infection are the very cells in the immune system that are usually mobilized to fight virus infections, macrophages and the T cells of the CD4 type. At first, the immune system seems to hold the HIV in check. But, over the long haul, the virus wins. Two features account for its success. First, the retroviral reverse transcriptase makes many errors when it copies RNA into DNA, producing many mutant viruses, some of which evade the immune system. Second, some of the cells infected by HIV are very long lived, allowing the virus to remain in the body integrated into host DNA for a long time. Over a period of time, typically ranging from two to ten years, the immune system gradually breaks down. This deterioration comes about directly from continued HIV infection of new cells, and indirectly because of the toxicity of free viral proteins and factors produced by infected cells, which affect other immune system cells and cells of the central nervous system as well. As a result of the HIV-induced deficiencies in the immune response, infected individuals become susceptible to infection by a wide variety of other microorganisms, many of which are little threat to people with normal immunity. Most AIDS victims die because they ultimately succumb to such "opportunistic" infections.

We now have very sensitive ways of detecting the amount of virus present in the blood of individuals infected with HIV, allowing us to observe the intensity of the battle fought between the virus and the immune system. Two independent research teams, those of David Ho at the Aaron Diamond AIDS Research Center in New York City and George Shaw at the University of Alabama, asked what the body had to do to maintain the status quo in the face of an infection. They started with AIDS patients whose numbers of T-cells and viruses in the blood were stable over time, administered an antiviral drug, and monitored the rate of decline of virus and rate of increase in T-cell counts after treatment. They calculated that, each day, the bodies of these AIDS patients destroy on the order of a billion viruses and that the viruses, in turn, destroy about 40 million T-cells. Normally, these T-cells live for years, but in infected individuals these cells survive for only about a day, showing how hard the body works to maintain the stable numbers of T-cells observed in these AIDS patients.

Even though the body destroys a billion viruses every day, the amount of virus in the blood remains very high, implying that people infected with HIV produce an extraordinary number of viruses each day. The high reproductive rate of the virus drives the pathogenic processes and amplifies the problem of the high error rate of reverse transcriptase. During replication, mutations rapidly accumulate in HIV, some of which will affect protease or reverse transcriptase, the targets of the antiviral drugs. Even if only a very small fraction of the virus is resistant to a drug, this high reproductive rate will allow these resistant viruses to take over the population. It is because of the rapidity with which such mutant viruses can take over the population that most anti-HIV therapies now use a cocktail of several potent antiviral drugs. Viruses that are resistant to any one of the drugs may be present in an infected individual, but it is unlikely that mutants resistant to all of the drugs in such a cocktail will preexist. Furthermore, no new mutants will arise, so long as the replication of the virus is held in check.

HIV is one of the most intensely studied viruses in history because it is such a serious health threat. Through study of HIV, we have learned much about retrovirus biology and the activities of the immune system, but we have not yet conquered AIDS. Moreover, our approaches thus far raise medical and ethical issues. These drugs are now prolonging the lives of AIDS victims in the developed countries of North America and Europe; however, the long-term outcome of this treatment is uncertain. Because some of the cells infected with HIV are long-lived, it has been estimated that treatment with cocktails of antiviral drugs will have to continue for five to seven years to be sure that all of the viral reservoirs in an infected individual are gone. Because some of these drugs have serious side effects, it is not yet certain how such prolonged treatment will be tolerated. Moreover, the powerful antiviral drugs that target reverse transcriptase and protease are not generally available to victims in the developing world because of expense and complexities of administration. If we are ever to control the AIDS pandemic, better treatments and, more important, an effective and inexpensive vaccine will have to be developed. But, the route to success is not yet clear and could come out of apparently unrelated studies. As was the case with study of the oncogenic retroviruses and cancer, our increased knowledge of HIV has provided important insights into the causes of AIDS but also sobering views of the magnitude of this problem.

FURTHER READING

Baltimore, D. (1976) Viruses, polymerases, and cancer. Science 192, 632–36.

Baltimore, D. (1995) Discovery of the reverse transcriptase. FASEB J. 9, 1660–63.

Bishop, J. M. (1982) Oncogenes. Sci. Am. 246, 80–92.

Temin, H. M. (1976) The DNA provirus hypothesis. Science 192, 1075–80.

GENETIC ENGINEERING:
GENES AND PROTEINS ON DEMAND

THE SOCIAL CONTEXT FOR BASIC RESEARCH IN BIOLOGY

The phenomenal progress of molecular biologists in understanding the replication, expression, and evolution of genes has depended on a social structure that supports such a basic research enterprise. Molecular biology is a world-wide pursuit but has flourished most successfully in the United States. Beyond sheer numbers, the success of American molecular biology has resulted from two special social foundations: the structure of the university system; and generous financial support from the U.S. government, especially through grants from the National Institutes of Health. Since these grants are reviewed by scientists and award is based on merit, they provide a tangible incentive for the successful pursuit of science. Together, these have provided the key elements—intellectual independence and financial support—required for any scientific endeavor.

The intellectual independence of beginning professors at American universities empowers these young scientists and fuels the scientific enterprise with a continual sense of new energy, excitement, and passion for discovery. Because independence is coupled with financial support provided by the NIH peer reviewed grant system, new faculty are free to strike out in novel directions and make their own unique contributions to the direction of science. This situation is not generally found elsewhere in the world, where a small number of senior professors reign supreme, in charge of both research directions and funding. Rather than change the university system, European countries have sought to solve the problem by developing research structures outside the universities. The approach of the research institute has met with notable success in a few cases, such as the MRC Lab in Cambridge, England; however, in most instances the research institute has eventually tended to mimic the hierarchical structure of the universities.

A BIOLOGICAL REVOLUTION FROM THE UNPREDICTABLE

Intellectual freedom remains a principle that the public generally supports, despite periodic assaults led by political figures eager to impose their particular view of a "moral" society. However, public financial support for basic research represents a complex and often misunderstood policy issue. Why should public funds be spent on biological research to understand phage λ when so many human diseases cause incredible suffering? What technological benefits could conceivably derive from studying how bacteria defend themselves from foreign DNA? Why can't research be directed toward clearly defined societal goals, such as curing cancer and AIDS? The answer is simple but difficult to communicate.

A highly directed attack on a problem becomes possible only when the correct approach becomes clearly defined. Backed by misguided scientific and political figures, a specially funded "War on Cancer" was declared some 30 years ago. The proclamation was about as useful as declaring a "War on Earthquakes." The solution to most societal problems related to biology has depended (and will continue to depend) on scientists who are attempting to understand basic biological mechanisms. The research described in this chapter represents a particularly good example of a biological revolution that has changed our approach to public health but was fueled by research from several different areas, each of which was seemingly unrelated to public health concerns.

HITCHHIKING GENES AND MULTIPLYING VECTORS

The essential concept of "genetic engineering," by which we mean manipulating DNA to create novel genetic combinations, is so simple that the reader will no doubt say, "Of course, it is obvious," just as Lise Jacob commented when her husband François first conceived of the operon idea. Suppose that you want to know the organization and expression of your favorite gene and obtain some of

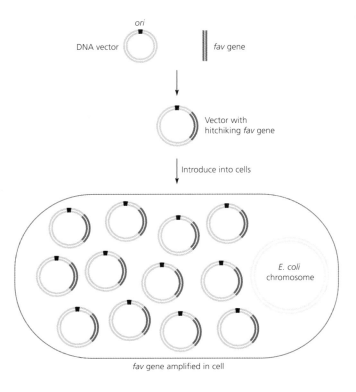

FIGURE 10–1. A crucial component of genetic engineering is a piece of DNA, or "vector," that has its own origin of replication (*ori*) so it can replicate in a cell independently of the bacterial chromosome. When a gene (*fav*) is spliced into this DNA vector and introduced into bacteria, the hitchhiking gene is amplified many times over when the vector replicates.

its protein product ("Fav," for *fav*orite protein). These goals can be expedited by inserting the DNA of the *fav* gene into a vector DNA molecule capable of replicating independently from the host chromosome, because it has its own origin of replication (*ori*). Two commonly used vectors are phage DNA (e.g. λ) or small circular DNAs called plasmids (like the F factor of Chapter 2). After transfer into an appropriate host cell, the vector DNA replicates from its own origin, consequently amplifying the hitchhiking foreign *fav* DNA. So now you have plenty of *fav* DNA to study its genetic structure and cellular expression (e.g., by DNA-RNA hybridization). If the vector DNA carries a powerful promoter that transcribes in the proper direction, the *fav* gene will be expressed and Fav protein produced. You do need a way to identify your *fav* gene DNA in the first place, but we will come to that later. This process for amplifying a particular piece of DNA is often termed "molecular cloning" because replication produces many identical copies from a single union of vector and foreign DNA segment (Figure 10–1).

We have just defined the basic needs of genetic engineering: first, a vector DNA that replicates independently of the host chromosome; second, a way to prepare foreign DNA segments and join them to the vector; third, a means of introducing the modified vector into cells to replicate; and fourth, a technique for identifying the foreign DNA segment of interest. The technical fusion of these elements into workable genetic engineering occurred between 1972 and 1975 through the efforts of several research groups, centered at Stanford. But the intellectual and experimental roots went back more than a decade earlier to some biological puzzles about phage growth.

BIOLOGICAL ROOTS:
PHAGE WITH BACTERIAL GENES AND COHESIVE SITES

The λgal phage provided the earliest example of a hitchhiking gene to pique the interest of molecular biologists. The λ that carried galactose genes was initially identified by Esther and Joshua Lederberg and Larry Morse, and its genetic structure was determined by Allan Campbell and Werner Arber in the late 1950's. Indeed, Campbell's fascination with explaining λ's peculiar acquisition of bacterial genes provided one impetus for his model of how λ integrates into the host chromosome (see Figure 5–6). The explicit model of gene regulation proposed by Jacob and Monod enlarged the fan club for phage carrying bacterial genes other than gal. A number of us realized that the molecular analysis of bacterial gene regulation depended on the ability to isolate the DNA of an operon. The improper excision of λ had already "cloned" the gal operon (see Figure 5–6); likewise, improper excision of phage 80, whose attachment site was very near the trp gene, created φ80trp. Jon Beckwith and Ethan Signer used further genetic trickery to make a φ80lac phage, which became the source of lac operon DNA for molecular studies. Not surprisingly, gal, lac and trp were the first bacterial operons to be intensively studied. Phage were not the only independently replicating DNA molecules to acquire hitchhiking genes. Jacob, Ed Adelberg, and Yukinori Hirota isolated variants of the F plasmid that carried the lac or gal operons, but these were not used for early molecular studies because it was very difficult to isolate pure F plasmid DNA.

The Campbell model explained integration, excision, and occasional λgal formation with wonderful clarity (see Figure 5–6). But

FIGURE 10–2.   Regions of DNA at either end of the phage λ chromosome, called cohesive (or *cos*) sites, are able to stick to each other or to ends of other λ chromosomes. Interaction between *cos* sites on the same λ chromosome results in circular DNA, and several phage chromosomes can be joined together when *cos* sites from different λ chromosomes interact.

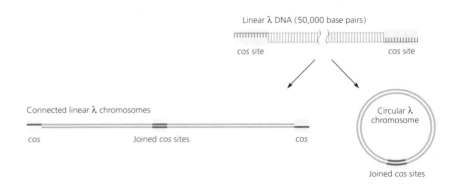

Al Hershey

the proposed mechanism required an intracellular circle, and the λ DNA injected into the cell from the phage was linear. How did λ DNA become a circle? Phage λ had invented a clever DNA-joining mechanism—"cohesive sites" on its ends. Solving the puzzle of cohesion brought scientists close to the brink of genetic engineering.

Al Hershey inferred the existence of cohesive sites on λ DNA from a brilliant set of centrifuge experiments, which I initially learned about in my first meeting with him in 1963. To me, Hershey was a semi-mythical hero who had shown that genes were DNA. He was also a famous recluse who worked by himself at Cold Spring Harbor and seldom came to meetings, even those held at Cold Spring Harbor. I had been told that his idea of an overly long conversation was "Pass the pipette can," when he ran out of these measuring gadgets. However, I found Hershey to be extremely friendly—he obviously took a real interest in the entry of new people with new ideas into his field. He later organized and edited the first λ book, presenting the multiple currents of λ research in a unified document.

Hershey came to lambdology from study of the T2 and T4 phages because he was interested in the structure of phage genomes. He had found that λ DNA molecules had a most interesting property: they would attach to other λ DNA molecules in solution to form molecules that were the size of several single λ DNA molecules. He also noted that the ends of a single DNA molecule could bind to each other, resulting in the formation of a circle. Both of these types of molecules could be distinguished from linear λ DNA molecules because they moved differently during centrifuga-

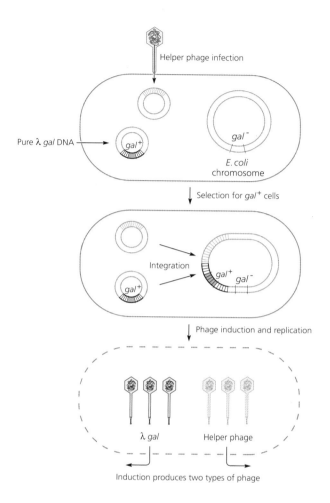

Helper phage infection

Pure λ *gal* DNA

gal⁺

gal⁻

E. coli chromosome

Selection for *gal*⁺ cells

Integration

gal⁺ *gal*⁻

gal⁺

Phage induction and replication

λ *gal*

Helper phage

Induction produces two types of phage

FIGURE 10–3. Kaiser and Hogness found that they could transfer genes into the *E. coli* chromosome with pure DNA from the λ*gal* phage. Direct gene transfer, or "transformation," required that the introduction of pure DNA be accompanied by infection with a "helper phage." Cells incorporating λ*gal* DNA are selected by requiring transformed *gal*⁻ cells to grow on medium with galactose. Because λ gal DNA does not integrate efficiently by itself, such cells were "double lysogens," having both λ*gal* DNA and helper phage DNA integrated into the bacterial chromosome. When the cells able to grow on galactose are induced to make phage by irradiation with UV, both λ and λ*gal* phage are produced.

tion. A conservative type, Hershey called these DNA-joining elements cohesive sites to avoid an interpretive name that implied a mechanism. However, from the properties of the joining reaction, he had already guessed that the cohesive (*cos*) sites at each end of the linear λ molecule were short regions of single-strand DNA whose base sequence was complementary to each other (Figure 10–2).

Dale Kaiser converted the notion of cohesive sites into biochemical reality by a beautiful series of experiments in the mid-1960's. After his experiments in Paris on the regulation of lysogenization by phage λ, Kaiser had joined Arthur Kornberg's

Dale Kaiser

Biochemistry Department. His intent was to compare the genetic map of λ to the physical structure of its DNA. With Dave Hogness, Kaiser found that pure λgal DNA itself could enter cells and convert gal⁻ to gal⁺ bacteria, provided that a normal λ "helper" phage infected the host bacteria at the same time (Figure 10–3 top, middle). Although direct transfer of λ DNA into *E. coli* was less efficient than infection by λ, the Kaiser-Hogness experiments revived interest in carrying out direct gene transfer into *E. coli*. "Transformation" of other bacterial species with pure DNA was a widely studied phenomenon, but many efforts to transform (or directly transfer) pure DNA into *E. coli* had been unsuccessful. Kaiser and Hogness also learned that cells transformed with pure λ DNA could execute a lytic infection, producing many virus particles. Efficient transformation by λgal DNA and efficient phage production by λ DNA turned out to require that the DNA molecules have cohesive sites, making their study a priority.

Kaiser and Hans Strack probed the nature of the cohesive sites with the Stanford battery of enzymes, using transformation by λgal DNA as a biological assay for the presence of cohesive ends on the phage DNA. These experiments would have been difficult to carry out anywhere else. Arthur Kornberg and Bob Lehman not only provided purified enzymes free of contaminating nucleases but also provided knowledge of the specificity and requirements of these enzymes, ensuring that these difficult experiments had a maximal chance of success. Treatment of λgal DNA with DNA polymerase prevented transformation. They inferred that, by adding nucleotides to the 3' end, DNA polymerase converted the 5' single-strand extension into duplex DNA, thereby removing the cohesive site (Figure 10–4a). Incubating polymerase-treated DNA with an exonuclease that removed bases from the 3' end of DNA restored transformation by converting the duplex DNA back into the 5' single-strand extension (Figure 10–4b). From those and other experiments, Kaiser and Strack concluded that the cohesive site consisted of a 5' single-strand extension of the duplex λ DNA molecule. Encouraged by this conclusion, Kaiser decided to use DNA polymerase to determine the base sequence of the single-strand cohesive site, withholding one or more dNTP precursors to limit the reaction. Using this approach, which presaged the more general Sanger method (see Figure 5–30), Ray Wu and Kaiser found that the cohesive sites consisted of 12 bases of complementary single-strand DNA (Figure 10–4c).

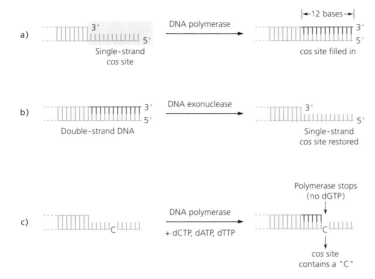

a) Single-strand *cos* site — DNA polymerase → *cos* site filled in |←12 bases→|

b) Double-strand DNA — DNA exonuclease → Single-strand *cos* site restored

c) — DNA polymerase + dCTP, dATP, dTTP → Polymerase stops (no dGTP) / *cos* site contains a "C"

FIGURE 10–4. *cos* sites were defined using the enzymes DNA polymerase and DNA exonuclease. Polymerase added nucleotides to the 3' end of the cos site, making the region double-stranded and no longer cohesive (*a*). An exonuclease that chews up nucleotides from the 3' end of the DNA removed the bases that polymerase added, regenerating the single-strand *cos* site (*b*). The sequence of the single-strand region was determined by withholding nucleotides from polymerase: if the absence of a specific nucleotide halted the addition reaction, then the base at the stopped position in the *cos* site must be complementary to the missing nucleotide (*c*).

In the fall of 1969, Peter Lobban, a graduate student with Kaiser, was faced with a standard task for prospective Ph.D. candidates— preparation of a theoretical research proposal to be presented to his examination committee. The purpose of such research proposals is to assess the students' ability to think creatively about a topic that is distinct from their ongoing lab work (which generally depends heavily on input from the thesis adviser). Usually those proposals constitute a relatively minor variation on previously published work, but in Lobban's case the proposal was an intellectual breakthrough. Lobban synthesized the insights from the work de-scribed above into a novel and detailed "game plan" for genetic engineering.

In essence, Lobban proposed that investigators could produce λ phages carrying any desired piece of DNA in a series of test–tube reactions based on the ones that had created the λ*gal* phage. In λ*gal*, the galactose genes replace some phage genes. Since phage λ can hold only slightly more than one λ-sized piece of DNA, adding the *gal* genes to one side of the prophage meant deleting the head and tail genes from the other side (see Figure 5–9). Mimicking λ*gal*, Lobban proposed that a 15,000 base-pair piece of foreign DNA could be added in place of head and tail genes (between the left end

Peter Lobban

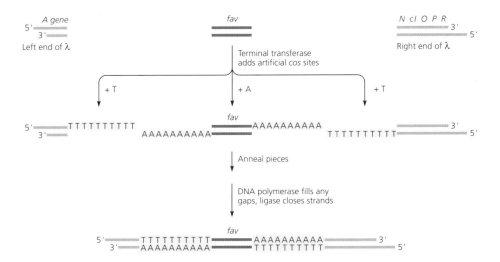

FIGURE 10–5. Lobban proposed that virtually any gene could be isolated by inserting it into the λ chromosome in place of the phage head and tail genes, using a series of test-tube reactions to mimic the creation of λgal. To make pieces of λ DNA and the DNA of a favorite gene (fav), Lobban proposed to shear these DNA's. He would then stitch them together by adding cohesive sites (complementary single strand regions) to the ends of λ and fav using the enzyme terminal transferase. Following ligation, the engineered DNA would be introduced into cells and amplified by replication.

of λ and the right half of λ DNA). A slightly modified version of Lobban's original diagram is presented in Figure 10–5.

The Lobban scheme proposed joining a piece of foreign DNA (*fav*) to λ by creating "internal cohesive sites" between the foreign DNA and phage DNA, using an enzyme called terminal transferase (Figure 10–5, middle). The notion of using terminal transferase was inspired by a brief comment in a talk by another graduate student, Tom Broker. Discovered by Fred Bollum some years earlier, terminal transferase adds nucleotides to one end of a DNA molecule (the 3′ end). Because the enzyme would use any triphosphate substrate presented to it, a short "homopolymer" of A's or T's could be added. These homopolymers of complementary bases would constitute artificial cohesive sites, allowing the junction of the three DNA segments in the proper orientation to make a phage carrying foreign DNA analogous to λ*gal*. Creation of this DNA would then be completed by a filling-in reaction with DNA polymerase, followed by strand closure with DNA ligase (Figure 10–5, bottom). (The actual scheme was slightly more complicated, because an additional exonuclease enzyme was needed to expose the 3′ end in a single-strand DNA, which is the substrate for terminal transferase.)

The λ DNA molecule with foreign DNA produced in a test tube should behave in every respect like λ*gal* DNA in the Kaiser-Hogness transformation experiment. In that experiment, both helper DNA and λ*gal* DNA integrated at the λ attachment site, producing "double lysogens" (see Figure 10–3, middle). When induced to lytic growth, both λ and λ*gal* phage were recovered (see Figure 10–3, bottom). Likewise, Lobban's λ should be integrated along with helper λ DNA at the λ attachment site and be recovered as a phage upon induction to lytic growth. This step provides a way to increase the number of copies or "amplify" the artificial DNA molecule created in the test tube. Moreover, just as λ*gal* could be selected by transforming *gal⁻* cells, so λ DNA carrying any other gene can be selected by transforming cells mutant in that gene, and then selecting those cells that had regained the wild type function of that gene. The only question remaining was the source of the small pieces of foreign DNA that would be inserted into λ. Kaiser, Hogness, and colleagues had already shown that pieces of λ DNA active in transformation could be prepared by the Hershey technique of shear forces (very rapid stirring). The preparation of bacterial DNA that was smaller than such pieces of λ could be accomplished by the more vigorous shearing technique of sonic oscillation.

The idea of joining DNA segments by cohesive sites became the guiding principle for the development of genetic engineering. However, it turned out that other enzymes were better suited for creating cohesive ends than the terminal transferase reaction envisioned by Lobban.

BIOLOGICAL ROOTS:
RESTRICTION ENZYMES FOR PIECES OF DNA

The appropriate engineering tool to make and rejoin specific DNA pieces turned out to be made by Nature—restriction enzymes. Some of this class of DNA-cutting enzymes cleave at specific sites, allowing the preparation of specific DNA fragments; as noted earlier, this technological advance provided for a vast leap in our ability to analyze the arrangement of bases in a genome and to study DNA transcription and RNA splicing (see Chapters 5 and 6). Some restriction enzymes, made in Heaven for genetic engineers, are even so thoughtful as to leave small cohesive sites, allowing the simple joining of DNA segments.

Before 1970, work on the "restriction" or cutting of foreign DNA by bacterial cells constituted a research topic that interested only a few people in the world. Now, many thousands of scientists

FIGURE 10–6.

E. coli host strain	Phage name	Phage progeny able to grow in	
		C	K-12
C	λ • C	Yes	No
K-12	λ • K	Yes	Yes

use these "restriction enzymes" daily, and the commercial production of the enzymes alone is a multi-million dollar enterprise. The four principal protagonists who gave the technology to the world came from very different backgrounds but were united in their determination to track biological phenomena to the biochemistry of DNA. Not so remarkably, all four came from the phage world: Werner Arber and Matt Meselson from the "old school"; Ham Smith and Dan Nathans from the younger generation.

Salvador Luria and other early phage workers were intrigued by a phenomenon termed "restriction and modification." Phage grown in one bacterial host often failed to grow on different bacterial strains ("restriction"). But the rare progeny phage that were produced in the restrictive host had become "modified" so that they now grew normally on the previously unfriendly host. The modification was not a heritable change, because the entire cycle could be repeated. For example, λ phage grown on the C strain of *E. coli* (λ·C) were restricted in the K-12 strain (the standard for most work). However, the rare λ phage that managed to grow in the K-12 strain now had "K" modification (λ·K). These λ·K phage grew normally on both C and K-12; however, after growth on C, these λ·C phage were again restricted in K-12. Thus, the K-12 strain knew how to mark its own genes for preservation, but eliminate invading genes from another distantly related strain (Figure 10–6).

Working at the University of Geneva, Werner Arber defined the molecular basis of restriction and modification. Arber started his career on phage as a graduate student with Edward Kellenberger, studying the growth of mutants of phage λ by electron microscopy. He became interested in λ*gal* phage through a visit by Jean Weigle, and, along with Campbell, demonstrated that these viruses were defective in lytic growth; they required "helper" phage to supply the viral genes missing from their own chromosome. After a postdoc-

Werner Arber

toral stint, Arber returned to Geneva in 1960 at Kellenberger's invitation, intending to work on radiation biology. Instead, he became fascinated with restriction and modification.

The prevailing view at the time Arber started was that foreign DNA was "restricted" because the host cells did not have the proper protein to protect it. By a clever combination of genetics and radiation biology, Arber could argue strongly that DNA was "marked" for protection by a change in the DNA itself rather than by a protective protein. In 1962, Arber and Daisy Dussoix found that λ DNA was degraded in a restricting host. So, the restrictive agent was probably a nuclease with a remarkable ability to distinguish the heritage of a DNA molecule. How could the nuclease tell whether the phage had been made in a protective or nonprotective host? One clue was that the protective modification was lost when the phage replicated in a different host. So the modified DNA must carry some kind of a protective tag added only by the protective host. During a visit by Arber to Berkeley, Gunther Stent pointed out that methyl groups were added to DNA at a limited number of sites; importantly, the location of methyl groups was known to vary among bacterial species. (This occurs because the enzymes that add these methyl groups to DNA differ among the different bacteria.) From this clue and some indirect experiments, Arber guessed correctly that the modification consisted of the judicious addition of methyl groups to protect those sites attacked by the restricting nuclease. Since the "methyl-modified" target sites were no longer recognized by the nuclease, the DNA was no longer degraded and so was protected. Later studies showed that the modification and restriction functions were different activities present in a single complex enzyme.

Meselson and Bob Yuan provided the initial biochemical characterization of a restriction enzyme in 1968. They identified the restriction enzyme from *E. coli* K-12 by its ability to attack λ·C modified DNA but not λ·K modified DNA. The purified enzyme cleaved λ·C DNA into about five pieces, but did not touch λ·K DNA. This result verified *in vivo* studies and also indicated that the nuclease had a high degree of specificity in choosing sites. Molecular biologists applauded Meselson's brilliant extension of his previous work on intracellular DNA transactions, but there was little interest in these enzymes outside of the restriction-modification field.

Ham Smith found himself working on restriction enzymes by surprise, but his success did not come as a surprise to me. I had known Smith from his postdoctoral research on phage P22 with

Ham Smith

Restriction enzymes cut DNA at symmetric recognition sites

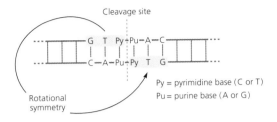

FIGURE 10–7.

Mike Levine, where he had carried out experiments closely similar to those going on in my lab with λ. All by himself, Ham seemed to be figuring out regulation and prophage integration while giving low key, "aw shucks" talks that made it all sound easy.

When Smith joined the faculty at Johns Hopkins, he intended to work on the biochemistry of genetic recombination. He decided to use the transformation of a bacterial strain with pure DNA as an assay for a successful recombination event accomplished in a test-tube reaction. Smith selected the bacterium *Hemophilus influenzae* for these studies because transformation of *Hemophilus* was being investigated in the nearby group of Roger Herriott. He also decided to use extracts from *Hemophilus* for the recombination reaction because, unlike *E. coli* extracts, *Hemophilus* extracts did not appear to have nuclease activities that degraded its own DNA.

At one point, Smith and his student Kent Wilcox added P22 DNA to the *Hemophilus* extract; the previously harmless extract now degraded the foreign DNA. This observation implied that the extract contained a restriction enzyme; Smith decided to purify the enzyme to "teach himself biochemistry." The purified *Hemophilus* nuclease behaved like the restriction enzyme from the *E. coli* K-12; it produced only a limited number of breaks in its target DNA. Smith decided to verify that this new enzyme, dubbed endonuclease R, recognized a very specific sequence by determining the DNA sequence immediately adjacent to the site of cutting. The methodology then available for determining DNA sequence could yield only two bases from much labor. Smith borrowed and made the necessary enzymes to get his two nucleotides. With a new postdoctoral associate, Tom Kelly, he then developed a way to determine the four nucleotides on each side of the break. The conclusion was clear; restriction enzyme R cleaved at a specific DNA site. The cleavage site exhibited "rotational" symmetry; when the sequence was rotated about a central base, it was the same on both sides (Figure 10–7). In this first example of a symmetric DNA sequence, Kelly and Smith correctly interpreted this property as a way for a symmetric, dimeric protein to use both subunits for DNA recognition.

Because of the explicit biochemical demonstration that the restriction enzyme cleaved at a specific DNA sequence, Smith's work underscored the potential utility of restriction enzymes for DNA work. Long before the data were published, Smith had recruited one eager apostle by mail: his fellow faculty member Dan Nathans, who had worked for several years with the RNA phage but had become intrigued with DNA tumor viruses while teaching about them in a class. These viruses replicate in animal cells and in special situations can induce tumors in the infected host. To learn about SV40 and polyoma virus, he went to spend a sabbatical year with Ernest Winocour at the Weizmann Institute in Israel in 1968. During this time, Smith wrote Nathans a letter about his evidence for cutting at a specific sequence. The letter arrived at an ideal time; Nathans knew enough about the animal viruses to realize the limitations in current experimental approaches, and his background helped him grasp immediately the potential value of specific pieces of DNA. He arrived back at Johns Hopkins with radioactive SV40 DNA, eager to start cutting it up.

Dan Nathans

The DNA of SV40 had been studied extensively with physical techniques by Jerry Vinograd at Cal Tech. At 5,200 bases in length, the duplex DNA was intriguingly small, with a fascinating closed circular topology. But an effective way to carry out genetic mapping by recombination did not exist, and physical mapping of transcripts on a closed circle looked like a nightmare. In his previous work, Nathans had used a segment of the viral RNA to help map genes on the RNA phage. Based on this experience, he was convinced that specific pieces of SV40 DNA would allow him to map both the genes and the transcripts of the virus.

Nathans' initial experiments with SV40 DNA clearly demonstrated the potential of restriction enzymes for mapping SV40 DNA. He and Kathy Danna found that treatment of SV40 DNA with Smith's enzyme yielded eleven specific fragments (Figure 10–8a). Upon further study, Danna and Nathans realized that their original cleavage pattern resulted from a mixture of two different restriction enzymes, later called *Hind*II and *Hind*III (for the bacterial source, *Hemophilus influenzae* strain d). (Smith had missed the *Hind*III restriction enzyme because his substrate for the reaction, T7 DNA, did not contain a site for cleavage by *Hind*III.) By cleaving first with one enzyme and then with the other, Danna and Nathans could order the cleavage fragments with respect to each other.

Soon, other restriction enzymes became available to amplify the analysis. The first restriction enzyme purified by Meselson and Yuan, the K–12 restriction enzyme, turned out not to have a specific cleavage site. Such "Type I" restriction enzymes recognize a specific DNA site but then randomly cut the DNA nearby. But Smith's enzyme was of a different type. "Type II" restriction enzymes cleave

FIGURE 10–8. Nathans cut SV40 DNA into eleven pieces (A through I) with the *Hind*II and *Hind*III restriction enzymes. Using the two enzymes sequentially, he was able to figure out the order of the fragments in the uncut virus (*a*). He used these restriction fragments to locate temperature-sensitive ("ts") mutants of the SV40 virus. Nathans infected cells with a heteroduplex made from one complete strand of the mutant virus and a strand from one of the restriction fragments of the wild-type virus. This was converted into a complete double-strand virus in the cell and then replicated. If the fragment covered the mutation, the cell would give rise to normal virus, which could grow at high temperature (*b*). This procedure allowed Nathans to place mutations on the ordered restriction fragment map of the SV40 chromosome. By hybridizing early and late RNA to the restriction fragments, Nathans also determined when regions of the chromosome were transcribed during the viral life cycle (*c*).

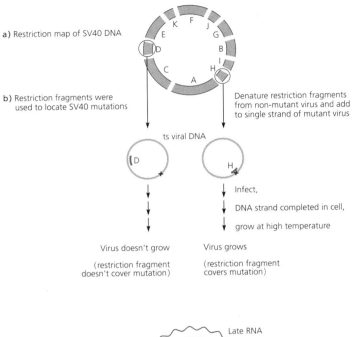

a) Restriction map of SV40 DNA

b) Restriction fragments were used to locate SV40 mutations

Denature restriction fragments from non-mutant virus and add to single strand of mutant virus

ts viral DNA

Infect,

DNA strand completed in cell,

grow at high temperature

Virus doesn't grow

(restriction fragment doesn't cover mutation)

Virus grows

(restriction fragment covers mutation)

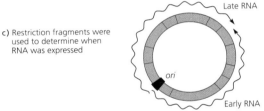

c) Restriction fragments were used to determine when RNA was expressed

Late RNA

Early RNA

ori

DNA at a specific site that is located within the binding site of the restriction enzyme.

Nathans applied his specific DNA pieces generated by the restriction enzymes to a series of important experiments on SV40. He used the restriction fragments to generate a genetic map of SV40. Viral DNA can be taken up by animal cells. When the viral DNA was added to cells growing in petri dishes, the virus particles that were produced could be visualized as holes (plaques) in the lawn of cells (see Figure 9–1). When the DNA came from a virus with a temperature-sensitive mutation, no plaques were visible at high temperature. Nathans and his collaborators made a heteroduplex between a complete single DNA strand from the mutant virus and a single complementary strand of one of the restriction fragments made from a wild type virus. This "hybrid" DNA was introduced into cells, and the cellular replication machinery then completed the second strand. If the mutant viruses had acquired the wild-type gene, virus production was normal at high temperature. By carrying out this experiment repetitively with each of the restriction fragments and many mutants, each mutation was assigned to a single restriction fragment. In this way, the eleven ordered restriction fragments of SV40 DNA were converted to an ordered genetic map of SV40 (Figure 10–8b). Further, by hybridizing viral RNA made in cells after infection to these same DNA fragments, the DNA source of the early and late viral RNA transcripts was determined (Figure 10–8c). (Similar experiments, carried out at Cold Spring Harbor, have already been described; see Figure 6–25.) Finally, Nathans and associates were able to use restriction fragments to locate the SV40 origin of DNA replication. Restriction fragments of DNA were spectacularly useful.

THE ENGINEERED JOINING OF DNA MOLECULES

There are two key concepts in practical genetic engineering: artificially joining DNA molecules; and amplifying the newly created DNA combination (by a replicating vector). The remarkable exam proposal of Lobban contained both concepts: joining by "A–T tailing" with terminal transferase and amplification by phage growth following induction. The initial demonstration of a controlled DNA-joining reaction for an engineering purpose resulted from experiments carried out at Stanford by Lobban in Dale Kaiser's lab and by David Jackson and Robert Symons in Paul Berg's group. Lobban was excited by his exam concept and proposed following this idea for his doctoral thesis. Kaiser encouraged this plan. As a demonstration of the joining reaction by artificial cohesive sites, Lobban decided to join two DNA molecules of phage P22, which lacks cohesive ends of its own. P22 had the further advantage that its ends were generated randomly from the circular molecule; if the joining reaction worked here, it should work on any piece of DNA. Joining P22 DNAs with polyA and polyT "tails" would give an easily recognized product, a double-length DNA molecule. The A–T tailing scheme required a variety of precisely controlled enzyme reactions that took some time to work out; the development of a successful joining reaction was achieved in 1972 (Figure 10–9).

Paul Berg came to genetic engineering from a quest for a way to

Paul Berg

FIGURE 10–9. Linear DNA from either SV40 virus (Jackson) or P22 phage (Lobban) was modified with exonuclease to make single strand ends (*a*). These ends were treated with terminal transferase to create artificial cohesive sites (*b*), then annealed and closed with DNA polymerase and ligase. This process created circular molecules twice as large as either of the starting linear DNAs (*c*).

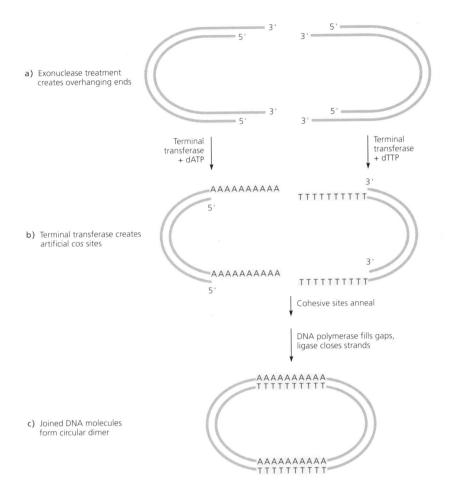

a) Exonuclease treatment creates overhanging ends

Terminal transferase + dATP

Terminal transferase + dTTP

b) Terminal transferase creates artificial *cos* sites

Cohesive sites anneal

DNA polymerase fills gaps, ligase closes strands

c) Joined DNA molecules form circular dimer

use animal viruses to carry genes into mammalian cells. As a new recruit to animal virus work, Berg was seeking a niche in "what seemed like the impenetrable wall separating prokaryotic and eukaryotic molecular biology." Berg had moved into his new field from a highly successful career in prokaryotic (bacterial) research. After joining Kornberg at Washington University as a postdoc, he became a faculty member in the same department and then moved to Stanford with Kornberg in 1960. In the late 1950's, Berg had been a co-discoverer of the transfer of amino acids to tRNA (Chapter 2), and he had worked for a number of years on characterizing the enzymes responsible for making the amino acid·tRNA complex (and also on RNA polymerase and cell-free protein synthesis). Berg was bounding restlessly among a variety of research areas when he decided in 1967 to spend a sabbatical year learning about DNA tumor viruses with Renato Dulbecco at the Salk Institute in San Diego (inspired in part by a comparison between phage λ and tumor viruses in a course given by Kaiser).

Berg returned to Stanford to convert his research program to the newly burgeoning field of the SV40 virus. At about the same time, Jim Watson, with the help of Joe Sambrook (also from Dulbecco's lab), started the Cold Spring Harbor animal virus group, and Nathans was in Israel thinking about using restriction enzymes in SV40 DNA. The development of restriction technology blew open the door to the molecular biology of animal viruses, and Berg joined in applying this new resource to SV40. Berg had developed a particular fascination with trying to use SV40 to carry genes into animal cells. SV40 did occasionally package host DNA but not efficiently enough to constitute a practical approach. Besides, how could such an event be detected? Kenichi Matsubara, a postdoc with Kaiser, had derived a plasmid from λ, called λdvgal, which was the plasmid analog of λgal. Rather than integrating into the bacterial chromosome, this plasmid freely replicated as a circular DNA molecule in E. coli. Berg decided that the gal operon might work as a test system for the transfer of genes into animal cells by SV40. The first problem was how to join the SV40 and λdvgal circular DNAs. Berg decided the best approach was to open the circles and join the two DNAs with artificial cohesive sites.

Jackson, a newly arrived postdoc in Berg's lab, set out to perform the DNA opening and joining reactions. By great good fortune, the perfect restriction enzyme for the DNA-opening job had just appeared on the scene. Working with Herb Boyer at the nearby University of California–San Francisco, Bob Yoshimori had purified a new restriction enzyme that cut at a specific DNA site (a type II restriction enzyme). This restriction enzyme was named EcoRI because the enzyme was specified by an "R-factor" plasmid of E. coli. John Morrow, a student in Berg's group, found that EcoRI made a single specific cleavage in SV40, opening the circle into a full-length linear DNA. Miraculously, EcoRI also cleaved λdvgal at a single site.

Jackson began his DNA-joining studies by producing polyA and polyT tails on linear SV40 DNA cut with EcoRI. In experiments done concurrently with those of Lobban on P22, Jackson used the tailing reaction to make SV40 circles that contained two SV40

DNA molecules (dimers). The joining reaction was essentially that diagrammed in Figure 10–9 and in the "Lobban exam" diagram (see Figure 10–5), except that two additional trimming reactions worked out by Lobban (using exonucleases that chew either at the 5' or 3' end of DNA) were required to produce the appropriate DNA substrates first for terminal transferase and then for DNA polymerase.

The completed, covalently joined molecules of SV40 or P22 were characterized as double-length by their sedimentation rate in centrifuge experiments and by direct visualization by electron microscopy. As expected, the production of double-length molecules required the A-T cohesive site; joining did not occur between two DNAs with A-tails only or T-tails only. Jackson and Symons then went on to use the same technology to pair SV40 DNA with λdv*gal* DNA. The Stanford experiments showed that an efficient artificial joining of DNA molecules could be achieved by test-tube reactions. At about this same time, Vittorio Sgaramella in Gobind Khorana's lab had shown that the DNA ligase of phage T4 could join DNA molecules without cohesive sites, but the process was inefficient and did not initially arouse much interest.

The paper describing the work by Jackson, Symons, and Berg was submitted to the *Proceedings of the National Academy of Sciences* in the summer of 1972 and published in the fall. The publication sounded a gong for genetic engineering; the content of the paper was notable, and Berg attracted attention by his energetic, outgoing, optimistic, and aggressive personality. Berg suggested to Kaiser that Lobban and Kaiser publish their work concurrently with his; however, Kaiser wanted to publish a more complete story. Kaiser represented an opposite personality type to Berg—shy, cautious, retiring, self-effacing—and he liked to see a problem unfold to a conclusion with no loose ends. The work by Lobban and Kaiser was submitted somewhat later as a long paper to the *Journal of Molecular Biology*. Unfortunately, the leisurely publication rate of that journal meant that the Lobban and Kaiser paper did not appear until a year after the Berg paper. By this time, A-T tailing had been bypassed by the much simpler reaction of cutting and joining through restriction enzymes, and Lobban's key early insights have not been generally recognized. Lobban left Stanford for a postdoctoral stay in Toronto, where he worked on animal cell genetics. He then sought an academic position, proposing to do genetic engineering with mammalian genes. A prophet ahead of his time, Lobban did not find a suitable job in molecular biology, so he returned to Stanford for a master's in electrical engineering and went on to pursue a successful career in biomedical instrumentation.

The work with artificial cohesive ends got everyone's attention. But, as Berg recalls, the experiments were considered a spectacular technical tour de force rather than a generally useful technology; the execution of the joining reaction involved complicated, multi-step reactions with enzymes that were not available to most molecular biologists. Moreover, from the point of view of Berg's long-term goal of gene transfer with SV40, there were no isolated mammalian genes to use.

The technological situation changed very quickly, largely through the efforts of two young professors at Stanford, Ron Davis and Stan Cohen. Davis, Boyer, and colleagues figured out that the miracle enzyme *Eco*RI produced cohesive sites all by itself, and Cohen developed a plasmid vector system to carry genes. Davis had done his doctoral work with Norman Davidson, applying electron microscopy to the physical chemistry of DNA. Davis and Janet Mertz were using electron microscopy to visualize the length of restriction fragments. They discovered that SV40 DNA cut by *Eco*RI behaved in a bizarre way; at low temperature, the DNA did not remain as a unit length linear SV40 molecule but produced either circles or linear dimers (molecules the size of two SV40's put together), just like the linear λ molecules studied by Hershey (see Figure 10–2). But at higher temperatures, the cleaved SV40 DNA remained as the expected linear monomer. These observations suggested that, when *Eco*RI cleaved DNA, it made a "staggered break," cutting each strand of the DNA duplex at a slightly different position. This would leave very short, easily melted cohesive sites. Only at low temperature would these sites be stable enough to generate the observed circles and dimers.

By a remarkable combination of thinking from physical chemistry and data from electron microscopy, Mertz and Davis determined that, after cutting, the *Eco*RI duplex DNA site "melted" to single strands at a temperature of 6°C. This weak interaction corresponded to an estimated cohesive site length of four to six bases. They showed, moreover, that different DNAs could be joined together by first cleaving them with *Eco*RI, mixing the DNA's together, and adding DNA ligase to join them into a single DNA molecule. In the paper from their work, Mertz and Davis pointed out that "Any two DNAs with RI endonuclease cleavage sites can be 'recombined' at their restriction sites by the sequential action of

Ron Davis

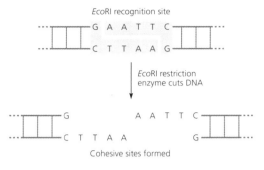

The *Eco*RI restriction enzyme makes staggered cuts, producing cohesive ends

*Eco*RI recognition site

```
        G A A T T C
        C T T A A G
```

↓ *Eco*RI restriction enzyme cuts DNA

```
        G              A A T T C
        C T T A A              G
```

Cohesive sites formed

FIGURE 10–10.

Stan Cohen

RI endonuclease and DNA ligase." The Berg group had spent months of effort putting artificial cohesive sites on DNA ends that already had natural joining sequences, albeit small ones!

All of the *Eco*RI for the Stanford work had come from a preparation of the enzyme by Yoshimori and Boyer. In the meantime, Boyer had joined with Howard Goodman to characterize the DNA sequence of the cleavage site for the miracle enzyme. Their work with Joe Hedgpeth showed that *Eco*RI recognized a six-base sequence; by making staggered breaks in the DNA, the cleavage reaction generated a four-base cohesive site (Figure 10–10). This can be contrasted to the cleavage reaction of *Hind*II, which Ham Smith had determined to give a "flush break" (see Figure 10–7), without cohesive sites.

DNA CLONING: AMPLIFICATION OF FOREIGN DNA BY A PLASMID VECTOR

Stan Cohen arrived at Stanford in 1968 as a beginning professor in the Department of Medicine. He had been introduced to molecular biology by working with phage λ during a postdoctoral stay with Jerry Hurwitz. Trained as a physician, Cohen initially planned to combine medicine and molecular biology, though he eventually chose a basic research career in the Genetics Department. For his first research project on a medically important topic, Cohen decided to study a class of plasmids that carried genes for antibiotic resistance. Some of these "resistance transfer plasmids" encoded proteins that blocked the action of several antibiotics; thus, the presence of these plasmids in a pathogenic bacterial population constituted a major problem for antibiotic therapy. The gathering of multiple genes for antibiotic resistance on a single plasmid represented a peculiar genetic phenomenon. He proposed to study the genetic structures that allowed them all to be present on one plasmid. (It turned out that the special mobility conferred by transposable elements is responsible for this phenomenon [see Figure 7–18].)

During his postdoctoral work, Cohen had prepared fragments of λ DNA by shearing it with rapid mixing. He was attracted to the idea of a similar physical analysis of plasmid structure to understand the antibiotic resistance problem. One major stumbling-block existed: there was no known way to put the plasmid segments back into *E. coli*. Fortunately, Morton Mandel had just found that a calcium treatment of the bacteria allowed effective uptake of λ DNA (without the helper phage of the Kaiser-Hogness experiments). Lobban, in turn, honed this technique with an eye to his own work. Cohen then successfully used the calcium technique to return plasmid DNA to *E. coli*, defining the biological assay system needed for his studies. A physical approach to plasmid structure now became possible. With Annie Chang, Cohen sheared the large plasmid DNA of the R-factor and isolated a set of smaller plasmids; the rare transformation events (where plasmids had been transferred to host bacteria) could be identified by selecting bacteria that had acquired resistance to an antibiotic carried on the R-factor.

Although successful, the shearing approach was limited. Only those fragments physically linked to the replication origin could be recovered by shearing because the other fragments would have no way of replicating in the host cell. A way to join any two plasmid fragments would be immensely valuable. Cohen knew about Lobban's work because he had occupied "guest space" in the Biochemistry Department when he had first arrived at Stanford. But the A-T tailing methodology was clearly extremely difficult. Then, in the fall of 1972, Cohen heard Boyer speak at a meeting on plasmids and learned about the cohesive sites provided by the *Eco*RI enzyme. The route to plasmid engineering became clear: restriction enzymes provided a way to obtain defined pieces of DNA and then to join them in desired arrays.

Cohen suggested to Boyer that they collaborate by using Boyer's *Eco*RI enzyme to analyze plasmid structure. Cohen searched for a plasmid DNA that had only a single *Eco*RI site located in a nonessential target sequence. A small plasmid, pSC101, produced by shear treatment of the original R-factor DNA turned out to have the ideal properties. *Eco*RI cut the plasmid at only a single site that left both the replication origin and the gene for resistance to the antibiotic tetracycline intact (Figure 10–11, top). In the first experiments, they were able to demonstrate joining between two DNA's cut with *Eco*RI: the pSC101 plasmid carrying tetracycline resistance, and a segment from another plasmid carrying kanamycin, a different antibiotic resistance gene. Now, the kanamycin gene no longer needed to come along with its original replication origin; it had become a hitchhiker on the pSC101 vector (Figure 10–11, middle and bottom). The pSC101 plasmid would clearly serve as a replicating vector into which other DNAs might be inserted. These experiments introduced practical genetic engineering. In principle,

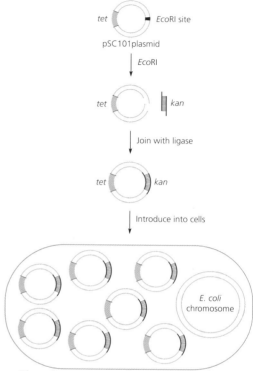

Plasmid vectors and restriction enzymes made DNA cloning a reality

When cells are grown on media containing tetracycline and kanamycin, only cells with correct plasmid will grow

FIGURE 10–11.   By shearing a large plasmid, Stan Cohen made a small plasmid vector ("pSC101") that carried a single *Eco*RI site and a gene for resistance to the antibiotic tetracycline (*tet*). He realized that he could stitch any foreign DNA into this vector by using the restriction enzyme, *Eco*RI, to make cohesive ends. He tested this idea by cutting both pSC101 and a DNA fragment carrying the kanamycin resistance gene (*kan*) with *Eco*RI, ligating the DNA's together, and then introducing them into *E. coli*. By growing the bacteria on media containing both tetracycline and kanamycin, he selected cells with a hybrid vector: pSC101 carrying the hitchhiking *kan* gene.

any piece of DNA could be inserted as a hitchhiker into the replicating vector DNA.

Would the procedure work for DNAs that were foreign to *E. coli*? Perhaps there were fundamental species "barriers" at the DNA level. Cohen and Chang initially addressed this question by successfully adding another antibiotic resistance gene to pSC101 from *Staphylococcus aureus*, a bacterium far distant from *E. coli* in evolutionary terms. There are actually two protection mechanisms that limit acquisition of foreign DNA by cells: restriction enzymes; and a mismatch recognition system that aborts poorly matched recombinants. However, both of these systems were bypassed when DNAs were joined in the test tube. When the artificially joined DNAs were returned to cells, the use of appropriate mutant strains disabled these two protection mechanisms, and the recombinant was obtained.

To demonstrate the general value of the plasmid technology, Cohen and Boyer then collaborated with John Morrow on the experiment that made genetic engineering a household word. To study eukaryotic gene regulation, Don Brown and associates had prepared the DNA coding for ribosomal RNA from the African toad. Brown had sent some of this DNA to Morrow, who was finishing his doctoral thesis with Berg before beginning postdoctoral work with Brown. Morrow, Boyer, and Cohen decided to introduce the ribosomal DNA from the toad into the pSC101 vector and amplify the DNA in *E. coli*. The experiment worked; replicated toad DNA was isolated from *E. coli* and cut out of pSC101 by *Eco*RI. Although the transformed *E. coli* bacteria had not learned to hop, they nevertheless attracted a great deal of attention; eukaryotic genes could clearly be amplified in *E. coli*!

SAFETY CONCERNS AND REGULATORY POLICY

Genetic engineering could produce new combinations of genes that had never appeared in the natural environment. The possible health and environmental hazards of the new technology became a highly discussed topic during the mid-1970's. Those discussions resulted in the development of formal guidelines for the formation and propagation of the test-tube recombinant molecules. The establishment of a regulatory policy represented an unusual, perhaps unique, instance in the history of science, in which scientists acted to restrain their own research in response to possible, but undocumented hazards.

What were the concerns about the new genetic technology? The essence of the perceived problem involved the formation of novel genetic combinations by a process that bypassed the normal requirements for species similarity in natural systems: gene transfer (sex); and subsequent joining of the genes of the two parents by genetic recombination. Although frogs become princes in fairy tales, genetic exchange between frogs and humans presumably does not happen in the real world. The specific concerns of most scientists focused on two possible "genetic accidents": the human acquisition of *E. coli* bacteria with an engineered animal gene producing a highly toxic compound (such as a growth-controlling hormone); or the production of a more virulent or more tumorigenic human virus. An additional potential health problem would have been the creation of plasmids with new combinations of antibiotic resistance genes that were not already found in nature. A more vague general worry that bothered a few people was the notion that a breakdown of species barriers (especially between prokaryotes and eukaryotes) might have catastrophic but unpredictable consequences in terms of genetic instability or bizarre gene expression. Molecular biologists were therefore faced with a complex issue. On the one hand, genetic engineering possessed clearly defined potential for science and society; on the other, there were conceivable hazards with the new technology. What should be done, if anything, to protect people from the possible dangers?

Berg became the central figure in the ensuing debate and move toward regulation. In 1972, he had planned with Mertz to introduce his λdv *gal*–SV40 recombinant into *E. coli* to look for SV40 gene expression. Mertz discussed the experiment during a visit to Cold Spring Harbor and drew a highly negative reaction. Robert Pollack and others at Cold Spring Harbor argued that introducing tumor virus genes into *E. coli*, a normal intestinal inhabitant of humans, constituted a very dangerous procedure. Berg decided not to do the experiment. (The planned experiment could not have worked because the *Eco*RI junction disrupted the λ gene needed for plasmid replication, but this fact was not known at the time.) Ensuing discussion led to a small and inconclusive meeting on the possible hazards of working with viruses able to cause cancer in animals. Genetic engineering had not yet become a major enterprise.

In 1973, Boyer presented his collaborative work with Cohen at the Gordon Conference on Nucleic Acids. Based on concerns expressed about possible future hazards, Dieter Söll and Maxine

Singer, the meeting organizers, wrote to Philip Handler and John Hogness, the presidents of the National Academy of Sciences and the National Institute of Medicine, respectively, requesting that "the Academies establish a study committee to consider this problem and to recommend specific actions or guidelines, should that seem appropriate." The letter was also published in the journal *Science*.

Berg was asked to organize the requested "Committee on Recombinant DNA." He gathered a group of seven scientists who met at MIT in the summer of 1974. The meeting resulted in an open letter to the scientific community containing four proposals. First, two classes of experiments should be voluntarily deferred: (1) construction of new bacterial plasmids that might result in the introduction of genes for bacterial toxins or antibiotic resistance into bacterial strains that did not already carry such genes; (2) linkage of segments from animal viruses to bacterial plasmids or other viral DNAs. Second, plans to link fragments of animal cell DNAs to plasmid or phage vectors "should be carefully weighed." Third, the director of the NIH was asked to establish an advisory committee to evaluate hazards and to devise experimental guidelines. Fourth, an international meeting should be convened to consider the matter further. In essence, the letter proposed postponing certain experiments until more information was available about potential hazards and protective approaches. The letter was signed by all seven attendees at the meeting; four scientists already heavily involved in genetic engineering also signed to express their support. The committee of course had no power to implement its proposals, but the signers represented a "high clout" group.

The international conference proposed in the committee letter was held in February 1975 at the Asilomar Conference Center in Pacific Grove, California (the meeting became known as the "Asilomar Conference"). Although the meeting turned out to be contentious, a provisional set of regulatory guidelines emerged at the end. The adoption of any policy at all represented a notable achievement, largely guided by the efforts of Berg and Sydney Brenner. The proposed guidelines endeavored to define the possible level of risk and to match that perceived risk with appropriate containment of the recombinant DNA. Both "physical" and "biological" containment procedures were suggested. "Physical containment" was already standard practice for work with pathogenic organisms: gloves, protective clothing, and special "hoods" with airflow designed to direct any escaped droplets away from the experimenter. "Biological containment" represented a new idea: to use enfeebled bacteria and vectors unable to grow outside of a defined experimental environment. Biological containment provided a more stringent and mistake-proof block to the accidental transfer of the gene cloning systems to humans.

The tricky question clearly was how to define the potential hazard of a given experiment, and any guess was bound to be controversial. Genetic engineering technology had the power to revolutionize research on animal viruses and eukaryotic genomes, but the cloning of these genes possessed the highest potential risk. For example, the random cloning of DNA from warm-blooded vertebrates was assigned to a moderate containment, enfeebled vector category. Therefore, these experiments could not be done until the enfeebled vectors were developed. However, given the pressures, the appropriate plasmid and phage vectors and bacterial hosts appeared quickly. The Asilomar recommendations became the basis for a formal proposal by an NIH committee, and ultimately for a set of "NIH guidelines" that were mandatory for NIH-funded experiments (most biological research). Universities, research institutes, and other funding agencies in the United States and abroad adopted these guidelines as operating procedures for recombinant DNA.

My impression is that most scientists at the time accepted the guidelines as a reasonable starting point. However, there were strongly held opinions either that no regulation was needed or that no gene cloning research should be done at all until the risks were thoroughly evaluated. Watson had signed and strongly urged the original committee letter that started the whole process; however, at Asilomar and thereafter, he strongly opposed regulatory protocols. On the other side, the volatile and literary Erwin Chargaff wrote in a letter to *Science*, "Have we the right to counteract, irreversibly, the evolutionary wisdom of millions of years, in order to satisfy the ambition and the curiosity of a few scientists?"

Over the intervening years, there has been no evidence of any unexpected hazard associated with gene cloning experiments, and the guidelines have become considerably less restrictive. In a recent discussion, Berg commented, "We overestimated the risks, but we had no data as a basis for deciding, and it was sensible to choose the prudent approach." He also pointed out that the initially stringent guidelines had generated public confidence and defused a number of emerging controversies. I agree with Berg's point of view. I was also strongly pro-regulation at the time, because I felt that biological scientists had the social responsibility to exercise caution with unknown gene arrangements. In addition, many nonscientists were genuinely frightened by some popular presentations of the new technology. I appeared on a local television show with a call-in segment. Most of the callers had interpreted novel gene combinations as producing new creatures, and I felt that explaining the regulatory efforts helped a great deal. Some scientists (often with the gift of hindsight) have asked me, "How could you be for regulating a technology with a vast potential benefit, and no defined hazard?" My answer is that benefits are often perceived before hazards. Consider the automobile. The benefits were obvious at the beginning, but who envisaged cities choking in automotive smog? I consider the regulatory efforts to be an historic achievement for biological science and an exceptional personal contribution by Paul Berg.

GENETIC ENGINEERING: A GIANT LEAP FOR EUKARYOTIC MOLECULAR BIOLOGY

In the early 1970's, there was great interest but little progress in understanding genetic organization and gene expression in eukaryotic organisms. DNA-DNA hybridization studies had established that many DNA sequences were represented once in the genome, and that some odd sequences were multiply represented. In addition, there was steady and sometimes spectacular progress in understanding the expression of animal virus genes (Chapter 6, Figures 6–24 to 6–27). But no counterpart existed to the genetic and biochemical studies that could be performed in bacteria. The central problem in carrying out detailed regulatory and molecular studies in eukaryotic organisms was isolating a cellular gene to study. Below I recount, in very abbreviated fashion, the successive approaches that have led to our current facility in studying such problems. As will be seen, it is largely the advent of genetic engineering that changed this scientific landscape.

The initial isolation of eukaryotic genes was accomplished by Don Brown and by Max Birnsteil at the University of Edinburgh and their collaborators, but the approach was clearly not a general one for other genes. The African toad produces an enormous number of copies of the ribosomal RNA genes; these genes also possess a distinct base composition. These properties were exploited to isolate the DNA for the 18S and 28S large rRNA, and for the 5S small rRNA.

What about isolating eukaryotic genes coding for proteins? Here, the genes of animal viruses defined one route for studying such genes. In addition, the mRNA of particular cellular genes could be isolated because of their great abundance in certain cell types (e.g., the developing red cell produces mainly hemoglobin mRNA). In these cases, gene regulation could be studied by direct analysis of mRNA levels. But isolating the DNA of most cellular genes, which was a prerequisite for a study of their regulation, seemed very remote indeed.

The development of genetic engineering suddenly let the eukaryotic molecular biologists through the looking glass into which they had been peering sadly for some time; the genes were almost within their grasp. The first gene sequences to be cloned were not chromosomal DNAs, but DNA copies of the abundant mRNAs already isolated (called cDNAs, for complementary to RNA). The initial techniques for making cDNAs depended on the polyA "tail" present on eukaryotic mRNA (see Figure 6–24). Phil Leder had developed a procedure to purify such mRNAs by adding cellular RNA to a column of short polydT DNA sequences. The polyA-containing mRNA bound to the column, whereas other RNAs did not. The pure mRNA was then freed from the column by weakening the association of polyT and polyA. A short polydT DNA was then added to the mRNA, but this time it was used as a primer for a special DNA polymerase that would copy RNA (reverse transcriptase; see Figure 9–3), resulting in a DNA-RNA duplex molecule. The RNA was then removed and the single-strand DNA could be converted to duplex DNA, which was then attached to a plasmid vector. Although an exciting achievement, the cDNA clones lacked important information: they contained only the coding sequence of the gene and not the transcriptional regulatory region that is usually located upstream of the gene itself.

This limitation did not last for long. Scientists quickly figured out how to use these cDNA clones or the pure mRNA itself to "fish out" the chromosomal DNA segment containing this gene. The conceptual breakthrough was the idea of making a "library" of a particular genome by cloning random segments of the genome into a vector. Then, DNA-DNA hybridization could be used to identify the clones that carried your favorite cDNAs. The best libraries were made in phage λ using exactly the strategy outlined by Lobban (see Figure 10–5), except that the "artificial cohesive sites" were made by cutting with a restriction enzyme rather than by adding

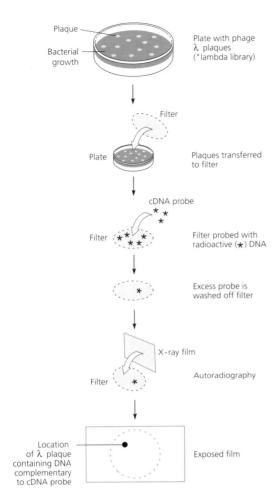

"Libraries" were the key to isolating a single gene

Plaque

Bacterial growth

Plate with phage λ plaques ("lambda library)

Filter

Plate

Plaques transferred to filter

cDNA probe

Filter

Filter probed with radioactive (★) DNA

Excess probe is washed off filter

X-ray film

Filter

Autoradiography

Location of λ plaque containing DNA complementary to cDNA probe

Exposed film

FIGURE 10–12.   A genetic library consisted of many phage, each of which carried a different bit of DNA from another, nonphage species. Collectively, this population of phage contained the entire genome of the other species. When the phage library was mixed with bacteria and plated on agar in petri dishes, the individual phage grew up and formed "plaques" in the surrounding layer of bacterial growth (see Figure 3–4). The phage in these plaques could be transferred onto filter disks cut to fit on top of the petri dishes. Then, the filters were covered with a radioactive (*) cDNA probe that was complementary to a particular gene. This probe would bind to the plaque that carried complementary sequence but be washed off of the rest of the filter. The location of the probe was detected by autoradiography with X-ray film (see Figure 5–27).

complementary AT tails. A collection of λ phage, each carrying a single random DNA segment of the genome constituted a library. A "lambda library" of a particular genome was used to infect *E. coli*, taking care that only one phage would infect any particular bacteria. These infected cells were mixed with large numbers of uninfected *E. coli* cells and plated on agar plates. Each phage produced a plaque, containing millions of phage, visible as a hole in the layer of bacterial growth (Figure 10–12, top). A unique segment of the eukaryotic genome was amplified in each plaque. Because eukaryotic genomes were so large, many plaques would have to be analyzed to find one that would hybridize to a particular cDNA. Tens of thousands of individual plaques could be scanned simultaneously by transferring their DNA from the plaque to a filter. This was simply accomplished by overlaying the filter on the agar plate (Figure 10–12, middle). Each filter was hybridized with a radioactive cDNA and then exposed to photographic film. The λ clones with complementary DNA appeared as black dots on the film (Figure 10–12, bottom). Clearly, it was important to have very large petri dishes so that many λ plaques could be scanned at once. Some laboratories even took to using cafeteria trays for these experiments. Libraries of human, mouse, Drosophila, and yeast genomes were quickly established by Tom Maniatis at Cal Tech and by Ron Davis at Stanford. In the collaborative spirit that permeated this field on the edge of a new scientific era, these libraries of billions of virus particles in a test tube were shipped around the world to different labs. With so many people joining the pursuit for eukaryotic genes, progress was very rapid indeed.

The rate of progress in this emerging field was clearly dependent upon the further development of recombinant DNA techniques. The exponential development of these methods resulted from the technical wizardry of the early scientists using these techniques, fueled by pivotal contributions from organic chemistry, engineering, and physics. It is now possible to use a protein in the test tube to isolate the gene that encodes it, without obtaining the cDNA. The amino acid sequence of the protein can be obtained from mass spectrometry and used to infer the sequence of a stretch of DNA that could encode for the peptide (by determining possible codons from the genetic code [see Figure 2–11]). Meanwhile, following up on the early work of Gobind Khorana and Keichii Itakura, organic chemists developed techniques to synthesize DNA, and engineers automated this process. So, armed with possible DNA sequences for your protein, an inexpensive mix of DNA probes that could potentially base-pair with your gene is ordered and used to screen a library.

Even more remarkable is the polymerase chain reaction (PCR) that skips the library step altogether. Conceived of by Kary Mullis, a researcher working at the Cetus Corporation in Emeryville, California, with contributions from many others, this technique uses the principles of DNA replication and base-pairing to replicate one specific portion of total cellular DNA in a test-tube reaction. A short DNA sequence complementary to one strand of the sequence flanking each side of the gene is hybridized to denatured cellular DNA (Figure 10–13, top). DNA polymerase then extends these primers, using the denatured cellular DNA as a template. At the end of one cycle of extension, a single copy of the gene has been made (Figure 10–13, middle). The key insight was to realize that this copy can be amplified many times if the DNA is denatured and subjected to the same procedure over and over again. Then, most synthesis would come from the short gene-sized pieces made in the previous round of synthesis (Figure 10–13, bottom). PCR became feasible when thermostable DNA polymerases were identified. These DNA polymerases, isolated from organisms that live at temperatures close to that of boiling water, retain activity even at the high temperatures necessary to denature DNA. The repeated cycles of DNA denaturation, annealing, and synthesis in a PCR reaction allow the sequence of a specific gene to be amplified a million-fold. This level of amplification allows for gene cloning directly from the total chromsomal DNA of an organism.

THE INDUSTRIAL AND MEDICAL RAMIFICATIONS OF GENETIC ENGINEERING

Genetic engineering was created from investigations on three rather obscure phenomena seemingly far removed from any practical human health concerns: phage carrying bacterial genes, λ ends that stuck together, and restriction of foreign DNA by certain bacterial hosts. The technology that resulted from these investigations has revolutionized many aspects of our life.

Modern biotechnology was born and raised in California with its foundation in two laboratory techniques: first, the process of plasmid DNA transformation in *E. coli*, and, second, the use of restriction enzymes. Any DNA cut with a particular restriction enzyme

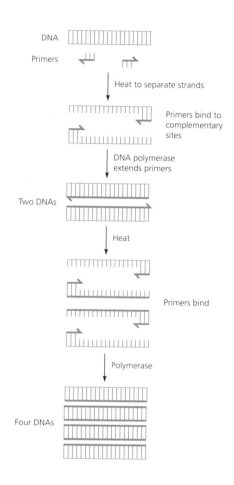

Polymerase chain reaction (PCR) can amplify a single piece of DNA many times over

FIGURE 10–13. By using short DNA primers that can bind to complementary sites on a piece of DNA, the stretch of DNA between the primer sites can be duplicated by DNA polymerase. In this procedure, the DNA strands are separated by heating. The primers then bind and the polymerase extends the primers, creating two identical DNA duplexes. The cycle can be repeated by heating the mixture, allowing the primers to rebind. Because the DNA polymerase used in this reaction remains active after heating, the cycle of duplication can be repeated as many times as desired.

can be joined to another DNA molecule cut with the same restriction enzyme via the overlapping ("sticky") ends generated from the staggered cleavage reaction. Since the DNA fragments need to be matched only over a few bases at their 3′ or 5′ ends to be joined, the particular source of the DNA is no longer of any consequence. Bacterial, plant, human, fungal, or even totally synthetic DNA can be fused and introduced into a host for expression. Hybrid or chimeric DNA became the norm.

Restriction enzymes became a business. A little more than 20 years after the fundamental work of Arber, Smith, and Nathans in identifying and using the first restriction enzymes, hundreds of these endonucleases have been identified (and they keep coming). Their discovery has spawned a business based on making and selling packaged kits of restriction enzymes to sustain research using genetic engineering in academia and industry.

More important, biotechnology itself has become an industry. The foresight and entrepreneurial nature of a few investors were responsible for seeding and nurturing the industry; in retrospect, it is hard to imagine this happening anywhere else but in California. DNA manipulation was the new industrial revolution: Genentech, the first company explicitly founded to pursue genetic engineering, opened in 1977; by this time Cetus was also pursuing recombinant molecular research. Genex (no longer in existence), Biogen, Amgen, and others rapidly followed. This industry has grown enormously and many shrewd investors have profited.

Biotechnology also changed the traditional pharmaceutical industry and revolutionized medicine. Apart from the use of insulin to treat diabetes, proteins had not been used as drugs. Genetic engineering made drugs out of proteins, a concept that had not been anticipated. The use of restriction enzymes to splice eukaryotic genes into prokaryotic expression vectors has permitted the production of protein drugs for medicine. This field started with the cloning of the human insulin gene into bacteria (and yeast) and the production of large quantities of human insulin, all but replacing the use of porcine or bovine insulin. Next, somatostatin and a succession of other biologically active proteins have been expressed in *E. coli* and other hosts and purified to clinical grade purity. These include human growth hormone, interferons, colony-stimulating factors, and a variety of cytokines.

There were many pioneers in the technical transition of molecular biology to industrial biotechnology. Two of the best known,

David Goedell (Genentech) and Charles Weissman (Biogen), have been recognized for advancing the use of *E. coli* as a production host for human proteins such as interferon, by developing the art of "cutting and splicing." During a period of just ten years, *E. coli* K-12 moved from being a laboratory model for studies of microbial function to an industrial microorganism. It is now used as a microbial factory to generate hundreds of grams of purified mammalian proteins (a wild dream in the mid-1970's), with a market value of several billion dollars. Equally important was the development of eukaryotic hosts for producing those proteins that contained certain modifications made only by eukaryotic cells. *Saccharomyces cerevisiae* is used for the production of hepatitis B vaccine, and mammalian cell lines are used for the production of erythropoietin and other glycosylated proteins.

Restriction endonucleases have also changed the field of medical diagnostics. The length of restriction fragments within a particular gene has been found to vary among individuals. This "restriction fragment length polymorphism" (RFLP) has been employed for identifying breast cancer genes and microbial pathogens, tracking wild life, settling paternity suits, and identifying lost people and criminals. Although the diagnostic use of these enzymes is gradually being replaced in some cases by direct gene amplification and rapid nucleic acid sequencing, the applications of endonucleases still pervade the field, surpassing all expectations for their utility.

The use of restriction endonucleases in these numerous processes is now largely routine and their seminal role largely forgotten, but academic and industrial biology will never be the same since the discovery of the precise way in which these bacterial enzymes cleave DNA molecules.

FURTHER READING

Davies, J., and W. S. Reznikoff, eds. (1992) *Milestones in Biotechnology: Classic Papers on Genetic Engineering.* Boston: Butterworth-Heinemann.

Gilbert, W., and L. Villa-Komaroff (1980) Useful proteins from recombinant bacteria. Sci. Am. 242, 74–94.

Marx, J. L. (1988) Multiplying genes by leaps and bounds. Science 240, 1408–10.

Mullis, K. B. (1990) The unusual origin of the polymerase chain reaction. Sci. Am. 262, 56–65.

AFTERWORD

It was Berkeley, 1970. The tear gas from the Vietnam War demonstrations had dissipated from People's Park, though only recently. Students were refocusing on their classwork. At the extreme uphill end of the Berkeley campus, beneath a grove of pungent eucalyptus trees, stood Wendell Stanley Hall. In a first-floor classroom, wearing a shockingly orange tie-dyed T-shirt and a headband, Hatch Echols was applying chalk to blackboard, illustrating a bacteriophage lambda experiment in a first-year graduate course on molecular biology.

As a student in that class, I was awestruck by the stories of DNA, RNA, and protein being offered by Hatch and his colleagues. A chemistry major at Grinnell College and now a chemistry Ph.D. student at Berkeley, I had studied these macromolecules as chemical and biochemical entities. Now new life was being breathed into the subject as these molecules were connected to biological processes. Hatch's often passionate descriptions of key ideas and experiments further enlivened the subject. The molecular biology revolution, already ongoing for more than a decade at that point, had somehow escaped my attention the first time around. Not this time.

The same personality that brought molecular biology to life for me in 1970 has done it again through the pages of this volume. What is special to me is not just the concepts and the experiments that gave rise to them, but the way Hatch has woven in the personalities of the scientists involved. At times he steps aside to provide a brief characterization of one of the protagonists. At other times, the personalities are illuminated by the style of the experiments through which they tackle problems. In some cases, hypotheses emerge slowly from data painstakingly collected and analyzed. At the other extreme, a flash of inspiration reveals an attractive solution to a problem, which is then quickly subjected to an incisive test. And of course the very success of our field derives from this diversity of styles: we make mistakes, but different sorts of people make different mistakes in the self-correcting process we call science.

In writing the preceding paragraph, I found myself unable to use the past tense when talking about Hatch. To say that he lives on through this book seems like a trite, overused sentiment. Yet no one who knew him and now hears him talking again in this book, his book, can possibly feel otherwise.

This living memorial did not simply happen through its own creative energy. Nor did it come to fruition solely through the efforts of the many who Carol Gross so generously acknowledges in her Preface, though their contributions were of course important. The real driving force was Carol herself, who contributed to the book at so many levels—writing unfinished sections, editing, working through the illustrations, and finding an enthusiastic editor. Hers was a classic labor of love, love both for the author and for the story he so much wanted to share.

Hatch has left us with a highly readable, personalized account of molecular biology past. It may be fitting to end this volume with a few thoughts on what the future might hold for the field. The information age is finally intruding on molecular biology and not at all subtly. The various genome sequencing projects, the determination and analysis of vast numbers of macromolecular structures, and the availability of biochips to analyze patterns of gene expression have already changed profoundly the way in which molecular biologists ask and answer questions. This bioinformatics revolution is as yet so young that we perhaps all underestimate the impact it will have in the next decades.

Yet there are some skills that today's students and practitioners of molecular biology cannot learn by surfing the web: the ability to state an idea clearly, to devise a rigorous test, to interpret results with self-skepticism. The challenge for the next generation of molecular biologists will be to embrace these values and apply them to the vast array of structural, genomic, and functional-genomic information found in ever-expanding databases. If the field is to remain vibrant, then the sort of chalk-in-hand, passionate exploration of ideas which lit up that Berkeley classroom 30 years ago will be just as important as ever.

Tom Cech

TIMELINE

IDEAS AND TECHNOLOGY		EXPERIMENTS
Beadle & Tatum propose "one gene–one enzyme" hypothesis.	1941	
	1944	Avery demonstrates that bacterial genes can be transferred as free DNA.
	1947	McClintock obtains evidence for transposable genetic elements in maize.
Lederberg distinguishes *lac*$^+$ and *lac*$^-$ bacteria on agar indicator plates.	1948	
	1949	Lethal effects of UV light on bacteria found to be reversible.
"Pauling Principles" for structure of proteins set forward.	1951	λ prophage discovered
Plaque assay for animal viruses developed.	1952	Hershey & Chase show that DNA carries genetic information.
Watson & Crick propose the double helical structure of DNA. *E. coli* strains having efficient gene transfer developed.	1953	Zamecnik achieves protein synthesis in a test-tube reaction

Gamow formulates the coding problem.   1954

1955   Ribosomes shown to be site of protein synthesis.

"Adaptation" found to require new enzyme synthesis; renamed "induced enzyme synthesis."

1956   DNA synthesis *in vitro*.

1957   Genes shown to be linear arrays by genetic mapping of *rII* mutations.

tRNA shown to be protein synthesis intermediate.

Key regulatory genes (*cI, cII, cIII*) controlling λ lysis/lysogeny decision identified.

Morphological transformation assay for quantitative studies of RNA tumor viruses developed.   1958   DNA replication shown to be semiconservative.

DNA polymerase I purified.

Stable diploid strains of *E. coli* developed for testing dominance.   1959   Pardee, Jacob, and Monod publish key experiments leading to repressor model of enzyme induction (PaJaMa experiment).

Jacob and Monod publish operon model for gene regulation and declare all gene regulation to be negative.   1960   First high-resolution structure of a protein.

Discovery of RNA polymerase.

1961   mRNA identified as intermediate in protein synthesis, is complementary to DNA.

First step in solving genetic code biochemically: polyU codes for polyPhe.

Genetic proof of a DNA triplet code.

Evidence for recombination by breakage and rejoining in phage.

Campbell proposes that λ lysogeny involves integration of phage DNA into chromosomal DNA.

1962

DNA shown to be degraded in "restrictive" host, leading to discovery of restriction enzymes.

Heat shock response discovered in Drosophila.

Temin proposes idea of reverse transcription.

1963

Visualization of DNA replication forks by electron microscopy shows both strands replicated in same direction.

λ found to have cohesive ends.

Hall develops first filter-binding assay to measure RNA-DNA hybrids.

Pyrimidine dimer found to be a major product of UV damage.

Conditional lethal mutants introduced for phage.

Holliday proposes model for mechanism of genetic recombination.

1964

1965

Nonsense chain termination codons, nonsense suppressor genes identified.

Key *E. coli* recombination genes *recA, B, & C* identified.

Solution of genetic code completed.

tRNA shown to have cloverleaf structure.

First evidence for insertion sequences (IS elements) in *E. coli*.

Crick proposes "wobble hypothesis" for codon-anticodon recognition.

1966

ara operon shown to be positively regulated (experiments from 1962 to 1966).

Lac repressor purified.

N-formyl methionine shown to be initiator amino acid for protein synthesis.

N and Q genes defined as sequential positive regulators of λ development.

uvrA, B, & C genes identified; required for repair of UV damage.

	1967	λ repressor purified, shown to bind DNA.
		Purified Lac repressor binds DNA containing *lac* operator.
	1968	Many genes shown to be involved in DNA replication (1968–1971).
		Isolation of λ integration-defective (*int⁻*) mutants.
		Discovery of site-specific recombination.
		Okazaki demonstrates discontinuous replication of one strand of DNA duplex.
		λ cohesive ends shown to be single-strand regions with complementary sequences.
		First biochemical characterization of a restriction enzyme.
Lobban first proposes method for joining DNA ends *in vitro*.	1969	Ribosomes found to select the beginning of genes.
		σ subunit of *E. coli* RNA polymerase identified.
		Discovery of Rho, *E. coli* transcription termination factor.
		DNA Pol I mutant isolated.
Ca⁺⁺ treatment allows *E. coli* to take up foreign DNA.	1970	*lac* operon shown to be under positive as well as negative control.
		First site of restriction enzyme cleavage determined.
		Reverse transcriptase described.
		Nonsense codons found to terminate normal proteins.

Witkin's and Radman's studies lead to the idea that all UV responses are part of single multi-gene pathway, SOS response.

Kornberg and Brutlag develop soluble extract system that will replicate a small DNA phage.

DNA fragments first joined in test tube to form "recombinant" DNA.

Purification of eukaryotic mRNA on polydT column.

Construction of first plasmids useful for molecular cloning.

Shine & Dalgarno propose sequence for rRNA-mRNA base-pairing.

Two-dimensional gels described by O'Farrell.

Yanofsky proposes mechanism for regulation of *trp* operon by attenuation.

1971 cII, cIII shown to be both positive regulators of λ*cI* gene & negative regulators of lytic genes in λ lysis/lysogeny decision.

Purified λ repressor inhibits transcription.

DNA polymerase III shown to be respon-sible for DNA replication in *E. coli*.

Demonstration that mRNA + nascent polypeptide move relative to ribosome (translocation).

1972 Cleavage by *Eco*RI restriction enzyme shown to generate cohesive ends.

1973 Heat shock in fruit flies results in new synthesis of a few proteins.

Sequence of *lac* operator determined, and found to be symmetric.

1974 λ N protein shown to act by anti-termination.

First eukaryotic genes cloned into *E. coli* plasmid.

Proteins required for DNA replication defined *in vitro* using single-stranded phage as model systems (1974–1977).

1975 λ DNA integrated into host DNA *in vitro*.

1976 RSV oncogene *src* shown to be originally a cellular gene.

The prokaryotic promoter is defined.

DNA sequencing methods developed by Maxam & Gilbert, Sanger.	1977	Mu phage shown to replicate by transposition.
		Splicing of discontinuous RNA segments in eukaryotic genes discovered.
DNase I "footprinting" developed to identify DNA sequences contacted by proteins.	1978	*E. coli* heat shock response first described.
"Libraries" of eukaryotic genes made in λ.		*recA* mediates cleavage of λ repressor.
		Initial work suggesting rRNA has vital role in ribosomal function.
Shapiro proposes mechanism for Mu transposition & replication.	1979	*In vitro*, RecA shown to initiate Holliday recombination.
Retroviruses proposed to integrate by transposition.	1980	Some RNAs shown to be capable of self-splicing.
		λ repressor & Cro shown to act as molecular switch in lysis/lysogeny decision.
	1981	First AIDS cases described.
		Replication starting from the origin of a duplex DNA is achieved in a test tube.
		Structure solved for first DNA binding proteins, Cro & CRP. Helix-turn-helix DNA binding motif identified.
	1982	Reactive component in λ integration found to be large nucleoprotein complex (intasome).
		λ Q protein shown to act as anti-terminator.
		Experiments show λ lysis/lysogeny decision mediated by amount of cII, which is regulated by control of cII synthesis & stability.

1983   HIV identified as cause of AIDS.

RNase P shown to be a ribozyme, or catalytic RNA.

mRNA splicing *in vitro*.

Mu transposition *in vitro*.

A nucleoprotein structure implicated in site-
specific recombination.

1984   Regulatory factor for *E. coli* heat shock
shown to be alternate σ factor for RNA
polymerase.

Heat shock proteins found to be universal.

RecA shown to induce SOS response by
mediating LexA autodigestion.

1985   OriC-dependent DNA replication *in vitro*
achieved with purified proteins.

Nucleoprotein complex at λ origin of replication visualized by electron microscopy.

PCR amplification technique developed.   1986

1987   Drosophila heat shock regulator purified;
is transcription factor.

Retroviral integration in a test-tube reaction.

16S RNA   See ribosome.

30S subunit   See ribosome.

50S subunit   See ribosome.

70S ribosome   See ribosome.

A site   See ribosome.

A-T tailing   A-T tailing was the original method pioneered by Peter Lobban for joining two pieces of DNA in the test tube. Cohesive sites were created by adding short "tails" of A and T respectively to the two DNA molecules. This was similar in concept to adding bits of velcro to the ends of the molecules, as the A and T tails were complementary and allowed the two DNA molecules to stick together.

active site   The active site of a protein is the site within its three-dimensional structure that is responsible for specific molecular associations and chemical transformations.

amino acid   Amino acids are the subunits of a protein that are joined together through peptide bonds to form a long chain, or polymer. Each of the 20 different amino acids used in proteins have chemically distinct properties.

amplification   Gene amplification is the generation of multiple copies of a gene by replication of the vector into which the gene is spliced. Genes can also be amplified with a PCR reaction.

anticodon   The 3-base sequence in a tRNA that recognizes a specific codon in mRNA is called the anticodon.

antitermination   The expression of certain genes is regulated by preventing RNA polymerase from stopping at a site upstream of the gene, a process known as antitermination. The λ genes N and Q act as antiterminator proteins.

assay   Assay is a scientific term for a way of measuring or testing something. For example, the commonly used assay for whether a cake is done is to poke it with a toothpick.

asymmetric replication   Asymmetric replication is the mode of DNA replication used to copy the DNA of certain single-stranded phage in which only one of the two DNA strands is copied.

ATP   Adenosine triphosphate (ATP) is a molecule used by the cell to store energy.

attachment sites   Attachment sites are the specific sites on the λ and *E. coli* chromosomes that are required for the λ integration reaction. The phage attachment site is large, requiring some 240 bp of DNA and having seven binding sites for Int protein and three for IHF; the bacterial attachment site is small, only about 20 bp, with 2 Int binding sites.

attenuator   An attenuator is a regulatory site located downstream of the promoter but preceding the genes of an operon, at which transcription is regulated by antitermination (see Figure 5–21).

autoradiography   Autoradiography is the blackening on an X-ray film caused by the decay of a radioactive element.

β*-galactosidase*   The enzyme β-galactosidase, specified by the *lacZ* gene, splits the sugar lactose into two simpler sugars, glucose and galactose.

bacteria   Bacteria are microscopic single-celled organisms. They are prokaryotes, having a simple cell organization with no separate nucleus, in contrast to the eukaryotic cells of higher organisms that have a membrane-bound nucleus enclosing the chromosomes. Bacteria multiply rapidly by splitting in two.

bacteriophage   Bacteriophage, often referred to as "phage," are viruses that infect bacteria. Phage have served as model systems for genetic and molecular analysis because their small genomes and simple structure make them (relatively) easy to study, while they demonstrate the same fundamental genetic and molecular mechanisms as larger creatures with more complex and differentiated cells.

base   Within the chain of nucleotides that makes a DNA or RNA strand, the bases are the chemical components that differ from one nucleotide to the next. The four bases adenine (A), guanine (G), thymine (T), and cytosine (C) are used in DNA; in RNA, adenine, guanine, uracil (U), and cytosine are used. Adenine and guanine are members of the chemical class known as purines, whereas thymine, cytosine, and uracil are pyrimidines.

base-pair   The bases in one strand of DNA are paired with partner (or complementary) bases in the other strand. These base-pairs constitute the equivalent subunits from which the duplex structure of DNA can be formed (see Figure 1–5).

branch point   When DNA strands are exchanged between two DNA duplexes, as in general genetic recombination, the point at which the DNA strand branches (like a tree branch) away from the duplex is referred to as the branch point (Figure 7–3).

branch point migration   In general recombination, one DNA strand from each parental DNA duplex is broken. Each broken strand pairs with the complementary parental strand of the other DNA molecule, forming a heteroduplex region. This heteroduplex region can then enlarge, or migrate, a phenomenon known as branch point migration, or branch migration (Figure 7–3).

cAMP   Cyclic AMP (cAMP) nucleotide is a small molecule involved in regulation of certain genes.

catalysis   The term catalysis refers to the acceleration of chemical reactions. Within the cell, catalysis is generally carried out by proteins called enzymes.

cell-free extract   A cell-free extract is the intracellular "soup" remaining after cells are broken (for instance by grinding) and cell debris is removed by low-speed centrifugation. Cell-free extracts provide a way of studying a biochemical process without having completely purified components.

Central Dogma   The idea that information flows from DNA to RNA to protein was termed the "Central Dogma" by Francis Crick.

centrifugation   Centrifugation is a way of separating the components of a mixture by rapid spinning. Heavier (or denser) components are forced farther from the center by centrifugal force, just as the riders on an amusement park ride are spun outward. In simple centrifugation, components are separated into two fractions: a pellet on the bottom of the centrifuge tube, and a "supernatant" (everything left in the liquid). In density gradient centrifugation, the centrifugal force causes a dense solution (of salt or sugar) to form a gradient, becoming more dense at the bottom of the centrifuge tube than at the top. Other components in the solution migrate to a position at which their density matches that of the surrounding solution (analogous to floating on very salty water).

chaperone   Heat shock proteins have been called molecular chaperones because of their functions in guiding newly synthesized proteins to fold and in accompanying proteins to facilitate the assembly of monomer units into active multimers.

chromosome   A chromosome is a DNA molecule containing the genes of an organism in a linear array. *E. coli* has one chromosome, which is circular. Eukaryotes have multiple chromosomes, in which the DNA is associated with specialized chromosomal proteins.

cI   The λ *cI* gene encodes a repressor able to turn off all the genes of lytic development. The cI repressor is required for maintaining lysogeny.

cII   The λ gene *cII* may be considered a "master regulator" of the decision to establish lysogeny. cII is a positive activator of the transcription of several genes necessary for lysogeny. It also activates transcription of a small RNA that prevents expression of the Q protein, required for lytic growth.

cIII   The λ gene *cIII* is required for λ to establish lysogeny. The cIII protein stabilizes the cII protein, increasing the amount of *cII* in the cell.

cis-*dominant*   A *cis*-dominant mutation effects only adjacent genes on the same chromosome (*cis* is from the Latin for *on this side of*).

clear plaque mutant   See plaque, λ *c* mutation.

cloning   In molecular cloning, a particular piece of DNA is isolated from its genomic source and attached to a vector DNA molecule capable of replicating independently of the host chromosome. In this way many copies of a specific piece of DNA may be made, allowing scientists to more easily study and manipulate specific genes or to produce large amounts of proteins of interest.

closed complex   RNA polymerase binds to promoters in several steps. It first recognizes the correct DNA sequence from the outside of the helix while the DNA remains base-paired; this is referred to as the closed complex.

codon   A three base (triplet) sequence of DNA specifies a single amino acid, just as a telephone area code specifies the region being called. This triplet coding unit is known as a codon.

cohesive ends   See cohesive sites.

cohesive sites Cohesive sites were first identified in phage λ as sites at each end of the linear λ molecule (called *cos* sites) that stuck together to allow the molecule to circularize, analogous to pieces of velcro on the ends of the molecule. These were determined to be short regions of single-strand DNA whose base sequences were complementary. The idea of joining DNA segments by cohesive sites (sometimes referred to as "sticky ends") became the guiding principle for the development of genetic engineering.

cointegrate In some types of transposition, the transposon itself is replicated to produce two copies, each linked to the new and old target sites; the product of this reaction is called a "cointegrate" because donor and recipient DNA molecules are fused (see Figure 7–18).

complementary The bases that can pair between two strands of nucleic acid are referred to as complementary. Adenine and thymine (or uracil in RNA) are complementary, as are cytosine and guanine. Complementarity of base-pairs is the means by which nucleic acids (for instance, the two strands of a DNA double helix) recognize one another.

complementation assay One approach to identifying proteins important in a process such as DNA replication is to first identify the genes involved by finding mutants defective in the process. The protein products of these genes can then be purified using an *in vitro* complementation assay: an extract from a mutant cell is used in the test tube to provide all enzymes but the one defective in the mutant. This missing function is supplied by a fraction from a normal cell (see Figure 4–14).

complementation test A complementation test is a genetic test used to determine whether two mutations are in the same or different genes (See Figure 3–9). In a complementation test, scientists ask whether two mutant pieces of DNA can each provide the functions missing from the other (or complement each other). In an everyday example, if the battery in one car and the brakes in another were defective, the functional battery could be used to "complement" the functional brakes to make a working car.

conditional-lethal Conditional-lethal mutations are mutations that allow the mutant phage or cells to grow under one set of conditions but not another. One type of conditional-lethal mutation is the temperature-sensitive mutation, which causes cells to stop growing at high temperature.

constitutive When a gene product is produced all the time, rather than in response to induction, this is referred to as constitutive expression. For example, Niagara Falls could be thought of as constitutive, whereas many waterfalls in Yosemite National Park are induced by rain or melting snow.

cooperative binding Some DNA-binding proteins, such as cI protein, bind to two or more DNA sites in a row. These proteins effectively increase their affinity to the DNA by binding to each other as well as to the DNA. This enhanced binding is called cooperative binding.

co-repressor A co-repressor is an environmental signal (or effector) that is required to convert a repressor to its active state.

Cro λ Cro protein is a negative regulator of the lysogenic pathway of λ development. Cro binds to the same operators as does cI.

CRP The catabolite receptor protein (CRP) is a protein that binds cAMP and acts as a positive regulator of transcription for certain operons. CRP is also called CAP (for catabolite activator protein).

cut and paste Some transposons use a "cut and paste" mechanism of transposition. The transposon is completely excised from its surrounding DNA by transposase. Transposase then makes staggered cuts in the target DNA and joins the transposon to this new DNA site.

degenerate The triplet DNA code is referred to as "degenerate," meaning that a given amino acid may be specified by more than one triplet codon. For example, both the codons UUU and UUC specify the amino acid phenylanine. Serine can be specified by any of four triplet codons.

density gradient centrifugation See centrifugation.

diploid An organism that has two copies of every gene is diploid (humans are diploid). A cell can also be diploid for a portion of its genome, or for a single gene.

discontinuous replication At a replication fork, one strand of DNA is synthesized continuously in the 5'–3' direction. The other strand of DNA, which has the opposite orientation in chemical terms, is synthesized by discontinuous replication: short DNA fragments are synthesized in a direction opposite to the direction of overall chain growth and then joined together by DNA ligase (see Figures 4–7 and 4–8).

DNA Deoxyribonucleic acid (DNA) is the hereditary material—genes are DNA. DNA is composed of long chains or "strands" of nucleotide subunits that contain the sugar deoxyribose. An intact DNA molecule is double-stranded, or "duplex": it is composed of two nucleotide strands wound together in a helical structure, much like two interlocked spiral staircases. See also nucleotide; base; base-pair; complementary.

DNA ligase DNA ligase is a DNA-joining enzyme. During replication, it is responsible for stitching together the DNA segments on the lagging strand into a continuous chain.

DNA polymerase DNA polymerase is an enzyme that synthesizes DNA. Three DNA polymerases have been identified in *E. coli*; Pol III is the polymerase responsible for chromosomal DNA replication.

DNA replication DNA replication is the copying, or duplication, of the genome to make two copies.

DNA sequencing The order, or sequence, of nucleotides in a segment of DNA can be determined (akin to reading out the letters in a written message) by techniques known as "DNA sequencing."

DnaG DnaG protein, specified by the gene *dnaG*, is the primase used in replication of the *E. coli* chromosome.

DNase I DNaseI is an enzyme that degrades DNA.

duplex DNA See DNA.

EcoRI *Eco*RI is a restriction enzyme, or restriction endonuclease, that cuts at a specific DNA sequence and leaves a staggered break, or "sticky ends." It was named *Eco*RI because the enzyme was specified by an "R-factor" plasmid of *E. coli*.

effector An effector is an environmental signal, positive or negative, that controls the activity of a regulatory protein.

elongation factor See translation factor.

end-labeled DNA fragments may be "end-labeled" by linking a radioactive phosphate to one end. End-labeling of DNA is essential to the Maxam-Gilbert DNA sequencing method and has been widely used in studying protein-DNA interactions (see Figures 5–28 and 5–29).

endonuclease An endonuclease is an enzyme that cuts DNA or RNA like a scissors at sites within the molecule.

enzymatic adaptation   The ability to produce enzymes in response to an environmental need, now referred to as "induced enzyme synthesis" or induction, was originally called enzymatic adaptation. See induction.

enzyme   Enzymes are proteins that accelerate, or catalyze, chemical reactions. Because of the ability of enzymes to catalyze chemical reactions, some reactions occur in fractions of seconds in the cells of our body that would not occur within our lifetimes with the free molecules in a test tube.

equivalent subunits   A key element of the Pauling Principles for the structure of proteins was that the amino acid subunits would be used in the same way in the structure. This principle was successfully extended to the structure of DNA by Jim Watson and Francis Crick, who determined that the equivalent subunits of DNA are not the individual nucleotides, but the base-pairs A-T and G-C (Figure 1–5).

eukaryote   A eukaryote is an organism whose cells have a membrane-bound nucleus enclosing the chromosomes (in contrast to the prokaryotes, or bacteria, which lack a nucleus).

excision   Excision is the removal of integrated viral, plasmid, or transposon sequences from the host chromosome.

exonuclease   An exonuclease is an enzyme capable of degrading DNA or RNA starting from the end of a strand.

extract   See cell-free extract.

F factor   The F factor is the *E. coli* mating factor, a nonchromosomal piece of circular DNA that encodes the proteins necessary to transfer itself from an F^+ to an F^- bacterial cell. F^+ bacteria are "male" strains capable of transferring the F factor, while F^- bacteria are "female" strains able to act as recipients.

F^+/F^- bacteria   See F factor.

fmet-tRNA$_f$ A special tRNA, formyl methionine tRNA or fmet-tRNA$_f$, is used for the initiation of protein synthesis in *E. coli*. Fmet-tRNA$_f$ establishes the reading frame for a protein by selecting the correct AUG as a start codon (see Figure 6–4).

formyl methionine   Formyl methionine is a variant of the amino acid methionine, in which an additional chemical group termed formyl is linked to the N-terminal end of the amino acid. Formyl methionine is used to start amino acid chains by fmet-tRNA$_f$ (see fmet-tRNA$_f$; Figure 6–3).

fractionation   Fractionation is a traditional biochemical approach to understanding a biological process, in which each active component is separated and purified in the test tube.

gel electrophoresis   Gel electrophoresis is a technique used to separate DNA or protein molecules by size. An electrical field is set up across a gel, a jello-like matrix. Small molecules move rapidly through the gel in the electrical field, whereas larger molecules move more slowly.

gene   A gene is a unit of heredity. Genes are now known to be segments of DNA, which are arranged in a linear array along a chromosome.

gene expression   Gene expression is the transfer of biological information from DNA to protein: an RNA copy of the gene (mRNA) is made by RNA polymerase; the mRNA binds to the ribosome; at the ribosome, the codons of the mRNA are translated into amino acid sequence to make a protein.

general recombination   General recombination, sometimes called genetic recombination, is the mode of genetic exchange studied by classical genetics. Here, new combinations of genes are created when homologous regions of two chromosomes exchange segments. Thus if one copy of a chromosome carries the A⁻ and B⁺ genes, and the other the A⁺ and B⁻ genes, recombination may result in the two new combinations (or recombinants) A⁻B⁻ and A⁺B⁺ (see Figure 2–2).

genetic code   The genetic code is the mechanism for storing information in DNA, the way in which the sequence of the bases in a gene determines the amino acid sequence of a protein. In the genetic code, a sequence of three bases (or "codon") specifies a particular amino acid. The code is thus sometimes referred to as a triplet code.

genetic engineering   Genetic engineering is the manipulation of DNA by scientists to create novel genetic combinations. Currently, genetic engineering is widely used in basic research as a means of studying gene function and in medicine to economically generate protein drugs such as human growth hormone, interferons, and vaccines.

genetic mapping   Genetic mapping is a technique by which the order and relative separation of mutations is inferred from the frequency with which two mutant DNAs exchange segments by genetic recombination to restore a normal gene. The farther two mutations are from each other on the chromosome, the more frequently recombination will occur between them (see Figure 2–2).

genetic recombination   See general recombination.

genome   The genome of an organism is the complete set of genetic information stored within its DNA.

genotype   The genotype of an organism is its array of units of heredity, or genes.

glucose effect   When fed a mixture of sugars including glucose, bacteria use the glucose first and do not waste energy and materials making the enzymes required for the other sugars. So long as glucose is present, operons such as lactose, arabinose, and maltose are not induced efficiently by their respective sugars. This is termed the "glucose effect."

gratuitous inducer   A compound which causes production of a particular protein is an inducer of that protein. Inducers of β-galactosidase that are not substrates of the enzyme are termed "gratuitous inducers."

GTP   The guanine nucleotide with three phosphates plays several roles in the cell. It is both a precursor for RNA synthesis and a required "cofactor" for protein synthesis, where the conversion from GTP to GDP mediates conformational changes in translation factors.

haploid   Cells or organisms that have only one copy of each gene are haploid.

heat shock   See heat shock response.

heat shock proteins   Heat shock proteins are the set of proteins produced in response to a heat shock. Many of the heat shock proteins function to refold or to facilitate the degradation of proteins damaged by the heat treatment—a molecular cleanup function. Heat shock proteins appear to be critically involved in protein folding and assembly and in the translocation of proteins across membranes under normal conditions as well. (See also heat shock response.)

heat shock response   High temperature is a common environmental stress for all organisms and a potentially dangerous one because the proteins that make cells work are especially sensitive to damage from elevated temperature. All cells react to a heat shock (a sudden increase in temperature) with a coordinated, multigene regulatory response that produces elevated amounts of a class of proteins termed the heat shock proteins. This response is called the heat shock response.

helicase   A helicase is an enzyme that separates the DNA double helix into single strands.

helix-turn-helix motif   The helix-turn-helix motif is a secondary structure found in many different DNA-binding proteins. In a helix-turn-helix, one α-helix is positioned to touch the bases of the DNA binding site and an adjacent α-helix lies along the sugar-phosphate backbone. The interaction can be visualized as a thumb-and-forefinger grip by the protein on a standard DNA double helix (See Figure 5–34.)

heteroduplex   In a heteroduplex segment of DNA the two strands come from two different parental DNA molecules. Heteroduplex regions are formed during the process of DNA recombination.

heterozygosis   After genetic recombination a region of the DNA can be found to be heteroduplex. When heteroduplexes have the wild type form of the gene (allele) on one DNA strand and the mutant allele on the other DNA strand, this phenomenon is known as heterozygosis.

Hfr   High frequency of recombination (Hfr) *E. coli* strains are "male" strains in which the mating factor F has been inserted into the bacterial chromosome. During mating, nearby chromosomal genes are transferred into the recipient strain along with F factor DNA. Different Hfr strains have the F factor inserted at different places in the chromosome (see Figure 3–7).

HIV-1   HIV-1, often referred to simply as HIV (for human immunodeficiency virus), is the retrovirus that causes the disease AIDS.

hnRNA   The term hnRNA, or heterogeneous nuclear RNA, was originally used to describe very long RNA molecules found in the nucleus of eukaryotic cells. This hnRNA was later determined to be the unspliced precursors to eukaryotic mRNAs (see RNA splicing).

Holliday model   In the Holliday model for general genetic recombination, one DNA strand from each parental DNA duplex is broken. The two broken strands then exchange places, pairing with the complementary strand on the other DNA molecule. This cross-like intermediate is "resolved" by cutting with a resolving nuclease, producing recombinants (see Figure 7–3).

holoenzyme   Holoenzyme is a term that refers to a multi-subunit enzyme (for instance DNA polymerase or RNA polymerase) with all subunits present.

homologous   Regions of DNA with identical or very similar sequence are referred to as homologous.

hybridization   When duplex DNA is separated into single strands by heating, the two complementary strands will reassociate if allowed to cool gradually. If RNA copies of the DNA, or complementary DNA fragments from another source, are included in the mixture, then the RNA or added DNA will bind to the complementary DNA sequences during the cooling process, forming DNA-RNA or DNA-DNA hybrids.

hydrogen bond   A hydrogen bond is a weak chemical interaction ("bond") between different sections of a molecule, created by sharing a hydrogen atom. Hydrogen bonds are responsible for much of the folded structure of proteins and for the base-pairing of DNA.

IHF   "Integrative host factor" (IHF) is an *E. coli* protein that acts along with Int and Xis in the integration and excision of λ.

immunity   Prophage immunity refers to the ability of the λ prophage to prevent lytic growth of an infecting λ (see Figure 3–3).

in vitro   *In vitro* is from the Latin for "in glass" and refers to any biochemical reaction taking place outside of a living cell.

in vivo   *In vivo* is from the Latin for "in life" and means "within a living organism."

induced   See induction.

induction   Many proteins are produced, or induced, only in response to conditions in which they are needed. This process is referred to as induction. Induction of protein synthesis may be seen as analogous to a front porch light on a motion sensor, which is turned on in response to the presence of a visitor. In prophage induction, a prophage replicating harmlessly along with the host cell genome is induced to the lytic state, in which it multiplies rapidly and lyses the host cell.

initiation codon   The initiation, or start, codon in the mRNA signals the start of the coding sequence for a protein, just as a capitalized word signals the start of a sentence. The initiation codon is usually the triplet sequence AUG, which encodes methionine. The initiator AUG defines the set of mRNA codons used for a given protein; correct initiation is said to define the "reading frame" of the protein.

initiation factor   See translation factor.

insertion element   Insertion elements, or IS elements, are transposable elements found in *E. coli* that were originally identified through their ability to cause mutations by inserting in or near genes.

Int   The λ Int protein is required for integrating and excising the λ prophage.

intasome   The reactive component in λ integration was found to be a nucleoprotein complex called the intasome. In the intasome, DNA containing the phage attachment site is wound around multiple copies of the Int and IHF proteins.

integration   Integration is the reaction in which viral, plasmid, or transposon DNA is inserted into the host chromosome. Bacteriophage λ is integrated into the bacterial chromosome by recombination between unique sites in the phage and bacterial DNA (see also site-specific recombination), whereas Mu phage integrates at random sites by a reaction called transposition.

intron   The expression of nearly all genes of higher eukaryotes depends on splicing together RNA sequences that are not contiguous in the DNA. The noninformational spliced-out segments are called intervening sequences or "introns."

IPTG   IPTG is a gratuitous (nonmetabolized) inducer of the *lac* operon.

isotope   An isotope is one of several possible forms of a chemical element, each differing in their number of neutrons. Both radioactive isotopes and "heavy" isotopes have been used as a way of "labeling" molecules of interest in biological experiments, as they differ from the forms found most commonly in nature and can easily be detected.

λ*c mutations*   λ*c* mutations, or clear plaque mutations, result in phage that are defective in the lysogenic response. (These produce clear plaques in a lawn of bacteria on an agar dish; see plaque.)

lacI   *lacI* is the gene for the *lac* repressor (LacI). A *lacI* ⁻ mutation causes *E. coli* to produce abundant β-galactosidase all the time (constitutively), rather than only in response to induction.

lacZ   *lacZ* is the *E. coli* gene coding for the protein β-galactosidase.

lagging strand   In DNA replication, the DNA chain growing overall in the 3′ to 5′ direction and elongated discontinuously (in the 5′ to 3′ direction) is called the lagging strand.

leading strand   In DNA replication, the DNA chain growing in the 5′ to 3′ direction and elongated continuously from a single RNA primer is called the leading strand.

LexA   The LexA protein is the master regulator of the bacterial SOS response, binding to multiple operator sites to turn off the multiple SOS genes. Cleavage of LexA, catalyzed by RecA, is the key event inducing SOS-regulated genes.

library   A "library," in molecular cloning terminology, is a set of clones in which an organism's complete genome is represented. These clones can be searched for a gene of interest by a variety of techniques. One form of library is the lambda library, in which the vector is λ phage.

long terminal repeat (LTR)   Each end of the proviral DNA of a retrovirus has identical sequences termed LTRs for "long terminal repeats." The LTR segments contain the promoter for making viral RNA, and serve to direct integration.

lysogenic   Lysogenic bacteria carry a phage genome in a latent, quiescent state. Such cells can be induced to produce phage particles.

lysogeny   Lysogeny refers to the presence of a phage genome in a bacterial cell in a harmonious, symbiotic state in which the viral genome replicates along with that of the host.

lytic   In the lytic state, bacteriophage multiply rapidly and lyse the host cell.

missense   Missense mutations are mutations that change one amino acid to another in the sequence of a protein (causing the amino acid sequence to have a different meaning, or "mis-sense").

morphological transformation   When cultured cells are infected with a tumor virus such as RSV, the virus can induce a "morphological transformation" (or cell transformation) causing the cells to have a different shape, and to gain certain characteristics of tumor cells. The viral alteration of the cells represents the cell culture equivalent of malignant transformation to tumor cells in an animal host.

mRNA   Messenger RNA (mRNA) is an RNA copy of a gene. mRNA molecules carry the genetic information from DNA to ribosome and serve as the template for the assembly of amino acids.

Mu phage   Mu (short for "mutator") phage is an *E. coli* phage that acts as a huge transposon. Mu can insert at many places throughout the host chromosome; insertion often disrupts a gene and thereby induces a mutation.

mutagenesis   Mutagenesis is the process of treating cells with chemicals, UV light, or X-rays to create mutations in the DNA.

mutagenic   A mutagenic agent is a treatment that causes mutations. Mutagenic agents include various chemicals, UV light, and X-rays.

mutation   A mutation is an alteration in the DNA sequence. Mutations often cause production of altered (or mutant) proteins.

N   N protein, encoded by the λ *N* gene, is a positive regulator of λ early gene transcription. N allows transcription of these genes by over-riding termination signals in front of the genes.

negative interference   In studying phage recombination, the frequency of two genetic exchanges very close to each other was found to be much higher than would be predicted from the frequency of single exchanges. This phenomenon was called "negative interference," to signify that having one exchange in a region seemed to encourage rather than inhibit a second exchange in the same region.

negative regulation   In negative regulation, a regulatory molecule turns OFF the genes of an operon under the appropriate conditions, such as in the absence of an inducer or in the presence of a co-repressor. Inactivating the regulator by mutation results in constitutive expression of the genes in the operon.

nonsense codons   Nonsense codons, also called stop codons, specify the termination of an amino acid sequence, like a stop sign for the ribosome.

nonsense mutations   Nonsense mutations are mutations in which a triplet codon specifying an amino acid is altered to one directing the ribosome to stop translation (a "stop" codon); thus nonsense mutations cause translation to terminate during synthesis of a protein.

nucleases   Nucleases are enzymes that degrade DNA or RNA. They include endonucleases, which cut like scissors at internal sites, and exonucleases, which nibble in from the end of the nucleic acid.

nucleoprotein complex   Specialized DNA–protein complexes termed nucleoprotein complexes, in which the DNA is wound around multiple DNA-binding proteins, are used in recombination and integration reactions, the initiation of DNA replication, and to promote distant interactions in transcriptional regulation. In this way, DNA-bound proteins do not have to act in a linear array but can interact over distance, bending or winding the intervening DNA.

nucleotide   Both DNA and RNA are long chains of subunits called nucleotides. Each nucleotide consists of three chemical parts, a base, a sugar, and a phosphate. Nucleotides with one phosphate are referred to as monophosphates, those with two phosphates diphosphates, and those with three phosphates triphosphates.

nus *genes*   *nus* stands for *N* utilization substance; Nus proteins interact with RNA polymerase during elongation to influence the termination response of the enzyme.

O   O protein, encoded by the λ *O* gene, is an initiation factor for λ replication.

O^c *mutation*   A mutation in the *lac* operator that causes constitutive synthesis of the adjacent *lac* operon. O^c mutations are *cis*-dominant.

oncogene   An oncogene is a gene capable of causing cancer. Many oncogenes are mutated forms of normal genes (proto-oncogenes) carried in the animal cell genome.

open complex   When RNA polymerase binds to a promoter, it first forms a closed complex and then partially unwinds the promoter DNA. This second step is referred to as the open complex.

operator   An operator, or O site, is the DNA site at which a repressor binds to prevent transcription of an operon.

operon   A group of adjacent genes transcribed from the same promoter and usually serving a related function, is known as an operon.

oriC   *oriC* is the origin of replication for the *E. coli* genome.

origin of replication   The site at which DNA replication initiates is the origin of replication. The *E. coli* genome has one origin of replication, called *oriC*.

oriT   *oriT* (or origin of transfer) is the site on the *E. coli* F factor at which replication begins for transfer of the F factor into the recipient cell during bacterial mating.

P   P protein, encoded by the λ P gene, is an initiation factor for λ replication.

P site   See ribosome.

pathway   The series of steps in a biological process are referred to as a pathway. The term pathway may be most familiar to students of biochemistry in the guise of metabolic pathways, in which molecule A is converted into molecule B which is then converted into molecule C, and so on.

PCR   PCR, or polymerase chain reaction, is a technique for making unlimited numbers of copies of a specific piece of DNA, a bit like a molecular photocopy machine.

peptide   The term peptide refers to a polymer of a small number (2–100) of amino acids linked together by peptide bonds; if the molecule has greater than 100 amino acids it is usually referred to as a polypeptide or a protein.

peptide bond   The chemical bond linking amino acids within a polypeptide chain is referred to as a peptide bond. See polypeptide.

peptidyl transferase   The peptidyl transferase activity of the ribosome is responsible for linking each new amino acid to the end of the growing peptide chain.

permease   A permease is an enzyme required for transport of a compound into the cell.

phage   See bacteriophage.

phage λ (lambda)   λ is a temperate *E. coli* phage that has served as a model system for the study of gene regulation.

phenotype   The phenotype of an organism is the expression of the genes in the observable traits of the organism, such as green eyes or curly tail.

plaque   A commonly used assay for phage growth is the plaque assay. Bacteria are spread on an agar dish, forming a uniform "lawn." Where phage infection occurs, a small circular hole in the lawn, called a plaque, is seen. Infection by virulent phage produces clear plaques, as there are no survivors of infection, whereas infection by temperate phage produces turbid plaques because surviving bacteria grow within the ring of lysed cells (see Figure 3–4). This same assay has been adapted to measure the growth of animal viruses (see Figure 9–1).

plasmid   A plasmid is a small circular DNA molecule that has its own origin of replication and can replicate in a cell independently of the bacterial chromosome.

pleiotropic A pleiotropic mutation affects the activity of many genes. For example, a mutation inactivating a positive regulatory protein turns off expression of all of the genes in an operon, a pleiotropic negative phenotype.

Pol III holoenzyme The complete DNA polymerase Pol III is a large multi-protein ensemble called Pol III holoenzyme. In addition to the subunits originally identified as the original minimal replication enzyme Pol III, the holoenzyme includes proteins required to recognize primers efficiently and to copy very long stretches of DNA without leaving the template, a property called processivity.

polar When a nonsense mutant terminates polypeptide chains in the middle of a protein, not only is enzyme activity from the mutant gene lost but in many cases enzyme activity from downstream genes in the operon is reduced as well. Because downstream but not upstream genes are affected, nonsense mutations are termed polar, meaning having direction (as an electric field). This phenomenon occurs because a failure of protein synthesis can lead to premature termination of RNA chains (see Figure 5–16).

polyA tail The 3′ end of a eukaryotic mRNA molecule carries a chain of adenines called the polyA tail.

polymerization When many similar subunits are linked together to form a polymer, the process is called polymerization. In molecular biology, polymerization commonly refers to the synthesis of DNA by DNA polymerase or RNA by RNA polymerase.

polynucleotide phosphorylase Polynucleotide phosphorylase, or PNPase, was briefly considered to be the enzyme responsible for RNA synthesis in the cell; in fact, its biological role is probably the reverse reaction of RNA degradation. In a test tube, PNPase synthesizes random polymers from a mix of diphosphate substrates (ADP, GDP, CDP, UDP).

polypeptide A polypeptide is a long chain, or polymer, of amino acids joined together by peptide bonds (see Figure 1–1). A protein may be composed of one or more polypeptides.

polyU polyU is a (synthetic) RNA in which all of the bases are uridine. The discovery that polyU encoded polyphenylalanine was the initial breakthrough in deciphering the genetic code.

positive regulation In a positive regulatory system, the regulator turns ON the genes of an operon when the appropriate inducer is supplied. Inactivating the regulator by mutation results in loss of ability to express the genes in the operon.

post-transcriptional Post-transcriptional regulation is regulation of gene expression at steps subsequent to the synthesis of RNA, including the translation of the mRNA, the stability of either the mRNA or protein, or the activity of the protein product.

precursor A precursor to a molecule is a compound used in the synthesis of the molecule. The nucleotides with three phosphates (dATP, dGTP, dCTP, and dTTP) are precursors for DNA.

primary structure The order of the amino acid subunits in a protein is known as the "primary structure" of the protein.

primase DNA primase (DnaG protein) is the enzyme that synthesizes the primers for *E. coli* DNA replication.

primer The primer strand is an existing strand of DNA or RNA to which DNA polymerase adds new bases.

promoter A promoter is the site at which RNA polymerase binds to the DNA in order to start transcription of a gene.

prophage The prophage state, also referred to as the lysogenic state, is the quiescent state in which a bacteriophage genome is integrated into the host chromosome and replicates along with that of the host cell. The latent viral genome is termed a prophage.

protein Proteins are composed of amino acids joined together through peptide bonds to form long chains, or polypeptides. A protein may be composed of one or more polypeptides. A key characteristic of proteins is their capacity for unique recognition and molecular association.

proteolysis Within the cell, proteins have only a finite lifespan. The breakdown of proteins is referred to as proteolysis, or proteolytic degradation, and the enzymes that accomplish this function as proteolytic enzymes.

provirus Just as λ phage knows how to integrate into the host chromosome to become an inactive prophage, retroviruses integrate into animal cell chromosomes and then replicate along with the host DNA as a provirus.

pulse-label In a pulse-label experiment, cells are "labeled" with a radioactive precursor of some molecule (for example DNA or protein) for a short period. In this way only the DNA or protein being synthesized at that particular time is labeled and can be visualized.

purine See base.

pyrimidine See base.

pyrimidine dimer The major lesion in DNA induced by UV is a stable chemical linkage between two adjacent pyrimidine bases, referred to as a pyrimidine dimer.

Q Q protein, encoded by the λ Q gene, is a positive regulator of the λ late genes. Q acts as an anti-terminator, over-riding signals to stop transcription in front of the late genes.

reading frame Three overlapping sets of triplet codons, or reading frames, are possible within the messenger RNA for a gene. The correct reading frame is the set of triplet codons that are actually used for a given protein and is defined by the position of the start codon (see Figure 2–12).

RecA protein RecA protein is a central player in general genetic recombination in *E. coli*, being required for pairing homologous duplexes and for strand transfer. Biochemical studies have shown that a RecA-coated single-strand of DNA can find homology in duplex DNA and begin the branch migration reaction (see Figure 7–6). RecA also promotes the cleavage of cI, LexA, and UmuD when cells experience DNA damage, thus initiating the SOS response.

RecBCD RecBCD enzyme, the product of the *recB*, *C*, and *D* genes, generates the single strand necessary for loading RecA onto DNA. It travels down the DNA, unwinding duplex DNA and degrading one strand until it reaches a Chi site, where it loads RecA onto that strand (see Figure 7–7).

recombinant A recombinant gene is the product of genetic recombination. For example, a wild-type recombinant may result from recombination, or exchange of DNA, between two mutant copies of a gene if each has a mutation in a different location.

recombination   Recombination is the process of rearranging DNA, and along with mutation, is a source of genetic diversity in a biological population. See also general recombination; site-specific recombination; transposition.

regulatory gene   The role of a regulatory gene is to control the activity of other genes.

release factor   See translation factor.

replication fork   The replication fork is the place where the two DNA strands separate, like a fork in the road, so that each strand can be replicated by DNA polymerase.

replicative form (RF)   In the life cycle of the single-stranded phages φX174 and M13, the parental single-strand (SS) DNA is copied to a double-strand (DS) form, which then replicates to make more double-strand circles, termed RF for replicative form.

repressor   A repressor is a gene product that functions to turn off the expression of other genes. The repressor functions by binding to the operator site of an operon.

restriction and modification   In a phenomenon termed restriction and modification, foreign DNA introduced into a bacterial host is cut by restriction enzymes (and thus "restricted" from growing). In contrast, DNA grown in the "restrictive" host is modified (by the addition of methyl groups) so that it is not cut.

restriction enzyme   Restriction enzymes are DNA endonucleases whose biological role is to digest foreign DNA that enters a bacteria, thus "restricting" its influence on the host cell. Some of this class of DNA-cutting enzymes cleave at specific sites. Restriction enzymes are now widely used as a molecular scissors to cut DNA into

specific fragments, which can then be stitched together in new combinations in the test tube.

restriction fragment   DNA pieces generated by cutting with restriction enzymes are termed restriction fragments.

retrotransposon   A retrotransposon is a transposon whose movement in the genome requires reverse transcription of an RNA copy of the transposon followed by integration of the DNA copy. Retrotransposons have their own reverse transcriptases.

retrovirus   The "retrovirus" (backwards virus) is an RNA virus that copies the information in its RNA into DNA upon entering the cell, integrates this DNA into the host genome, and finally copies the DNA into RNA to make new viruses. This is contrary to the direction of information flow in the Central Dogma (DNA to RNA to protein).

reverse transcriptase   Reverse transcriptase is an enzyme that copies RNA into DNA (in the reverse of the general direction of information flow from DNA to RNA to protein). First discovered in retroviruses, reverse transcriptase is also encoded by retrotransposons.

ribosome   Ribosomes, originally termed "ribonucleoprotein particles," are the site of protein synthesis in the cell. The intact ribosome (referred to as 70S, for its rate of sedimentation in a high-speed centrifuge) can be separated into two subunits, the 30S and 50S subunits. Together these subunits contain >50 proteins and 3 RNA species (5S, 16S, and 23S). Several functional sites have been defined in the ribosome, including a P site (for peptide binding), an A site (for accepting the next tRNA), and an E site (for exit). (See Figures 6–9, 6–10, 6–11, and 6–16.)

ribozyme The term "ribozyme" refers to an RNA molecule with catalytic activity, or an "RNA enzyme." Ribozymes exhibit the hallowed characteristics of a protein enzyme—exceptional specificity and rate acceleration.

R-loop mapping R-loop mapping was a technique developed by Ron Davis and Norman Davidson for mapping RNA to its DNA source. Under appropriate conditions, a homologous RNA could form a DNA-RNA hybrid, termed an "R-loop," within otherwise duplex DNA (see Figure 6–26).

RNA Ribonucleic acid (RNA) is, like DNA, a long chain of nucleotide subunits. Unlike DNA, RNA contains the sugar ribose and is primarily found in single-strand form. The major role of RNA in the cell is in the transfer of biological information from DNA to protein, in the form of messenger RNA (mRNA), transfer RNA (tRNA), and ribosomal RNA (rRNA). See also mRNA; tRNA; rRNA.

RNA phage RNA phage are bacteriophage having an RNA genome. RNA phage were used as a packaged, naturally occurring source of mRNA by scientists studying protein synthesis.

RNA polymerase RNA polymerase is the enzyme that synthesizes RNA copies of genes using a DNA template (or "transcribes" DNA).

RNA processing The RNA slicing events that make the "mature" rRNA and tRNA are collectively termed RNA processing.

RNA splicing In eukaryotes, the DNA coding sequence of many genes is discontinuous, requiring a process called RNA splicing to remove the extraneous information and assemble a functional mRNA capable of specifying a complete protein.

RNase H RNase H is an RNA degrading activity that degrades only RNA in an RNA-DNA hybrid (hence the H designation, for hybrid). RNase H is a second enzymatic activity of reverse transcriptase.

RNase P tRNAs in *E. coli* are produced as longer precursor molecules. These are processed to their mature size in part through cleavage by endonuclease RNase P. This enzyme has both a protein and an RNA component, with the RNA component responsible for catalysis.

RNase III RNase III is the central processing enzyme for *E. coli* rRNA and also processes certain mRNA species.

Rous Sarcoma Virus (RSV) The Rous sarcoma virus (RSV for short) is a retrovirus capable of causing tumors in chickens. A great many of the features of retroviruses have been discovered by studying RSV. Because of its role in the study of retroviruses the Rous virus has become one of the most famous viruses in scientific history.

rRNA Ribosomal RNA (rRNA) is the RNA found stably associated with ribosomes.

σ (sigma) The σ subunit of RNA polymerase is the subunit that allows the enzyme to recognize and bind to promoters. The major σ subunit of *E. coli* RNA polymerase is now referred to as σ^{70} (for its molecular mass of 70,000) to distinguish it from several other "minor" sigma factors. Each sigma factor directs RNA polymerase to a different set of promoters (see σ^{32}).

σ^{32} σ^{32}, encoded by the *rpoH* (*htpR*) gene, is a special sigma factor for RNA polymerase, directing the core transcription apparatus to the promoters of the heat shock genes so that their expression can be coordinately regulated.

salvational response   A salvational response is a coordinated set of changes within the cell in response to an external environmental challenge.

secondary structure   The amino acid chain of a protein follows a certain spatial contour; for instance, it may be coiled or be aligned in sheets. This is referred to as the secondary structure of the protein. Within a single-strand RNA or DNA molecule, some regions of the nucleotide chain can base-pair with one another. This is referred to as the secondary structure of the nucleic acid.

self-splicing   Certain RNA molecules, of which Tetrahymena rRNA is an example, undergo self-splicing; that is, the RNA molecule is capable of catalyzing its own splicing, without the help of protein enzymes.

semi-conservative replication   In replication of the *E. coli* genome, each strand serves as the template for a new strand. This mode of replication is called "semi-conservative," meaning that the newly replicated molecule is composed of one parent and one daughter strand—only one of the two parental DNA strands is conserved in the product of replication (see Figure 4–2).

Shine-Dalgarno sequence   The Shine-Dalgarno sequence is a sequence upstream of the initiator AUG in mRNA that base-pairs with a region at the end of the 16S ribosomal RNA, allowing the ribosome to recognize the initiator AUG (see Figures 6–18 and 6–19).

site-specific recombination   Site-specific recombination is recombination that occurs between unique, specific sequences on both recombining DNA's and joins regions of DNA with little or no homology. λ integration is a well-studied example of site-specific recombination.

snRNPs   The spliceosome is composed of smaller RNA-protein structures, termed snRNPs (pronounced "snurps") for small nuclear ribonucleoprotein particles.

SOS   *E. coli* responds to DNA damage with a complex, coordinated set of changes in gene expression termed the SOS response. In the SOS response, cell division is blocked, DNA repair mechanisms are induced, and the mutation rate is increased.

specificity   The specificity of a protein is its capacity for unique recognition and molecular association, a key property of proteins.

spliceosome   The machinery of RNA splicing is a huge RNA-protein complex analogous to a ribosome, termed the spliceosome. This apparatus is nearly as large as a ribosome, containing five different RNAs and more than 50 proteins.

Ssb   Ssb is a single-strand binding protein that helps replication by coating single-strand DNA, thereby preventing complementary regions from pairing with each other.

staggered break   A staggered break, or staggered cut, is created when each strand of the DNA duplex is cut at a slightly different position. This leaves very short, easily melted cohesive sites. A staggered break is created by λ Int protein in the λ integration and excision reactions, by transposases during transposition, and by many restriction enzymes during their cleavage reactions.

stem and loop   When two regions close to one another in an RNA molecule are complementary, they may base-pair, forming a "stem" (the base-paired region itself) and a "loop." (The stem and loop structure looks rather like a lollipop; see Figure 5–20.) Such RNA secondary structures are often involved in gene regulation.

substrate   The term substrate refers to the target molecule of enzyme catalysis; the molecule upon which an enzyme acts to accelerate a chemical transition.

subunit   Many enzymes, such as DNA polymerase and RNA polymerase, are composed of multiple polypeptide chains, or subunits. The term subunit also refers to the repeating unit that makes up a larger molecule: a protein is composed of amino acid subunits joined together, and DNA is composed of nucleotide subunits joined together.

suppressor tRNA   A suppressor tRNA is a mutant tRNA that overcomes the effect of nonsense mutations by "misreading" the nonsense codon as a sense codon. Instead of stopping translation, the suppressor tRNA inserts an amino acid to permit translation of a full length protein.

SV40   SV40 (simian virus 40) is a DNA tumor virus that has been extensively studied as a model for understanding both tumor virus function and mammalian molecular biology.

symmetric replication   In symmetric DNA replication, each of the two parental strands is replicated in the same direction at about the same time. The E. coli chromosome is replicated symmetrically.

T cells   T cells are a type of immune system cell, one of whose functions is to fight viral infections. T cells are one of the major targets for infection by HIV-1.

temperate phage   Temperate phage are capable of achieving a harmonious, symbiotic state (the prophage state) in which the viral genome replicates along with that of the host (which is referred to as lysogenic) (see Figure 3–3).

temperature-sensitive   A temperature-sensitive (ts) mutant can grow at lower but not at higher temperature. For example, in dnats mutants, replication proceeds normally at lower temperature but stops at elevated temperature.

template   In DNA or RNA synthesis, the template DNA strand is the strand that is copied by DNA polymerase to give rise to a daughter DNA strand or by RNA polymerase to produce an RNA copy.

terminal transferase   Terminal transferase is an E. coli enzyme that adds nucleotides to one end of a DNA molecule.

terminator   A terminator is the site at which an RNA transcript is terminated. There are two types of terminators: intrinsic terminators, which have a characteristic RNA structure recognized by RNA polymerase on its own, and rho-dependent terminators, which require a special termination protein called rho.

tertiary structure   The overall three-dimensional folded structure of a protein is known as its tertiary structure.

T4   T4 is a virulent bacteriophage originally favored for molecular biological study by Max Delbrück.

thymidine   Thymidine (the base thymine plus deoxyribose sugar) is a precursor for DNA but not for RNA.

tif   tif mutant E. coli (named for "temperature-induced filamentation") are constitutive for the SOS response at high temperature—all of the SOS functions are "on." The tif mutation is in the recA gene and results in an altered form of RecA, the positive regulator of the SOS response.

topoisomerase   Topoisomerase is an enzyme that removes the supertwists from the DNA double helix.

transcription RNA polymerase synthesizes an RNA copy of a gene using the DNA as a template; this process is referred to as transcription.

transformation Bacterial genes can be transferred into new cells as free DNA in a process known as "transformation." The term "transformation" is used in a different sense when studying tumor viruses; see morphological transformation.

translation When mRNA is used as instructions for the synthesis of a protein, the language of nucleotides is translated into the language of amino acids. Thus protein synthesis in the cell is called translation.

translation factor Translation factors are proteins that are required for protein synthesis in addition to the ribosome. These include three initiation factors (IF1, IF2, and IF3), three elongation factors (EF-Tu, EF-Ts, and EF-G), and three factors required to release the finished amino acid chain (RF1, RF2, and RF3).

translocation During the translation of mRNA, a tRNA bearing an amino acid is brought in to a part of the ribosome called the A site, where it binds to the next triplet codon in the mRNA. For translation to proceed, the mRNA must then move with respect to the ribosome, putting a new triplet codon in position to receive a tRNA and making the A site is available for the next amino acid. This movement is termed translocation.

transposase Transposase is the enzyme responsible for movement (transposition) of a transposon and is carried by the transposon itself.

transposition DNA transposition is the movement of transposable genetic elements (transposons) from one region of a genome to another.

transposon A transposable genetic element, or transposon, in its simplest incarnation is a DNA unit with a self-directed capacity to move from one region of a genome to another.

triplet binding assay The triplet (or trinucleotide) binding assay of Phil Leder and Marshall Nirenberg was a rapid binding assay in which ribosome-mRNA-tRNA complexes but not free tRNAs were retained on a special type of filter (analogous to a mini coffee filter). This assay was used to solve the genetic code by providing a series of synthetic mRNA's of only three bases—and determining which tRNA was bound to each (see Figure 2–10).

triplet code In a triplet DNA code, a linear sequence of three bases (known as a codon) specifies a single amino acid, just as a telephone area code specifies the region being called.

tRNA Transfer RNA (tRNA) is a set of small RNA molecules whose role in protein synthesis is to transfer amino acids to the ribosome for assembly into proteins.

unstable protein An unstable protein is one that is rapidly degraded in the normal cellular environment. The stability of some proteins is regulated—that is, the proteins may be more stable under certain cellular conditions.

UVR pathway The UVR pathway, involving the *uvrA*, *B*, and *C* genes, is a major component of DNA repair in *E. coli*. By excising T-T dimers, the Uvr pathway of excision repair saves about 99% of UV-irradiated *E. coli* from death. After the dimers are removed, repair DNA synthesis by DNA polymerase I fills the gaps (see Figure 8–12).

vector In molecular cloning, a gene of interest is spliced into a piece of DNA, called a vector, that can replicate in a cell independently of the bacterial chromosome (see Figure 10–1).

virulent phage Virulent phage are phage capable only of lytic growth, in which the host cell is killed.

virus A virus is an infectious agent composed of nucleic acid (DNA or RNA) and a protein coat. Viruses are parasites; that is, they are dependent upon the machinery of the host cell to replicate.

wild type A normal gene or organism is known as wild type (that is, the variety that is found in the wild).

wobble In tRNA-mRNA recognition, the criteria for base-pairing in the third position of the codon is relaxed. In addition to the Watson-Crick base-pairs of G-C and A-U, "wobble" base-pairs are also allowed: U and G can pair, and inosine, a G-like base found only in tRNA, can pair with U or A (see Figure 6–1).

Xis The λ Xis protein is required along with Int protein to excise the λ prophage.

X-ray crystallography X-ray crystallography is a technique used to determine the positions of the individual atoms in a molecule.

Chapter 1. Beginnings: Simplicity and Elegance, DNA and Protein

PROTEINS

Buchner shows that fermentation reactions can occur in cell extracts

Buchner, Eduard (1897) Alkoholische Gährung ohne Hefezellen [Vorläufige Mittheilung]. Berichte der Deutschen Chemischen Gesellschaft 30, 117–24. English translation in Gabriel, M. L., and S. Fogel, eds. (1955) *Great Experiments in Biology*. Englewood Cliffs, N.J.: Prentice-Hall.

Pauling proposes that structures of proteins derive from the the structural features of the amino acid subunits

Pauling, L., and R. B. Corey (1951) Configurations of polypeptide chains with favored orientations around single bonds: Two new pleated sheets. Proc. Natl. Acad. Sci. USA 37, 729–40.

Pauling and Corey propose α-helices and β-sheets

Pauling, L., R. B. Corey, and H. R. Branson (1951) The structure of proteins: Two hydrogen bonded helical configurations of the polypeptide chain. Proc. Natl. Acad. Sci. USA 37, 205–11.

Pauling, L., and R. B. Corey (1951) The pleated sheet, a new layer configuration of polypeptide chains. Proc. Natl. Acad. Sci. USA 37, 251–56.

Kendrew obtains crystal structure of myoglobin to high resolution

Kendrew, J. C., R. E. Dickerson, B. E. Strandberg, R. G. Hart, D. R. Davies, D. C. Phillips, and V. C. Shore (1960) Structure of myoglobin: A three-dimensional Fourier synthesis at 2-Å resolution. Nature 185, 422–27.

Perutz obtains structure of hemoglobin

Perutz, M. F., M. G. Rossman, A. F. Cullis, H. Muirhead, G. Will, and A. C. T. North (1960) Structure of haemoglobin: A three-dimensional fourier synthesis at 5.5-Å resolution, obtained by X-ray analysis. Nature 185, 416–22.

DNA

Watson and Crick propose structure of DNA

Watson, J. D., and F. H. C. Crick (1953) A structure of deoxyribose nucleic acid. Nature 171, 737–38.

X-ray work on DNA by Franklin and Wilkins

Wilkins, M. H. F., A. R. Stokes, and H. R. Wilson (1953) Molecular structure of deoxypentose nucleic acids. Nature 171, 738–40.

Franklin, R. E., and R. G. Gosling (1953) Molecular configuration of sodium thymonucleate. Nature 171, 740–41.

Chargaff's "rule": proportions of A and T and proportions of G and C are equal in DNA

Chargaff, E. (1950) Chemical specificity of nucleic acids and mechanism of their enzymic degradation. Experientia 6, 201–9.

Watson, "Suddenly I became aware . . ." and "The structure was too pretty not to be true"

Watson, J. (1968) *The Double Helix*, pp. 194, 205. New York: Atheneum.

Chapter 2. The Code for Life: DNA to Protein

GENES CODE FOR PROTEINS

"One gene–one enzyme" hypothesis of Beadle and Tatum

Beadle, G. W., and E. L. Tatum (1941) Genetic control of biochemical reactions in *Neurospora*. Proc. Natl. Acad. Sci. USA 27, 499–506.

Garrod proposes the idea of metabolic diseases

Garrod, Sir Archibald. (1923) *Inborn Errors of Metabolism*, 2d ed. London: Oxford University Press. 1st ed., 1909.

Beadle, "I have the impression . . ."

Beadle, G. W. (1996) Biochemical genetics: Some recollections," p. 31 in J. Cairns, G. S. Stent, and J. D. Watson, eds., *Phage and the Origins of Molecular Biology*. Cold Spring Harbor, N.Y.: Cold Spring Harbor Laboratory Press.

DNA carries genetic information

Avery, O. T., C. M. MacLeod, and M. McCarty (1944) Studies on the chemical nature of the substance inducing transformation of pneumococcal types. I. Induction of transformation by a deoxyribonucleic acid fraction isolated from *Pneumococcus* III. J. Exp. Med. 79, 137–58.

Hershey, A. D., and M. Chase (1952) Independent functions of viral protein and nucleic acid in growth of bacteriophage. J. Gen. Physiol. 39–56.

Zamecnik, "some of the doors . . ."

Zamecnik, P. C. (1969) An historical account of protein synthesis, with current overtones—a personalized view. Cold Spring Harbor Symp. Quant. Biol. 34, 9.

DEFINING THE CODING PROBLEM: THE INFORMATION PEOPLE

Formulation of the coding problem by Gamow

Gamow, G. (1954) Possible relation between deoxyribonucleic acid and protein structures. Nature 173, 318.

Caldwell and Hinshelwood paper
Caldwell, P. C., and C. Hinshelwood (1950) Some considerations on autosynthesis in bacteria. J. Chem. Soc. 4, 3156–59.

Presentation of sequence-matching concept by Dounce
Dounce, A. (1952) Duplicating mechanism for peptide chain and nucleic acid synthesis. Enzymologia 15, 251–58.

Amino acid sequence of insulin determined
Sanger, F., and H. Tuppy (1951) The amino-acid sequence in the phenylalanyl chain of insulin. 1 and 2. Biochem. J. 49, 463–80 and 481–90.
Sanger, F., and E. O. P. Thompson (1953) The amino-acid sequence in the glycyl chain of insulin. 1 and 2. Biochem. J. 53, 353–65 and 366–74.

Anfinsen demonstrates that proteins can refold in vitro
Anfinsen, C. B. (1973) Principles that govern the folding of protein chains. Science 181, 223–30.

Genetic mapping of rII mutations shows that gene is a linear array
Benzer, S. (1955) Fine structure of a genetic region in bacteriophage. Proc. Natl. Acad. Sci. 41, 344–54.

Genetic evidence for a triplet code
Crick, F. H. C., L. Barnett, S. Brenner, and R. J. Watts-Tobin (1961) General nature of the genetic code for proteins. Nature 192, 1227–32.

Sickle-cell hemoglobin has altered charge
Pauling, L., H. A. Itano, S. J. Singer, and I. C. Wells (1949) Sickle cell anemia, a molecular disease. Science 110, 543–48.

Sickle-cell hemoglobin contains an altered amino acid
Ingram, V. J. (1956) A specific chemical difference between the globins of normal human and sickle-cell anaemia haemoglobin. Nature 178, 792–94.

Sanger develops method to use proteases to break proteins into fragments
Sanger, F. (1988) Sequences, sequences, and sequences. Annu. Rev. Biochem. 57, 1–28.

Colinearity of gene and protein
Levinthal, C. (1959) Genetic and chemical studies with alkaline phosphatase of *Escherichia coli*. Brookhaven Symp. Biol. 12, 76–85.
Sarabhai, A. S., A. O. W. Stretton, S. Brenner, and A. Bolle (1964) Co-linearity of the gene with the polypeptide chain. Nature 201, 13–17.
Yanofsky, C., B. C. Carlton, J. R. Guest, D. R. Helinski, and U. Henning (1964) On the colinearity of gene structure and protein structure. Proc. Natl. Acad. Sci. USA 51, 266–72.

DNA TO RNA TO PROTEIN: THE BIOCHEMICAL PATHWAY PEOPLE
Early work by Caspersson and Brachet implicating role of RNA in protein synthesis
Caspersson, T., and J. Schultz (1939) Pentose nucleotides in the cytoplasm of growing tissues. Nature 143, 602–3.

Brachet, J. (1941) Le détection histochimique et al microdosage des acides pentose-nucléiques (tissus animaux-développement embryonnaire des amphibiens). Enzymologia Acta Biocatalytica 10, 87–96.

Brachet, "ribonucleoprotein particles . . ."
Brachet, J. (1946) Nucleic acids in the cell and the embryo. Symp. Soc. Exp. Biol. 1, 215.

Tracking of amino acids done initially by Zamecnik with animal cells and by Roberts with E. coli
McQuillen, K., R. B. Roberts, and R. J. Britten (1959) Synthesis of nascent protein by ribosomes in *E. coli* Proc. Natl. Acad. Sci. USA 45, 1437–47.

Zamecnik achieves some protein synthesis in a "cell-free" extract
Zamecnik, P. C. (1953) Incorporation of radioactivity from DL-leucine-1-C^{14} into proteins of rat liver homogenate. Fed. Proc. 12, 295.
Zamecnik, P. C., and E. B. Keller (1954) Relation between phosphate energy donors and incorporation of labeled amino acids into proteins. J. Biol. Chem. 209, 337–54.

Definitive demonstration using cell-free system that ribosomes are the sites of protein synthesis
Littlefield, J. W., E. B. Keller, J. Gross, and P. C. Zamecnik (1955) Studies on cytoplasmic ribonucleoprotein particles from the liver of the rat. J. Biol. Chem. 217, 111–23.

Zamecnik and Hoagland show that sRNA (now tRNA) is an intermediate in protein synthesis
Hoagland, M., P. C. Zamecnik, and M. L. Stephenson (1957) Intermediate reactions in protein biosynthesis. Biochm. Biophys. Acta 24, 215–16.
Hoagland, M. B., M. L. Stephenson, J. F. Scott, L. I. Hecht, and P. C. Zamecnik (1958) A soluble ribonucleic acid intermediate in protein synthesis. J. Biol. Chem. 231, 241–57.

Evidence for sRNA intermediate also obtained by Holley and Berg
Holley, R. W. (1957) An alanine-dependent, ribonuclease-inhibited conversion of AMP to ATP, and its possible relationship to protein synthesis. J. Amer. Chem. Soc. 79, 658–62.
Berg, P., and E. J. Ofengand (1958) An enzymatic mechanism for linking amino acids to RNA. Proc. Natl. Acad. Sci. USA 44, 78–86.

Crick's prediction of an adapter between nucleic acids and amino acids
Crick, F. H. C. (1958) On protein synthesis. Symp. Soc. Exp. Biol. 12, 138–63.

Review article of experiments by Holley, Berg, and others
Novelli, G. D. (1967) Amino acid activation for protein synthesis. Annu. Rev. Biochem. 36, 449–84.

Jacob proposes existence of mRNA
Pardee, A. B., F. Jacob, and J. Monod (1958) Sur l'expression et role des allèlles "inductible" et "constitutif" dans la syntheèse de la β-galactosidase chez les zygotes d'*Escherichia coli*. Comptes rendus des Academie des Sciences 246, 3125–27.

Pardee, A. B., F. Jacob, and J. Monod (1959) The genetic control and cytoplasmic expression of "inducibility" in the synthesis of β-galactosidase by *E. coli*. J. Mol. Biol. 1, 165–78.

Jacob, F., and J. Monod (1961) Genetic regulatory mechanisms in the synthesis of proteins. J. Mol. Biol. 3, 318–56.

Additional articles on the role of tRNA

Crick, F. H. C. (1966) Codon-anticodon pairing: The wobble hypothesis. J. Mol. Biol. 19, 548–55.

Chapeville, F., F. Lipmann, G. VonEhrenstein, B. Weisblum, W. J. Ray Jr., and S. Benzer (1962) On the role of soluble ribonucleic acid in coding for amino acids. Proc. Natl. Acad. Sci. USA 48, 1086–92.

CRACKING THE CODE: SYNTHETIC MRNA

E. coli *extracts for cell-free protein synthesis*

Lamborg, M. R., and Zamecnik, P. C. (1960) Amino acid incorporation into protein by extracts of *E. coli*. Biochimica et Biophysica Acta 206–11.

Tissières, A., D. Schlessinger, and F. Gros (1960) Amino acid incorporation into proteins by *Escherichia coli* ribosomes. Proc. Natl. Acad. Sci. USA 46, 1450–63.

Nirenberg's classic paper on protein synthesis

Nirenberg, M. W., and J. H. Matthaei (1961) The dependence of cell-free protein synthesis in *E. coli* upon naturally occurring or synthetic polyribonucleotides. Proc. Natl. Acad. Sci. USA 47, 1588–1602.

Polynucleotide phosphorylase discovered

Grunberg-Manago, M., and S. Ochoa (1955) Enzymatic synthesis and breakdown of polynucleotides; polynucleotide phosphorylase. J. Am. Chem. Soc. 77, 3165–66.

Khorana had begun to develop chemical methods for polynucleotide synthesis

Khorana, H. G. (1966–67) Polynucleotide synthesis and the genetic code. Harvey Lectures Series 1966–67, vol. 62. New York: Academic Press.

Khorana and Adler use RNA polymerase to copy synthetic DNA into RNA

Falaschi, A., J. Adler, and H. G. Khorana (1963) Chemically synthesized deoxypolynucleotides as templates for ribonucleic acid polymerase. J. Biol. Chem. 238, 3080–85.

Kaji demonstrates polyU directed association of tRNA-Phe with the ribosome

Kaji, A., and H. Kaji (1963) Specific interaction of soluble RNA with polyribonucleic acid induced polysomes. Biochem. Biophys. Res. Comm. 13, 186–92.

Leder and Nirenberg show that a trinucleotide can bind tRNA-Phe to the ribosome

Nirenberg, M., and P. Leder (1964) RNA codewords and protein synthesis: The effect of trinucleotides upon the binding of sRNA to ribosomes. Science 145, 1399–1407.

Leder and Nirenberg produce all trinucleotides by limited random synthesis with PNPase

Nirenberg, M. (1965) Protein synthesis and the RNA code. Harvey Lectures Series 1965, vol. 59, 155–85.

Khorana's group produces all trinucleotides by chemical synthesis
Khorana, H. G. (1966–67) Polynucleotide synthesis and the genetic code. Harvey Lectures Series 1966–67, vol. 62. New York: Academic Press.

THE GENETIC CODE: GOD HAS AN ORDERLY MIND
Brenner and Garen infer two of the three nonsense codons
Brenner, S., A. O. W. Stretton, and S. Kaplan (1965) Genetic code: The "nonsense" triplets for chain termination and their suppression. Nature 206, 994–98.
Weigert, M.G., and A. Garen (1965) Base composition of nonsense codons in *E. coli*: Evidence from amino acid substitutions at a tryptophan site in alkaline phosphatase. Nature 206, 992–94.
Garen, A. (1968) Sense and nonsense in the genetic code. Science 160, 149–59.

Proposal of "Central Dogma" by Crick
Crick, F. H. C. (1970) Central dogma of molecular biology. Nature 227, 561–63.

Chapter 3. Turning Genes On and Off: Genes that Control Other Genes

Lise Jacob, "Of course, it is obvious"
Taped interview with François Jacob

BACTERIAL GROWTH AND THE INDUCED SYNTHESIS OF β–GALACTOSIDASE
Lederberg isolates lac⁻ mutants on indicator medium
Lederberg, J. (1948) Gene control of β-galactosidase in *Escherichia coli*. Genetics 33, 617–18.

Lederberg introduces use of colorimetric substrate NPG
Lederberg, J. (1951) Genetic studies with bacteria, pp. 263–89 in L. C. Dunn, ed., *Genetics in the 20th Century*. New York: Macmillan.

Lactose needs a permease enzyme to transport it into the cell
Rickenberg, H.V., G. N. Cohen, G. Guttin, and J. Monod (1956) La galactoside-perméase d'*Escherichia coli*. Annales de l'Institut Pasteur 91, 829–57.

Cohn synthesizes chemical analogs of lactose
Cohn, M. (1980) In memoriam, pp. 1–9 in J. H. Miller and W. S. Reznikoff, eds., *The Operon*. Cold Spring Harbor, N.Y.: Cold Spring Harbor Laboratory Press.

Not all substrates are inducers, not all inducers are substrates
Monod, J., G. Cohen-Bazire, and M. Cohn (1951) Sur la biosynthèse de la β-galactosidase (lactase) chez *Escherichia coli*: La spécificité de l'induction. Biochim. Biophys. Acta 7, 585–99.

Spiegelman's plasmagene hypothesis
Spiegelman, S. (1946) Nuclear and cytoplasmic factors controlling enzymatic constitution. Cold Spring Harbor Symp. Quant. Biol. 11, 256–77.

Addition of inducer results in new synthesis of β-galactosidase

Rotman, B., and S. Spiegelman (1954) On the origin of the carbon in the induced synthesis of β-galactosidase in *Escherichia coli.* J. Bacteriol. 68, 419–29.

Hogness, D. S., M. Cohn, and J. Monod (1955) Studies on the induced synthesis of β-galactosidase in *Escherichia coli*: The kinetics and mechanism of sulfur incorporation. Biochim. Biophys. Acta 16, 99–116.

THE PROBLEM OF LYSOGENY AND PHAGE λ: THE JACOB GROUP
Lwoff establishes that lysogeny is real

Lwoff, A., and A. Gutmann (1950) Recherches sur un *Bacillus megathérium* lysogene. Ann. Inst. Pasteur (Paris) 78, 711–39. English translation in G. S. Stent, ed. (1960) *Papers on Bacterial Viruses.* Boston: Little, Brown.

Lwoff, A. (1953) Lysogeny. Bacteriol. Rev. 17, 269–337.

Lwoff, A. (1996) The prophage and I, pp. 88–99 in J. Cairns, G. S. Stent, and J. D. Watson, eds., *Phage and the Origins of Molecular Biology.* Cold Spring Harbor, N.Y.: Cold Spring Harbor Laboratory Press.

E. coli *K12 found to be lysogenic for phage* λ

Lederberg, E. M. (1951) Lysogenicity in *E. coli* K-12. Genetics 36, 560 (Abstract).

Lederberg, E. M., and J. Lederberg (1953) Genetic studies on lysogenicity in *Escherichia coli.* Genetics 38, 51–64.

B. megaterium *phage induced by UV*

Lwoff, A., L. Siminovitch, and N. Kjeldgaard (1950) Induction de la production de bacteriophages chez une bactérie lysogeène. Annales de l'Institut Pasteur 79, 815–59.

λ *prophage is inducible by UV irradiation*

Weigle, J., and M. Delbrück (1951) Mutual exclusion between an infecting phage and a carried phage. J. Bacteriol. 62, 301–18.

Mutagenesis is inducible

Weigle, J. (1953) Induction of mutations in a bacterial virus. Proc. Natl. Acad. Sci. USA 39, 628–36.

BACTERIAL SEX AND EROTIC INDUCTION
Review of early work on bacterial conjugation

Wollman, E.-L., F. Jacob, and W. Hayes (1956) Conjugation and genetic recombination in *Escherichia coli* K-12. Cold Spring Harbor Symp. Quant. Biol. 21, 141–62.

Lederberg and Tatum demonstrate genetic exchange in E. coli, *conjugation*

Lederberg, J., and E. L. Tatum (1946) Gene recombination in *E. coli.* Nature 158, 558.

Hayes develops "male" strains efficient in genetic exchange

Hayes, W. (1952) Recombination in *Bact. coli* K-12: Unidirectional transfer of genetic material. Nature 169, 118–19.

Hayes, W. (1953) The mechanism of genetic recombination in *E. coli.* Cold Spring Harbor Symp. Quant. Biol. 18, 75–93.

First Hfr isolated; significance not realized
Cavalli-Sforza, L. L. (1950) La sessualità nei batteri. Bolletino Istituto Sieroterapico Milanese 29, 281–89.

λ *genome often transmitted together with genes for galactose utilization*
Morse, M. L., E. M. Lederberg, and J. Lederberg (1956a) Transduction in *Escherichia coli* K-12. Genetics 41, 142–56.
Morse, M. L., E. M. Lederberg, and J. Lederberg (1956b) Transductional heterogenotes in *Escherichia coli*. Genetics 41, 758–79.

"Zygotic induction," λ *prophage shown to be integrated in* E. coli *chromosome*
Jacob, F., and E. Wollman (1954) Induction spontanée du développement du bactériophage λ au cours de la recombinaison génétique chez *E. coli* K-12. Comptes Rendus Acad. Sci. 239, 317–19.
Jacob, F., and E. L. Wollman (1956) Sur les processus de conjugaison et de recombination génetique chez *Escherichia coli*. I. L'induction par conjugaison ou induction zygotique. Ann. Inst. Pasteur 91, 486–510.

Interrupted mating experiments
Wollman, E. L., and F. Jacob (1955) Sur le mécanisme du transfert de matériel génétique au cours de la recombinaison chez *Escherichia coli* K-12. Comptes rendus Académie des Sciences 240, 2449–51.

E. coli *chromosome shown to be circular by mating experiments with different Hfr donors*
Jacob, F., and E. L. Wollman (1961) *Sexuality and the Genetics of Bacteria*. New York: Academic Press.

TURNING OFF RELATED GENES: THE ROAD TO THE REPRESSOR
AND THE OPERON
First picture of a phage in electron microscope
Luria, S. E., and T. F. Anderson (1942) The identification and characterization of bacteriophages with the electron microscope. Proc. Natl. Acad. Sci. USA 28, 127–30.

Regulators of lysogeny (cI, cII, cIII) identified
Kaiser, A. D. (1957) Mutations in a temperate bacteriophage affecting its ability to lysogenize *E. coli*. Virology 3, 42–61.

Regulatory genes of P22 identified
Levine, M. (1957) Mutations in the temperature phage P22 and lysogeny in *Salmonella*. Virology 3, 22–41.

Regulator of lac *gene expression,* lacI, *identified genetically*
Cohen-Bazire, G., and M. Jolit (1953) Isolement par sélection de mutants d'*Escherichia coli* synthétisant spontanément l'amylomaltase et al β-galactosidase. Annales de l'Institut Pasteur 84, 937–45.

Evidence for lactose permease
Rickenberg, H. V., G. N. Cohen, G. Buttin, and J. Monod (1956) La galactoside-perméase d'*Escherichia coli*. Annales de l'Institut Pasteur 91, 829–57.

Monod's model of an internal inducer

Monod, J. (1947) The phenomenon of enzymatic adaptation and its bearings on problems of genetics and cellular differentiation. Growth Symposium 11, 223–89.

PaJaMa experiments; a test of the internal inducer idea led to the repressor model

Pardee, A. B., F. Jacob, and J. Monod (1958) Sur l'expression et role des allèlles "inductible" et "constitutif" dans la syntheèse de la β-galactosidase chez les zygotes d'*Escherichia coli*. Comptes rendus des Academie des Sciences 246, 3125–27.

Pardee, A. B., F. Jacob, and J. Monod (1959) The genetic control and cytoplasmic expression of "inducibility" in the synthesis of β-galactosidase by *E. coli*. J. Mol. Biol. 1, 165–78.

Szilard proposes repressor model

According to Judson, Szilard visited the Pasteur in the winter of 1957–58 shortly after Pardee's first experiments. Szilard argued for activation by release of inhibition. Judson, H. F. (1979) *The Eighth Day of Creation: Makers of the Revolution in Biology*, pp. 395–98. New York: Simon & Schuster.

Jacob, "It seemed impossible . . ." and "ping-pong game . . ."

Taped interview with François Jacob

Genetic tests showed that lacI⁻ *was recessive to* lacI⁺

Jacob, F., and J. Monod (1961) Genetic regulatory mechanisms in the synthesis of proteins. J. Mol. Biol. 3, 318–56.

Jacob realizes that dominant operator mutations in λ *(*λ *vir mutants) exist*

Jacob, F., and E. L. Wollman (1953) Induction of phage development in lysogenic bacteria. Cold Spring Harbor Symp. Quant. Biol. 18, 101–21.

Isolation of F'lac

Jacob, F., and E. A. Adelberg (1959) Transfert de caractères génétique par incorporation au facteur sexuel d'*Escherichia coli*. Comptes rendus des Séances de l'académie des Sciences 249, 189–91.

Isolation of dominant, constitutive mutations for lac; *first proposal of operon model*

Jacob, F., D. Perrin, C. Sanchez, and J. Monod (1960) L'opéron: Group de gènes á expression coordonnée par un opératur. Comptes rendus des Academie des Sciences 250, 1727–29.

Jacob, F., and J. Monod (1961) Genetic regulatory mechanisms in the synthesis of proteins. J. Mol. Biol. 3, 318–56.

Genetic identification of the promoter

Scaife, J., and J. R. Beckwith (1966) Mutational alteration of the maximal level of *lac* operon expression. Cold Spring Harbor Symp. Quant. Biol. 31, 403–8.

THE DISCOVERY OF MESSENGER RNA

Volkin and Astrachan find unstable RNA fraction in T2 infected E. coli

Volkin, E., and L. Astrachan (1956) Phosphorus incorporation in *Escherichia coli* ribonucleic acid after infection with bacteriophage T2. Virology 2, 149–61.

Density-labeling procedure developed by Meselson and Stahl
Meselson, M., and F. W. Stahl (1958) The replication of DNA in *E. coli*. Proc. Natl. Acad. Sci. USA 44, 671–82.

Jacob, Brenner, and Meselson show that mRNA is the informational intermediate
Brenner, S., F. Jacob, and M. Meselson (1961) An unstable intermediate carrying information from genes to ribosomes for protein synthesis. Nature 190, 576–81.
Spiegelman, S. (1961) On the relation of informational RNA to DNA. Cold Spring Harbor Symp. Quant. Biol. 26, 75–90.

Uninfected E. coli *possessed an unstable RNA with the properties expected of mRNA*
Gros, F., W. Gilbert, H. Hiatt, C. Kurland, R. W. Risebrough, and J. D. Watson (1961) Unstable ribonucleic acid revealed by pulse labeling of *Escherichia coli*. Nature 190, 581–85.
Hayashi, M., and S. Spiegelman (1961) The selective synthesis of informational RNA in bacteria. Proc. Natl. Acad. Sci. USA 47, 1564–80.

WHAT IS THE REPRESSOR?
Jacob and Monod initially propose that repressor is an RNA molecule
Jacob, F., and J. Monod (1961) Genetic regulatory mechanisms in the synthesis of proteins. J. Mol. Biol. 3, 318–56.

Repressor shown to be a protein
Jacob, F., R. Sussman, and J. Monod (1962) Sur la nature du répresseur assurant l'immunité des bactéries lysogènes. Comptes rendus des Academie des Sciences 254, 4214–16.
Bourgeois, S., M. Cohn, and L. E. Orgel (1965) Suppression of and complementation among mutants of the regulatory gene of the lactose operon of *Escherichia coli*. J. Mol. Biol. 14, 300.

Monod, "Mel, we were never wrong"
Taped interview with Mel Cohn

POSITIVE AND NEGATIVE REGULATION: TWO WAYS TO RUN AN OPERON
Trp operon is negatively regulated
Cohen, G., and F. Jacob (1959) Sur la repression de la synthase des enzymes intervenant dans la formations du tryptophane chez *Escherichia coli*. Compt. Rend. 248, 3490–92.

Gal operon is also under negative regulation
Buttin, G. (1963a) Mécanismes régulateurs dans la biosynthèse des enzymes du métabolisme du galactose chez *Escherichia coli* K12. I. J. Mol. Biol. 7, 164–82.
Buttin, G. (1963b) Mécanismes régulateurs dans la biosynthèse des enzymes du métabolisme du galactose chez *Escherichia coli* K12. II. J. Mol. Biol. 7, 183–205.

Alkaline phosphatase synthesis under positive control
Garen, A., and H. Echols (1962) Genetic control of induction of alkaline phosphatase synthesis in *E. coli*. Proc. Natl. Acad. Sci. USA 48, 1398–1402.

AraC proposed to be a positive control factor

Helling, R. B., and R. Weinberg (1963) Complementation studies of arabinose genes in *Escherichia coli*. Genetics 48, 1397–1410.

Englesberg, E., J. Irr, J. Power, and N. Lee (1965) Positive control of enzyme synthesis by gene C in the L-arabinose system. J. Bacteriol. 90, 946–57.

Dove, Thomas, and Echols show that λ has two positive regulatory genes

Dove, W. F. (1966) Action of the lambda chromosome. I. Control of functions late in bacteriophage development. J. Mol. Biol. 19, 187–201.

Joyner, A., L. N. Isaacs, H. Echols, and W. S. Sly (1966) DNA replication and messenger RNA production after induction of wild-type lambda bacteriophage and lambda mutants. J. Mol. Biol. 19, 174–86.

Thomas, R. (1966) Control of development in temperature bacteriophages. I. Induction of prophage genes following heteroimmune superinfection. J. Mol. Biol. 22, 79–95.

Maltose operon under positive control

Schwartz, M. (1967) Expression phénotypique et localisation génétique de mutations affectant le metabolisme du maltose chez *Escherichia coli*. Ann. Inst. Pasteur 112, 673–702.

Schwartz, M. (1979) Another route, in A. Ullmann and A. Lwoff, eds., *The Origins of Molecular Biology: A Tribute to Jacques Monod*. New York: Academic Press.

Historical essay by Jon Beckwith

Beckwith, J. (1987) The operon: An historical account, in F. C. Neidhardt, J. L. Ingraham, K. B. Low, B. Magasanik, M. Schaechter, and H. E. Umbarger, eds., *Escherichia coli and Salmonella typhimurium: Cellular and Molecular Biology*. Washington, D.C.: ASM Press.

MULTI-OPERON REGULATION: THE GLUCOSE EFFECT
"Glucose effect" discovered by Monod

Monod, J. (1941) Sur un phénomène nouveau de croissance complexe dans les cultures bactériennes. Comptes rendus des Academiei des Sciences 212, 934–36.

Major basis of glucose control due to activity of positive regulatory protein

Emmer, M., B. deCrombrugghe, I. Pastan, and R. Perlman (1970) Cyclic AMP receptor protein of *E. coli*: Its role in the synthesis of inducible enzymes. Proc. Natl. Acad. Sci. USA 66, 480–87.

Zubay, G., D. O. Schwartz, and J. Beckwith (1970) The mechanism of activation of catabolite sensitive genes: A positive control system. Proc. Natl. Acad. Sci. USA 66, 104–10.

REGULATORY DIVERSITY: REVIEW AND PREVIEW
Feedback inhibition of enzyme activity

Yates, R. A., and A. B. Pardee (1956) Control of pyrimidine biosynthesis in *Escherichia coli* by a feedback mechanism. J. Biol. Chem. 221, 757–70.

Umbarger, H. E. (1956) Evidence for a negative-feedback mechanism in the biosynthesis of leucine. Science 123, 848.

Gerhart, J. C., and A. B. Pardee (1962) The enzymology of control by feedback inhibition. J. Biol. Chem. 237, 891–96.

Monod, J., J.-P. Changeux, and F. Jacob (1963) Allosteric proteins and cellular control systems. J. Mol. Biol. 6, 306–29.

Chapter 4. Replicating the Genome
DNA POLYMERASE REPLICATES DNA
Structure of DNA proposed by Watson and Crick
Watson, J. D., and F. H. C. Crick (1953) A structure of deoxyribose nucleic acid. Nature 171, 737–38.

Kornberg, "Bob, how would you like to switch . . ."
Taped interview with Arthur Kornberg

Purification of DNA polymerase
Kornberg, A., I. R. Lehman, M. J. Bessman, and E. S. Simms (1956) Enzymatic synthesis of deoxyribonucleic acid. Biochim. Biophys. Acta. 21, 197–98.
Lehman, I. R., M. J. Bessman, E. S. Simms, and A. Kornberg (1958) Enzymatic synthesis of deoxyribonucleic acid. I. Preparation of substrates and partial purification of an enzyme from *Escherichia coli.* J. Biol. Chem. 233, 163–70.
Lehman, I. R., S. B. Zimmerman, J. Adler, M. J. Bessman, E. S. Simms, and A. Kornberg (1958) Enzymatic synthesis of deoxyribonucleic acid. V. Chemical composition of enzymatically synthesized deoxyribonucleic acid. Proc. Natl. Acad. Sci. USA 44, 1191–96.

THE PROBLEMS OF REPLICATING A GENOME
DNA replication in E. coli *is semi-conservative*
Meselson, M., and F. W. Stahl (1958) The replication of DNA in *Escherichia coli.* Proc. Natl. Acad. Sci. USA 44, 671–82.

Cairns shows that DNA replication in vivo *is by a fork mechanism from very few origins*
Cairns, J. (1963a) The bacterial chromosome and its manner of replication as seen by autoradiography. J. Mol. Biol. 6, 208–13.
Cairns, J. (1963b) The chromosome of *Escherichia coli.* Cold Spring Harbor Symp. Quant. Biol. 28, 43–45.

Replication of λ *is bidirectional*
Schnös, M., and R. B. Inman (1970) Position of branch points in replicating λ DNA. J. Mol. Biol. 51, 61–73.

THE GENETICS OF REPLICATION
Identification of temperature-sensitive mutants with defects in DNA replication
Bonhoeffer, F., and H. Schaller (1965) A method for selective enrichment of mutants based on the high UV sensitivity of DNA containing 5-bromouracil. Biochem. Biophys. Res. Comm. 20, 93–97.
Kohiyama, M., D. Cousin, A. Ryter, and F. Jacob (1966) Mutants thermosensibles d'*Escherichia coli* K12. I. Isolement et characterisation rapide. Ann. Inst. Pasteur 110, 465–86.
Fangman, W. L., and A. Novick (1968) Characterization of two bacterial mutants with temperature-sensitive synthesis of DNA. Genetics 60, 1–17.
Hirota, Y., A. Ryter, and F. Jacob (1968) Thermosensitive mutants of *E. coli* affected in the processes of DNA synthesis and cellular division. Cold Spring Harbor Symp. Quant. Biol. 33, 677–93.
Carl, P. L. (1970) *Escherichia coli* mutants with temperature-sensitive synthesis of DNA. Mol. Gen. Genet. 109, 107–22.

Wechsler, J. A., and J. D. Gross (1971). *Escherichia coli* mutants temperature-sensitive for DNA synthesis. Mol. Gen. Genet. 113, 273–84.

Isolation of Pol I mutant

DeLucia, P., and J. Cairns (1969) Isolation of an *E. coli* strain with a mutation affecting DNA polymerase. Nature 224, 1164–66.

THE FINE STRUCTURE OF DNA REPLICATION

One DNA strand is replicated by a discontinuous mechanism

Okazaki, R. T., K. Okazaki, K. Sakabe, K. Sugimoto, and A. Sugino (1968) Mechanism of DNA chain growth. I. Possible discontinuity and unusual secondary structure of newly synthesized chains. Proc. Natl. Acad. Sci. USA 59, 598–605.

Discovery of DNA ligase

Gefter, M. L., A. Becker, and J. Hurwitz (1967) The enzymatic repair of DNA. I. Formation of circular λ DNA. Proc. Natl. Acad. Sci. USA 58, 240–47.

Gellert, M. (1967) Formation of covalent circles of lambda DNA by *E. coli* extracts. Proc. Natl. Acad. Sci. USA 57, 148–55.

Olivera, B. M., and I. R. Lehman (1967) Linkage of polynucleotides through phosphodiester bonds by an enzyme of *Escherichia coli*. Proc. Natl. Acad. Sci. USA 57, 1426–33.

Weiss, B., and C. C. Richardson (1967) Enzymatic breakage and joining of deoxyribonucleic acid, I. Repair of single-strand breaks in DNA by an enzyme system from *Escherichia coli* infected with T4 bacteriophage. Proc. Natl. Acad. Sci. USA 57, 1021–28.

Brutlag and Kornberg devise a soluble system to replicate small DNA phage and find that an RNA primer is used to initiate DNA synthesis

Brutlag, D., R. Schekman, and A. Kornberg (1971) A possible role for RNA polymerase in the initiation of M13 DNA synthesis. Proc. Natl. Acad. Sci. USA 68, 2826–29.

Wickner, W., D. Brutlag, R. Schekman, and A. Kornberg (1972) RNA synthesis initiates *in vitro* conversion of M13 DNA to its replicative form. Proc. Natl. Acad. Sci. USA 69, 965–69.

THE "REAL" REPLICATION ENZYME IS JUST LIKE POL I

DNA Pol III is the replication enzyme

Knippers, R. (1970) DNA polymerase II. Nature (London) 228, 1050–53.

Gefter, M., Y. Hirota, T. Kornberg, J. A. Wechsler, and C. Barnaux (1971) Analysis of DNA polymerases II and III in mutants of *Escherichia coli* thermosensitive for DNA synthesis. Proc. Natl. Acad. Sci. USA 68, 3150–53.

Wechsler, J. A., and J. D. Gross (1971) *Escherichia coli* mutants temperature-sensitive for DNA synthesis. Mol. Gen. Genet. 113, 273–84.

Kornberg and Gefter finish characterizing DNA Pol III

Kornberg, T., and M. L. Gefter (1974) Deoxyribonucleic acid synthesis in cell-free extracts. IV. Purification and catalytic properties of deoxyribonucleic acid polymerase III. J. Biol. Chem. 247, 5369–75.

A SIMPLE TASK THAT NEEDS MANY PROTEINS: PRIMING AND ELONGATING

φX174 shown to have a single-stranded, circular DNA chromosome

Sinsheimer, R. L. (1959) A single-stranded deoxyribonucleic acid from bacteriophage φX174. J. Mol. Biol. 1, 43–53.

φX174 replicates by an interesting sequential mechanism

Sinsheimer, R. L., R. Knippers, and T. Komano (1968) Stages in the replication of bacteriophage φX174 DNA *in vivo*. Cold Spring Harbor Symp. Quant. Biol. 33, 443–47.

φX174 synthesis dependent on dna ts *mutants*

Schekman, R., W. Wickner, O. Westergaard, D. Brutlag, K. Geider, L. L. Bertsch, and A. Kornberg (1972) Initiation of DNA synthesis: Synthesis of φX174 replicative form requires RNA synthesis resistant to rifampicin. Proc. Natl. Acad. Sci. USA 69, 2691–95.

Wickner, R. B., M. Wright, S. Wickner, and J. Hurwitz (1972) Conversion of φX174 and fd single-stranded DNA to replicative forms in extracts of *Escherichia coli*. Proc. Natl. Acad. Sci. USA 69, 3233–37.

Purification of proteins required for SS to RF conversion by M13, φX174, and G4

Wickner, S., and J. Hurwitz (1974) Conversion of φX174 viral DNA to double-stranded forms by *E. coli* proteins. Proc. Natl. Acad. Sci. USA 71, 4120–24.

Geider, K., and A. Kornberg (1974) Conversion of the M13 viral single strand to the double-stranded replicative forms by purified proteins. J. Biol. Chem. 249, 3999–4005.

Schekman, R., J. H. Weiner, A. Weiner, and A. Kornberg (1975) Ten proteins required for conversion of φX174 single-stranded DNA to duplex form *in vitro*. Resolution and reconstitution. J. Biol. Chem. 250, 5859–65.

Weiner, J. H., R. McMacken, and A. Kornberg (1976) Isolation of an intermediate which precedes dnaG RNA polymerase participation in enzymatic replication of bacteriophage φX174 DNA. Proc. Natl. Acad. Sci. USA 73, 752–56.

SPECIAL PROTEINS NEEDED FOR GENOME REPLICATION: STRETCHING, UNWINDING, UNTWISTING

Isolation of T4 Ssb and its role in T4 DNA replication

Alberts, B. M., F. J. Amodio, M. Jenkins, E. D. Gutmann, and F. L. Ferris (1969) Studies with DNA-cellulose chromatography. I. DNA-binding proteins from *Escherichia coli*. Cold Spring Harbor Symp. Quant. Biol. 33, 289–305.

Alberts, B. M. and L. Frey (1970) T4 bacteriophage gene 32: A structural protein in the replication and recombination of DNA. Nature (London) 227, 1313–18.

Purification of E. coli Ssb

Sigal, N., H. Delius, T. Kornberg, M. L. Gefter, and B. M. Alberts (1972) A DNA-unwinding protein isolated from *Escherichia coli*. Its interaction with DNA and DNA polymerases. Proc. Natl. Acad. Sci. USA 69, 3537–41.

First topoisomerase discovered

Wang, J. C. (1971) Interaction between DNA and an *Escherichia coli* protein ω. J. Mol. Biol. 55, 523–33.

Discovery of DNA helicases

Friedman, E. A., and H. O. Smith (1973) Production of possible recombination intermediates by an ATP-dependent DNase. Nature New Biol. 241, 54–58.

Abdel-Monem, M., H. Durwald, and H. Hoffmann-Berling (1976) Enzymic unwinding of DNA. II. Chain separation by an ATP-dependent DNA unwinding enzyme. Eur. J. Biochem. 65, 441–49.

Wilcox, K. W., and H. O. Smith (1976) Mechanism of DNA degradation by the ATP-dependent DNase from *Hemophilus influenzae* Rd. J. Biol. Chem. 251, 6127–34.

REPLICATING DUPLEX DNA
RF to SS replication by φX174 requires φX174 A protein and E. coli Rep protein
Eisenberg, S., J. Griffith, and A. Kornberg (1977) φX174 cistron A protein is a multifunctional enzyme in DNA replication. Proc. Natl. Acad. Sci. USA 74, 3198–3202.
Eisenberg, S. (1980) The role of gene A protein and *E. coli rep* protein in the replication of φX174 replicative form DNA. Proceedings Royal Society of London-Series B 210, 337–49.

Parallel studies on DNA replication by Alberts with T4 and Richardson with T7
Richardson, C. C. (1983) Bacteriophage T7: Minimal requirements for the replication of a duplex DNA molecule. Cell 33, 315–17.
Alberts, B. M. (1984) The DNA enzymology of protein machines. Cold Spring Harbor Symp. Quant. Biol. 49, 1–12.

STARTING GENOME REPLICATION: A SPECIAL STRUCTURE
Development of suppression analysis to study λ replication
Georgopoulos, C., and I. Herskowitz (1971) *Escherichia coli* mutants blocked in lambda DNA synthesis, in A. D. Hershey, ed., *The bacteriophage lambda.* Cold Spring Harbor, N.Y.: Cold Spring Harbor Laboratory Press.

Suppression analysis also used by Tomizawa
Tomizawa, J. (1971) Functional cooperation of genes O and P, in A. D. Hershey, ed., *The bacteriophage lambda.* Cold Spring Harbor, N.Y.: Cold Spring Harbor Laboratory Press.

E. coli oriC cloned
von Meyenburg, K., F. G. Hansen, L. D. Nielsen, and P. Jorgensen (1977) Origin of replication, *oriC*, of the *Escherichia coli* chromosome: Mapping of genes relative to *EcoRI* cleavage sites in the *oriC* region. Mol. Gen. Genet. 158, 101–9.
Yasuda, S., and Y. Hirota (1977) Cloning and mapping of the replication origin of *Escherichia coli*. Proc. Natl. Acad. Sci. USA 74, 5458–62.

OriC-dependent replication in cell extracts and with purified proteins
Fuller, R. S., J. M. Kaguni, and A. Kornberg (1981) Enzymatic replication of the origin of the *Escherichia coli* chromosome. Proc. Natl. Acad. Sci. USA 78, 7370–74.
Funnell, B. E., T. A. Baker, and A. Kornberg (1986) Complete enzymatic replication of plasmids containing the origin of the *Escherichia coli* chromosome. J. Biol. Chem. 261, 5616–24.

Biochemical analysis of λ replication
Wickner, S. (1979) DNA replication proteins of *E. coli* and phage λ. Cold Spring Harbor Symp. Quant. Biol. 43, 303–10.
Wold, M. S., J. B. Mallory, J. D. Roberts, J. H. LeBowitz, and R. McMacken (1982) Initiation of bacteriophage lambda DNA replication *in vitro* with purified lambda replication proteins. Proc. Natl. Acad. Sci. USA 79, 6176–80.
Zylicz, M., J. H. LeBowitz, R. McMacken, and C. Georgopoulos (1983) The DnaK protein of *Escherichia coli* possesses an ATPase and autophosphorylating activity and is essential in an *in vitro* DNA replication system. Proc. Natl. Acad. Sci. USA 80, 6431–35.

LeBowitz, J. H., M. Zylicz, C. Georgopoulos, and R. McMacken (1985) Initiation of DNA replication on single-stranded DNA templates catalyzed by purified replication proteins of bacteriophage lambda and *Escherichia coli*. Proc. Natl. Acad. Sci. USA 82, 3988–92.

Yamamoto, T., J. McIntyre, S. M. Sell, C. Georgopoulos, D. Skowyra, and M. Zylicz (1987) Enzymology of the pre-priming steps in lambda dv DNA replication *in vitro*. J. Biol. Chem. 262, 7996–99.

Liberek, K., C. Georgopoulos, and M. Zylicz (1988) Role of the *Escherichia coli* DnaK and DnaJ heat shock proteins in the initiation of bacteriophage lambda DNA replication. Proc. Natl. Acad. Sci. USA 85, 6632–36.

Alfano, C., and R. McMacken (1989a) Ordered assembly of nucleoprotein structures at the bacteriophage lambda replication origin during the initiation of DNA replication. J. Biol. Chem. 264, 10699–708.

Alfano, C., and R. McMacken (1989b) Heat shock protein-mediated disassembly of nucleo-protein structures is required for the initiation of bacteriophage lambda DNA replication. J. Biol. Chem. 264, 10709–18.

Mensa-Wilmot, K., R. Seaby, C. Alfano, M. C. Wold, B. Gomes, and R. McMacken (1989) Reconstitution of a nine-protein system that initiates bacteriophage lambda DNA replication. J. Biol. Chem. 264, 2853–61.

Zylicz, M., D. Ang, K. Liberek, and C. Georgopoulos (1989) Initiation of lambda DNA replication with purified host- and bacteriophage-encoded proteins: The role of the dnaK, dnaJ and grpE heat shock proteins. EMBO J. 8, 1601–08.

Robley Williams pioneered use of EM to see nucleoprotein structures

Williams, R. C., and R. C. Glaeser (1972) Ultrathin carbon support films for electron microscopy. Science 175, 1000–1001.

Williams, R. C. (1977) Use of polylysine for adsorption of nucleic acids and enzymes to electron microscope specimen films. Proc. Natl. Acad. Sci. USA 74, 2311–15.

Pathway of λ replication observed in electron microscope

Dodson, M., J. Roberts, R. McMacken, and H. Echols (1985) Specialized nucleoprotein structures at the origin of replication of bacteriophage λ: Complexes with λ O, λ P and *Escherichia coli* DnaB proteins. Proc. Natl. Acad. Sci. USA 82, 4678–82.

Dodson, M., H. Echols, S. Wickner, C. Alfano, K. Mensa-Wilmot, J. LeBowitz, J. D. Roberts, and R. McMacken (1986) Specialized nucleoprotein structures at the origin of replication of bacteriophage λ: Localized unwinding of duplex DNA by a six-protein reaction. Proc. Natl. Acad. Sci. USA 83, 7638–42.

Dodson, M., R. McMacken, and H. Echols (1989) Specialized nucleoprotein structures at origin of replication of bacteriophage λ. Protein association and dissociation reactions responsible for localized initiation of replication. J. Biol. Chem. 264, 10719–25.

Similar initiation pathway determined in parallel studies by the Kornberg group

Funnell, B. E., T. A. Baker, and A. Kornberg (1987) *In vitro* assembly of a prepriming complex at the origin of the *Escherichia coli* chromosome. J. Biol. Chem. 262, 10327–34.

Requirements for SV40 replication are similar to those of E. coli

Li, J. J., and T. J. Kelly (1984) Simian virus 40 DNA replication *in vitro*. Proc. Natl. Acad. Sci. USA 81, 6973–77.

Kelly, T. J. (1988) SV40 DNA replication. J. Biol. Chem. 263, 17889–92.

Hurwitz, J., F. B. Dean, A. D. Kwong, and S. H. Lee (1990) The *in vitro* replication of DNA containing the SV40 origin. J. Biol. Chem. 265, 18043–46.

Stillman, B. (1992) Initiation of chromosome replication in eukaryotic cells. Harvey Lectures Series 1992–93, vol. 88, 115–40.

Chapter 5. Making RNA from DNA

THE TRANSCRIPTION ENZYME: RNA POLYMERASE

Isolation of RNA polymerase

Weiss, S. B., and L. Gladstone (1959) A mammalian system for the incorporation of cytidine triphosphate into ribonucleic acid. J. Amer. Chem. Soc. 81, 4118.

Hurwitz, J., A. Bresler, and R. Diringer (1960) The enzymic incorporation of ribonucleotides into polyribonucleotides and the effect of DNA. Biochem. Biophys. Res. Comm. 3, 15–19.

Weiss, S. B. (1960) Enzymatic incorporation of ribonucleoside triphosphates into the interpolynucleotide linkages of ribonucleic acid. Proc. Natl. Acad. Sci. USA 46, 1020–30.

Huang, R. C., N. Maheshwari, and J. Bonner (1960) Enzymatic synthesis of RNA. Biochem. Biophys. Res. Comm. 3, 689.

Stevens, A. (1960) Incorporation of the adenine ribonucleotide into RNA by cell fractions from *E. coli*. B. Biochem. Biophys. Res. Comm. 3, 92.

Stevens, A. (1961) Net formation of polyribonucleotides with base compositions analogous to deoxyribonucleic acid. J. Biol. Chem. 236, PC43–PC45.

Chamberlin, M., and P. Berg (1962) Deoxyribonucleic acid-directed synthesis of ribonucleic acid by an enzyme from *Escherichia coli*. Proc. Natl. Acad. Sci. USA 48, 81–94.

THE PROPERTIES OF mRNA

Brenner and Jacob link "Volkin T2 RNA" to informational RNA

Brenner, S., F. Jacob, and M. Meselson (1961) An unstable intermediate carrying information from genes to ribosomes for protein synthesis. Nature 190, 576–81.

Volkin, E., and L. Astrachan (1956) Phosphorus incorporation in *Escherichia coli* ribonucleic acid after infection with bacteriophage T2. Virology 2, 149–61.

Nomura, Hall, and Spiegelman show that mRNA can be separated from rRNA

Nomura, M., B. D. Hall, and S. Spiegelman (1960) Characterization of the RNA synthesized in *Escherichia coli* after bacteriophage T2 infection. J. Mol. Biol. 2, 306–26.

DNA-RNA hybridization technique

Hall, B. D., and S. Spiegelman (1961) Sequence complementarity of T2-DNA and T2-specific RNA. Proc. Natl. Acad. Sci. USA 47, 137–46.

Marmur and Doty had shown that duplex DNA could be separated into single strands and then reassociate

Doty, P., J. Marmur, J. Eigner, and C. Schildkraut (1960) Strand separation and specific recombination in deoxyribonucleic acids: Physical chemical studies. Proc. Natl. Acad. Sci. 46, 461–76.

Hall uses nitrocellulose filter binding to detect RNA-DNA hybrids
Nygaard, A. D., and B. D. Hall (1963) A method for the detection of RNA-DNA complexes. Biochem. Biophys. Res. Comm. 12, 98–104.

Gros shows increased amounts of gal *and* lac *RNA after induction*
Attardi, G., S. Naono, J. Rouvière, F. Jacob, and F. Gros (1963) Production of messenger RNA and regulation of protein synthesis. Cold Spring Harbor Symp. Quant. Biol. 28, 363–72.

Spiegelman and associates showed rRNA and tRNA produced as transcripts from DNA
Giacomoni, D., and S. Spiegelman (1962) Origin and biologic individuality of the genetic dictionary. Science 138, 1328–31.
Yanofsky, S. A., and S. Spiegelman (1962) The identification of the ribosomal RNA cistron by sequence complementarity. I. Specificity of complex formation. Proc. Natl. Acad. Sci. USA 48, 1069–78.
Yanofsky, S. A., and S. Spiegelman (1962) The identification of the ribosomal RNA cistron by sequence complementarity. II. Saturation of and competitive interaction at the RNA cistron. Proc. Natl. Acad. Sci. USA 48, 1466–72.
Yanofsky, S. A., and S. Spiegelman (1963) Distinct cistrons for the two ribosomal RNA components. Proc. Natl. Acad. Sci. USA 49, 538–44.

Goodman and Rich also showed that tRNA produced as a transcript from DNA
Goodman, H. M., and A. Rich (1962) Formation of a DNA-soluble RNA hybrid and its relation to the origin, evolution, and degeneracy of soluble RNA. Proc. Natl. Acad. Sci. USA 48, 2101–9.

Spiegelman and coworkers showed that mRNA is unstable and rRNA stable
Hayashi, M., and S. Spiegelman (1961) The selective synthesis of informational RNA in bacteria. Proc. Natl. Acad. Sci. USA 47, 1564–80.

DEVELOPMENTAL PATHWAYS OF PHAGE
Campbell isolates nonsense mutants of λ *(originally called host-defective [hd] mutants)*
Campbell, A. (1961) Sensitive mutants of bacteriophage λ. Virology 14, 22–32.

Isolation of conditional-lethal mutants of T4
Epstein, R. H., A. Bolle, C. M. Steinberg, E. Kellenberger, E. Boy de la Tour, R. Chevalley, R. S. Edgar, M. Susman, G. H. Denhardt, and A. Lielausis (1963) Physiological studies of conditional lethal mutants of bacteriophage T4D. Cold Spring Harbor Symp. Quant. Biol. 28, 375–92.

Suppressor genes
Brenner, S., A. O. W. Stretton, and S. Kaplan (1965) Genetic code: the "nonsense" triplets for chain termination and their suppression. Nature 206, 994–98.
Weigert, M. G., and A. Garen (1965) Base composition of nonsense codons in *E. coli*: Evidence from amino acid substitutions at a tryptophan site in alkaline phosphatase. Nature 206, 992–94.
Garen, A. (1968) Sense and nonsense in the genetic code. Science 160, 149–59.

Temperature-sensitive phage mutants as another class of conditional-lethals
Horowitz, N. H., and U. Leupold (1951) Some recent studies bearing on the one gene-one enzyme hypothesis. Cold Spring Harbor Symp. Quant. Biol. 16, 65–72.
Edgar, R. S., and I. Lielausis (1964) Temperature sensitive mutants of bacteriophage T4D: Their isolation and genetic characterization. Genetics 49, 649–62.

LAMBDOLOGY: THE DEVELOPMENTAL BIOLOGY OF A FIELD
Jacob and Wollman propose that prophage DNA hooks onto the bacterial chromosome
Jacob, F., and E. Wollman (1961) *Sexuality and the Genetics of Bacteria*. New York: Academic Press.

Order of genes in λ is different in the prophage and the phage
Calef, E., and G. Licciardello (1960) Recombination experiments on prophage host relationships. Virology 12, 81–103.
Franklin, N. C., W. F. Dove, and C. Yanofsky (1965) The linear insertion of a prophage into the chromosome of *E. coli* shown by deletion mapping. Biochem. Biophys. Res. Commun. 18, 910–23.

Campbell model for λ integration into host DNA
Campbell, A. (1962) Episomes. Advances in Genetics 11, 101–45.

LAMBDOLOGY: THE DEVELOPMENTAL BIOLOGY OF A PHAGE
Echols finds that N, O, P, and Q mutants are defective in producing λ late mRNA
Joyner, A., L. N. Isaacs, H. Echols, and W. S. Sly (1966) DNA replication and messenger RNA production after induction of wild type bacteriophage and lambda mutants. J. Mol. Biol. 19, 174–86.

Dove independently concludes that Q was a positive regulator of late events in λ
Dove, W. F. (1966) Action of the lambda chromosome. I. Control of functions late in bacteriophage development. J. Mol. Biol. 19, 187–201.

Weigle maps late genes to the "left" end of λ
Weigle, J. (1966) Assembly of phage lambda *in vitro*. Proc. Natl. Acad. Sci. 55, 1462–65.

Skalka and Hershey isolate piece of λ DNA that contains the late genes
Hershey, A. D., E. Burgi, and C. I. Davern (1965) Preparative density-gradient centrifugation of the molecular halves of lambda DNA. Biochem. Biophys. Res. Comm. 18, 675–78.
Skalka, A. (1966) Regional and temporal control of genetic transcription in phage lambda. Proc. Natl. Acad. Sci. USA 55, 1190–95.

Echols and Skalka use DNA-RNA hybridization to figure out transcription
Skalka, A., B. Butler, and H. Echols (1967) Genetic control of transcription during development of phage λ. Proc. Natl. Acad. Sci. USA 58, 576–83.

Thomas and Radding independently show that N turns on other genes
Thomas, R. (1966) Control of development in temperature bacteriophages. I. Induction of prophage genes following heteroimmune superinfection. J. Mol. Biol. 22, 79–95.

Radding, C. M., and D. C. Shreffler (1966) Regulation of lambda exonuclease. II. Joint regulation of exonuclease and a new lambda antigen. J. Mol. Biol. 18, 251–61.

Operator sites for λ cI protein had been identified by isolating virulent mutants
Jacob, F., and E. L. Wollman (1953) Induction of phage development in lysogenic bacteria. Cold Spring Harbor Symp. Quant. Biol. 18, 101–21.

Szybalski et al. isolate each strand of λ chromosome
Taylor, K., Z. Hradecna, and W. Szybalski (1967) Asymmetric distribution of the transcribing regions on the complementary strands of coliphage lambda DNA. Proc. Natl. Acad. Sci. USA 57, 1618–25.

Single site to the right of Q is needed for all late gene transcription
Herskowitz, I., and E. R. Signer (1970) A site essential for expression of all late genes in bacteriophage λ. J. Mol. Biol. 47, 545–46.

LAMBDOLOGY AS A COMMUNITY
Proposal for a more cooperative scientific community
Echols, H. (1972) Scientific community [correspondence]. Nature 239, 476–77.

THE LAC AND λ REPRESSORS BIND TO OPERATOR DNA
Müller-Hill and Gilbert isolate Lac repressor
Gilbert, W., and B. Müller-Hill (1966) Isolation of the lac repressor. Proc. Natl. Acad. Sci. USA 56, 1891–98.

Ptashne prepared radoactively pure λ cI protein
Ptashne, M. (1967) Isolation of the λ phage repressor. Proc. Natl. Acad. Sci. USA 57, 306–13.

Host protein synthesis can be turned off by irradiating cells with UV
Hosoda, J., and C. Levinthal (1968) Protein synthesis of *Escherichia coli* infected with bacteriophage T4D. Virology 34, 709–27.

Jacob and Campbell isolate λ Ind⁻ mutant
Jacob, F., and A. Campbell (1959) Sur le système de répression assurant l'immunité chez les bactéries lysogènes. Comptes rendus des Academie des Sciences 248, 3219–21.

λ cI protein binds to λ DNA, but not to 434 DNA
Ptashne, M. (1967) Specific binding of the λ phage repressor to λ DNA. Nature 214, 232–34.

Lac repressor binds to DNA
Gilbert, W., and B. Müller-Hill (1967) The *lac* operator is DNA. Proc. Natl. Acad. Sci. 58, 2415–21.

Extracts of a λ lysogen inhibit transcription from λ DNA, but not from imm 434 DNA
Echols, H., L. Pilarski, and P. Y. Cheng (1968) *In vitro* repression of phage λ DNA transcription by a partially purified repressor from lysogenic cells. Proc. Natl. Acad. Sci. USA 59, 1016–23.

THE BIOCHEMISTRY OF RNA POLYMERASE

E. coli *RNA polymerase copies only one strand of T4 DNA*

Grau, O., A. Guha, E. P. Geiduschek, and W. Szybalski (1969) Transcription of the bacteriophage T4 template: Strand selection by *E. coli* RNA polymerase *in vitro*. Nature 224, 1105.

RNA polymerase contains multiple subunits

Zillig, W., K. Zechel, D. Rabussay, M. Schachner, V. S. Sethi, P. Palm, A. Heil, and W. Seifert (1970) On the role of different subunits of DNA-dependent RNA polymerase from *E. coli* in the transcription process. Cold Spring Harbor Symp. Quant. Biol. 35, 47–58.

Isolation of sigma unit of RNA polymerase

Burgess, R. R., A. A. Travers, J. J. Dunn, and E. K. Bautz (1969) Factor stimulating transcription by RNA polymerase. Nature 221, 43–46.

Sigma dissociates from RNA polymerase during elongation

Travers, A. A., and R. R. Burgess (1969) Cyclic re-use of the RNA polymerase sigma factor. Nature 222, 537–40.

RNA polymerase binds tightly at promoters and rapidly initiates transcripts

Zillig, W., K. Zechel, D. Rabussay, M. Schachner, V. S. Sethi, P. Palm, A. Heil, and W. Seifert (1970) On the role of different subunits of DNA-dependent RNA polymerase from *E. coli* in the transcription process. Cold Spring Harbor Symp. Quant. Biol. 35, 47–58.

Hinkle, D. C., and M. J. Chamberlin (1972) Studies of the binding of *Escherichia coli* RNA polymerase to DNA. I. The role of sigma subunit in site selection. J. Mol. Biol. 70, 157–85.

Hinkle, D. C., and M. J. Chamberlin (1972) Studies of the binding of *Escherichia coli* RNA polymerase to DNA. II. The kinetics of the binding reaction. J. Mol. Biol. 70, 187–95.

Chamberlin defined a multi-stage process of chain initation

Chamberlin, M. J. (1974) The selectivity of transcription. Annu. Rev. Biochem. 43, 721–75.

RNA polymerase by itself terminates transcription in the early region of T7

Millete, R. L., C. D. Trotter, P. Herrlich, and M. Schweiger (1970) *In vitro* synthesis, termination, and release of active messenger RNA. Cold Spring Harbor Symp. Quant. Biol. 35, 135–42.

RNA polymerase requires other proteins to terminate in early region of λ

Roberts, J. W. (1969) Termination factor for RNA polymerase. Nature 224, 1168–1174.

λ *repressor inhibits transcription from* p_L *and* p_R

Steinberg, R. A., and M. Ptashne (1971) *In vitro* repression of RNA synthesis by purified lambda phage repressor. Nature New Biol. 230, 76–80.

Wu, A. M., S. Ghosh, H. Echols, and W. G. Spiegelman (1972) Repression by the cI protein of phage λ: *In vitro* inhibition of RNA synthesis. J. Mol. Biol. 67, 407–21.

POSITIVE AND NEGATIVE REGULATORS OF THE LAC OPERON

Addition of cAMP "cures" glucose repression

Perlman, R. L., and I. Pastan (1968) Regulation of β-galactosidase synthesis in *Escherichia coli* by cyclic adenosine 3′, 5′-monophosphate. J. Biol. Chem. 243, 5420–27.

Addition of cAMP to Zubay's extract promotes transcription of lac operon

Chambers, D. A., and G. Zubay (1969) The stimulatory effect of cyclic adenosine 3′, 5′-monophosphate on DNA-directed synthesis of β-galactosidase in a cell-free system. Proc. Natl. Acad. Sci. USA 63, 118–22.

Isolation of CRP and cya mutants

Perlman, R. L., and I. Pastan (1969) Pleiotropic deficiency of carbohydrate utilization in an adenyl cyclase deficient mutant of *Escherichia coli*. Biochem. Biophys. Res. Commun. 37, 151–57.

Schwartz, D., and J. R. Beckwith (1970) Mutants missing a factor necessary for the expression of catabolite-sensitive operons in *E. coli* In J. Beckwith and D. Zipser, eds., *The Lac Operon* Cold Spring Harbor, N.Y.: Cold Spring Harbor Laboratory Press.

Purification of CRP

Emmer, M., B. de Crombrugghe, I. Pastan, and R. Perlman (1970) Cyclic AMP receptor protein of *E. coli*: Its role in the synthesis of inducible enzymes. Proc. Natl. Acad. Sci. USA 66, 480–87.

Zubay, G., D. Schwartz, and J. Beckwith (1970) Mechanism of activation of catabolite-sensitive genes: A positive control system. Proc. Natl. Acad. Sci. USA 66, 104–10.

Anderson, W. B., A. B. Schneider, M. Emmer, R. L. Perlman, and I. Pastan (1971) Purification and properties of the cyclic adenosine 3′, 5′-monophophate receptor protein which mediates cyclic adenosine 3′, 5′-monophosphate-dependent gene transcription in *Escherichia coli*. J. Biol. Chem. 246, 5929–37.

Majors, J. (1975) Specific binding of CAP factor to lac promoter DNA. Nature 256, 672–74.

Repression by LacI demonstrated in vitro

de Crombrugghe, B., B. Chen, W. Anderson, P. Nissley, M. Gottesman, and I. Pastan (1971) *Lac* DNA, RNA polymerase and cyclic AMP receptor protein, cyclic AMP, *lac* repressor and inducer are essential elements for controlled *lac* transcription. Nature New Biol. 231, 139–42.

Eron, L., and R. Block (1971) Mechanism of initiation and repression of *in vitro* transcription of the *lac* operon of *Escherichia coli*. Proc. Natl. Acad. Sci. USA 68, 1828–32.

THE IDEA OF ANTITERMINATION REGULATION: PROBLEMS IN PARADISE
"Transcription-translation coupling" model proposed to explain regulation of the his operon

Ames, B. N., and P. E. Hartmann (1963) The histidine operon. Cold Spring Harbor Symp. Quant. Biol. 28, 349–56.

Polar nonsense mutations prevent transcription of downstream genes in an operon

Imamoto, F. (1970) Evidence for premature termination of transcription of the tryptophan operon in polarity mutants of *Escherichia coli*. Nature 228, 232–35.

λ N does not activate early gene expression by activating new promoters

Luzzati, D. (1970) Regulation of λ exonuclease synthesis: Role of the N gene product and λ repressor. J. Mol. Biol. 49, 515–19.

Transcription activates the replication origin of λ

Dove, W. F., E. Hargrove, M. Ohashi, F. Haugli, and A. Guha (1969) Replicator activation in lambda. Japan J. Genet. 44, suppl. 1, 11–22.

Dove, W. F., H. Inokuchi, and W. F. Stevens (1971) Replication control in phage lambda, in A. D. Hershey, ed., *The Bacteriophage Lambda*. Cold Spring Harbor, N.Y.: Cold Spring Harbor Laboratory Press.

Roberts proposes that N protein blocks action of rho

Roberts, J. (1969) Termination factor for RNA polymerase. Nature 224, 1168–74.

REGULATED TERMINATION IN THE *trp* OPERON: THE ATTENUATOR
Stent's book says that everything fundamental is known in molecular genetics

Stent, G. S. (1969) *The Coming of the Golden Age: A View of the End of Progress*. Garden City, N.Y.: Natural History Press.

trpR identified

Cohen, G., and F. Jacob (1959) Sur la répression de la synthèse des enzymes intervenant dans la formations du tryptophane chez *Escherichia coli*. Compte Rendus 248, 3490–92.

In trpR mutant, trp *operon mRNA increases in the absence of tryptophan*

Baker, R., and C. Yanofsky (1972) Transcription initiation frequency and translational yield for the tryptophan operon of *Escherichia coli*. J. Mol. Biol. 69, 89–102.

Deletions downstream of the trp *promoter-operator region increase* trp *operon expression*

Jackson, E. N., and C. Yanofsky (1973) The region between the operator and first structural gene of the tryptophan operon of *Escherichia coli* may have a regulatory function. J. Mol. Biol. 76, 89–101.

Almost all transcription of the trp *operon* in vitro *terminates after 140 nucleotides*

Lee, F., C. L. Squires, C. Squires, and C. Yanofsky (1976) Termination of transcription *in vitro* in the *Escherichia coli* tryptophan operon leader region. J. Mol. Biol. 103, 383–93.

Bertrand and Yanofsky show differential regulation of different parts of trp *mRNA*

Bertrand, K., and C. Yanofsky (1976) Regulation of transcription termination in the leader region of the tryptophan operon of *Escherichia coli* involves tryptophan or its metabolic product. J. Mol. Biol. 103, 339–49.

Yanofsky predicts the secondary structure of the trp *attenuator region from its sequence*

Squires, C., F. Lee, K. Bertrand, C. L. Squires, M.J. Bronson, and C. Yanofsky (1976) Nucleotide sequence of the 5′ end of tryptophan messenger RNA of *Escherichia coli*. J. Mol. Biol. 103, 351–58.

Lee, F., and C. Yanofsky (1977) Transcription termination at the *trp* operon attenuators of *Escherichia coli* and Salmonella typhimurium: RNA secondary structure and regulation of termination. Proc. Natl. Acad. Sci. USA 74, 4365–69.

Mutation in Rho protein reduces polarity caused by nonsense mutations

Richardson, J. P., C. Grimley, and C. Lowery (1975) Transcription termination factor rho activity is altered in *Escherichia coli* with suA gene mutations. Proc. Natl. Acad. Sci. 72, 1725–28.

Ratner, D. (1976) Evidence that mutations in the suA polarity suppressing gene directly affect termination factor rho. Nature 259, 151–53.

TRANSCRIPTION TERMINATION AND HOW TO AVOID IT:
λ N AND Q PROTEINS
N protein reverses the effect of polar mutations in the trp *and* gal *operon*
    Adhya, S., M. Gottesman, and B. de Crombrugghe (1974) Release of polarity in *Escherichia coli* by gene N of phage λ: Termination and antitermination of transcription. Proc. Natl. Acad. Sci. USA 71, 2534–38.
    Franklin, N. C. (1974) Altered reading of genetic signals fused to the N operon of bacteriophage λ: Genetic evidence for modification of polymerase by the protein product of the N gene. J. Mol. Biol. 89, 33–48.

Forbes and Herskowitz provide genetic evidence for antitermination by λQ *protein*
    Forbes, D., and I. Herskowitz (1982) Polarity suppression by the Q gene product of bacteriophage lambda. J. Mol. Biol. 160, 549–69.

Isolation of nus mutants
    Friedman, D. I. (1971) A bacterial mutant affecting lambda development, in A. D. Hershey, ed., The Bacteriophage Lambda. Cold Spring Harbor, N.Y.: Cold Spring Harbor Laboratory Press.
    Georgopoulos, C. P. (1971) A bacterial mutant affecting N function, in A. D. Hershey, ed., The Bacteriophage Lambda. Cold Spring Harbor, N.Y.: Cold Spring Harbor Laboratory Press.
    Ward, D. F., and M. E. Gottesman (1981) The nus mutations affect transcription termination in *Escherichia coli*. Nature 292, 212–15.

Greenblatt uses extracts to identify an activity for λ *N protein*
    Greenblatt, J. (1972) Positive control of endolysin synthesis *in vitro* by the gene N protein of phage λ. Proc. Natl. Acad. Sci. USA 69, 3606–10.

Purification of NusA
    Greenblatt, J., and J. Li (1981) The nusA gene protein of *Escherichia coli.*: Its identification and a demonstration that it interacts with the gene N transcription anti-termination protein of bacteriophage lambda. J. Mol. Biol. 147, 11–23.

NusB, NusE, and NusG are part of the antitermination complex with N and NusA
    Mason, S. W., and J. Greenblatt (1991) Assembly of transcription elongation complexes containing the N protein of phage lambda and the *Escherichia coli* elongation factors NusA, NusB, NusG, and S10. Genes. Devel. 5, 1504–12.
    Sullivan, S. A., and M. E. Gottesman (1992) Requirement for *E. coli* NusG protein in factor-dependent transcription termination. Cell 68, 989–94.
    Das, A. (1993) Control of transcription termination by RNA-binding proteins. Annu. Rev. Biochem. 62, 893–930.

NusA binds to RNA polymerase in the absence of N
    Greenblatt, J., and J. Li (1981) Interaction of the sigma factor and the nusA gene protein of *E. coli* with RNA polymerase in the initiation-termination cycle of transcription. Cell 24, 421–28.

NusA alters the elongation rate and termination capacity of RNA polymerase

Schmidt, M. C., and M. J. Chamberlin (1984) Amplification and isolation of *Escherichia coli* nusA protein and studies of its effects on *in vitro* RNA chain elongation. Biochemistry 23, 197–203.

Schmidt, M. C., and M. J. Chamberlin (1987) NusA protein of *Escherichia coli* is an efficient transcription termination factor for certain termination sites. J. Mol. Biol. 195, 809–18.

Isolation of λ Q protein

Grayhack. E. J., and J. W. Roberts (1983) Purification of the bacteriophage lambda late gene regulator encoded by gene Q. J. Biol. Chem. 258, 9192–96.

Antitermination by Q protein requires NusA

Grayhack, E. J., X. J. Yang, L. F. Lau, and J. W. Roberts (1985) Phage lambda gene Q antiterminator recognizes RNA polymerase near the promoter and accelerates it through a pause site. Cell 42, 259–69.

THE REGULATOR–OPERATOR INTERACTION: HOW TO LOCATE A SITE ON DNA

Lac repressor proteins a 24 bp segment of DNA from DNase I digestion

Gilbert, W., and A. Maxam (1973) The nucleotide sequence of the lac operator. Proc. Natl. Acad. Sci. USA 70, 3581–84.

Maizels sequences the lac operator

Maizels, N. (1973) The nucleotide sequence of the lactose messenger ribonucleic acid transcribed from the UV5 promoter mutant of *E. coli*. Proc. Natl. Acad. Sci. USA 70, 3585–89.

lac operator is a symmetric sequence

Sadler, J. R., and T. F. Smith (1971) Mapping of the lactose operator. J. Mol. Biol. 62, 139–69.

Gierer points out that inverted repeats in DNA could form hairpins

Gierer, A. (1967) [On the function of deoxyribonucleic acids and the theory of the regulation of gene action] (German). Naturwissenschaften 54, 389–96.

A NEW APPROACH TO DNA SEQUENCE AND ITS RECOGNITION:
SINGLE-CLEAVAGE ANALYSIS

Arber, Smith, and Nathans show the existence and utility of restriction endonucleases

Arber, W., and D. Dussoix (1962) Host controlled modification of bacteriophage λ. J. Mol. Biol. 5, 18–36.

Smith, H. O., and K. W. Wilcon (1970) A restriction enzyme from Haemophilus influenzae. I. Purification and general properties. J. Mol. Biol. 51, 379–92.

Danna, K., and D. Nathans (1971) Specific cleavage of simian virus 40 DNA by restriction endonuclease of Hemophilus influenzae. Proc. Natl. Acad. Sci. USA 68, 2913–17.

Development of DNA sequencing by Maxam and Gilbert

Maxam, A. M., and W. Gilbert (1977) A new method for sequencing DNA. Proc. Natl. Acad. Sci. USA 74, 560–64.

DNase I footprinting

Galas, D. J., and A. Schmitz (1978) DNase I footprinting: a simple method for the detection of protein-DNA binding specificity. Nucleic Acids Res. 5, 3157–70.

DETERMINING DNA SEQUENCE BY DNA REPLICATION

Sanger, F., and A. R. Coulson (1975) A rapid method for determining sequences in DNA by primed synthesis with DNA polymerase. J. Mol. Biol. 94, 441–48.

Sanger, F., S. Nicklen, and A. R. Coulson (1977) DNA sequencing with chain-terminating inhibitors. Proc. Natl. Acad. Sci. USA 74, 5463–67.

THE MOLECULAR STUDY OF PROTEIN-DNA INTERACTIONS

Size of DNA protected by λ cI protein depends on the concentration of cI

Maniatis, T., and M. Ptashne (1973) Multiple repressor binding at the operators in bacteriophage λ. Proc. Natl. Acad. Sci. USA 70, 1531–1535.

Pirrotta V. (1973) Isolation of the operators of phage λ. Nature New Biol. 244, 13–16.

Maniatis sequences o_L and o_R operator regions

Maniatis, T., M. Ptashne, K. Backman, D. Kield, S. Flashman, A. Jeffrey, and R. Maurer (1975) Recognition sequences of repressor and polymerase in the operators of bacteriophage lambda. Cell 5, 109–13.

o_L and o_R consist of three contiguous operator sites

Ibid.

Pribnow box

Pribnow, D. (1975) Nucleotide sequence of an RNA polymerase binding site at an early T7 promoter. Proc. Natl. Acad. Sci. USA 72, 784–88.

Bases in -10 and -35 regions are contacted by RNA polymerase

Siebenlist, U., R. B. Simpson, and W. Gilbert (1980) *E. coli* RNA polymerase interacts homologously with two different promoters. Cell 20, 269–81.

RNA polymerase-promoter interaction

Dickson, R. C., J. Abelson, W. M. Barnes, and W. S. Reznikoff (1975) Genetic regulation: The Lac control region. Science 187, 27–35.

Gilbert, W. (1976) Starting and stopping sequences for the RNA polymerase, in R. Losick and M. Chamberlin, eds., RNA Polymerase. Cold Spring Harbor, N.Y.: Cold Spring Harbor Laboratory Press.

Siebenlist, U. (1979) RNA polymerase unwinds an 11-base pair segment of a phage T7 promoter. Nature 279, 651–52.

Straney, D. C., and D. M. Crothers. 1987. Comparison of the open complexes formed by RNA polymerase at the *Escherichia coli* lac UV5 promoter. J. Mol. Biol. 193, 279–92.

BUILDING A REGULATORY PROTEIN: A DNA-BINDING MOTIF

Structures of myoglobin and hemoglobin

Kendrew, J. C., R. E. Dickerson, B. E. Strandberg, R. G. Hart, D. R. Davies, D. C. Phillips, and V. C. Shore (1960) Structure of myoglobin: A three-dimensional Fourier synthesis at 2 Å resolution. Nature 185, 422–27.

Perutz, M. F., M. G. Rossman, A. F. Cullis, H. Muirhead, G. Will, and A. C. T. North (1960) Structure of haemoglobin. A three-dimensional fourier synthesis at 5.5-Å resolution, obtained by X-ray analysis. Nature 185, 416–22.

X-ray structures of regulatory proteins

Anderson, W. F., D. H. Ohlendorf, Y. Takeda, and B. W. Matthews (1981) Structure of the cro repressor from bacteriophage lambda and its interaction with DNA. Nature 290, 754–58.

McKay, D. B., and T. A. Steitz (1981) Structure of catabolite gene activator protein at 2.9 A resolution suggests binding to left-handed B-DNA. Nature 290, 744–49.

Pabo, C. O., and M. Lewis (1982) The operator-binding domain of lambda repressor: Structure and DNA recognition. Nature 298, 443–47.

Structures of cocrystals of cI and 434 cI and operator fragments

Aggarwal, A. K., D. W. Rodgers, M. Drottar, M. Ptashne, and S. C. Harrison (1988) Recognition of a DNA operator by the repressor of phage 434: A view at high resolution. Science 242, 899–907.

Jordan, S. R., and C. O. Pabo (1988) Structure of the lambda complex at 2.5 Å resolution: Details of the repressor-operator interactions. Science 242, 893–99.

von Hippel proposes that DNA-binding proteins "slide" along DNA

Berg, O. G., R. B. Winter, and P. H. von Hippel (1981) Diffusion-driven mechanisms of protein translocation on nucleic acids. I. Models and theory. Biochemistry 20, 6929–48.

THE REGULATOR–POLYMERASE INTERACTION: GENERAL RULES?

Adhya proposes "looping" model to explain repression of gal *operon*

Irani, M. H., L. Orosz, and S. Adhya (1983) A control element within a structural gene: The *gal* operon of *Escherichia coli*. Cell 32, 783–88.

Majumdar, A., and S. Adhya (1984) Demonstration of two operator elements in *gal*: *in vitro* repressor binding studies. Proc. Natl. Acad. Sci. USA 81, 6100–6104.

Schleif proposes looping at ara

Dunn, T. M., S. Hahn, S. Ogden, and R. F. Schleif (1984) An operator at –280 base pairs that is required for repression of araBAD operon promoter: Addition of DNA helical turns between the operator and promoter cyclically hinders repression. Proc. Natl. Acad. Sci. USA 81, 5017–20.

Models for activation at a distance

Echols, H. (1984) Specialized nucleoprotein structures in high-fidelity DNA transactions. BioEssays 1, 148–52.

Echols, H. (1986) Multiple DNA-protein interactions governing high-precision DNA transactions. Science 133, 1050–56.

Ptashne, M. (1986) Gene regulation by proteins acting nearby and at a distance. Nature 322, 697–701.

Kustu and coworkers show looping by NtrC

Su, W., S. Porter, S. Kustu, and H. Echols (1990) DNA-looping and enhancer activity: Association between DNA-bound NtrC activator and RNA polymerase at the bacterial glnA promoter. Proc. Natl. Acad. Sci. USA 87, 5504–8.

Wedel, A., D. S. Weiss, D. Popham, P. Droge, and S. Kustu (1990) A bacterial enhancer functions to tether a trancriptional activator near a promoter. Science 248, 486–90.

Geiduschek shows that a sliding clamp can activate transcription

Herendeen, D. R., G. A. Kassavetis, and E. P. Geiduschek (1992) A transcriptional enhancer whose function imposes a requirement that proteins track along DNA. Science 256, 1293–1303.

THE ANTICODON: BASE-PAIRS CAN WOBBLE

in vitro translation systems of Zamecnik and Tissières

Lamborg, M. R., and P. C. Zamecnik (1960) Amino acid incorporation into protein by extracts of *E. coli*. Biochim. Biophys. Acta 42, 206–11.

Tissières, A., D. Schlessinger, and F. Gros (1960) Amino acid incorporation into proteins by *Escherichia coli* ribosomes. Proc. Natl. Acad. Sci. USA 46, 1450–63.

Crick proposes "wobble hypothesis"

Crick. F. H. C. (1966) Codon-anticodon pairing: The wobble hypothesis. J. Mol. Biol. 19, 548–55.

A single tRNA can recognize more than one codon

Khorana, H. G. (1966–67) Polynucleotide synthesis and the genetic code. Harvey Lectures Series 1966–67, vol. 62. New York: Academic Press.

Söll, D., and U. L. RajBhandary (1967) Studies on polynucleotides. LXXVI. Specificity of transfer RNA for codon recognition as studied by amino acid incorporation. J. Mol. Biol. 29, 113–24.

BASE SEQUENCE OF TRNA: A THREE-LEAF CLOVER FOR HOLLEY

Sequence determined for yeast alanine tRNA and "clover-leaf" structure predicted

Holley, R. W., J. Apgar, G. A. Everett, J. T. Madison, M. Marquisee, S. H. Merrill, J. R. Penswick, and A. Zamir (1965) Structure of a ribonucleic acid. Science 147, 1462–65.

Sequence of yeast tyrosine-tRNA

Madison, J. T., G. A. Everett, and H. Kung (1966) Nucleotide sequence of a yeast tyrosine transfer RNA. Science 153, 531–34.

Zachau and colleagues sequence two yeast serine tRNAs

Zachau, H. G., D. Dutting, and H. Feldmann (1966) Nucleotide sequences of two serine-specific transfer ribonucleic acids. Angewandte Chemie 5, 422.

Crystal structure determined for phenylalanine tRNA

Kim, S. H., F. L. Suddath, G. J. Quigley, A. McPherson, J. L. Sussman, A. H. Wang, N. C. Seeman, and A. Rich (1974) Three-dimensional tertiary structure of yeast phenylalanine transfer RNA. Science 185, 435–40.

Robertus, J. D., J. E. Ladner, J. T. Finch, D. Rhodes, R. S. Brown, B. F. Clark, and A. Klug (1974) Structure of yeast phenylalanine tRNA at 3 A resolution. Nature 250, 546–51.

Recognition of tRNAs by aminoacyl-tRNA synthetases

Normanly, J., R. C. Ogden, S. J. Horvath, and J. Abelson (1986) Changing the identity of a transfer RNA. Nature 321, 213–19.

Hou, Y. M., and P. Schimmel (1988) A simple structural feature is a major determinant of the identity of a transfer RNA. Nature 333, 140–45.

McClain, W. H., and K. Foss (1988) Changing the identity of a tRNA by introducing a G-U wobble pair near the 3' acceptor end. Science 240, 793–96.

McClain, W. H. (1993) Transfer RNA identity. FASEB J. 7, 72–78.

MESSAGE DECODING: METHODS AND PROTAGONISTS

Isolation of phage f2

Loeb, T., and N. D. Zinder (1961) A bacteriophage containing RNA. Proc. Natl. Acad. Sci. USA 47, 282–89.

THE PROBLEM OF STARTING A PROTEIN

Proteins synthesized from the N-terminus to the C-terminus

Dintzis, H. M. (1961) Assembly of the peptide chains of hemoglobin. Proc. Natl. Acad. Sci. USA 47, 247–61.

N-formyl-methionine tRNA discovered

Marcker, K., and F. Sanger (1964) N-formyl-methionyl-s-RNA. J. Mol. Biol. 8, 835–40.

Marcker and Clark examined the decoding properties of f-met-tRNA

Clark, B. F. C., and K. A. Marcker (1966) The role of N-formyl-methionyl-tRNA in protein biosynthesis. J. Mol. Biol. 17, 394–406.

Waller finds that only 50% of E. coli *proteins begin with an N-terminal methionine residue*

Waller, J. (1963) The NH2-terminal residues of the proteins from cell-free extracts of *E. coli* J. Mol. Biol. 7, 483–96.

Proteins synthesized in vitro *begin with N-formyl methionine residue*

Adams, J. M., and M. R. Capecchi (1966) N-formylmethionyl-sRNA as the initiator of protein synthesis. Proc. Natl. Acad. Sci. USA 55, 147–55.

Webster, R. E., D. L. Engelhardt, and N. D. Zinder (1966) *in vitro* protein synthesis: Chain initiation. Proc. Natl. Acad. Sci. USA 55, 155–61.

ENDING A PROTEIN: THE VALUE OF NONSENSE

Nonsense mutations in the major head protein of phage T4 produce truncated protein products in nonsuppressing hosts

Sarabhai, A. S., A. O. W. Stretton, S. Brenner, and A. Bolle (1964) Colinearity of the gene with the polypeptide chain. Nature 201, 13–17.

Brenner and coworkers determine the sequence of two nonsense codons

Brenner, S., A. O. W. Stretton, and S. Kaplan (1965) Genetic code: The "nonsense" triplets for chain termination and their suppression. Nature 206, 994–98.

Garen and coworkers determine the sequence of nonsense codons in alkaline phosphatase

Weigert, M. G., and A. Garen (1965) Base composition of nonsense codons in *E. coli*: Evidence from amino acid substitutions at a tryptophan site in alkaline phosphatase. Nature 206, 992–94.

Garen, A. (1968) Sense and nonsense in the genetic code. Science 160, 149–59.

UAA, UAG, and UGA triplets do not bind tRNA in the trinucleotide binding assay

Nirenberg, M., and P. Leder (1964) The effect of trinucleotides upon the binding of sRNA to ribosomes. Science 145, 1399–1407.

Nirenberg, M., P. Leder, M. Bernfield, R. Brimacombe, J. Trupin, F. Rottman, and C. O'Neal (1965) RNA codewords and protein synthesis. VII. On the general nature of the RNA code. Proc. Natl. Acad. Sci. USA 53, 1161–68.

Söll, D., E. Ohtsuka, D. S. Jones, R. Lohrmann, H. Hayatsu, S. Nishimura, and H. G. Khorana (1965) Studies on polynucleotides. XLIX. Stimulation of the binding of aminoacyl-sRNA's to ribosomes by ribotrinucleotides and a survey of codon assignments for 20 amino acids. Proc. Natl. Acad. Sci. USA 54, 1378–85.

Further evidence that nonsense codons cause translation to terminate
Capecchi, M. R., and G. N. Gussin (1965) Suppression *in vitro*: Identification of a serine-sRNA as a "nonsense" suppressor. Science 149, 416–22.
Engelhardt, D. L., R. E. Webster, R. C. Wilhelm, and N. Zinder (1965) *in vitro* studies on the mechanism of suppression of a nonsense mutation. Proc. Natl. Acad. Sci. USA 54, 1791–97.

A mutant tyrosine tRNA with altered anticodon can recognize UAG codons
Capecchi, M. R., and G. N. Gussin (1965) Suppression *in vitro*: Identification of a serine-sRNA as a "nonsense" suppressor. Science 149, 416–22.
Goodman, H. M., J. Abelson, A. Landy, S. Brenner, and J. D. Smith (1968) Amber suppression: A nucleotide change in the anticodon of a tyrosine transfer RNA. Nature 217, 1019–24.

Initial verification that nonsense codons terminate normal proteins
Nichols, J. L. (1970) Nucleotide sequence from the polypeptide chain termination region of the coat protein cistron in bacteriophage R17 RNA. Nature 225, 147–51.
Last, J. A., W. M. Stanley Jr., M. Salas, M. B. Hille, A. J. Wahba, and S. Ochoa (1967) Translation of the genetic message. IV. UAA as a chain termination codon. Proc. Natl. Acad. Sci. USA 57, 1062–67.

PROTEIN TRANSLATION FACTORS:
TURNING THE GEARS IN PROTEIN SYNTHESIS
Purification of elongation factors for protein synthesis
Lucas-Lenard, J., and F. Lipmann (1966) Separation of three microbial amino acid polymerization factors. Proc. Natl. Acad. Sci. USA 55, 1562–66.

Purification of initiation factors
Brawerman, G., and J. M. Eisenstadt (1966) A factor from *E. coli* concerned with the stimulation of cell-free polypeptide synthesis by exogenous ribonucleic acid. II. Characteristics of the reaction promoted by the stimulation factor. Biochem. 5, 2784–89.
Revel, M., and F. Gros (1966) A factor from *E. coli* required for translation of natural messenger RNA. Biochem. Biophys. Res. Comm. 25, 124–32.
Stanley, W. M. Jr., M. Salas, A. J. Wahba, and S. Ochoa (1966) Translation of the genetic message: Factors involved in the initiation of protein synthesis. Proc. Natl. Acad. Sci. 56, 290–95.
Hershey, J. W., K. F. Dewey, and R. E. Thach (1969) Purification and properties of initiation factor f-1. Nature 222, 944–47.

Experiments of Nomura, Guthrie, and Lowry on ribosome dissociation
Nomura, M., and C. Lowry (1967) Phage f2 RNA-directed binding of formyl-methionyl-tRNA to ribosomes and the role of 30S ribosomal subunits in the initiation of protein synthesis. Proc. Natl. Acad. Sci. USA 58, 946–53.
Nomura, M., C. Lowry, and C. Guthrie (1967) The initiation of protein synthesis: Joining of the 50S ribosomal subunit to the initiation complex. Proc. Natl. Acad. Sci. USA 58, 1487–93.

Guthrie, C., and M. Nomura (1968) Initiation of protein synthesis: A critical test of the 30S subunit model. Nature 219, 232–35.

Evidence for ribosome dissociation cycle

Mangiarotti, G., and D. Schlessinger (1967) Polyribosome metabolism in *Escherichia coli*. II. Formation and lifetime of RNA molecules, ribosomal subunit couples and polyribosomes. J. Mol. Biol. 29, 395–418.

Kaempfer, R., M. Meselson, and H. Raskas (1968) Cyclic dissociation into stable subunits and re-formation of ribosomes during bacterial growth. J. Mol. Biol. 31, 277–89.

Kaempfer, R., and M. Meselson (1969) Studies of ribosomal subunit exchange. Cold Spring Harbor Symp. Quant. Biol. 34, 209–20.

USING THE RIBOSOME: GTP AND TRANSLOCATION
GTP is a required cofactor for protein synthesis

Keller, E. B., and P. C. Zamecnik (1956) The effect of guanosine diphosphate and triphosphate on the incorporation of labeled amino acids into proteins. J. Biol. Chem 221, 45–59.

First role determined for GTP was to bring the charged tRNA to the ribosome

Anderson, J. S., M. S. Bretscher, B. F. C. Clark, and K. A. Marcker (1967) A GTP requirement for binding initiation tRNA to ribosomes. Nature 215, 490–92.

Lengyel proposes that GTP hydrolysis changes the conformation of EF-Tu

Ono, Y., A. Skoultchi, J. Waterson, and P. Lengyel (1969) Peptide chain elongation: GTP cleavage catalysed by factors binding aminoacyl-transfer RNA to the ribosome. Nature 222, 645–48.

Peptidyl transferase activity of the ribosome discovered

Maden, B. E. H., R. R. Traut, and R. E. Monro (1968) Ribosome-catalyzed peptidyl transfer: The polyphenylalanine system. J. Mol. Biol. 35, 333–45.

"Particularly clear experiment" establishing the role of EF-G in protein synthesis

Erbe, R. W., M. M. Nau, and P. Leder (1969) Translation and translocation of defined RNA messengers. J. Mol. Biol. 39, 441–60.

Translocation reaction of ribosomes demonstrated

Gupta, S. L., J. Waterson, M. L. Sopori, S. M. Weissman, and P. Lengyel (1971) Movement of the ribosome along the messenger ribonucleic acid during protein synthesis. Biochemistry 10, 4410–21.

Thach, S. S., and R. E. Thach (1971) Translocation of messenger RNA and "accomodation" of fmet-tRNA. Proc. Natl. Acad. Sci. USA 68, 1791–95.

Translational release factors identified

Capecchi, M. R. (1967) Polypeptide chain termination *in vitro*: Isolation of a release factor. Proc. Natl. Acad. Sci. USA 58, 1144–51.

Caskey, C. T., R. Tomkins, E. Scolnick, T. Caryk, and M. Nirenberg (1968) Sequential translation of trinucleotide codons for initiation and termination of protein synthesis. Science 162, 135–38.

Capecchi, M. R., and H. A. Klein (1969) Characterization of three proteins involved in polypeptide chain termination. Cold Spring Harbor Symp. Quant. Biol. 34, 469–77.

There are multiple ribosomal proteins, early 1960's

Waller, J. P., and J. H. Harris (1961) Studies on the composition of the protein from *Escherichia coli* ribosomes. Proc. Natl. Acad. Sci. USA 47, 18–23.

Leboy, P. S., E. C. Cox, and J. G. Flaks (1964) The chromosomal site specifying a ribosomal protein in *Escherichia coli*. Proc. Natl. Acad. Sci. USA 52, 1367–74.

Waller, J. P. (1964) Fractionation of the ribosomal protein from *Escherichia coli*. J. Mol. Biol. 10, 319–36.

Separation of ribosomal proteins

Traut, R. R., P. B. Moore, H. Delius, H. Noller, and A. Tissières (1967) Ribosomal proteins of *E. coli* Demonstration of different primary structures. Proc. Natl. Acad. Sci. USA 57, 1294–1301.

Craven, G. R., P. Voynow, S. J. S. Hardy, and C. G. Kurland (1969) The ribosomal proteins of *E. coli* II. Chemical and physical characterization of the 30S ribosomal proteins. Biochemistry 8, 2906–15.

Hardy, S. J. S., C. G. Kurland, P. Voynow, and G. Mora (1969) The ribosomal proteins of *E. coli* I. Purification of the 30S proteins. Biochemistry 8, 2897–2905.

Later work by Wittmann defined 32 proteins in the 50S ribosomal subunit

Kaltschmidt, E., and H. G. Wittmann (1970b) Ribosomal proteins. XII. Number of proteins in small and large ribosomal subunits of *Escherichia coli* as determined by two-dimensional gel electrophoresis. Proc. Natl. Acad. Sci. USA 67, 1276–82.

Steitz identified A protein of phage R17

Steitz, J. A. (1968) Identification of the A protein as a structural component of bacteriophage R17. J. Mol. Biol. 33, 923–36.

Ribosome could protect mRNA from nucleases

Takanami, M., Y. Yan, and T. H. Jukes (1965) Studies on the site of ribosomal binding of f2 bacteriophage RNA. J. Mol. Biol. 12, 761–63.

Lengyel found conditions where mRNA binding to ribosomes depended on initiator tRNA

Kondo, M., G. Eggertsson, J. Eisenstadt, and P. Lengyel (1968) Ribosome formation from subunits: Dependence on formylmethionyl-transfer RNA in extracts from *E. coli*. Nature 220, 368–71.

Steitz finds AUG at beginning of the three genes of RNA phage

Steitz, J. A. (1969) Polypeptide chain initiation: Nucleotide sequences of the three ribosomal binding sites in bacteriophage R17 RNA. Nature 224, 957–64.

Shine and Dalgarno sequence

Shine, J., and L. Dalgarno (1974) The 3′-terminal sequence of *Escherichia coli* 16S ribosomal RNA: Complementarity to nonsense triplets and ribosome binding sites. Proc. Natl. Acad. Sci. USA 71, 1342–46.

Steitz shows that fragment of 16S rRNA pairs with the initiator region of phage RNA

Steitz, J. A., and K. Jakes (1975) How ribosomes select initiator regions in mRNA: Base pair formation between the 3' terminus of 16S rRNA and the mRNA during initiation of protein synthesis in *Escherichia coli*. Proc. Natl. Acac. Sci. USA 72, 4734–38.

Traub and Nomura reassemble 30S subunit from 16S rRNA and 21 ribosomal proteins
Traub, P., and M. Nomura (1968) Structure and function of *E. coli* ribosomes. V. Reconstitution of functionally active 30S ribosomal particles from RNA and proteins. Proc. Natl. Acad. Sci. USA 59, 777–84.

Work on 50S ribosomal subunit
Nomura, M., and V. A. Erdmann (1970) Reconstitution of 50S ribosomal subunits from dissociated molecular components. Nature 228, 774–78.
Nierhaus, K. H., and F. Dohme (1974) Total reconstitution of functionally active 50S ribosomal subunits from *Escherichia coli*. Proc. Natl. Acad. Sci. USA 71, 4713–17.
Nierhaus, K. H. (1982) Structure, assembly, and function of ribosomes. Curr. Top. Microbiol. Immunol. 97, 81–155.

Moore and Engelman neutron scattering studies on ribosomes
Moore, P. B., D. M. Engelman, and B. P. Schoenborn (1975) A neutron scattering study of the distribution of protein and RNA in the 30S ribosomal subunit of *Escherichia coli*. J. Mol. Biol. 91, 101–20.

Electron microscope studies on ribosomes by Lake and Stöffler
Lake, J. A. (1979) Ribosome structure and functional sites, in G. Chambliss, G. R. Craven, J. Davies, K. Davis, L. Kahan, and M. Nomura, eds., *Ribosomes: Structure, Function, and Genetics*. Baltimore, Md.: University Park Press.
Stöffler, G., R. Bald, B. Kastner, R. Lührmann, M. Stöffler-Meilicke, and G. Tischendorf (1979) Structural organization of the *Escherichia coli* ribosome and localization of functional domains, in G. Chambliss, G. R. Craven, J. Davies, K. Davis, L. Kahan, and M. Nomura, eds., *Ribosomes: Structure, Function, and Genetics*. Baltimore, Md.: University Park Press.

Work on 30S subunit by neutron scattering completed by Moore and colleagues in 1987
Capel, M. S., M. Kjeldgaard, D. M. Engelman, and P. B. Moore (1988) Positions of S2, S13, S16, S17, S19 and S21 in the 30S ribosomal subunit of *Escherichia coli*. J. Mol. Biol. 200, 65–87.

Later immuno-EM studies by the Lake and Stöffler groups on the 50S subunit
Walleczek, J., D. Schuler, M. Stöffler-Meilicke, R. Brimacombe, and G. Stöffler (1988) A model for the spatial arrangement of the proteins in the large subunit of the *Escherichia coli* ribosome. EMBO J. 7, 3571–76.
Lotti, M., M. Noah, M. Stöffler-Meilicke, and G. Stöffler (1989) Localization of proteins L4, L5, L20 and L25 on the ribosomal surface by immuno-electron microscopy. Mol. Gen. Genet. 216, 245–53.

Woese talk at 1979 ribosomes meeting in Madison, Wisconsin
Woese, C. R. (1979) Just so stories and Rube Goldberg machines: Speculations on the origin of the protein synthetic machinery, in G. Chambliss, G. R. Craven, J. Davies, K. Davis, L. Kahan, and M. Nomura, eds., *Ribosomes: Structure, Function, and Genetics*. Baltimore, MD: University Park Press.

Noller and colleagues determine sequence of 16S rRNA gene
Brosius, J., M. L. Palmer, P. J. Kennedy, and H. F. Noller (1978) Complete nucleotide sequence of a 16S ribosomal RNA gene from *Escherichia coli*. Proc. Natl. Acad. Sci. USA 75, 4801–05.

16S rRNA sequences important for tRNA binding were universally conserved

Woese, C. R., G. E. Fox, L. Zablen, T. Uchida, L. Bonen, K. Pechman, B. J. Lewis, and D. Stahl (1975) Conservation of primary structure in 16S ribosomal RNA. Nature 254, 83–86.

Woese, C. R., L. J. Magrum, R. Gupta, R. B. Siegel, D. A. Stahl, J. Kop, N. Crawford, J. Brosius, R. Gutell, J. J. Hogan, and H. F. Noller (1980) Secondary structure model for bacterial 16S ribosomal RNA: Phylogenetic, enzymatic, and chemical evidence. Nucl. Acids Res. 8, 2275–93.

Noller, H. F., and C. R. Woese (1981) Secondary structure of 16S ribosomal RNA. Science 212, 403–11.

Noller finds that he can destroy the ability of tRNA to bind to the ribosome by modifying only a few bases in rRNA

Noller, H. F., and J. B. Chaires (1972) Functional modification of 16S ribosomal RNA by kethoxal. Proc. Natl. Acad. Sci. USA 69, 3115–18.

Noller, H. F. (1974) Topography of 16S RNA in 30S ribosomal subunits. Nucleotide sequences and location of sites of reaction with kethoxal. Biochemistry 13, 4694–4703.

RNA can be an enzyme

Zaug, A. J., and T. R. Cech (1980) *in vitro* splicing of the ribosomal RNA precursor in nuclei of Tetrahymena. Cell 19, 331–38.

Cech, T. R., A. J. Zaug, and P. J. Grabowski (1981) *in vitro* splicing of the ribosomal RNA precursor of Tetrahymena: Involvement of a guanosine nucleotide in the excision of the intervening sequence. Cell 27, 487–96.

The RNA view of the ribosome

Noller, H. F., V. Hoffrath, and L. Zimniak (1992) Unusual resistance of peptidyl transferase to protein extraction procedures. Science 256, 1416–19.

Samaha, R. R., R. Green, and H. F. Noller (1995) A base pair between tRNA and 23S rRNA in the peptidyl transferase centre of the ribosome. Nature 377, 309–14. (Erratum: Nature 378, 419.)

Khaitovich, P., A. S. Mankin, R. Green, L. Lancaster and H. F. Noller (1999). Characterization of functionally active subribosomal particles from Thermus aquaticus. Proc. Natl. Acad. Sci. USA 96, 85-90.

Ban, N., P. Nissen, J. Hansen, P. B. Moore and T. A. Steitz (2000). The complete atomic structure of the large ribosomal subunit at 2.4Å resolution. Science 289, 905-20.

Nissen, P., J. Hansen, N. Ban, P. B. Moore, and T. A. Steitz (2000). The structural basis of ribosome activity in peptide bond synthesis. Science 289, 920-30.

RNA SLICING IN BACTERIA
Early papers on RNA processing

Dunn, J. J., ed. (1975) Processing of RNA. Report of a symposium held May 29–31, 1974. Biology Dept., Brookhaven National Laboratory, Upton, N.Y.

Smith and Brenner find that suppression results from a change in the anticodon

Goodman, H., J. Abelson, A. Landy, S. Brenner, and J. D. Smith (1968) Amber suppression: A nucleotide change in the anticodon of a tyrosine transfer tRNA. Nature 217, 1019–24.

Altman finds tRNA precursor molecules

Altman, S. (1971) Isolation of tyrosine tRNA precursor molecules. Nature New Biol. 229, 19–21.

Altman, S., and J. D. Smith (1971) Tyrosine tRNA precursor molecule polynucleotide sequence. Nature New Biol. 233, 35–39.

Robertson and Altman partially purify RNase P activity

Robertson, H. D., S. Altman, and J. D. Smith (1972) Purification and properties of a specific *Escherichia coli* ribonuclease which cleaves a tyrosine transfer ribonucleic acid precursor. J. Biol. Chem. 247, 5243–51.

RNase P is a ribozyme

Guerrier-Takada, C., K. Gardiner, T. Marsh, N. Pace, and S. Altman (1983) The RNA moiety of ribonuclease P is the catalytic subunit of the enzyme. Cell 35, 849–57.

Ts mutation in RNase P

Schedl, P., and P. Primakoff (1973) Mutants of *E. coli* thermosensitive for the synthesis of transfer RNA. Proc. Natl. Acad. Sci. USA 70, 2091–95.

Sakano, H., S. Yamada, T. Ikemura, Y. Shimura, and H. Ozeki (1974) Temperature sensitive mutants of *Escherichia coli* for tRNA synthesis. Nucleic Acids Res. 1, 355.

All of the tRNAs in E. coli *are probably produced as longer precursor molecules*

Schedl, P., P. Primakoff, and J. Roberts (1975) Processing of *E. coli* tRNA precursors. Brookhaven Symposia in Biology 26, 53–76.

Ikemura, T., Y. Shimura, H. Sakano, and H. Ozeki (1975) Precursor molecules of *Escherichia coli* transfer RNAs accumulated in a temperature-sensitive mutant. J. Mol. Biol. 96, 69–86.

The 3′ end of the mature tRNA is probably produced by one or more enzymes

Reviewed in: Abelson, J. (1979) RNA processing and the intervening sequence problem. Annu. Rev. Biochem. 48, 1035–69.

Reviewed in: Deutscher, M. P. (1984) Processing of tRNA in prokaryotes and eukaryotes. CRC Critical Reviews in Biochemistry 17, 45–71.

Finding of multiple T7 mRNA molecules

Siegel, R. B., and W. C. Summers (1970) The process of infection with coliphage T7. III. Control of phage-specific RNA synthesis *in vivo* by an early phage gene. J. Mol. Biol. 49, 115–23.

Dunn purifies activity responsible for cleaving T7 mRNA, which turns out to be RNase III

Dunn, J. J., and F. W. Studier (1973) T7 early mRNAs are generated by site-specific cleavages. Proc. Natl. Acad. Sci. USA 70, 1559–63.

RNase III previously isolated by Robertson and Zinder

Robertson, H. D., R. E. Webster, and N. D. Zinder (1968) Purification and properties of ribonuclease III from *Escherichia coli*. J. Biol. Chem. 243, 82–91.

RNaseIII mutant

Kindler, P., T. U. Keil, and P. H. Hofschneider (1973) Isolation and characterization of a ribonuclease 3 deficient mutant of *Escherichia coli*. Molec. Gen. Genet. 126, 53–59.

rRNA is unprocessed in RNase III mutant

Dunn, J. J., and F. W. Studier (1973) T7 early RNAs and *Escherichia coli* ribosomal RNAs are cut from large precursor RNAs *in vivo* by ribonuclease 3. Proc. Natl. Acad. Sci. USA 70, 3296–3300.

Nikolaev, N., L. Silengo, and D. Schlessinger (1973) Synthesis of a large precursor to ribosomal RNA in a mutant of *Escherichia coli*. Proc. Natl. Acad. Sci. USA 70, 3361–65.

Nikolaev, N., L. Silengo, and D. Schlessinger (1973) A role for ribonuclease 3 in processing of ribosomal ribonucleic acid and messenger ribonucleic acid precursors in *Escherichia coli*. J. Biol. Chem. 248, 7967–69.

Target sites for RNase III in phage T7 RNA

Oakley, J. L., and J. E. Coleman (1977) Structure of a promoter for T7 RNA polymerase. Proc. Natl. Acad. Sci. USA 74, 4266–70.

Robertson, H., E. Dickson, and J. Dunn (1977) A nucleotide sequence from a ribonuclease III processing site in bacteriophage T7. Proc. Natl. Acad. Sci. USA 74, 822–26.

Rosenberg, M., and R. A. Kramer (1977) Nucleotide sequence surround a ribonuclease III processing site in bacteriophage T7 RNA. Proc. Natl. Acad. Sci. USA 74, 984–88.

16S, 23S and 5S are in same transcript

Dunn, J. J., and F. W. Studier (1973) T7 early RNAs and *Escherichia coli* ribosomal RNAs are cut from large precursor RNAs in vivo by ribonuclease 3. Proc. Natl. Acad. Sci. USA 70, 3296–3300.

Nikolaev, N., L. Silengo, and D. Schlessinger (1973) Synthesis of a large precursor to ribosomal RNA in a mutant of *Escherichia coli*. Proc. Natl. Acad. Sci. USA 70, 3361–65.

Seven rRNA operons

Kenerley, M. E., E. A. Morgan, L. Post, L. Lindahl, and M. Nomura (1977) Characterization of hybrid plasmids carrying individual ribosomal ribonucleic acid transcription units of *Escherichia coli*. J. Bacteriol. 132, 931–49.

Kiss, A., B. Sain, and P. Venetianer (1977) The number of rRNA genes in *Escherichia coli*. FEBS Lett. 79, 77–79.

There are tRNAs between 16S and 23S sequences

Lund, E., J. E. Dahlberg, L. Lindahl, S. R. Jaskunas, P. P. Dennis, and M. Nomura (1976) Transfer RNA genes between 16S and 23S rRNA genes in rRNA transcription units of *E. coli*. Cell 7, 165–77.

Young and Steitz work on processing of tRNAs from rRNA transcripts, and they determine sequence of mature 16S rRNA

Young, R. A., and J. A. Steitz (1978) Complementary sequences 1700 nucleotides apart form a ribonuclease III cleavage site in *Escherichia coli* ribosomal precursor RNA. Proc. Natl. Acad. Sci. USA 75, 3593–97.

RNase III cuts within this duplex region

Ginsburg, D., and J. A. Steitz (1975) The 30S ribosomal precursor RNA from *Escherichia coli*. A primary transcript containing 23S, 16S, and 5S sequences. J. Biol. Chem. 250, 5647–54.

Doty infers existence of RNA duplexes from physical measurements
Doty, P., H. Boedtker, J. R. Fresco, R. Haselkorn, and M. Litt (1959) Secondary structure in ribonucleic acids. Proc. Natl. Acad. Sci. USA 45, 482–99.

RNA SPLICING IN EUKARYOTES
hnRNA found by Darnell and coworkers in 1960's
Darnell, J. E. (1968) Ribonucleic acids from animal cells. Bacteriol. Rev. 32, 262–90.

Eukaryotic mRNAs have a polyA tail
Edmonds, M., and M. G. Caramela (1969) The isolation and characterization of adenosine monophosphate-rich polynucleotides synthesized by Ehrlich ascites cells. J. Biol. Chem. 244, 1314–24.
Edmonds, M., M. H. Vaughan Jr., and H. Nakazato (1971) Polyadenylic acid sequences in the heterogeneous nuclear RNA and rapidly-labeled polyribosomal RNA of HeLa cells: Possible evidence for a precursor relationship. Proc. Natl. Acad. Sci. USA 68, 1336–40.
Darnell, J. E., L. Philipson, R. Wall, and M. Adesnik (1971) Polyadenylic acid sequences: Role in conversion of nuclear RNA into messenger RNA. Science 174, 507–10.

Eukaryotic mRNAs have 5' cap
Rottman, F., A. J. Shatkin, and R. P. Perry (1974) Sequences containing methylated nucleotides at the 5' termini of messenger RNAs: Possible implications for processing. Cell 3, 197–99.
Shatkin, A. J. (1976) Capping of eucaryotic mRNAs. Cell 9, 645–53.

Perry finds that both hnRNA and cytoplasmic RNA have 5' cap
Perry, R. P., and D. E. Kelley (1976) Kinetics of formation of 5' terminal caps in mRNA. Cell 8, 433–42.

Nathans shows the utility of restriction enzymes
Danna, K. J., G. H. Sack Jr., and D. Nathans (1973) Studies of simian virus 40 DNA. VII. A cleavage map of the SV40 genome. J. Mol. Biol. 78, 363–76.
Khoury, G., M. A. Martin, T. N. Lee, K. J. Danna, and D. Nathans (1973) A map of simian virus 40 transcription sites expressed in productively infected cells. J. Mol. Biol. 78, 377–89.

Temporal program for SV40 and adenovirus gene expression
Sambrook, J., B. Sugden, W. Keller, and P. A. Sharp (1973) Transcription of simian virus 40. III. Mapping of "early" and "late" species of RNA. Proc. Natl. Acad. Sci. USA 70, 3711–15.
Sambrook, J., P. A. Sharp, B. Ozanne, and V. Pettersson (1974) Studies on the transcription of simian virus 40 and adenovirus type 2. Basic Life Sciences 3, 167–79.

R-looping
Thomas, M., R. L. White, and R. W. Davis (1976) Hybridization of RNA to double-stranded DNA: Formation of R-loops. Proc. Natl. Acad. Sci. USA 73, 2294–98.

Davidson had already characterized the tail seen at 3' polyA end of RNA-DNA duplexes
Bender, W., and N. Davidson (1976) Mapping of poly(A) sequences in the electron microscope reveals unusual structure of type C on cornavirus RNA molecules. Cell 7, 595–607.

Berget and Sharp studies on RNA loops that led to recognition of RNA splicing
Berget, S. M., C. Moore, and P. A. Sharp (1977) Spliced segments at the 5′ terminus of adenovirus 2 late mRNA. Proc. Natl. Acad. Sci. USA 74, 3171–75.

Roberts and Gelinas work on late viral mRNAs, all have same 5′ sequence
Gelinas, R. E., and R. J. Roberts (1977) One predominant 5′-undecanucleotide in adenovirus 2 late messenger RNAs. Cell 11, 533–44.

Klessig also finds that two mRNAs for viral structural proteins have same sequence at 5′ end
Klessig, D. F. (1977) Two adenovirus mRNAs have a common 5′ terminal leader sequence encoded at least 10 kb upstream from their main coding regions. Cell 12, 9–21.

Chow and Broker find 5′ tails in R-looping experiments with viral mRNAs
Chow, L. T., J. M. Roberts, J. B. Lewis, and T. R. Broker (1977) A map of cytoplasmic RNA transcripts from lytic adenovirus type 2, determined by electron microscopy of RNA:DNA hybrids. Cell 11, 819–36.

Chow and Broker find that viral RNA hybridizes to four distinct regions of viral DNA
Chow, L. T., R. E. Gelinas, T. R. Broker, and R. J. Roberts (1977) An amazing sequence arrangement at the 5′ ends of adenovirus 2 messenger RNA. Cell 12, 1–8.

Klessig completes the case for splicing
Klessig, D. (1977) Two adenovirus mRNAs have a common 5′ terminal leader sequence encoded at least 10 kb upstream from their main coding regions. Cell 12, 9–21.

Evidence for splicing in other systems
Hozumi, N., and S. Tonegawa (1976) Evidence for somatic rearrangement of immunoglobulin genes coding for variable and constant regions. Proc. Natl. Acad. Sci. USA 73, 3628–32.
Breathnach, R., J. L. Mandel, and P. Chambon (1977) Ovalbumin gene is split in chicken DNA. Nature 270, 314–19.
Jeffreys, A. J., and R. A. Flavell (1977) The rabbit beta-globin gene contains a large insert in the coding sequence. Cell 12, 1097–1108.
Tilghman, S. M., D. C. Tiemeier, F. Polsky, M. H. Edgell, J. G. Seidman, A. Leder, L. W. Enquist, B. Norman, and P. Leder (1977) Cloning specific segments of the mammalian genome: Bacteriophage lambda containing mouse globin and surrounding gene sequences. Proc. Natl. Acad. Sci. USA 74, 4406–10.
Berk, A. J., and P. A. Sharp (1978) Spliced early mRNAs of simian virus 40. Proc. Natl. Acad. Sci. USA 75, 1274–78.
Lai, E. C., S. L. Woo, A. Dugiaczyk, J. F. Canterall, and B. W. O'Malley (1978) The ovalbumin gene: Structural sequences in native chicken DNA are not contagious. Proc. Natl. Acad. Sci. USA 75, 2205–9.
Reviewed in: Breathnach, R., and P. Chambon (1981) Organization and expression of eucaryotic split genes coding for proteins. Annu. Rev. Biochem. 50, 349–83.
Reviewed in: Abelson, J. (1979) RNA processing and the intervening sequence problem. Annu. Rev. Biochem. 48, 1035–69.

THE BIOCHEMISTRY OF SPLICING: LARIATS ON A SPLICEOSOME

Development of cell-free systems for splicing

Green, M. R., T. Maniatis, and D. A. Melton (1983) Human beta-globin pre-mRNA synthesized *in vitro* is accurately spliced in Xenopus oocyte nuclei. Cell 32, 681–94.

Hernandez, N., and W. Keller (1983) Splicing of *in vitro* synthesized messenger RNA precursors in HeLa cell extracts. Cell 35, 89–99.

Padgett, R. A., S. F. Hardy, and P. A. Sharp (1983) Splicing of adenovirus RNA in a cell-free transcription system. Proc. Natl. Acad. Sci. USA 80, 5230–34.

Padgett, R. A., M. M. Konoraska, P. J. Grabowski, S. F. Hardy, and P. A. Sharp (1984) Lariat RNAs as intermediates and products in the splicing of messenger RNA precursors. Science 225, 898–903.

Ruskin, B., A. R. Krainer, T. Marianis, and M. R. Green (1984) Excision of an intact intron as a novel lariat structure during pre-mRNA splicing *in vitro*. Cell 38, 317–31.

RNA events in the splicing pathway, reviewed in:

Green, M. R. (1986) PRE-mRNA splicing. Annu. Rev. Genet. 20, 671–708.

Moore, M. J., C. C. Query, and P. A. Sharp (1993) Splicing of precursors to mRNA by the spliceosome, in R. F. Gesteland and J. F. Atkins, eds., *The RNA World*. Cold Spring Harbor, N.Y.: Cold Spring Laboratory Press.

Branched structure previously detected in cells by Wallace and Edmonds

Wallace, J. C., and M. Edmonds (1983) Polyadenylated nuclear RNA contains branches. Proc. Natl. Acad. Sci. 80, 950–54.

Lerner and Steitz isolate snRNPs with human antibodies

Lerner, M. R., and J. A. Steitz (1979) Antibodies to small nuclear RNAs complexed with proteins are produced by patients with systemic lupus erythrematosus. Proc. Natl. Acad. Sci. USA 75, 5495–99.

Padgett, R. A., S. M. Mount, J. A. Steitz, and P. A. Sharp (1983) Splicing of messenger RNA precursors is inhibited by antisera to small nuclear ribonucleoprotein. Cell 35, 101–7.

Guthrie uses a phylogenetic comparison to show that certain sequences are conserved

Brow, D. A., and C. Guthrie (1988) Spliceosomal RNA U6 is remarkably conserved from yeast to mammals. Nature 334, 213–18.

Base-pairing shown between U2 snRNA and the branch site, and between U1 and U6 and the 5′ splice site, using site-directed mutagenesis and complementary changes

Parker, R., P. G. Siciliano, and C. Guthrie (1987) Recognition of the TACTAAC box during mRNA splicing in yest involves pairing to the U2-like snRNA. Cell 49, 229–39.

Siciliano, P. G., and C. Guthrie (1988) 5′ splice site selection in yeast: Genetic alterations in base-pairing with U1 reveal additional requirements. Genes & Devel. 2, 1258–67.

Madhani, H. D., and C. Guthrie (1992) A novel base-pairing interaction between U2 and U6 snRNAs suggests a mechanism for the catalytic activation of the splicesome. Cell 71, 803–17.

Lesser, C. F., and C. Guthrie (1993) Mutations in U6 snRNA that alter splice site specificity: Implications for the active site. Science 262, 1982–88.

Photochemical crosslinking of RNA:RNA interactions during splicing

Sawa, H., and J. Abelson (1992) Evidence for a base-pairing interaction between U6 small nuclear RNA and 5′ splice site during the splicing reaction in yeast. Proc. Natl. Acad. Sci. USA 89, 11269–73.

Wassarman, D. A., and J. A. Steitz (1992) Interactions of small nuclear RNAs with precursor messenger RNA during *in vitro* splicing. Science 257, 1918–25.

Newman, A. J., S. Teigelkamp, and J. D. Beggs (1995) snRNA interactions at 5′ and 3′ splice sites monitored by photoactivated crosslinking in yeast spliceosomes. RNA 1, 968–80.

Association of snRNPs in the larger spliceosome shown by centrifugation experiments

Brody, E., and J. Abelson (1985) The "spliceosome": Yeast pre-messenger RNA associates with a 40S complex in a splicing-dependent reaction. Science 228, 963–67.

Frendewey, D., and W. Keller (1985) Stepwise assembly of a pre-mRNA splicing complex requires U-snRNPs and specific intron sequences. Cell 42, 355–67.

Grabowski, P. J., S. R. Seiler, and P. A. Sharp (1985) A multicomponent complex is involved in the splicing of messenger RNA precursors. Cell 42, 345–53.

Mechanism of tRNA splicing in eukaryotes

Knapp, G., R. C. Ogden, C. Peebles, and J. Abelson (1979) Splicing of yeast tRNA precursors: Structure of the reaction intermediates. Cell 18, 37–45.

Peebles, C. L., R. C. Ogden, G. Knapp, and J. Abelson (1979) Splicing of yeast tRNA precursors: A two-stage reaction. Cell 18, 27–35.

Greer, C. L., C. L. Peebles, P. Gegenheimer, and J. Abelson (1983) Mechanism of action of a yeast RNA ligase in tRNA splicing. Cell 32, 537–46.

RNA PROCESSING WITHOUT PROTEINS: CATALYTIC RNA
Cech uses R-looping to identify transcription unit for rRNA in Tetrahymena

Cech, T. R., and D. C. Rio (1979) Localization of transcribed regions on extrachromosomal ribosomal RNA genes of Tetrahymena thermophila by R-loop mapping. Proc. Natl. Acad. Sci. USA 76, 5051–55.

The RNA was splicing itself

Zaug, A. J., and T. R. Cech (1980) *in vitro* splicing of the ribosomal RNA precursor in nuclei of Tetrahymena. Cell 19, 331–38.

Precursor rRNA made in vitro *using cloned DNA and purified RNA polymerase*

Kruger, K., P. J. Grabowski, A. J. Zaug, J. Sands, D. E. Gottschling, and T. R. Cech (1982) Self-splicing RNA: Autoexcision and autocyclization of the ribosomal RNA intervening sequence of Tetrahymena. Cell 31, 147–57.

Cech et al. suggest that the precursor RNA itself carries the catalytic activity

Cech, T. R., A. J. Zaug, and P. J. Grabowski (1981) *in vitro* splicing of the ribosomal RNA precursor of Tetrahymena: Involvement of a guanosine nucleotide in the excision of the intervening sequence. Cell 27, 487–96.

In 1986, Cech and coworkers engineer the RNA to act as a true catalyst

Zaug, A. J., and T. R. Cech (1986) The intervening sequence RNA of Tetrahymena is an enzyme. Science 231, 470–76.

Additional papers on self-splicing introns

Garriga, G., and A. M. Lambowitz (1984) RNA splicing in Neurospora mitochondria: Self-splicing of a mitochondrial intron *in vitro*. Cell 39, 631–41.

Gott, J. M., D. A. Shub, and M. Belfort (1986) Multiple self-splicing introns in bacteriophage T4: Evidence from autocatalytic GTP labeling of RNA *in vitro*. Cell 47, 81–87.

RNase P

Kole, R., and S. Altman (1979) Reconstitution of RNase P activity from inactive RNA and protein. Proc. Natl. Acad. Sci. USA 76, 3795–99.

Guerrier-Takada, C., K. Gardiner, T. Marsh, N. Pace, and S. Altman (1983) The RNA moiety of ribonuclease P is the catalytic subunit of the enzyme. Cell 35, 849–57.

James, B. D., G. J. Olsen, J. S. Liu, and N. R. Pace (1988) The secondary structure of ribonuclease P RNA, the catalytic element of a ribonucleoprotein enzyme. Cell 52, 19–26.

Reich, C., G. J. Olsen, B. Pace, and N. R. Pace (1988) Role of the protein moiety of ribonuclease P, a ribonucleoprotein enzyme. Science 239, 178–81.

Chapter 7. DNA on Its Own: Genetic Recombination

GENERAL RECOMBINATION AND THE HOLLIDAY MODEL

Early work on genetic recombination in phage (T-even phages)

Hershey, A. D. (1947) Spontaneous mutations in bacterial viruses. Cold Spring Harbor Symp. Quant. Biol. 11, 67–77.

Delbrück, M., and W. T. Bailey (1947) Induced mutations in bacterial viruses. Cold Spring Harbor Symp. Quant. Biol. 11, 33–37.

Hershey, A. D., and R. Rotman (1949) Genetic recombination between host-range and plaque-type mutants of bacteriophage in single bacterial cells. Genetics, Princeton, 34, 44–71.

Hershey, A. D., and M. Chase (1952a) Genetic recombination and heterozygosis in bacteriophage. Cold Spring Harbor Symp. Quant. Biol. 16, 471–79.

Levinthal, C. (1954) Recombination in phage T2: Its relationship to heterozygosis and growth. Genetics, Princeton, 39, 169–84.

Chase, M., and A. H. Doermann (1958) High negative interference over short segments of the genetic structure of bacteriophage T4. Genetics 43, 332–53.

Weigle, J., M. Meselson, and K. Paigen (1959) Density alterations associated with transducing ability in the bacteriophage lambda. J. Mol. Biol. 1, 379–86.

Evidence for breakage and rejoining mechanism for recombination in λ

Meselson, M., and J. J. Weigle (1961) Chromosome breakage accompanying genetic reconstruction in bacteriophage. Proc. Natl. Acad. Sci. USA 47, 857–68.

Holliday model for recombination

Whitehouse, H. L. K. (1963) A theory of crossing-over by means of hybrid deoxyribonucleic acid. Nature, London, 199, 1034–40.

Holliday, R. (1964) A mechanism for gene conversion in fungi. Genet. Res. 5, 282–304.

HOLLIDAY INTERMEDIATES IN CELLS

Meselson and Fox infer the existence of long heteroduplex regions of DNA

Meselson, M. (1967) The molecular basis of genetic recombination, in R. A. Brink, ed., *Heritage from Mendel*. Madison: University of Wisconsin Press.

White, R. L., and M. S. Fox (1974) On the molecular basis of high negative interference. Proc. Natl. Acad. Sci. USA 71, 1544–48.

White, R. L., and M. S. Fox (1975) Genetic heterozygosity in unreplicated bacteriophage λ recombinants. Genetics 81, 33–50.

Fox, M. S., C. S. Dudney, and E. J. Sodergren (1979) Heteroduplex regions in unduplicated bacteriophage lambda recombinants. Cold Spring Harbor Symp. Quant. Biol. 43, Pt. 2, 999–1007.

Holliday-type DNA molecules extracted from cells infected with φX174

Thompson, B. J., C. Escarmis, B. Parker, W. C. Slater, J. Doniger, I. Tessman, and R. C. Warner (1975) Figure-8 configuration of dimers of S13 and φX174 replicative form DNA. J. Mol. Biol. 91, 409–19.

GENES AND ENZYMES IN BACTERIAL RECOMBINATION

Genetic identification of recA

Clark, A. J., and A. D. Margulies (1965) Isolation and characterization of recombination-deficient mutants of *Escherichia coli* K-12. Proc. Natl. Acad. Sci. USA 53, 451–58.

Clark, Howard-Flanders, and colleagues defined recB *and* recC *genes*

Clark, A. J., and A. D. Margulies (1965) Isolation and characterization of recombination-deficient mutants of *Escherichia coli* K-12. Proc. Natl. Acad. Sci. USA 53, 451–58.

Howard-Flanders, P., and L. Theriot (1966) Mutants of *Escherichia coli* K-12 defective in DNA repair and in genetic recombination. Genetics 53, 1137–50.

Genetic identification of the RecF pathway by Clark

Horii, Z., and A. J. Clark (1973) Genetic analysis of the RecF pathway to genetic recombination in *Escherichia coli* K-12: Isolation and characterization of mutants. J. Mol. Biol. 80, 327–44.

Additional work on genes involved in bacterial recombination

Konrad, E. B. (1977) Method for the isolation of *Escherichia coli* mutants with enhanced recombination between chromosomal duplications. J. Bacteriol. 130, 167–72.

THE RecA PROTEIN AND BRANCH MIGRATION

Purification of RecA protein

McEntee, K., J. E. Hesse, and W. Epstein (1976) Identification and radiochemical purification of the *recA* protein of *Escherichia coli* K-12. Proc. Natl. Acad. Sci. USA 73, 3973–83.

Weinstock, G. M., K. McEntee, and I. R. Lehman (1979) ATP-dependent renaturation of DNA catalyzed by the RecA protein of *E. coli* Proc. Natl. Acad. Sci. USA 76, 126–30.

Craig and Roberts develop λ *repressor cleavage assay for RecA protein*

Roberts, J. W., C. W. Roberts, and N. L. Craig (1978) *Escherichia coli* recA gene product inactivates phage λ repressor. Proc. Natl. Acad. Sci. USA 75, 4714–18.

Independently, Ogawa demonstrates that RecA is a DNA-dependent ATPase
Ogawa, T., H. Wabiko, T. Tsurimoto, T. Horii, H. Masukata, and H. Ogawa (1978) Characteristics of purified recA protein and the regulation of its synthesis *in vivo*. Cold Spring Harbor Symp. Quant. Biol. 43, 909–16.

Biochemical studies on role of RecA in recombination
McEntee, K., G. M. Weinstock, and I. R. Lehman (1979) Initiation of general recombination catalyzed *in vitro* by the recA protein of *Escherichia coli*. Proc. Natl. Acad. Sci. USA 76, 2615–19.

Shibata, T., C. DasGupta, R. P. Cunningham, and C. M. Radding (1979) Purified *E. coli recA* protein catalyses homologous pairing of superhelical DNA and single-stranded fragments. Proc. Natl. Acad. Sci. USA 76, 1638–42.

DasGupta, C., T. Shibata, R. P. Cunningham, and C. M. Radding (1980) The topology of homologous pairing promoted by RecA protein. Cell 22, 437–46.

Cox, M. M., and I. R. Lehman (1981) RecA protein of *Escherichia coli* promotes branch migration, a kinetically distinct phase of DNA strand exchange. Proc. Natl. Acad. Sci. USA 78, 3433–37.

West, S. C., E. Cassuto, and P. Howard-Flanders (1981) Heteroduplex formation by RecA protein: Polarity of strand exchange. Proc. Natl. Acad. Sci. USA 78, 6149–53.

Radding, C. M. (1982) Homologous pairing and strand exchange in genetic recombination. Annu. Rev. Genetics 16, 405–37.

RuvA and RuvB proteins promote branch migration in vitro
Tsaneva, I. R., B. Müller, and S. C. West (1992) ATP-dependent branch migration of Holliday junctions promoted by the RuvA and RuvB proteins of *E. coli* Cell 69, 1171–80.

RuvC protein resolves Holliday junctions in vitro
Connolly, B., C. A. Parsons, F. E. Benson, H. J. Dunderdale, G. J. Sharples, R. G. Lloyd, and S. C. West (1991) Resolution of Holliday junctions *in vitro* requires the *Escherichia coli ruvC* gene product. Proc. Natl. Acad. Sci. USA 88, 6063–67.

Additional references for branch migration
Thompson, B. J., M. N. Camien, and R. C. Warner (1976) Kinetics of branch migration in double-stranded DNA. Proc. Natl. Acad. Sci. USA 73, 2299–2303.

Hsieh, P., C. S. Camerini-Otero, and D. Camerini-Otero (1990) Pairing of homologous DNA sequences by proteins: Evidence for three-stranded DNA. Genes and Development 4, 1951–63.

Panyutin, I. G., and P. Hsieh (1994) The kinetics of spontaneous DNA branch migration. Proc. Natl. Acad. Sci. USA 91, 2021–25.

THE RecBC PROTEIN: UNWINDING AND NICKING
Isolation of RecBC based on its nuclease activity on duplex DNA
Goldmark, P. J., and S. Linn (1970) An endonuclease activity from *E. coli* absent from certain rec⁻ strains. Proc. Natl. Acad. Sci. USA 67, 434–41.

Goldmark, P. J., and S. Linn (1972) Purification and properties of the RecBC DNase of *Escherichia coli* K-12. J. Biol. Chem. 247, 1849–60.

RecBCD is also a DNA helicase

Taylor, A., and G. R. Smith (1980) Unwinding and rewinding of DNA by the RecBC enzyme. Cell 22, 447–57.

Muskavitch, K. M. T., and S. Linn (1982) A unified mechanism for the nuclease and unwinding activites of the recBC enzyme of *Escherichia coli*. J. Biol. Chem. 257, 2641–48.

Chi sites

Lam, S. T., M. M. Stahl, K. D. McMilin, and F. W. Stahl (1974) Rec-mediated recombinational hot spot activity in bacteriophage lambda. II. A mutation which causes hot spot activity. Genetics 77, 425–33.

McMilin, K. D., M. M. Stahl, and F. W. Stahl (1974) Rec-mediated recombinational hot spot activity in bacteriophage lambda. I. Hot spot activity associated with spi⁻ deletions and bio substitutions. Genetics 77, 409–23.

Stahl, F. W., J. M. Craseman, and M. M. Stahl (1975) Rec-mediated recombinational hot spot activity in bacteriophage lambda. III. Chi mutations are site-mutations stimulating rec-mediated recombination. J. Mol. Biol. 94, 203–12.

Stahl, F. W., and M. M. Stahl (1977) Recombination pathway specificity of Chi. Genetics 86, 715–25.

Stahl, F. W., L. C. Thomason, I. Siddiqui, and M. M. Stahl (1990) Further tests of a recombination model in which χ removes the RecD subunit from the RecBCD enzyme in *Escherichia coli*. Genetics 126, 519–33.

Anderson, D. G., and S. C. Kowalczykowski (1997) The recombination hot spot χ is a regulatory element that switches the polarity of DNA degradation by the RecBCD enzyme. Genes Dev. 11, 571–81.

GENERAL RECOMBINATIONAL PATHWAYS: THE FUTURE
Yeast Rad51 protein

Sung, P. (1994) Catalysis of ATP-dependent homologous DNA pairing and strand exchange by yeast Rad51 protein. Science 265, 1241–43.

RecA homolog found in Archaea

Ridder, R., R. Marquadt, and K. Esser (1991) Molecular cloning and characterization of the *recA* gene of Methylomonas clara and construction of a *recA* deficient mutant. Applied Microbiol. & Biotech. 35, 23–31.

Seitz, E. M., J. P. Brockman, S. J. Sandler, A. J. Clark, and S. C. Kowalczykowski (1998) RadA protein is an archaeal RecA protein homolog that catalyzes DNA strand exchange. Genes Devel. 12, 1248–53.

BRCA1 and BRCA2 proteins interact with human Rad51 protein

Mizuta, R., J. M. LaSalle, H. L. Cheng, A. Shinohara, H. Ogawa, N. Copeland, N. A. Jenkinds, M. Lalande, and F. W. Alt (1997) RAB22 and RAB163/mouse BRCA2: Proteins that specifically interact with the Rad51 protein. Proc. Natl. Acad. Sci. USA 94, 6927–32.

Scully, R., J. Chen, A. Plug, Y. Xiao, D. Weaver, J. Feunteun, T. Ashley, and D. M. Livingston (1997) Association of BRCA1 with Rad51 in mitotic and meiotic cells. Cell 88, 265–75.

Katagiri, T., H. Saito, A. Shinohara, H. Ogawa, N. Kamada, Y. Nakamura, and Y. Miki (1998) Multiple possible sites of BRCA2 interacting with DNA repair protein Rad51. Genes, Chromosomes & Cancer 21, 217–22.

SITE-SPECIFIC RECOMBINATION AND THE CAMPBELL MODEL

Campbell model of prophage integration

Campbell, A. (1962) The episomes. Adv. Genet. 11, 101–45.

Intracellular λ DNA is a circle

Young, E. T. II, and R. L. Sinsheimer (1964) Novel intra-cellular forms of lambda DNA. J. Mol. Biol. 10, 562–64.

Bode, V. C., and A. D. Kaiser (1965) Changes in the structure and activity of λ DNA in a superinfected immune bacterium. J. Mol. Biol. 14, 399–417.

Dove, W. F., and J. J. Weigle (1965) Intracellular state of the chromosome of bacteriophage lambda. I. The eclipse of infectivity of the bacteriophage DNA. J. Mol. Biol. 12, 620–29.

Gellert, M. (1967) Formation of covalent circles of lambda DNA by *E. coli* extracts. Proc. Natl. Acad. Sci. USA 57, 148–55.

Bacterial genes contiguous to prophage genes in the chromosome

Franklin, N. C., W. F. Dove, and C. Yanofsky (1965) The linear insertion of a prophage into the chromosome of *E. coli* shown by deletion mapping. Biochem. Biophys. Res. Commun. 18, 910–23.

Signer initially proposes site-specific recombination model

Signer, E., and J. Beckwith (1966) Transposition of the lac region of *Escherichia coli*. III. The mechanism of attachment of bacteriophage φ80 to the bacterial chromosome. J. Mol. Biol. 22, 33–51.

Isolation of λ int⁻ mutants

Gingery, R. and H. Echols (1967) Mutants of bacteriophage λ unable to integrate into the host chromosome. Proc. Natl. Acad. Sci. USA 58, 1507–14.

Zissler, J. (1967) Integration-negative (int) mutants of phage λ. Virology 31, 189.

Gottesman, M. E., and M. B. Yarmolinsky (1968) Integration-negative mutants of bacteriophage lambda. J. Mol. Biol. 31, 487–505.

Isolation of b2 mutants, deletion mutants that fail to integrate into chromosome

Kellenberger, G., M. L. Zichichi, and J. Weigle (1960) Mutations affecting the density of bacteriophage λ. Nature London 187, 161–62.

Campbell, A. M. (1965) The steric effect in lysogenization by bacteriophage lambda. II. Chromosomal attachment of the b2 mutant. Virology 27, 340–45.

Isolation of λ red⁻ mutants

Echols, H., and R. Gingery (1968) Mutants of bacteriophage λ defective in vegetative genetic recombination. J. Mol. Biol. 34, 239–49.

Signer, E. R., and J. Weil (1968) Recombination in bacteriophage lambda. I. Mutants deficient in general recombination. J. Mol. Biol. 34, 261–71.

Localization of region of λ needed for site-specific recombination

Echols, H., R. Gingery, and L. Moore (1968) Integrative recombination function of bacteriophage lambda: Evidence for a site-specific recombination enzyme. J. Mol. Biol. 34, 251–60.

Weil, J., and E. R. Signer (1968) Recombination in bacteriophage λ. II. Site-specific recombination promoted by the integration system. J. Mol. Biol. 34, 273–79.

PROPHAGE EXCISION AND EXCISIVE RECOMBINATION
Isolation of excision-defective mutants
Guarneros, G., and H. Echols (1970) New mutants of bacteriophage λ with a specific defect in excision from the host chromosome. J. Mol. Biol. 47, 565–74.

Gottesman argues for region of sequence identity in the phage and bacterial att sites
Shulman, M., and M. Gottesman (1973) Attachment site mutants of bacteriophage lambda. J. Mol. Biol. 81, 461–82.

Guerrini argued for differences in regions flanking the attachment sites
Guerrini, F. (1969) On the asymmetry of λ integration sites. J. Mol. Biol. 46, 523–42.

Echols tested prediction that excision reaction requires Xis
Echols, H. (1970) Integrative and excisive recombination by bacteriophage λ: Evidence for an excision-specific recombination protein. J. Mol. Biol. 47, 575–83.

THE BIOCHEMISTRY OF SITE-SPECIFIC RECOMBINATION
in vitro *recombination reaction for Int*
Nash, H. A. (1975) Integrative recombination of bacteriphage lambda DNA *in vitro*. Proc. Natl. Acad. Sci. USA 72, 1072–76.

Mizuuchi and Nash switched to identifying the recombinant molecules directly
Mizuuchi, K., and H. A. Nash (1976) Restriction assay for integrative recombination of bacteriophage λ DNA *in vitro*: Requirement for closed circular DNA substrate. Proc. Natl. Acad. Sci. USA 73, 3524–28.

The Mizuuchis develop intermolecular assay for λ integrase
Mizuuchi, K., and M. Mizuuchi (1979) Integrative recombination of bacteriophage lambda: Study of the intermolecular reaction. Cold Spring Harbor Symp. Quant. Biol. 43 pt 2, 1111–14.

Discovery of DNA gyrase
Gellert, M., K. Mizuuchi, M. H. O'Dea, and H. A. Nash (1976) DNA gyrase: An enzyme that introduces superhelical turns into DNA. Proc. Natl. Acad. Sci. USA 73, 3872–76.

The substrate DNA in integrase reaction must be covalently closed
Mizuuchi, K., M. Gellert, and H. A. Nash (1978) Involvement of supertwisted DNA in integrative recombination of bacteriophage lambda. J. Mol. Biol. 121, 375–92.

Bacterial chromosome and plasmid DNA are negatively supercoiled
Worcel, A., and E. Burgi (1972) On the structure of the folded chromosome of *Escherichia coli*. J. Mol. Biol. 71, 127–47.
Gellert, M., K. Mizuuchi, M. H. O'Dea, H. Ohmori, and J. Tomizawa (1979) DNA gyrase and DNA supercoiling. Cold Spring Harbor Symp. Quant. Biol. 43, 35–40.
Cozzarelli, N. (1980) DNA gyrase and the supercoiling of DNA. Science 207, 953–60.
Gellert, M. (1981) DNA topoisomerases. Annu. Rev. Biochem. 50, 879–910.

Int carries out a DNA topoisomerase reaction
Kikuchi, Y., and H. A. Nash (1979) Nicking-closing activity associated with bacteriophage λ *int* gene product. Proc. Natl. Acad. Sci. USA 76, 3760–64.

Miller and Friedman isolate host mutants defective in λ integration and excision
Miller, H. I., A. Kikuchi, H. A. Nash, R. A. Weisberg, and D. I. Friedman (1979) Site-specific recombination of bacteriophage λ: The role of host gene products. Cold Spring Harbor Symp. Quant. Biol. 43, 1121–26.

DNA–PROTEIN INTERACTIONS IN SITE-SPECIFIC RECOMBINATION
Landy and Ross determine the DNA sequences of att *sites*
Landy, A., and W. Ross (1977) Viral integration and excision: Structure of the lambda *att* sites. Science 197, 1147–60.

Phage att *site is about 240 bp whereas bacterial* att *site is only about 20 bp, and the Mizuuchis determine the site of DNA recombination within the att sites*
Mizuuchi, K., R. Weisberg, L. Enquist, M. Mizuuchi, M. Buraczynska, C. Foeller, P.-L. Hsu, W. Ross, and A. Landy (1981) Structure and function of the phage λ *att* site: Size, Int-binding sites, and location of the crossover point. Cold Spring Harbor Symp. Quant. Biol. 45, 429–37.

Ross and Landy map Int binding sites
Ross, W., A. Landy, Y. Kikuchi, and H. Nash (1979) Interaction of Int protein with specific sites on lambda att DNA. Cell 18, 297–307.
Ross, W., and A. Landy (1982) Bacteriophage λ Int protein recognizes two classes of sequence in the phage att site: Characterization of arm-type sites. Proc. Natl. Acad. Sci. USA 79, 7724–28.
Ross, W., and A. Landy (1983) Patterns of lambda Int recognition in the regions of strand exchange. Cell 33, 261–72.

Landy, "the phage attachment site . . ."
Personal communication.

IHF also binds to phage att *site*
Craig, N. L., and H. A. Nash (1984) *E. coli* integration host factor binds to specific sites in DNA. Cell 39, 707–16.

Nucleoprotein complex at λ att *site visualized*
Better, M., C. Lu, R. C. Williams, and H. Echols (1982) Site-specific DNA condensation and pairing mediated by the int protein of bacteriophage λ. Proc. Natl. Acad. Sci. 79, 5837–41.

Role of Xis in excision recombination visualized
Better, M., S. Wickner, J. Auerbach, and H. Echols (1983) Role of the Xis protein of bacteriophage lambda in a specific reactive complex at the attR prophage attachment site. Cell 32, 161–68.

Nash discovers that there is always a "left-over" supercoil after integration
Pollock, T. J., and H. A. Nash (1983) Knotting of DNA caused by a genetic rearrangement. Evidence for a nucleosome-like structure in site-specific recombination with bacteriophage lambda. J. Mol. Biol. 170, 1–18.

Landy and Nash add evidence for long-distance "cooperativity" in DNA binding

Richet, E., P. Abcarian, and H. A. Nash (1986) The interaction of recombination proteins with supercoiled DNA: Defining the role of supercoiling in lambda integrative recombination. Cell 46, 1011–21.

Thompson, J. F., L. M. de Vargas, S. E. Skinner, and A. Landy (1987) Protein-protein interactions in a higher order structure direct lambda site-specific recombination. J. Mol. Biol. 195, 481–93.

Int recombination reaction occurs sequentially

Kitts, P. A., and H. A. Nash (1987) Homology-dependent interactions in phage λ site-specific recombination. Nature 329, 346–48.

Nunes-Duby, S. E., L. Matsumoto, and A. Landy (1987) Site-specific recombination intermediates trapped with suicide substrates. Cell 50, 779–88.

First molecular characterization of the higher-order nucleoprotein complexes at att *site*

Kim, S., L. Moitoso de Vargas, S. E. Nunes-Düby, and A. Landy (1990) Mapping of a higher order protein-DNA complex: Two kinds of long-range interactions in λ attL. Cell 63, 773–81.

The properties of λ integrase provide a basis for intasome structures

Moitoso de Vargas, L., C. A. Pargellis, N. M. Hasan, E. W. Bushman, and A. Landy (1998) Autonomous DNA binding domains of λ integrase recognize two different sequence families. Cell 54, 923–29.

OTHER SITE-SPECIFIC RECOMBINASES
Hin system of Salmonella

Zieg, J., M. Silverman, M. Hilmen, and M. Simon (1977) Recombinational switch for gene expression. Science 196, 170–72.

Van de Putte, P., and N. Goosen (1992) DNA inversions in phages and bacteria. Trends in Genetics 8, 457–62.

Cre-Lox system of phage P1

Austin, S., M. Ziese, and N. Sternberg (1981) A novel role for site-specific recombination in maintenance of bacterial replicons. Cell 25, 729–36.

Yeast Flp site-specific recombinase

Broach, J. R., V. R. Guarascio, and M. Jayaram (1982) Recombination with the yeast plasmid 2 micron circle is site-specific. Cell 29, 227–34.

TRANSPOSITION AND REPLICATIVE SITE-SPECIFIC RECOMBINATION
McClintock discovers maize transposable genetic elements (first published 1950–52)

McClintock, B. (1952) Chromosome organization and gene expression. Cold Spring Harbor Symp. Quant. Biol. 16, 13–47.

McClintock, B. (1961) Some parallels between gene control systems in maize and in bacteria. Am. Naturalist 95, 265–77.

McClintock, B. (1965) The control of gene action in maize. Brookhaven Symp. Biol. 18, 162–84.

Sherratt and Falkow provide first evidence for cointegrate transposition intermediates

Gill, R., F. Heffron, G. Dougan, and S. Falkow (1978) Analysis of sequences transposed by complementation of two classes of transposition-deficient mutants of Tn*3*. J. Bacteriol. 136, 742–56.

Grindley, N. D., and D. J. Sherratt (1979) Sequence analysis at IS*1* insertion sites: Models for transposition. Cold Spring Harbor Symp. Quant. Biol. 43, 1257–61.

THE IS INSERTION ELEMENTS
Some gal3 *revertants express the* gal *operon constitutively*

Hill, C. W., and H. Echols (1969) Properties of a mutant blocked in inducibility of messenger RNA for the galactose operon of *Escherichia coli*. J. Mol. Biol. 19, 38–51.

Test of insertion idea by CsCl gradient

Jordan, E., H. Saedler, and P. Starlinger (1968) 0° and strong-polar mutations in the *gal* operon are insertions. Mol. Gen. Genet. 102, 353–63.

Shapiro, J. A. (1969) Mutations caused by the insertion of genetic material into the galactose operon of *Escherichia coli*. J. Mol. Biol. 40, 93–105.

Malamy finds that a number of polar mutations in the lac operon are insertions

Malamy, M. H. (1970) Some properties of insertion mutations in the lac operon, in J. R. Beckwith and D. Zipser, eds., *The Lactose Operon*. Cold Spring Harbor, N.Y.: Cold Spring Harbor Laboratory Press.

Development of heteroduplex analysis

Davis, R. W., and N. Davidson (1968) Electron-microscopic visualization of deletion mutations. Proc. Natl. Acad. Sci. USA 60, 243–50.

Westmoreland, B. D., W. Szybalski, and H. Ris (1969) Mapping of deletions and substitutions in heteroduplex DNA molecules of bacteriophage lambda by electron microscopy. Science 163, 1343–48.

Starlinger and Szybalski use heteroduplex mapping to look at insertion mutants

Hirsch, H. J., H. Saedler, and P. Starlinger (1972) Insertion mutations in the control region of the galactose operon of *E. coli* II. Physical characterization of the mutations. Mol. Gen. Genet. 115, 266–76.

Malamy, M. H., M. Fiandt, and W. Szybalski (1972) Electron microscopy of polar insertions in the lac operon of *Escherichia coli*. Mol. Gen. Genet. 119, 207–22.

MU: A PHAGE THAT DOES NOT INTEGRATE LIKE LAMBDA
Initial characterization of Mu by Taylor

Taylor, A. L. (1963) Bacteriophage-induced mutation in *Escherichia coli*. Proc. Natl. Acad. Sci. USA 50, 1043–51.

Mu lysogens have a high frequency of spontaneous mutations near the site of integration

Taylor, A. L. (1963) Bacteriophage-induced mutation in *Escherichia coli*. Proc. Natl. Acad. Sci. USA 50, 1043–51.

There are multiple, independent Mu insertions in the lacZ *gene*

Bukhari, A. I., and D. Zipser (1972) Random insertion of Mu-1 DNA within a single gene. Nature (London) 264, 580–83.

Daniell, E., R. Roberts, and J. Abelson (1972) Mutations in the lactose operon caused by bacteriophge Mu. J. Mol. Biol. 69, 1–8.

Order of genetic markers is the same in the Mu prophage as in linear phage DNA

Abelson, J., W. Boram, A. I. Bukhari, M. Faelen, M. Howe, M. Meylay, A. L. Taylor, A. Toussaint, P. van de Putte, G. C. Westmaas, and C. A. Wijffelman (1973) Summary of the genetic mapping of prophage Mu. Virology 54, 90–92.

Wijffelman, C. A., G. C. Westmaas, and P. van de Putte (1973) Similarity of vegetative map and prophage map of bacteriophage Mu-1. Virology 54, 125–34.

Mu DNA has variable ends

Daniell, E., J. Abelson, J. S. Kim, and N. Davidson (1973) Heteroduplex structures of bacteriophage Mu DNA. Virology 51, 237–39.

Waggoner, B. T., N. S. González, and A. L. Taylor (1974) Isolation of heterogeneous circular DNA from induced lysogens of bacteriophage Mu-1. Proc. Natl. Acad. Sci. 71, 1255–59.

Schröder, W., E. G. Bade, and H. Deluis (1974) Participation of *E. coli* DNA in the replication of temperature bacteriophage Mu-1. Virology 60, 534–42.

The junctions between a Mu prophage and host DNA are maintained after induction

Ljungquist, E., and A. I. Bukhari (1977) State of prophage Mu DNA upon induction. Proc. Natl. Acad. Sci. USA 74, 3143–47.

Replicating Mu molecules had branch points and Y-structures

Schröder, W., E. G. Bade, and H. Deluis (1974) Participation of *E. coli* DNA in the replication of temperature bacteriophage Mu-1. Virology 60, 534–42.

Waggoner, B. T., N. S. González, and A. L. Taylor (1974) Isolation of heterogeneous circular DNA from induced lysogens of bacteriophage Mu-1. Proc. Natl. Acad. Sci. 71, 1255–59.

Waggoner, B. T., M. L. Pato, and A. L. Taylor (1977) Characterization of covalently closed circular DNA molecules isolated after bacteriophage Mu induction, in A. I. Bukhari, J. A. Shapiro, and S. Adhya, eds., *DNA Insertion Elements, Plasmids and Episomes*. Cold Spring Harbor, N.Y.: Cold Spring Harbor Laboratory Press.

Direct repeats flanking site of insertion of IS elements and phage Mu

Calos, M. P., L. Johnsrud, and J. H. Miller (1978) DNA sequence at the integration sites of the insertion element IS1. Cell 13, 411–18.

Grindley, N. D. F. (1978) IS1 insertion generates duplication of a nine base-pair sequence at its target site. Cell 13, 419–26.

Allet, B. (1979) Mu insertion duplicats a 5 base-pair sequence at the host insertion site. Cell 16, 123–29.

Kahmann, R., and D. Kamp (1979) Nucleotide sequences of the attachment sites of bacteriophage Mu DNA. Nature 280, 247–50.

Mechanism to generate short duplications at the ends of IS insertions

Grindley, N. D., and D. J. Sherratt (1979) Sequence analysis at IS insertion sites: Models for transposition. Cold Spring Harbor Symp. Quant. Biol. 43, 1257–61.

Shapiro model for Mu transposition and replication

Shapiro, J. A. (1979) Molecular model for the transposition and replication of bacteriophage Mu and other transposable elements. Proc. Natl. Acad. Sci. USA 76, 1933–37.

IDENTIFYING THE MU COMPONENTS NEEDED FOR TRANSPOSITION

Mu A and B genes essential for replication and phage growth

Faelen, M., O. Huisman, and A. Toussaint (1978) Involvement of phage Mu-1 early functions in Mu mediated chromosomal rearrangements. Nature 271, 580–82.

O'Day, K. J., D. W. Schultz, and M. M. Howe (1978) A search for integration deficient mutants of bacteriophage Mu, in D. Schlessinger, ed., *Microbiology 1978*. Washington, D.C.: ASM Publications.

Mini-Mus

Faelen, M., A. Résibois, and A. Toussaint (1979) Mini-Mu: An insertion element derived from temperate phage Mu-1. Cold Spring Harbor Symp. Quant. Biol. 43, 1169–77.

Chaconas, G., F. J. de Bruijn, M. J. Casadaban, J. R. Lupski, T. J. Kwoh, R. M. Harshey, M. S. Du Bow, and A. I. Bukhari (1981) *in vitro* and *in vivo* manipulations of bacteriophage Mu DNA: Cloning of Mu ends and construction of mini-Mu's carrying selectable markers. Gene 13, 37–46.

A UNIFYING MODEL FOR TRANSPOSITION

Mizuuchi develops in vitro *transposition assay for Mu*

Mizuuchi, K. (1983) *in vitro* transposition of bacteriophage Mu: A biochemical approach to a novel replication reaction. Cell 35, 785–94.

Mizuuchi shows that transposition intermediate has structure predicted by Shapiro model

Craigie, R., and K. Mizuuchi (1985) Mechanism of transposition of bacteriophage Mu: Structure of a transposition intermediate. Cell 41, 867–76.

Miller, J. L., and G. Chaconas (1986) Electron microscopic analysis of *in vitro* transposition intermediates of bacteriophage Mu DNA. Gene 48, 101–8.

Chapter 8. Regulating the Regulators: Developmental and Salvational Decisions

THE LYSIS–LYSOGENY DECISION BY PHAGE λ:

CII AS THE MASTER REGULATOR

Kaiser established the critical features of the lambda lysogenic pathway in 1957

Kaiser, A. D. (1957) Mutations in a temperate bacteriophage affecting its ability to lysogenize *Escherichia coli*. Virology 3, 42–61.

Kaiser, A. D., and F. Jacob (1957) Recombination between related temperate bacteriophages and the genetic control of immunity and its lysogenizing properties. Virology 4, 509–21.

Eisen argues that cII and cIII turned on synthesis of the cI protein

Eisen, H., P. Brachet, L. Pereira da Silva, and F. Jacob (1970) Regulation of repressor expression in lambda. Proc. Natl. Acad. Sci. USA 66, 855–62.

Echols argues that cII and cIII delay expression of lytic genes

McMacken, R., N. Mantei, B. Butler, A. Joyner, and H. Echols (1970) Effect of mutations in the c2 and c3 genes of bacteriophage lambda on macromolecular synthesis in infected cells. J. Mol. Biol. 49, 639–55.

Direct measurements of cI protein

Echols, H., and L. Green (1971) Establishment and maintenance of repression by bacteriophage lambda: The role of the cI, cII, and cIII proteins. Proc. Natl. Acad. Sci. USA 68, 2190–94.

Reichardt, L., and A. D. Kaiser (1971) Control of lambda repressor synthesis. Proc. Natl. Acad. Sci. USA 68, 2185–89.

cy regulatory mutant defines p_{RE}, *the site of action of cII and cIII proteins*

Brachet, P., and R. Thomas (1969) Mapping and functional analysis of y and c II mutants. Mutat. Res. 7, 257–60.

Maintenance promoter for cI synthesis, p_{RM}, *defined*

Yen, K. M., and G. N. Gussin (1973) Genetic characterization of a *prm⁻* mutant of bacteriophage λ. Virology 56, 300–312.

int can be transcribed from separate promoter, p_I

Shimada, K., and A. Campbell (1974) Int-constitutive mutants of bacteriophage lambda. Proc. Natl. Acad. Sci. USA 71, 237–41.

Efficient Int production depended on cII

Katzir, N., A. Oppenheim, M. Belfort, and A. B. Oppenheim (1976) Activation of the lambda *int* gene by the *cII* and *cIII* gene products. Virology 74, 324–31.

cII and cIII stimulated Int expression but not production of Xis

Chung, S., and H. Echols (1977) Positive regulation of integrative recombination by the *cII* and *cIII* genes of bacteriophage lambda. Virology 79, 312–19.

cII and cIII negatively regulate expression of the lytic genes by activating p_{aQ}

Ho, Y. S., and M. Rosenberg (1985) Characterization of a third, *cII*-dependent, coordinately activated promoter on phage lambda involved in lyosgenic development. J. Biol. Chem. 260, 11838–44.

Hoopes, B. C., and W. R. McClure (1985) A *cII*-dependent promoter is located within the Q gene of bacteriophage lambda. Proc. Natl. Acad. Sci. USA 82, 3134–38.

cIII not required for establishment of lysogeny in E. coli hfl *mutants*

Belfort, M., and D. L. Wulff (1971) A mutant of *Escherichia coli* that is lyosgenized with high frequency, in A. D. Hershey, ed., *The Bacteriophage Lambda*. Cold Spring Harbor, N.Y.: Cold Spring Harbor Laboratory Press.

Shimatake and Rosenberg purify cII and show it functions in vitro

Shimatake, H., and M. Rosenberg (1981) Purified lambda regulatory protein cII positively activates promoters for lysogenic development. Nature 292, 128–32.

DNA sequence analysis identifies a conserved sequence in both cII-activated promoters
Wulff, D. L., and M. Rosenberg (1983) Establishment of repressor synthesis, in R. W.
Hendrix, J. W. Roberts, F. W. Stahl, and R. A. Weisberg, eds., *Lambda II*. Cold Spring Harbor,
N.Y.: Cold Spring Harbor Laboratory Press.

Rosenberg shows that cII protein binds to sequence present in cII-activated promoters
Ho, Y. S., D. L. Wulff, and M. Rosenberg (1983) Bacteriophage lambda protein cII binds promoters on the opposite face of the DNA helix from RNA polymerase. Nature 304, 703–8.

REGULATING THE REGULATOR: CONTROL OF CII ACTIVITY
Epp observes that cII is an unstable protein
Epp, C. D. (1978) Early protein synthesis and its control in bacteriophage lambda. Ph.D.
diss., University of Toronto, Ontario, Canada.

Both hfl *and* cIII *regulate cII stability*
Hoyt, M. A., D. M. Knight, A. Das, H. I. Miller, and H. Echols (1982) Control of phage
lambda development by stability and synthesis of cII protein: Role of the viral cIII and host
hflA, *himA* and *himD* genes. Cell 31, 565–73.

Mutations in genes encoding IHF protein affect the frequency of lysogenization
Miller, H. I., A. Kikuchi, H. A. Nash, R. A. Weisberg, and D. I. Friedman (1979) Site-specific
recombination of bacteriophage λ: The role of host gene products. Cold Spring Harbor
Symp. Quant. Biol. 43, 1121–26.

MOLECULAR DECISION MAKING: PRIMARY AND SECONDARY SWITCHES
Johnson analyzes the binding of cI and Cro to their operator sites
Johnson, A., B. J. Meyer, and M. Ptashne (1978) Mechanism of action of the cro protein of
bacteriophage λ. Proc. Natl. Acad. Sci. USA 75, 1783–87.
Johnson, A. D., B. J. Meyer, and M. Ptashne (1979) Interactions between DNA-bound re-
pressors govern regulation by the λ phage repressor. Proc. Natl. Acad. Sci. USA 76, 5061–65.
Johnson, A. D., C. O. Pabo, and R. T. Sauer (1980) Bacteriophage lambda repressor and cro
protein: Interactions with operator DNA. Methods enzymol. 65, 839–56.
Reviewed in: Ptashne, M., A. Jeffrey, A. D. Johnson, R. Maurer, B. J. Meyer, C. O. Pabo, T. M.
Roberts, and R. T. Sauer (1980) How the lambda repressor and cro work. Cell 19, 1–11.

Meyer studies the effects of cI and Cro on transcription from p_L *and* p_R
Meyer, B., R. Maurer, and M. Ptashne (1980) Gene regulation at the right operator (o_R) of
bacteriophage lambda. II. o_{R1}, o_{R2}, and o_{R3}: Their roles in mediating the effects of repressor
and cro. J. Mol. Biol. 139, 163–94.

Takeda and Echols also note the transcription switch between cI and Cro
Takeda, Y., A. Folkmanis, and H. Echols (1977) Cro regulatory protein specified by bacteriophage lambda. Structure, DNA-binding, and repression of RNA synthesis. J. Biol. Chem.
252, 6177–83.

"Retroregulation" of Int synthesis
Schindler, D., and H. Echols (1981) Retroregulation of the int gene of bacteriophage
lambda: Control of translation completion. Proc. Natl. Acad. Sci. USA 78, 4475–79.

THE HEAT SHOCK RESPONSE AND MOLECULAR CHAPERONES
Discovery of the heat shock response
Ritossa, F. M. (1962) A new puffing pattern induced by a temperature shock and DNP in *Drosophila*. Experientia 18, 571–73.

Induced chromosomal puffs in the salivary gland chromosomes of Drosophila
Ritossa, F. M. (1962) A new puffing pattern induced by a temperature shock and DNP in *Drosophila*. Experientia 18, 571–73.
Ritossa, F. (1996) Discovery of the heat shock response. Cell Stress & Chaperones 1, 97–98.

Tissières shows that synthesis of heat shock proteins is induced by temperature upshift
Tissières, A., H. K. Mitchell, and U. M. Tracy (1974) Protein synthesis in salivary glands of *Drosophila* melanogaster: Relation to chromosome puffs. J. Mol. Biol. 84, 389–98.

Heat shock RNA hybridizes to the chromosomal puffs
McKenzie, S. L., S. Henikoff, and M. Meselson (1975) Localization of RNA from heat-induced polysomes at puff sites in *Drosophila* melanogaster. Proc. Natl. Acad. Sci. USA 72, 1117–21.
Spradling, A., S. Penman, and M. L. Pardue (1975) Analysis of *Drosophila* mRNA by in situ hybridization: Sequences transcribed in normal and heat shocked cultured cells. Cell 4, 395–404.

First Drosophila heat shock gene cloned
Craig, E. A., B. J. McCarthy, and S. C. Wadsworth (1979) Sequence organization of two recombinant plasmids containing genes for the major heat shock-induced protein in D. melanogaster. Cell 16, 575–88.

REGULATION AND FUNCTION OF HEAT SHOCK PROTEINS IN *E. COLI*
Two-dimensional electrophoresis
O'Farrell, P. H. (1975) High-resolution two-dimensional electrophoresis of proteins. J. Biol. Chem. 250, 4007–21.

Induction of heat shock proteins in E. coli *after shift to 42°C*
Lemaux, P. G., S. L. Herendeen, P. L. Bloch, and F. C. Neidhardt (1978) Transient rates of synthesis of individual polypeptides in *E. coli* following temperature shifts. Cell 13, 427–34.
Yamamori, T., K. Ito, Y. Nakamura, and T. Yura (1978) Transient regulation of protein synthesis in *Escherichia coli* upon shift-up of growth temperature. J. Bacteriol. 134, 1133–40.

Yura shows that induction of heat shock protein synthesis depends on induced transcription
Yamamori, T., and T. Yura (1980) Temperature-induced synthesis of specific proteins in *Escherichia coli*: Evidence for transcriptional control. J. Bacteriol. 142, 843–51.

Identification of positive regulator of the E. coli *heat shock response*
Neidhardt, F. C., and R. A. VanBogelen (1981) Positive regulatory gene for temperature-controlled proteins in *Escherichia coli*. Biochem. Biophys. Res. Commun. 100, 894–900.
Yamamori, T., and T. Yura (1982) Genetic control of heat-shock protein synthesis and its bearing on growth and thermal resistance in *Escherichia coli* K-12. Proc. Natl. Acad. Sci. USA 79, 860–64.

A Ts lethal mutation prevents production of several proteins at elevated temperature
Cooper, S., and T. Ruettinger (1975) A temperature sensitive nonsense mutation affecting the synthesis of a major protein of *Escherichia coli* K-12. Mol. Gen. Genet. 139, 167–76.
Cooper, S., and T. Ruettinger (1975) Temperature dependent alteration in bacterial protein composition. Biochem. Biophys. Res. Commun. 62, 584–86.

Georgopoulos identifies several E. coli *mutants defective in lytic growth of lambda*
Georgopoulos, C. P., and I. Herskowitz (1971) *Escherichia coli* mutants blocked in lambda DNA synthesis, in A. D. Hershey, ed., *The Bacteriophage Lambda*. Cold Spring Harbor, N.Y.: Cold Spring Harbor Laboratory Press.
Georgopoulos, C. P., R. W. Hendrix, S. R. Casjens, and A. D. Kaiser (1973) Host participation in bacteriophage lambda head assembly. J. Mol. Biol. 76, 45–60.

GroEL is identical to the largest heat shock protein
Neidhardt, F. C., T. A. Phillips, R. A. VanBogelen, M. W. Smith, Y. Georgalis, and A. R. Subramanian (1981) Identity of the B56.5 protein, the A-protein, and the *groE* gene product of *Escherichia coli*. J. Bacteriol. 145, 513–20.

DnaK is the second most abundant heat shock protein
Georgopoulos, C., K. Tilly, D. Drahos, and R. Hendrix (1982) The B66.0 protein of *Escherichia coli* is the product of the *dnaK*$^+$ gene. J. Bacteriol. 149, 1175–77.
Reviewed in: Neidhardt, F. C., and R. A. VanBogelen (1987) Heat shock response, in F. C. Neidhardt, J. L. Ingraham, K. B. Low, B. Magasanik, M. Schaechter, and H. E. Umbarger, eds., *Escherichia coli and "Salmonella typhimurium" Cellular and Molecular Biology*. Washington, D.C.: American Society for Microbiology.

Further work shows that DnaJ, GrpE, and GroES are also heat shock proteins
Reviewed in: Neidhardt, F. C., and R. A. VanBogelen (1987) Heat shock response, in F. C. Neidhardt, J. L. Ingraham, K. B. Low, B. Magasanik, M. Schaechter, and H. E. Umbarger, eds., *Escherichia coli and "Salmonella typhimurium" Cellular and Molecular Biology*. Washington, D.C.: American Society for Microbiology.

DnaK was very similar to the Hsp70 class of eukaryotic heat shock proteins
Bardwell, J. C. A., and E. A. Craig (1984) Major heat shock gene of Drosophila and the *Escherichia coli* heat-inducible dnaK gene are homologous. Proc. Natl. Acad. Sci. USA 81, 848–52.

The hsp90 class also has a bacterial homologue
Bardwell, J. C., and E. A. Craig (1987) Eukaryotic Mr 83,000 heat shock protein has a homologue in *Escherichia coli*. Proc. Natl. Acad. Sci. USA 84, 5177–81.

REGULATION OF THE HEAT SHOCK RESPONSE:
NEW PROMOTER RECOGNITION BY RNA POLYMERASE
The synthesis of heat shock proteins is not affected by thermal inactivation of σ^{70}
Gross, C. A., A. D. Grossman, H. Liebke, W. Walter, and R. R. Burgess (1984) Effects of the mutant sigma allele *rpoD800* on the synthesis of specific macromolecular components of the *Escherichia coli* K12 cell. J. Mol. Biol. 172, 283–300.

Grossman and Gross show that the heat shock activator is the product of the htpR *gene*
Grossman, A. D., J. W. Erickson, and C. A. Gross (1984) The htpR gene product of *E. coli* is a sigma factor for heat-shock promoters. Cell 38, 383–90.

Sequence of htpR gene
Landick, R., V. Vaughn, E. T. Lau, R. A. VanBogelen, J. W. Erickson, and F. C. Neidhardt (1984) Nucleotide sequence of the heat shock regulatory gene of *E. coli* suggests its protein product may be a transcription factor. Cell 38, 175–82.

A consensus sequence for the promoters of heat shock genes
Cowing, D. W., J. C. Bardwell, E. A. Craig, C. Woolford, R. W. Hendrix, and C. A. Gross (1985) Consensus sequence for *Escherichia coli* heat shock gene promoters. Proc. Natl. Acad. Sci. USA 82, 2679–83.

The existence of alternate sigmas is first demonstrated by Losick
Losick, R., and J. Pero (1981) Cascades of sigma factors. Cell 25, 582–84.

Almost all bacteria have more than one sigma factor
Wösten, M. M. S. M. (1998) Eubacterial σ-factors. FEMS Microbiol. Rev. 22, 127–50.

REGULATION OF THE REGULATOR:
POST-TRANSCRIPTIONAL CONTROL OF σ^{32}
The synthesis of heat shock proteins correlates with the level of σ^{32}
Straus, D. B., W. A. Walter, and C. A. Gross (1987) The heat shock response of *E. coli* is regulated by changes in the concentration of σ^{32}. Nature 329, 348–51.

Level of σ^{32} *is controlled by regulating its synthesis and its stability*
Straus, D. B., W. A. Walter, and C. A. Gross (1987) The heat shock response of *E. coli* is regulated by changes in the concentration of σ^{32}. Nature 329, 348–51.
Tilly, K., J. Spence, and C. Georgopoulos (1989) Modulation of stability of the *Escherichia coli* heat shock regulatory factor sigma. J. Bacteriol. 171, 1585–89.
Morita, M.T., M. Kanemori, H. Yanagi and T. Yura (2000). Dynamic interplay between antagonistic pathways controlling the sigma 32 level in *Escherichia coli*. Proc. Natl Acad. Sci. USA 97, 5860-5.

Lambda cIII protein provokes the heat shock response by stabilizing σ^{32}
Bahl, H., H. Echols, D. B. Straus, D. Court, R. Crowl, and C. P. Georgopoulos (1987) Induction of the heat shock response of *E. coli* through stabilization of σ^{32} by the phage lambda cIII protein. Genes. Devel. 1, 57–64.

D'Ari, Bukau, and Ogura show that both cII and σ^{32} *are degraded by HflB*
Herman, C., T. Ogura, T. Tomoyasu, S. Hiraga, Y. Akiyama, K. Ito, R. Thomas, R. D'Ari, and P. Bouloc (1993) Cell growth and lambda phage development controlled by the same essential *Escherichia coli* gene, *ftsH/hflB*. Proc. Natl. Acad. Sci. USA 90, 10861–65.
Herman, C., D. Thevenet, R. D'Ari, and P. Bouloc (1995) Degradation of σ^{32}, the heat shock regulator in *Escherichia coli*, is governed by HflB. Proc. Natl. Acad. Sci. USA 92, 3516–20.

Tomoyasu, T., J. Gamer, B. Bukau, M. Kanemori, H. Mori, A. J. Rutman, A. B. Oppenheim, T. Yura, K. Yamanaka, H. Niki, et al. (1995) *Escherichia coli* FtsH is a membrane-bound, ATP-dependent protease which degrades the heat-shock transcription factor σ^{32}. EMBO J. 14, 2551–60.

Shotland, Y., S. Koby, D. Teff, N. Mansur, D. A. Oren, K. Tatematsu, T. Tomoyasu, M. Kessel, B. Bukau, T. Ogura, and A. B. Oppenheim (1997) Proteolysis of the phage lambda *c*II regulatory protein by FtsH (HflB) of *Escherichia coli*. Mol. Microbiol. 24, 1303–10.

dnaK, dnaJ, *and* grpE *mutants alter the heat shock response*

Tilly, K., N. McKittrick, M. Zylicz, and C. Georgopoulos (1983) The *dnaK* protein modulates the heat-shock response of *Escherichia coli*. Cell 34, 641–46.

Straus, D., W. Walter, and C. A. Gross (1990) DnaK, DnaJ, and GrpE heat shock proteins negatively regulate heat shock gene expression by controlling the synthesis and stability of σ^{32}. Genes Devel. 4, 2202–9.

Reviewed in: Craig, E. A., and C. A. Gross (1991) HSP70—The cellular thermometer? TIBS. 16, 135–40.

Another mode of regulation of σ^{32}—*inactivation of* σ^{32} *at low temperature*

Straus, D. B., W. A. Walter, and C. A. Gross (1989) The activity of σ^{32} is reduced under conditions of excess heat shock protein production in *Escherichia coli*. Genes Devel. 3, 2003–10.

DnaK and DnaJ bind to σ^{32}

Gamer, J., H. Buhard, and B. Bukau (1992) Physical interaction between heat shock proteins DnaK, DnaJ, and GrpE and the bacterial heat shock transcription factor σ^{32}. Cell 69, 833–42.

Liberek, K., T. P. Galitski, M. Zylicz, and C. Georgopoulos (1992) The DnaK chaperone modulates the heat shock response of *Escherichia coli* by binding to the σ^{32} transcription factor. Proc. Natl. Acad. Sci. USA 89, 3516–20.

THE HEAT SHOCK RESPONSE IN EUKARYOTIC CELLS
Identification of promoter region of Drosophila *heat shock genes*

Pelham, H. R. (1982) A regulatory upstream promoter element in the *Drosophila* hsp 70 heat-shock gene. Cell 30, 517–28.

HSE is the regulatory sequence

Pelham, H. R., and M. Bienz (1982) A synthetic heat-shock promoter element confers heat-inducibility on the herpes simplex virus thymidine kinase gene. EMBO J. 1, 1473–77.

Wu and Parker identify a protein that binds to the HSE

Parker, C. S., and J. Topol (1984) A *Drosophila* RNA polymerase II transcription factor binds to the regulatory site of an hsp 70 gene. Cell 37, 273–83.

Wu, C. (1984) Activating protein factor binds *in vitro* to upstream control sequences in heat shock gene chromatin. Nature 311, 81–84.

Heat shock factors are purified by Wu and Pelham

Wu, C., S. Wilson, B. Walker, I. Dawid, T. Paisley, V. Zimarino, and H. Ueda (1987) Purification and properties of *Drosophila* heat shock activator protein. Science 238, 1247–53.

Sorger, P. K., and H. R. Pelham (1987) Purification and characterization of a heat-shock element binding protein from yeast. EMBO J. 6, 3035–41.

Hsp70 genes in yeast (at least ten Hsp70 cognates)

Craig, E. A. (1990) Regulation and function of the HSP70 multigene family of *Saccharomyces cerevisiae*, in R. I. Morimoto, A. Tissières, and C. Georgopoulos, eds., *Stress Proteins in Biology and Medicine*. Cold Spring Harbor, N.Y.: Cold Spring Harbor Laboratory Press.

Reviewed in: Craig, E. A. (1989) Essential roles of 70kDa heat inducible proteins. Bioessay 11, 48–52.

BiP, an Hsp70 homolog, participates in assembly of the polypeptide chains of antibodies

Haas, I. G., and M. Wabl (1983) Immunoglobulin heavy chain binding protein. Nature 306, 387–89.

Gething, M. J., K. McCammon, and J. Sambrook (1986) Expression of wild-type and mutant forms of influenza hemagglutinin: The role of folding in intracellular transport. Cell 46, 939–50.

Munro, S., and H. R. Pelham (1986) An Hsp-70 like protein in the ER: Identity with the 78 kd glucose-regulated protein and immunoglobulin heavy chain binding protein. Cell 46, 291–300.

Haas, I. G., and T. Meo (1988). cDNA cloning of the immunoglobulin heavy chain binding protein. Proc. Natl. Acad. Sci. USA 85, 2250–54.

Hsp104 is required for the disaggregation of certain protein complexes after heat shock

Parsell, D. A., A. S. Kowal, M. A. Singer, and S. Lindquist (1994) Protein disaggregation mediated by heat-shock protein Hsp104. Nature 372, 475–78.

Hsp60 participates in transport and folding of proteins in subcellular organelles

Sigler, P. B., Z. Xu, H. S. Rye, S. G. Burston, W. A. Fenton, and A. L. Horwich (1998). Structure and function in GroEL-mediated protein folding. Annu Rev. Biochem 67, 581–608.

Cheng, M. Y., F. U. Hartl, J. Martin, R. A. Pollock, F. Kalousek, W. Neupert, E. M. Hallberg, R. L. Hallberg, and A. L. Horwich (1989) Mitochondrial heat-shock protein hsp60 is essential for assembly of proteins imported into yeast mitochondria. Nature 337, 620–25.

Ostermann, J., A. L. Horwich, W. Neupert, F. U. Hartl (1989) Protein folding in mitochondria requires complex formation with hsp60 and ATP hydrolysis. Nature 341, 125–30.

Hendrick, J. P., and F.-U. Hartl (1993) Molecular chaperone functions of heat-shock proteins. Annu. Rev. Biochem. 62, 349–84.

Proposal that molecular chaperones participate in folding newly synthesized polypeptides

Chappell, T. G., W. J. Welch, D. M. Schlossman, K. B. Palter, M. J. Schlesinger, and J. E. Rothman (1986). Uncoating ATPase is a member of the 70 kilodalton family of stress proteins. Cell 45 (1), 3–13.

REGULATED DNA REPAIR AND MUTAGENESIS

In the 1920's, X-rays were shown to produce mutations in Drosophila and maize

Muller, H. J. (1927) Artificial transmutation of the gene. Science 66, 84–87.

Stadler, L. J. (1928) Mutations in barley induced by X-rays and radium. Science 68, 186–87.

Evidence for DNA UV-damage repair mechanisms in bacteria

Kelner, A. (1949) Effect of visible light on the recovery of *Streptomyces griseus conidia* from ultraviolet irradiation injury. Proc. Natl. Acad. Sci. USA 35, 73–79.

Roberts, R. R., and E. Aldous (1949) Recovery from ultraviolet irradiation in *Escherichia coli*. J. Bacteriol. 57, 363–75.

Witkin found evidence for "dark repair" mechanisms

Witkin, E. M. (1956) Time, temperature, and protein synthesis: A study of ultraviolet-induced mutation in bacteria. Cold Spring Harbor Symp. Quant. Biol. 21, 123–40.

THE UVR PATHWAY OF EXCISION REPAIR
UV-induced pyrimidine dimers found in isolated DNA

Wacker, A., H. Dellweg, and E. Lodemann (1961) Strahlenchemische Veränderung der Nucleinsäuren. Angew. Chem. 73, 64–65.

Wacker, A., H. Dellweg, and D. Jacherts (1962) Thymin-demerisierung und Überlebensrate bei Bakterien. J. Mol. Biol. 4, 410–12.

UV-induced pyrimidine dimers found in E. coli *genome*

Setlow, R. B., P. A. Swenson, and W. L. Carrier (1963) Thymine dimers and inhibition of DNA synthesis by ultraviolet irradiation of cells. Science 142, 1464–66.

Hill isolates bacterial mutants that are highly sensitive to UV

Hill, R. F. (1958) A radiation-sensitive mutant of *Escherichia coli*. Biochim. Biophys. Acta 30, 636–37.

Hill, R. F. (1965) Ultraviolet-induced lethality and reversion to prototrophy in *Escherichia coli* strains with normal and reduced dark repair ability. Photochemistry & Photobiology 4, 563–68.

Thymine dimers cut out of DNA during repair, but not in some of Hill's mutants

Boyce, R. P., and P. Howard-Flanders (1964) Release of ultraviolet light-induced thymine dimers from DNA in *E. coli*. Proc. Natl. Acad. Sci. USA 51, 293–97.

Setlow, R. B., and W. L. Carrier (1964) The disappearance of thymine dimers from DNA: An error-correcting mechanism. Proc. Natl. Acad. Sci. USA 51, 226–30.

Isolation of uvrA, B, *and* C *genes*

Howard-Flanders, P., R. P. Boyce, and R. Theriot (1966) Three loci in *Escherichia coli* K-12 that control the excision of thymine dimers and certain other mutagen products from host or phage DNA. Genetics 53, 1119–36.

Hanawalt demonstrates repair DNA synthesis to fill gap left by dimer excision

Pettijohn, A. R., and P. C. Hanawalt (1964) Evidence for repair replication of ultraviolet damaged DNA in bacteria. J. Mol. Biol. 9, 395–410.

Cooper, P. K., and P. C. Hanawalt (1972) Role of DNA polymerase I and the *rec* system in excision-repair in *Escherichia coli*. Proc. Natl. Acad. Sci. USA 69, 1156–60.

Hanawalt, P. (1975) Molecular mechanisms involved in DNA repair. Genetics 79, 179–97.

recA-MEDIATED RECOMBINATIONAL REPAIR
recA *mutants were sensitive to the lethal effects of UV*

Clark, A. J., and A. D. Margulies (1965) Isolation and characterization of recombination-deficient mutants of *Escherichia coli* K-12. Proc. Natl. Acad. Sci. USA 53, 451–58.

recA uvr *double mutant much more sensitive to killing by UV than the single mutants*

Howard-Flanders, P., and R. P. Boyce (1966) DNA repair and genetic recombination: Studies on mutants of *Escherichia coli* defective in these processes. Radiat. Res. 6 (Suppl.), 156–84.

Pyrimidine dimers block DNA replication

Rupp, W. D., and P. Howard-Flanders (1968) Discontinuities in the DNA synthesized in an excision-defective strain of *Escherichia coli* following ultraviolet irradiation. J. Mol. Biol. 31, 291–304.

Proteins utilized by φX174 to initiate DNA synthesis are used in restart replication

Zavitz, K. H., and K. J. Marians (1991) Dissecting the functional role of PriA protein-catalysed primosome assembly in *Escherichia coli* DNA replication. Mol. Microbiol. 5, 2869–73.

Masai, H., T. Asai, Y. Kubota, K. Arai, and T. Kogoma (1994) *Escherichia coli* PriA protein is essential for inducible and constitutive stable DNA replication. EMBO J. 13, 5335–45.

Cox, M. M. (1998) A broadening view of recombinational DNA repair in bacteria. Genes to Cells 3, 65–78.

Reviewed in: Marians, K. J. (1999) PriA: At the crossroads of DNA replication and recombination. Progress in Nucleic Acid Research and Molecular Biology 63, 39–67.

THE SOS RESPONSE TO DNA DAMAGE

Hill, Witkin, and Bridges show that excision repair of UV damage is error free

Hill, R. F. (1965) Ultraviolet-induced lethality and reversion to prototrophy in *Escherichia coli* strains with normal and reduced dark repair ability. Photochemistry & Photobiology 4, 563–68.

Bridges, B. A., and R. J. Munson (1966) Excision-repair of DNA damage in an auxotrophic strain of *E. coli* Biochem. Biophys. Res. Commun. 22, 268–73.

Witkin, E. M. (1966) Radiation-induced mutations and their repair. Science 152, 1345–53.

Witkin, E. M. (1967) Mutation-proof and mutation-prone modes of survival in derivatives of *Escherichia coli* B differing in sensitivity to ultraviolet light. Brookhaven Symp. Biol. 20, 17–55.

Witkin and Bridges propose "error-prone" type of repair replication past the lesion site

Bridges, B. A., R. E. Dennis, and R. J. Munson (1967) Differential induction and repair of ultraviolet damage leading to true reversions and external suppressor mutations of an ochre codon in *Escherichia coli* B/r WP2. Genetics 57, 897–908.

Witkin, E. M. (1967) Mutation-proof and mutation-prone modes of survival in derivatives of *Escherichia coli* B differing in sensitivity to ultraviolet light. Brookhaven Symp. Biol. 20, 17–55.

Witkin found that recA *mutants fail to exhibit UV-induced mutagenesis*

Witkin, E. M. (1969) The mutability toward ultraviolet light of recombination-deficient strains of *Escherichia coli*. Mutation Res. 8, 9–14.

The same physiological conditions induced both prophage and filamentation

Witkin, E. M. (1967) The radiation sensitivity of *Escherichia coli* B: A hypothesis relating filament formation and prophage induction. Proc. Natl. Acad. Sci. USA 57, 1275–79.

Jacob isolates a bacterial mutant that induces lambda at high temperature

Goldthwait, D., and F. Jacob (1964) Sur le mécanisme de l'induction du développement du prophage chez les bactéries lysogènes. C. R. Acad. Sci. Paris D 259, 661–64.

Filament formation is also induced at high temperature in the mutant

Kirby, E. P., F. Jacob, and D. A. Goldthwait (1967) Prophage induction and filament formation in a mutant strain of *Escherichia coli*. Proc. Natl. Acad. Sci. USA 58, 1903–10.

Radman proposes that RecA and lexA regulate "SOS replication"

Defais, M., P. Fauquet, M. Radman, and M. Errera (1971) Ultraviolet reactivation and ultraviolet mutagenesis of λ in different genetic systems. Virology 43, 495–503.

tif *mutants also show "Weigle mutagenesis" at high temperature*

Castellazzi, M., J. George, and G. Buttin (1972) Prophage induction and cell division in *E. coli* I. Further characterization of the thermosensitive mutation tif-1 whose expression mimics the effect of UV irradiation. Mol. Gen. Genet. 119, 139–52.

UV mutagenesis induced in tif *mutants at high temperature*

Witkin, E. M. (1974) Thermal enhancement of ultraviolet mutability in a tif-1 *uvrA* derivative of *Escherichia coli* B/r: Evidence that ultraviolet mutagenesis depends upon an inducible function. Proc. Natl. Acad. Sci. USA 71, 1930–34.

Development of the idea of an "SOS response"

Witkin, E. M. (1973) Ultraviolet mutagenesis in bacteria: The inducible nature of error-prone repair. An. Acad. Bras. Cienc. 45 (Suppl.), 188–92.

Radman, M. (1974) Phenomenology of an inducible mutagenic DNA repair pathway in *Escherichia coli*: SOS repair hypothesis, in L. Prokash, F. Sherman, M. Miller, C. Lawrence, and H. W. Tabor, eds., *Molecular and Environmental Aspects of Mutagenesis*. Springfield, Ill.: Charles C. Thomas Publisher.

Radman, M. (1975) SOS repair hypothesis: Phenomenology of an inducible DNA repair which is accompanied by mutagenesis, in P. Hanawalt and R. B. Setlow, eds., *Molecular Mechanisms for Repair of DNA*, part A. New York: Plenum Press.

The tif *mutation is in the* recA *gene*

McEntee, K. (1977) Protein X is the product of the *recA* gene of *Escherichia coli*. Proc. Natl. Acad. Sci. USA 74, 5275–79.

Gudas, L. J., and D. W. Mount (1977) Identification of the *recA* (Tif) gene product of *Escherichia coli*. Proc. Natl. Acad. Sci. USA 74, 5280–84.

lexA *mutations also cause UV-sensitivity, blocked UV mutagenesis, and filamentation*

Witkin, E. M. (1967) Mutation-proof and mutation-prone modes of survival in derivatives of *Escherichia coli* B differing in sensitivity to ultraviolet light. Brookhaven Symp. Biol. 20, 17–55.

UV-sensitive lexA *mutations were dominant to* lexA$^+$

Mount, D. W., K. B. Low, and S. J. Edmiston (1972) Dominant mutations (lex) in *Escherichia coli* which affect radiation sensitivity and frequency of ultraviolet light-induced mutations. J. Bacteriol. 112, 886–93.

Isolation of knockout mutations in lexA

Mount, D. W. (1977) A mutant of *Escherichia coli* showing constitutive expression of the lysogenic induction and error-prone DNA repair pathways. Proc. Natl. Acad. Sci. USA 74, 300–304.

THE BIOCHEMISTRY OF SOS REGULATION:
INDUCTION BY REPRESSOR CLEAVAGE

Roberts and Roberts observed that prophage induction led to cleavage of cI protein

Roberts, J. W., and C. W. Roberts (1975) Proteolytic cleavage of bacteriophage lambda repressor in induction. Proc. Natl. Acad. Sci. USA 72, 147–51.

RecA protein mediates the cleavage of cI protein

Roberts, J. W., C. W. Roberts, and N. L. Craig (1978) *Escherichia coli* recA gene product inactivates phage lambda repressor. Proc. Natl. Acad. Sci. USA 75, 4714–18.

Little and coworkers worked out the pathway of LexA cleavage

Little, J. W., S. H. Edmiston, L. Z. Pacelli, and D. W. Mount (1980) Cleavage of the *Escherichia coli lexA* protein by the *recA* protein. Proc. Natl. Acad. Sci. USA 77, 3225–29.

Under appropriate in vitro *conditions, cI and LexA proteins can cleave themselves*

Little, J. W. (1984) Autodigestion of *lexA* and phage lambda repressors. Proc. Natl. Acad. Sci. USA 81, 1375–79.

LexA binds to promoter regions of SOS-regulated genes and blocks transcription

Brent, R., and M. Ptashne (1981) Mechanism of action of the *lexA* gene product. Proc. Natl. Acad. Sci. U.S.A. 78, 4204–8.

Little, J. W., D. W. Mount, and C. R. Yanisch-Perron (1981) Purified lexA protein is a repressor of the *recA* and *lexA* genes. Proc. Natl. Acad. Sci. USA 78, 4199–4203.

Technique for gene fusions utilizing lacZ *gene carried by phage Mu*

Casadaban, M. J., and S. N. Cohen (1980) Lactose genes fused to exogenous promoters in one step using a Mu-*lac* bacteriophage: *In vivo* probe for transcriptional control sequences. Proc. Natl. Acad. Sci. USA 76, 4530–33.

Kenyon and Walker identify genes whose expression was induced by DNA damage

Kenyon, C. J., and G. C. Walker (1980) DNA-damaging agents stimulate gene expression at specific loci in *Escherichia coli*. Proc. Natl. Acad. Sci. USA 77, 2819–23.

Lu and Echols find that RecA also binds to dsDNA that contains pyrimidine dimers

Lu, C., R. H. Scheuermann, and H. Echols (1986) Capacity of RecA protein to bind preferentially to UV lesions and inhibit the editing subunit (epsilon) of DNA polymerase III: A possible mechanism for SOS-induced targeted mutagenesis. Proc. Natl. Acad. Sci. USA 83, 619–23.

INDUCED MUTATION IN THE SOS RESPONSE

Identification of umuC *and* umuD *genes*

Kato, T., and Y. Shinoura (1977) Isolation and characterization of mutants of *Escherichia coli* deficient in induction of mutations by ultraviolet light. Mol. Gen. Genet. 156, 121–31.

Expression of umuC *and* umuD *is induced by the SOS response*

Bagg, A., C. J. Kenyon, and G. C. Walker (1981) Inducibility of a gene product required for UV and chemical mutagenesis in *Escherichia coli*. Proc. Natl. Acad. Sci. USA 78, 5749–53.

Elledge, S. J., and G. C. Walker (1983) Proteins required for ultraviolet light and chemical mutagenesis: Identification of the products of the *umuC* locus of *Escherichia coli*. J. Mol. Biol. 164, 175–92.

SOS mutagenesis requires the RecA-mediated cleavage of the UmuD protein
Burckhardt, S. E., R. Woodgate, R. H. Scheuermann, and H. Echols (1988) UmuD mutagenesis protein of *Escherichia coli*: Overproduction, purification, and cleavage by RecA. Proc. Natl. Acad. Sci. USA 85, 1811–15.
Nohmi, T., J. R. Battista, L. A. Dodson, and G. C. Walker (1988) RecA-mediated cleavage activates UmuD for mutagenesis: Mechanistic relationship between transcriptional derepression and posttranslational activation. Proc. Natl. Acad. Sci. USA 85, 1816–20.
Shinagawa, H., H. Iwasaki, T. Kato, and A. Nakata (1988) RecA protein-dependent cleavage of UmuD protein and SOS mutagenesis. Proc. Natl. Acad. Sci. USA 85, 1806–10.

in vitro *demonstration that RecA, UmuC/UmuD′, and pol III interact at the UV lesion*
Rajagopalan, M., C. Lu, R. Woodgate, M. O'Donnell, M. F. Goodman, and H. Echols (1992) Activity of the purified mutagenesis proteins UmuC, UmuD′, and RecA in replicative bypass of an abasic DNA lesion by DNA polymerase III. Proc. Natl. Acad. Sci. USA 89, 10777–81.
Maor-Shoshani, A., N. B. Reuven, G. Tomer and Z. Livneh (2000). Highly mutagenic replication by DNA polymerase V (UmuC) provides a mechanistic basis for SOS untargeted mutagenesis. Proc Natl Acad Sci USA 18, 565-70
Tang, M., P. Pham, X. Shen, J. S. Taylor, M. O'Donnell, R. Woodgate and M. F. Goodman (2000). Roles of *E. coli* DNA polymerases IV and V in lesion-targeted and untargeted SOS mutagenesis. Nature 404, 1014-8.

Chapter 9. Making DNA from RNA: The Strange Life of the Retrovirus

QUANTITATIVE ANIMAL VIROLOGY: THE PLAQUE AND FOCUS ASSAYS
Technique developed by Earle for growing cells on solid surface
Shannon, J. E., W. R. Earle, and H. K. Waltz (1952) Massive tissue cultures prepared from whole chick embryos planted as a cell suspension on glass substrate. J. Natl. Cancer Inst. 13, 349–55.

Dulbecco uses technique developed by Earle to develop a plaque assay
Dulbecco, R. (1952) Production of plaques in monolayer tissue cultures by single particles of an animal virus. Proc. Natl. Acad. Sci. USA 38, 747–52.

Discovery of Rous sarcoma virus
Rous, P. (1911) A sarcoma of the fowl transmissible by an agent separable from the tumor cells. J. Exp. Med. 13, 397–411.

Duesberg, "The Rous virus has led to five Nobel prizes . . ."
Personal communication.

"Morphological transformation" of chicken cells infected by RSV
Manaker, R. A., and V. Groupé (1956) Discrete foci of altered chicken embryo cells associated with Rous sarcoma virus in tissue culture. Virology 2, 838–40.

Focus assay for RSV developed by Temin and Rubin
Temin, H. M., and H. Rubin (1958) Characteristics of an assay for Rous sarcoma virus and Rous sarcoma cells in tissue culture. Virology 6, 669–88.

Experiments on radiation sensitivity of RSV growth
Rubin, H., and H. M. Temin (1959) A radiological study of cell-virus interaction in the Rous sarcoma. Virology 7, 75–91.
Levinson, W., and H. Rubin (1966) Radiation studies of avian tumor viruses and of Newcastle Disease Virus. Virology 28, 533–42.

Temin carries out experiments that encourage his belief in the proviral form of RSV
Temin, H. M., and H. Rubin (1959) A kinetic study of infection of chick embryo cells *in vitro* by Rous sarcoma virus. Virology 8, 209–22.
Temin, H. M. (1960a) The control of cellular morphology in embryonic cells infected with Rous sarcoma virus *in vitro*. Virology 10, 182–97.
Temin, H. M. (1960b) Infection of chick cells by Rous sarcoma virus *in vitro*. Ph.D. thesis, California Institute of Technology.
Temin, H. M. (1967) In *The Molecular Biology of Viruses*, p. 709. New York: Academic Press.

RSV packages RNA into its virus particles
Crawford, L. V., and E. M. Crawford (1961) The properties of Rous sarcoma virus purified by density gradient centrifugation. Virology 13, 227–32.

MAKING DNA FROM RNA
Virus production is blocked by an inhibitor of transcription
Temin, H. (1963) The effects of actinomycin D on growth of Rous sarcoma virus *in vitro*. Virology 20, 577–82.

Virus growth is also prevented by general inhibitors of DNA replication
Bader, J. P. (1964) The role of deoxyribonucleic acid in the synthesis of Rous Sarcoma Virus. Virology 22, 462–68.
Bader, J. P. (1965) The requirement for DNA synthesis in growth of Rous sarcoma and Rous-associated viruses. Virology 26, 253–61.
Temin, H. M. (1967) Studies on carcinogenesis by avian sarcoma virus. V. Requirement for new DNA synthesis and for cell division. J. Cell. Physiol. 69, 53–64.
McDonnell, J. P., A.-C. Garapin, W. E. Levinson, N. Quintrell, L. Fanshier, and J. M. Bishop (1970) DNA polymerases of Rous sarcoma virus: Delineation of two reactions with actinomycin. Nature 228, 433–35.

Temin found the "early experiments completely convincing"
Taped interview with Howard Temin

lac repressor originally identified as an RNA molecule by inhibitor experiments
Jacob, F., and J. Monod (1961) Genetic regulatory mechanisms in the synthesis of proteins. J. Mol. Biol. 3, 318–56.

Mizutani and Temin found evidence that RSV could replicate without new protein synthesis
Temin, H. M., and S. Mizutani (1970) RNA-dependent DNA polymerase in virions of Rous sarcoma virus. Nature 226, 1211–13.

Isolation of reverse transcriptase

Baltimore, D. (1970) Viral RNA-dependent DNA polymerase. Nature 226, 1209–11.

Temin, H. M., and S. Mizutani (1970) RNA-dependent DNA polymerase in virions of Rous sarcoma virus. Nature 226, 1211–13.

THE PATH FROM VIRUS TO PROVIRUS AND BACK AGAIN

Ts mutant of RSV that did not transform at high temperature

Martin, G. S. (1970) Rous sarcoma virus: A function required for maintenance of the transformed state. Nature 227, 1021–23.

Deletion mutants of RSV that failed to transform

Hanafusa, H., and T. Hanafusa (1966) Determining factor in the capacity of Rous sarcoma virus to induce tumors in mammals. Proc. Natl. Acad. Sci. USA 55, 532–38.

Duff, R. G., and P. K. Vogt (1969) Characteristics of two new avian tumor virus subgroups. Virology 39, 18–30.

Vogt, P. K. (1971) Spontaneous segregation of nontransforming viruses from cloned sarcoma viruses. Virology 46, 939–46.

The essential RSV genome defined

Duesberg, P. H. (1968) Physical properties of Rous sarcoma virus RNA. Proc. Natl. Acad. Sci. USA 60, 1511–18.

Duesberg, P. H., and P. K. Vogt (1970) Differences between the ribonucleic acids of transforming and nontransforming avian tumor viruses. Proc. Natl. Acad. Sci. USA 67, 1673–80.

Beemon, K., P. Duesberg, and P. Vogt (1974) Evidence for crossing-over between avian tumor viruses based on analysis of viral RNAs. Proc. Natl. Acad. Sci. USA 71, 4254–58.

Billeter, M. A., J. T. Parsons, and J. M. Coffin (1974) The nucleotide sequence complexity of avian tumor virus RNA. Proc. Natl. Acad. Sci. USA 71, 3560–64.

Detection of proviral DNA

Varmus, H. E., R. A. Weiss, R. R. Friis, W. E. Levinson, and J. M. Bishop (1972) Detection of avian tumor virus-specific nucleotide sequences in avian cell DNA's. Proc. Natl. Acad. Sci. USA 69, 20–24.

Hill, M., and J. Hillova (1972) Virus recovery in chicken cells tested with Rous sarcoma cell DNA. Nature New Biol. 237, 35–39.

Various forms of RSV DNA were identified: linear, circular, and integrated

Varmus, H. E., P. K. Vogt, and J. M. Bishop (1973) Integration of deoxyribonucleic acid specific for Rous sarcoma virus after infection of permissive and nonpermissive hosts. Proc. Natl. Acad. Sci. USA 70, 3067–71.

Varmus, H. E., R. V. Guntaka, C. T. Deng, and J. M. Bishop (1974) Synthesis, structure and function of avian sarcoma virus-specific DNA in permissive and nonpermissive cells. Cold Spring Harbor Symp. Quant. Biol. 39, pt. 2, 987–96.

Retroviral RNA contains 20 base direct repeat at each end

Haseltine, W. A., A. M. Maxam, and W. Gilbert (1977) Rous sarcoma virus genome is terminally redundant: The 5′ sequence. Proc. Natl. Acad. Sci. USA 74, 989–93.

Schwartz, D. E., P. C. Zamecnik, and H. L. Weith (1977) Rous sarcoma virus is terminally redundant: The 3′ sequence. Proc. Natl. Acad. Sci. USA 74, 994–98.

The existence of LTRs in proviruses

Hughes, S. H., P. R. Shank, D. H. Spector, H. J. Kung, J. M. Bishop, H. E. Varmus, P. K. Vogt, and M. L. Breitman (1978) Proviruses of avian sarcoma virus are terminally redundant, co-extensive with unintegrated linear DNA and integrated at many sites. Cell 15, 1397–1410.

Shank, P. R., S. H. Hughes, H. J. Kung, J. E. Majors, N. Quintrell, R. V. Guntaka, J. M. Bishop, and H. E. Varmus (1978) Mapping unintegrated avian sarcoma virus DNA: Termini of linear DNA bear 300 nucleotides present once or twice in two species of circular DNA. Cell 15, 1383–95.

Sabran, J. L., T. W. Husa, C. Yeater, A. Kaji, W. S. Mason, and J. M. Taylor (1979) Analysis of integrated avian RNA tumor virus DNA in transformed chicken, duck and quail fibroblasts. J. Virology 29, 170–78.

THE WILD RIDE OF REVERSE TRANSCRIPTASE

Primer for reverse transcriptase is a tRNA

Dahlberg, J. E., R. C. Sawyer, J. M. Taylor, A. J. Faras, W. E. Levinson, H. M. Goodman, and J. M. Bishop (1974) Transcription of DNA from the 70S RNA of Rous sarcoma virus. I. Identification of a specific 4S RNA which serves as a primer. J. Virology 13, 1126–33.

Faras, A. J., J. E. Dahlberg, R. C. Sawyer, F. Harada, J. M. Taylor, W. E. Levinson, J. M. Bishop, and H. M. Goodman (1974) Transcription of DNA from the 70S RNA of Rous sarcoma virus. II. Structure of a 4S RNA primer. J. Virology 13, 1134–42.

Primer binds close to the 5' end of the RNA

Reviewed in: Taylor, J. M. (1977) An analysis of the role of tRNA species as primers for the transcription into DNA of RNA tumor virus genomes. Biochim. Biophys. Acta 473, 57–71.

Discovery of RNase H

Hausen, P., and H. Stein (1970) Ribonuclease H: An enzyme degrading the RNA moiety of DNA-RNA hybrids. Eur. J. Biochem. 14, 278–83.

Mölling, K., D. P. Bolognesi, H. Bauer, W. Busen, H. W. Plassman, and P. Hausen (1971) Association of viral reverse transcriptase with an enzyme degrading the RNA moiety of RNA-DNA hybrids. Nature New Biol. 234, 240–43.

Completion of the right LTR and synthesis of left LTR

Varmus, H. E., S. Heasley, H.-J. Kung, H. Oppermann, V. C. Smith, J. M. Bishop, and P. R. Shank (1978) Kinetics of synthesis, structure and purification of avian sarcoma virus-specific DNA made in the cytoplasm of acutely infected cells. J. Mol. Biol. 120, 55–82.

Mitra, S. W., S. Goff, E. Gilboa, and D. Baltimore (1979) Synthesis of a 600-nt long plus strand DNA (plus strong stop DNA) by virions of Moloney murine leukemia virus. Proc. Natl. Acad. Sci. USA 76, 4355–59.

Gilboa, E., S. Goff, A. Shields, F. Yoshimura, S. Mitra, and D. Baltimore (1979a) *in vitro* synthesis of a 9 kbp terminally redundant DNA carrying the infectivity of Moloney murine leukemia virus. Cell 16, 863–74.

Gilboa, E., S. W. Mitra, S. Goff, and D. Baltimore (1979b) A detailed model of reverse transcription and tests of crucial aspects. Cell 18, 93–100.

Swanstrom, R., H. E. Varmus, and J. M. Bishop (1981) The terminal redundancy of the retroviral genome facilitates chain elongation by reverse transcriptase. J. Biol. Chem. 256, 1115–21.

Boone, L. R., and A. M. Skalka (1981) Viral DNA synthesized *in vitro* by avian retrovirus particles permeabilized with melittin. I. Kinetics of synthesize and size of minus- and plus-strand transcripts. J. Virol. 37, 109–16.

Boone, L. R., and A. M. Skalka (1981) Viral DNA synthesized *in vitro* by avian retrovirus particles permeabilized with melittin. II. Evidence for a strand displacement mechanism in plus-strand synthesis. J. Virol. 37, 117–26.

Reviewed in: Telesnitsky, A., and S. P. Goff (1997) Reverse transcriptase and the generation of retroviral DNA, in J. M. Coffin, S. H. Hughes, and H. E. Varmus, eds., *Retroviruses*. Cold Spring Harbor, N.Y.: Cold Spring Harbor Laboratory Press.

THE RETROVIRUS AND ITS CANCER GENE

A summary of Temin's early work arguing for idea of cancer genes

Temin, H. M. (1964) Nature of the provirus of rous sarcoma. Natl. Cancer Inst. Mono. 17, 557–70.

Experiments define the src *gene at one end of the viral RNA*

Wang, L.-H., P. Duesberg, K. Beemon, and P. K. Vogt (1975) Mapping RNase T1-resistant oligonucleotides of avian tumor virus RNAs: Sarcoma-specific oligonucleotides are near the poly(A) end and oligonucleotides common to sarcoma and transformation-defective viruses are at the poly(A) end. J. Virol. 16, 1051–70.

Coffin, J. M., and M. A. Billeter (1976) A physical map of the Rous sarcoma virus genome. J. Mol. Biol. 100, 293–318.

Virogene-oncogene hypothesis for cancer

Huebner, R. J., and G. J. Todaro (1969) Oncogenes of RNA tumor viruses as determinants of cancer. Proc. Natl. Acad. Sci. USA 64, 1087–94.

Chicken cells contain inactive provirus; all other avian species tested have only a src *gene*

Stehelin, D., H. E. Varmus, J. M. Bishop, and P. K. Vogt (1976) DNA related to the transforming gene(s) of avian sarcoma viruses is present in normal avian DNA. Nature 260, 170–73.

Homology between src *DNA and DNA from mammalian cells*

Schartl, M., and A. Barnekow (1982) The expression in eukaryotes of a tyrosine kinase which is reactive with pp60-vsrc antibodies. Differentiation 23, 109–14.

Other cellular proto-oncogenes identified

Reviewed in: Varmus, H. E. (1984) The molecular genetics of cellular oncogenes. Annu. Rev. Genet. 18, 553–612.

How retroviruses capture cellular proto-oncogenes

Herman, S. A., and J. M. Coffin (1987) Efficient packaging of readthrough RNA in ALV: Implications for oncogene transduction. Science 236, 845–48.

Zhang, J., and H. M. Temin (1993) Rate and mechanism of nonhomologous recombination during a single cycle of retroviral replication. Science 259, 234–38.

How retroviruses without oncogenes can also cause cancer

Hayward, W. S., B. G. Neel, and S. M. Astrin (1981) Activation of a cellular onc gene by promoter insertion in ALV-induced lymphoid leukosis. Nature 290, 475–80.

Neel, B. G., W. S. Hayward, H. L. Robinson, J. Fang, and S. M. Astrin (1981) Avian leukosis virus-induced tumors have common proviral integration sites and synthesize discrete new RNAs: Oncogenesis by promoter insertion. Cell 23, 323–34.

Payne, G. S., J. M. Bishop, and H. E. Varmus (1982) Multiple arrangements of viral DNA and an activated host oncogene in bursal lymphomas. Nature 295, 209–14.

RETROVIRUSES AS TRANSPOSONS

DNA sequence analysis suggested retroviral integration uses transposition

Dhar, R., W. L. McClements, L. W. Enquist, and G. F. Vande Woude (1980) Nucleotide sequences of integrated Moloney sarcoma provirus long terminal repeats and their host and viral junctions. Proc. Natl. Acad. Sci. USA 77, 3937–41.

Ju, G., and A. M. Skalka (1980) Nucleotide sequence analysis of the long terminal repeat (LTR) of avian retroviruses: Structural similarities with transposable elements. Cell 22, 379–86.

Shimotohno, K., S. Mizutani, and H. M. Temin (1980) Sequence of retrovirus provirus resembles that of bacterial transposable elements. Nature (London) 285, 550–54.

Hughes, S. H., A. Mutschler, J. M. Bishop, and H. E. Varmus (1981) A Rous sarcoma provirus is flanked by short direct repeats of a cellular DNA sequence present in only one copy prior to integration. Proc. Natl. Acad. Sci. USA 78, 4299–4303.

An endonuclease activity associated with preparations of reverse transcriptase

Golomb, M., and D. P. Grandgenett (1979) Endonuclease activity of purified RNA directed DNA polymerase from AMV. J. Biol. Chem. 254, 1606–13.

Leis, J., G. Duyk, S. Johnson, M. Longiaru, and A. Skalka (1983) Mechanism of action of the endonuclease associated with the αβ and ββ forms of avian RNA tumor virus reverse transcriptase. J. Virol. 45, 727–39.

Integrase coding sequence identified

Donehower, L. A., and H. E. Varmus (1984) A mutant murine leukemia virus with a single missense codon in *pol* is defective in a function affecting integration. Proc. Natl. Acad. Sci. USA 81, 6461–65.

Panganiban, A. T., and H. M. Temin (1984) The retrovirus pol gene encodes a product required for DNA integration: Identification of a retrovirus *int* locus. Proc. Natl. Acad. Sci. USA 81, 7885–89.

Schwartzenberg, P., J. Colicelli, and S. P. Goff (1984) Construction and analysis of deletion mutations in the pol gene of Moloney murine leukemia virus: A new viral function required for productive infection. Cell 37, 1043–52.

Development of an in vitro *integration assay*

Brown, P. O., B. Bowerman, H. E. Varmus, and J. M. Bishop (1987) Correct integration of retroviral DNA *in vitro*. Cell 49, 347–56.

Purified retroviral integration protein executes breaking and joining reaction

Craigie, R., T. Fujiwara, and F. Bushman (1990) The IN protein of Moloney murine leukemia virus processes the viral DNA ends and accomplishes their integration *in vitro*. Cell 62, 829–37.

Katz, R. A., G. Merkel, J. Kulkosky, J. Leis, and A. M. Skalka (1990) The avian retorviral IN protein is both necessary and sufficient for integrative recombination *in vitro*. Cell 63, 87–95.

Retroviral insertion intermediates have structure similar to that of phage Mu

Fujiwara, T., and K. Mizuuchi (1988) Retroviral DNA integration: Structure of an integration intermediate. Cell 54, 497–504.

Brown, P. O., B. Bowerman, H. E. Varmus, and J. M. Bishop (1989) Retroviral integration: Structure of the initial covalent product and its precursor, and a role for the viral IN protein. Proc. Natl. Acad. Sci. USA 86, 2525–29.

THE HIV RETROVIRUS
Montagnier and colleagues isolate HIV

Barré-Sinoussi, F., J. C. Chermann, F. Rey, M. T. Nugeyre, S. Chamaret, J. Gruest, C. Dauguet, C. Axler-Blin, F. Vézinet-Brun, C. Rouzioux, W. Rozenbaum, and L. Montagnier (1983) Isolation of a T-lymphotrophic retrovirus from a patient at risk for acquired immune deficiency syndrome (AIDS). Science 220, 868–71.

Gallo and coworkers establish that HIV causes AIDS

Gallo, R. C., P. S. Sarin, E. P. Gelmann, M. Robert-Guroff, E. Richardson, V. S. Kalyanaraman, D. Mann, G. D. Sidhu, R. E. Stahl, S. Zolla-Pazner, J. Leibowitch, and M. Popovic (1983) Isolation of a human T-cell leukemia virus in acquired immune deficiency syndrome (AIDS). Science 220, 865–67.

HIV tat and rev genes

Rabson, A. B., and B. J. Graves (1997) Synthesis and processing of viral RNA, in J. M. Coffin, S. H. Hughes, and H. E. Varmus, eds., *Retroviruses*. Cold Spring Harbor, N.Y.: Cold Spring Harbor Laboratory Press.

The first report of the in vitro *activity of AZT against HIV*

Mitsuya, H., K. J. Weinhold, P. A. Furman, M. H. St. Clair, S. N. Lehrman, R. C. Gallo, D. Bolognesi, D. W. Barry, and S. Broder (1985) 3′-Azido-3′-deoxythymidine (BW A509U): An antiviral agent that inhibits the infectivity and cytopathic effect of human T-lymphotrophic virus type III/lymphadenopathy-associated virus *in vitro*. Proc. Natl. Acad. Sci. USA 82, 7096–7100.

AZT and other drugs in AIDS therapy

Reviewed in: Mitsuya, H., R. Yarchoan, and S. Broder (1990) Molecular targets for AIDS therapy. Science 249, 1533–44.

Studies by research teams of Ho and Shaw

Wei, X., S. K. Ghosh, M. E. Taylor, V. A. Johnson, E. A. Emini, P. Deutsch, J. D. Lifson, S. Bonhoeffer, M. A. Nowak, B. H. Hahn, M. S. Saag, and G. M. Shaw (1995) Viral dynamics in human immunodeficiency virus type 1 infection. Nature 373, 117–22.

Perelson, A. S., A. U. Neumann, M. Markowitz, J. M. Leonard, and D. D. Ho (1996) HIV-1 dynamics *in vivo*: Virion clearance rate, infected cell life-span, and viral generation time. Science 271, 1582–86.

Ogg, G. S., X. Jin, S. Bonhoeffer, P. R. Dunbar, M. A. Nowak, S. Monard, J. P Segal, Y. Cao, S. L. Rowland-Jones, V. Cerundolo, et al. (1998) Quantification of HIV-1-specific cytotoxic T lymphocytes and plasma load of viral RNA. Science 279, 2103–6.

Chapter 10. Genetic Engineering: Genes and Proteins on Demand

BIOLOGICAL ROOTS: PHAGE WITH BACTERIAL GENES AND COHESIVE SITES

λ*gal isolated by Lederberg and Lederberg*

Morse, M. L., E. M. Lederberg, and J. Lederberg (1956) Transduction in *Escherichia coli*. Genetics 41, 142–56.

Morse, M. L., E. M. Lederberg, and J. Lederberg (1956) Transductional heterogenotes in *Escherichia coli*. Genetics 41, 758–79.

Genetic structure of λgal *determined*

Arber, W., G. Kellenberg, and J. Weigle (1957) La défectuosité du phage λ transducteur. Schweiz. Z. Allgem. Pathol. Bakteriol. 20, 659–65.

Campbell, A. (1957) Transduction and segregation in *Escherichia coli* K12. Virology 4, 366–84.

Isolation of φ*80* trp

Matsushiro, A. (1963) Specialized transduction of tryptophan markers in *Escherichia coli* K12 by bacteriophage φ80. Virology 19, 475–82.

Isolation of φ*80* lac

Beckwith, J. R., and E. R. Signer (1966) Transposition of the *lac* region of *Escherichia coli*. I. Inversion of the *lac* operon and transduction of *lac* by φ80. J. Mol. Biol. 19, 254–65.

Isolation of F'lac and F'gal

Hirota, Y. (1959) Mutants of the sex factor in *Escherichia coli* K12. Genetics 44, 515.

Jacob, F., and E. A. Adelberg (1959) Transfert de caractères génétique par incorporation au facteur sexual d'*Escherichia coli*. Comptes Rendus Acad. Sci. 249, 189–91.

Hirota, Y., and P. H. A. Sneeth (1961) F' and F mediated transduction in *Escherichia coli* K12. Jpn. J. Genetics 36, 307–18.

Hershey inferred the existence of cohesive sites on λ *DNA*

Hershey, A. D., E. Burgi, and L. Ingraham (1963) Cohesion of DNA molecules isolated from phage λ. Proc. Natl. Acad. Sci. USA 49, 748–55.

Kaiser and Hogness show direct transfer of λ *DNA into* E. coli *requires cohesive sites*

Kaiser, A. D., and D. S. Hogness (1960) The transformation of *Escherichia coli* with deoxyribonucleic acid isolated from bacteriophage λdg. J. Mol. Biol. 2, 392–415.

Experiments by Kaiser and Strack on cohesive ends

Strack, M. B., and A. D. Kaiser (1965) On the structure of the ends of λ DNA. J. Mol. Biol. 12, 36–49.

Kaiser and Wu sequence the cohesive end site

Wu, R., and A. D. Kaiser (1968) Structure and base sequence in the cohesive ends of phage lambda. J. Mol. Biol. 33, 523–37.

Lobban Ph.D. dissertation

Lobban, P. (1972) An enzymatic method for end-to-end joining of DNA molecules. Ph.D. diss., Stanford University.

Discovery of terminal transferase

Bollum, F. J. (1974) Terminal deoxynucleotidyl transferase, in P. D. Boyer, ed., *The enzymes*, 3rd ed., vol. 10. New York: Academic Press.

Hershey technique of shear forces

Hershey, A. D., E. Burgi, and L. Ingraham (1962) Sedimentation coefficient and fragility under hydrodynamic shear as measures of molecular weight of the DNA of phage T5. Biophys. J. 2, 423–31.

BIOLOGICAL ROOTS: RESTRICTION ENZYMES FOR PIECES OF DNA
Arber defines the molecular basis of restriction and modification

Arber, W., and D. Dussoix (1962) Host controlled modification of bacteriophage λ. J. Mol. Biol. 5, 18–36.

Arber, W., S. Hattman, D. Dussoix (1963) On the host-controlled modification of bacteriophage λ. Virology 21, 30–35.

λgal *is defective in lytic growth*

Arber, W., G. Kellenberg, and J. Weigle (1957) La défectuosité du phage λ transducteur. Schweiz. Z. Allgem. Pathol. Bakteriol. 20, 659–65.

Campbell, A. (1957) Transduction and segregation in *Escherichia coli* K12. Virology 4, 366–84.

Initial biochemical characterization of a restriction enzyme

Meselson, M., and R. Yuan (1968) DNA restriction enzyme from *E. coli*. Nature 217, 1111–13.

Restriction enzyme purified from Haemophilus influenzae

Smith, H. O., and K. W. Wilcox (1970) A restriction enzyme from *Haemophilus influenzae*. I. Purification and general properties. J. Mol. Biol. 51, 379–91.

Smith and Kelly determine DNA sequence adjacent to cleavage site

Kelly, T. J., and H. O. Smith (1970) A restriction enzyme from *Haemophilus influenzae*. II. Base sequence recognition site. J. Mol. Biol. 51, 393–409.

Physical studies by Vinograd on SV40 DNA

Bauer, W., and J. Vinograd (1968) The interaction of closed circular DNA with intercalative dyes. I. The superhelix density of SV40 DNA in the presence and absence of dye. J. Mol. Biol. 33, 141–71.

Bauer, W., and J. Vinograd (1970) Interaction of closed ciruclar DNA with intercalative dyes. II. The free energy of superhelix formation in SV40 DNA. J. Mol. Biol. 47, 419–35.

Nathans and Danna digest SV40 DNA with the H. influenzae *endonuclease*

Danna, K., and D. Nathans (1971) Specific cleavage of simian virus 40 DNA by restriction endonuclease of *Hemophilus influenzae*. Proc. Natl. Acad. Sci. USA 68, 2913–17.

Nathans used restriction digests to generate genetic map of SV40

Danna, K. J., G. H. Sack Jr., and D. Nathans (1973) Studies of simian virus 40 DNA. VII. A cleavage map of the SV40 genome. J. Mol. Biol. 78, 363–76.

Nathans, D., S. P. Adler, W. W. Brockman, K. J. Danna, T. N. Lee, and G. H. Sack Jr. (1974) Use of restriction endonucleases in analyzing the genome of simian virus 40. Fed. Proc. 33, 1135–35.

Heteroduplexes between mutant and wild-type DNA fragments
Lai, C. J., and D. Nathans (1974) Mapping temperature-sensitive mutants of simian virus 40: Rescue of mutants by fragments of viral DNA. Virology 60, 466–75.

Restriction fragments used to map early and late SV40 transcripts
Khoury, G., M. A. Martin, T. N. Lee, K. J. Danna, and D. Nathans (1973) A map of simian virus 40 transcription sites expressed in productively infected cells. J. Mol. Biol. 78, 377–89.

SV40 origin of replication located
Danna, K. J., and D. Nathans (1972) Bidirectional replication of simian virus 40 DNA. Proc. Natl. Acad. Sci. USA 69, 3097–3100.

THE ENGINEERED JOINING OF DNA MOLECULES
Berg, "the impenetrable wall . . ."
Taped interview with Paul Berg

Plasmid derivative of λ, λdvgal, developed
Matsubara, K., and A. D. Kaiser (1968) Lambda *dv*: An autonomously replicating DNA fragment. Cold Spring Harbor Symp. Quant. Biol. 33, 769–75.
Berg, D. E., D. A. Jackson, and J. E. Mertz (1974) Isolation of a λ *dv* plasmid carrying the bacterial *gal* operon. J. Virol. 14, 1063–69.

EcoRI cleaves single sites in both SV40 and λdvgal
Morrow, J. F., and P. Berg (1972) Cleavage of simian virus 40 DNA at a unique site by a bacterial restriction enzyme. Proc. Natl. Acad. Sci. USA 69, 3365–69.

T4 DNA ligase could join DNA molecules without cohesive ends
Sgaramella, V., J. H. van de Sande, and H. G. Khorana (1970) Studies on polynucleotides. C. A novel joining reaction catalyzed by the T4-polynucleotide ligase. Proc. Natl. Acad. Sci. USA 67, 1468–75.

Successful joining of DNA molecules in the test tube
Jackson, D. A., R. H. Symons, and P. Berg (1972) Biochemical method for inserting new genetic information into DNA of Simian Virus 40: Circular SV40 DNA molecules containing λ phage genes and the galactose operon of *Escherichia coli*. Proc. Natl. Acad. Sci. USA 69, 2904–9.
Lobban, P. E., and A. D. Kaiser (1973) Enzymatic end-to end joining of DNA molecules. J. Mol. Biol. 78, 453–71.

EcoRI generates cohesive ends
Hedgpeth, J., H. M. Goodman, and H. W. Boyer (1972) DNA nucleotide sequence restricted by the RI endonuclease. Proc. Natl. Acad. Sci. USA 69, 3448–52.
Mertz, J. E., and R. W. Davis (1972) Cleavage of DNA by RI restriction endonuclease generates cohesive ends. Proc. Natl. Acad. Sci. USA 69, 3370–74.

DNA CLONING: AMPLIFICATION OF FOREIGN DNA BY A PLASMID VECTOR
Calcium treatment allows E. coli *to take up λ DNA*
Mandel, M., and A. Higa (1970) Calcium-dependent bacteriophage DNA infection. J. Mol. Biol. 53, 159–62.

Calcium treatment used to transform bacteria with plasmid DNA
Cohen, S. N., A. C.Y. Chang, and L. Hsu (1972) Nonchromosomal antibiotic resistance in bacteria: Genetic transformation of *Escherichia coli* by R-factor DNA. Proc. Natl. Acad. Sci. USA 69, 2110–14.

R-factor DNA sheared to isolate set of smaller plasmids
Cohen, S. N., and A. C. Chang (1973) Recircularization and autonomous replication of a sheared R-factor DNA segment in *Escherichia coli* transformants. Proc. Natl. Acad. Sci. USA 70, 1293–97.

pSC101 developed as cloning vector
Cohen, S. N., A. C.Y. Chang, H. W. Boyer, and R. B. Helling (1973) Construction of biologically functional bacterial plasmids *in vitro*. Proc. Natl. Acad. Sci. USA 70, 3240–44.

S. aureus *DNA cloned into pSC101*
Chang, A. C., and S. N. Cohen (1974) Genome construction between bacterial species *in vitro*: Replication and expression of *Staphylococcus* plasmid genes in *Escherichia coli*. Proc. Natl. Acad. Sci. USA 71, 1030–34.

*Brown prepares DNA coding for ribosomal RNA from African toad (*Xenopus laevis*)*
Brown, D. D., P. C. Wensink, and E. Jordan (1971) Purification and some characteristics of 5S DNA from *Xenopus laevis*. Proc. Natl. Acad. Sci. USA 68, 3175–79.
Brown, D. D. (1973) The isolation of genes. Scientific American 229, 20–29.

Toad DNA cloned into pSC101
Morrow, J. F., S. N. Cohen, A. C. Chang, H. W. Boyer, H. M. Goodman, and R. B. Helling (1974) Replication and trancription of eukaryotic DNA in *Escherichia coli*. Proc. Natl. Acad. Sci. 71, 1743–47.

SAFETY CONCERNS AND REGULATORY POLICY
Letter to scientific community from "Committee on Recombinant DNA"
Berg, P., D. Baltimore, H. W. Boyer, S. N. Cohen, R. W. Davis, D. S. Hogness, D. Nathans, R. Roblin, J. D. Watson, S. Weissman, and N. D. Zinder (1974) Potential biohazards of recombinant DNA molecules. Science 185, 303.

Report from 1975 Asilomar Conference on recombinant DNA
Berg, P., D. Baltimore, S. Brenner, R. O. Roblin 3rd, and M. F. Singer (1975) Asilomar conference on recombinant DNA molecules. Science 188, 991–94.
Berg, P., D. Baltimore, S. Brenner, R. O. Roblin, and M. F. Singer (1975) Summary statement of the Asilomar conference on recombinant DNA molecules. Proc. Natl. Acad. Sci. USA 72, 1981–84.

Chargaff's letter to Science *opposing cloning*
Chargaff, E. (1976) On the dangers of genetic meddling. Science 192, 938–40.

GENETIC ENGINEERING: A GIANT LEAP FOR EUKARYOTIC MOLECULAR BIOLOGY
Initial isolation of eukaryotic genes
Birnstiel, M. L., H. Wallace, J. Sirlin, and M. Fischberg (1966) Localization of the ribosomal DNA complements in the nucleolar organizer region of *Xenopus laevis*. Nat. Cancer Inst. Mono. 23, 431–47.

Birnstiel, M., J. Speirs, I. Purdom, K. Jones, and U. E. Loening (1968) Properties and composition of the isolated ribosomal DNA satellite of *Xenopus laevis*. Nature 219, 454–63.

Brown, D. D., P. C. Wensink, and E. Jordan (1971) Purification and some characteristics of 5S DNA from *Xenopus laevis*. Proc. Natl. Acad. Sci. USA 68, 3175–79.

Brown, D. D. (1973) The isolation of genes. Scientific American 229, 20–29.

Leder develops procedure to purify mRNAs with a polyT-DNA column

Aviv, H., and P. Leder (1972) Purification of biologically active globin messenger RNA by chromatography on oligothymidylic acid-cellulose. Proc. Natl. Acad. Sci. USA 69, 1408–12.

Synthesis of cDNA using polydT primer and reverse transcriptase

Ross, J., H. Aviv, E. Scolnick, and P. Leder (1972) *in vitro* synthesis of DNA complementary to purified rabbit globin mRNA. Proc. Natl. Acad. Sci. USA 69, 264–68.

Libraries of human, mouse, Drosophila, and yeast genomes made

Thomas, M., J. R. Cameron, and R. W. Davis (1974) Viable molecular hybrids of bacteriophage λ and eukaryotic DNA. Proc. Natl. Acad. Sci. USA 71, 4579–83.

Benton, W. D., and R. W. Davis (1977) Screening λgt recombinant clones by hybridization to single plaques *in situ*. Science 196, 180–82.

Lawn, R. M., E. F. Fritsch, R. C. Parker, G. Blake, and T. Maniatis (1978) The isolation and characterization of linked delta- and beta-globin genes from a cloned library of human DNA. Cell 15, 1157–74.

Maniatis, T., R. C. Hardison, E. Lacy, J. Lauer, C. O'Connell, D. Quon, G. K. Sim, and A. Efstratiadis (1978) The isolation of structural genes from libraries of eucaryotic DNA. Cell 15, 687–701.

Polymerase chain reaction developed

Mullis, K., F. Faloona, S. Scharf, R. Saiki, G. Horn, and H. Erlich (1986) Specific enzymatic amplification of DNA *in vitro*: The polymerase chain reaction. Cold Spring Harbor Symp. Quant. Biol. 51, pt 1, 263–73.

THE INDUSTRIAL AND MEDICAL RAMIFICATIONS OF GENETIC ENGINEERING
Restriction fragment length polymorphism

Botstein, D., R. L. White, M. Scolnick, and R. W. Davis (1980) Construction of a genetic linkage map in man using restriction fragment length polymorphisms. Am. J. Human Genet. 32, 314–31.

NAME INDEX

Takanami, M., 191
Takeda, Y., 159, 269
Tatum, E. L., 12, 45, 135
Taylor, J., 306, 308
Taylor, A. L. (Larry), 252
Temin, H. M.: and focus assay, 299, **299**; identifies retroviral *int* coding sequence, 313; personality of, 300; and provirus idea, 299–300; and retroviral oncogenes, 308, 312; and reverse transcription, 300–301, **301**
Tessman, I., 226
Thach, R. E. (Bob), 180
Thomas, R., 62, 115, 117, 263
Tilly, K., 280
Tissières, A., 27, 190, 274
Todaro, G. J., 309
Tomizawa, J., 97
Tompkins, G., 28
Torriani, A., 38, 60
Toussaint, A., 252, 255, 256
Traub, P., 194–195
Traut, R. R., 190
Travers, A. A., 126

Van Bogelen, R. A., 276
van de Putte, P., 252–253
Vande Woude, G. F., 313
Varmus, H. E., 303–308, 310, 312, 313

Vinograd, J., 331
Vogt, P., 302, 308, 310
Volkin, E., 57, 103–104
von Hippel, P., 162

Wacker, A., 283
Walker, G. C., 291–292, 293, **292**
Wallace, J. C., 212
Waller, J.-P., 173, 189–190
Wang, J. C., 91
Warner, R. C. (Bob), 226
Watson, J. D.: characterization of Crick, 13–14; and DNA structure, 6, 8, 65; as director of Cold Spring Harbor Labs, 204–205, 335; and identification of mRNA, 58; opposition to regulation of genetic engineering, 343; realization of base-pair equivalence, 7; scientific style, 173, 205; and translation, 173, 174, 176, 183, 189–190
Weigle, J., 43, 114, 222–223, 328, **223**
Weil, J., 235–236
Weisberg, R. A., 242
Weissbach, H., **184**
Weiss, S., 104
Weissman, C., 348
Welch, W. J. (Bill), 282
West, S. C., 230
Whitehouse, H. L. K., 223–224

Wickner, W. (Bill), 84
Wickner, R., 84
Wickner, S. H., 84, 99
Wilcox, K. W., 330
Wilkins, M. H. F., 7
Williams, R. C., 98–99
Witkin, E. M., 282–283, 287–289
Wittmann, H. G., 190, 195–196
Woese, C. R., 197–198
Wollman, E. L., 41–43, 45–47, 51, 54–55, 111
Wu, C., 281
Wu, R., 324
Wulff, D. L., 265

Yamamori, T., 276
Yanofsky, C., 20, 136–141, **137**, **139**
Yarmolinsky, M. B., 235–236
Yoshimori, R. (Bob), 335, 338
Young, R. A., 202
Yuan, R. (Bob), 329
Yura, T., 275–276, 279

Zachau, H. G., 170
Zamecnik, P., 14–15, 22–23, 27, 183, **22**
Zillig, W., 126, 127
Zinder, N. D., 171–172, 174, 176, 201
Zipser, D., 252
Zissler, J., 235–236
Zubay, G., 129–130

SUBJECT INDEX

References to figures and tables are given in **boldface.**

16S RNA, 190; base-pairing with mRNA (Shine-Dalgarno sequence) 192–194, 197, **193**, **194**; evolutionarily conserved structure, 198, **198**; maturation of, 202, **202**; sequence of, 197–198, **198**. *See also* ribosomal RNA, ribosome, protein synthesis

23S RNA, 190; maturation of, 202, **202**; peptidyl transferase activity, 198. *See also* protein synthesis, ribosomal RNA, ribosome

acridine yellow, 18, **18**

active site, 7–8

adenine, 6, **8**

adenovirus, 205; and splicing, 204, 209–210, **210**; temporal program for growth defined, 206, **207**

affinity chromatography, 143–144, **144**

AIDS, 314–317

alkaline phosphatase, regulation of, 60–61

allosteric regulation, 64

α-helix 4, **5**, **6**, 158–162

amber mutations, 109, 133, 176–178, **133, 177**

amino acids: activation of, 22–23, **24**; directionality of, 172; pathway to protein, 23, **22, 24** (*see also under* protein synthesis); in protein primary structure, 2, **3**; regulation of synthesis, 140–141; specified by triplet code, **15, 18, 32** (*see also under* genetic code)

aminoacyl tRNA synthetase, 23, **24**

antibiotic resistance: and plasmids, 338, 339–340; and transposable elements, 338, 259–260

antibodies, to snRNPs, 213–214

antibody diversity, 260

antibody mRNA, splicing in, 211

anti-sense RNA, regulation by, 265

antitermination: attenuation, 136–141, **137, 139**; mechanism of, 142–145; by N protein, 128, 134–135, 142–145, 271–272, **135, 272**; by Q protein, 142, 143, 145

ara operon (arabinose operon), 62–63, 164

ATP, 11, 183

attenuation, 136–141, **137, 139, 140**

autoradiography, 150, **150**

Bacillus megaterium, lysogeny of, 42

Bacillus subtilis: alternate sigma subunits in, 278–279; RNase P of, 218

bacteriophage. *See* phage

Bacteriophage Lambda, The (Hendrix, *et al.*), 110

base-pairing: in DNA structure, 7–8, **8, 9**; and genome fidelity, 168; RNA:RNA ("wobble"), 167–168, 193, **168**

β-galactosidase: assays for, 38, **38, 39**; constitutive expression of, 51–52; function of, 11; gratuitous inducers of, 39; reasons for study, 38; regulation of, 36, 39–40, 51–56, 63, 129–131, **131**. See also *lac* operon

β-galactosides. *See* lactose analogs

β-sheet, 4, 159, 162, **5, 6**

biochemistry, 13, 66

biotechnology, 348–349; Amgen, 348; Biogen, 348, 349; Cetus, 347, 348; Genentech, 348, 349; Genex, 348

biotin gene, 239–240

blender experiment, 46

branch migration, 207, 224, 226, 230, 231, **225, 229**

BRCA1 and BRCA2 proteins, 233

breast cancer, 233

cI gene, 44, 50–51; regulation of, 263–264, **263, 264**

cI repressor: cleavage of, 290–291; competition with Cro as a transcriptional switch, 269–271; cooperativity of, 156, 270; helix-turn-helix motif, 160; inhibits transcription, 125–126, 128–129, **129**; operator DNA binding, 125, 155–156, 160, 269–271, **125, 155**; operator sites, 116, 155–156, **116, 155**; purification of, 122–124, **124**; role of, 116, 134–135, 262, 268–271, **116**; in zygotic induction, 46

cII gene, 50–51

cII protein: is a transcriptional activator, 266; regulation of, 266–268, 280, **267, 269**; role in lysis-lysogeny decision, 262–266, 271, 268–269, **263, 264, 265, 269**

cIII gene, 50–51

cIII protein: regulates cII stability, 266–268; required for estabishment of lysogeny, 262–265; stabilizes σ^{32}, 280

calcium, bacterial transformation and, 338–339

cAMP, 130–131

cAMP receptor protein (Crp or CAP), 63, 130–132, **131**, 159–160

CAP. *See* cAMP receptor protein

catabolite repression (glucose effect), 62–63, 129–132. *See also* cAMP receptor protein cDNA clones, 345–346

cell division, blockage of, 286

cell metabolism, 11–12

Central Dogma, 34, 296, 300

chaperones, 273–274, 277, 281–282. *See also* heat shock proteins

chromatin, 164–165

chromatography, column, 169

chromosome, bacterial, 45–48, 71–74, **45, 48**

chromosome, replication of. *See* DNA replication

chromosome puffs, *Drosophila,* 274

cis-dominance, 55, **55**

clear-plaque mutations. *See* λ mutants: *c* mutations

cloning, molecular: concept of, 319–20, **320**; of eukaryotic genes, 274, 344–347, **346**; phage variants as precursor of, 107; vectors for, 320

coding problem, the, 13, 15–16. *See also* genetic code

codon, as coding unit, 15. *See also* genetic code

cohesive sites: created by "A–T tailing," 326–327, 336, **326, 334**; created by restriction enzymes, 337–338, **338,** 345–346; in λ, 233, 322–324, **322, 325**; and molecular cloning, 325–327, 333, 335–338, 345–346, **326, 334**

colicins, 194

colinearity hypothesis, 19–21, **19**

complementarity, in DNA structure, 7–8, 65

complementation assay, 85–86, **86**

conjugation, bacterial, 45–48, 52, 226, **45, 47, 48**

consensus sequence, 193

cooperative binding: and cI repressor, 156, 270; and Ssb, 90

co-repressor, 60

counter-current distribution, 169

Cre recombinase, 246

Cro protein: competition with cI, 269–271; DNA recognition, 160, **161**; essential for lytic response, 269; operator binding, **155,** 156, 269–271; purification of, 159; structure of, 158–160, **159, 161**. *See also* λ lysis-lysogeny decision; λ lysogeny; λ lytic response

cross-linking, chemical 196

Crp. *See* cAMP receptor protein

cya gene, 130

cytosine, 6, **8**

data, discrepancies in, 54

degeneracy, in genetic code, 15, 32

deletion, by site-specific recombination, 245–247

density gradient centrifugation, 70, 222–223, **223**

density labeling, 57–58, 70

developmental processes, 49, 108–110, 261–262

diploidy, in *E. coli,* 45, 47, 52, 55, **55**

DMS (dimethyl sulfoxide), 150–153, **153**

DNA: base composition, 7, 24, 67; carries genetic information, 2, 12–13, 102

dnaA gene, 75, 96–97

DnaA protein, 100

dnaB gene, 75; mutations in, 97, **96, 97**

DnaB protein, 88, **89**; is *E. coli* helicase, 94, 96, 99–100; and λ replication, 94, 96–97, 99, **96, 100**; at origin of replication, 96–97, 99–100, **95, 100**

dnaC gene, 75, 84, 97

DnaC protein, 88, 100, **89**

DNA damage, 88, 282–283, 285; repair of (*see* DNA repair); response to (*see* SOS response)

dnaE gene, 75, 84; encodes DNA polymerase III subunit, 87–88

DNA end-labeling, 149–152, **151, 153**

dnaG gene, 75, 84

DnaG protein: is a primase, 88, **88**; and replication, 88, 94, **88, 89, 95**

DNA gyrase, 241

dnaJ mutants, 276. *See also* DnaJ protein

DNA joining: cohesive sites and, 322–327; publications regarding, 336; and restriction enzymes, 336–338; using ligase, 326, 336; using terminal transferase ("A–T tailing"), 333, 335–336, 339, **334**

DnaJ protein: function of, 277; and heat shock, 277, 280; and replication initiation, 99–101, **100**. *See also dnaJ* mutants

dnaK mutants, 276, 280

DnaK protein: function of, 277; and heat shock, 277, 280; and replication initiation, 99–101, **100**

DNA ligase: discovery of, 78; and genetic engineering, 78, 336, 338; in lagging strand synthesis, 81, **81**

DNA looping, and gene regulation, 163–164, **163**

DNA methylation, 329

dnaN gene, 75, 88

DNA polymerase I: editing function, 82; is essential, 82; direction of replication, 73; function of, 76; genetics, 75–77; problems with, 73, 75; purification and properties of, 68–69, 77 **69**; and repair DNA synthesis, 284; requirements, 81; role in progress of field

DNA polymerase II, 81

DNA polymerase III, 77, 81–82, **82,** 294; holoenzyme, 87–88

DNA polymerases, 323–324, 326, 336; direction of replication, 80; and double-strand DNA, 90; editing function, 82; predictions for "real" polymerase, 77; reverse transcriptase, 301; role of, 66–67; thermostable, 347; UmuC/UmuD', 294

dnaQ gene, 88

DNA repair, 282–283; and breast cancer 233; and DNA polymerase I, 76; by error-prone replication, 287; excision, 283–284, **284**; recombinational, 231–233, 284–285, **286**; Uvr pathway, 283–285, 292, **284**. *See also* SOS response

DNA replication, 65, 68, **69**; asymmetric, 83, 93; bidirectional, 73; biochemical approach to, 66–68; chain elongation during, 86–87, 92–93; discontinuous, 77–81, 82, **77, 78, 95**; and DNA damage, 88, **89**; fine structure of, 77–81; genetic analysis of, 74–77, 81, 96–97, **75**; genomic, visualized (Cairns' experiments), 69–73, **71, 72**; helicase requirement of, 91, 93, 94, 99–100, **91, 93, 95**; and induced mutagenesis, 294; initiation of, 75, 95–101, **95, 100** (*see also* origin of replication); interplay between biochemical and cellular approaches to, 69; *in vitro,* 68, 82–89, 92–95, 97–98; key questions regarding, 69, 74, 77; lagging strand, 80, 94–95, **81**; of λ phage, 73, 94, 96–99, 101, 115, **100, 115** (*see also* λ replication); leading strand, 80, 94–95, **81**; nucleoprotein complexes in, 96, 98–101, **100**; *oriC*-dependent, 97–98, 100; priming of, 68, 79–80, 86–88, 94, 96, **80, 81, 95**; processivity of, 87, 89; protein requirements of, 68, 79, 80–101; purification of components, 84–86, **85**; replication forks, 71–73, 90, **71, 72, 73**; restart of, after recombinational repair, 285; role of DNA ligase in, 81, **81**; role of Ssb in, 87–90, 93, 99–100, **90**; rolling circle, 93, **92**; semiconservative, 69–70; single-strand

DNA replication (continued)
DNA phage RF to SS replication, 79–80, 93–94, **80, 92, 93, 94**; single-strand DNA phage SS to RF (SS to DS) replication, 83, 86–88, 92–93, **87, 88, 89**; speed of, 95; supertwists and, 91, **91**; symmetric, 71, 73, 77, 81, **71, 73**; T4 and T7 phage, 94; template for, 68–70, **69**; topoisomerase requirement of, 91, 93, **91**; two views of, 66–67; unwinding during, 67, 90–92, **91, 93, 94**. *See also under names of individual genes and gene products.*

DNA Replication (Arthur Kornberg), 66
Dnase, DNA replication assays and, 67
DNase I footprinting, 146, 148–153, **146, 153**; and cI and Cro operator binding, 155, 269; and cII binding, 266; and RNA polymerase binding, 156
DNA sequence: determination of, 146, 148–154, **151, 154**; motifs in, 160
DNA sites, probability of uniqueness, 147–148, **147**
DNA structure, 6–8, **7, 9**; altered by transcriptional activators, 163, **163**; implications of, 8, 15, 65, 66; and protein binding, 148, 162; stability of duplex, 90
DNA supercoiling, 91, 163, 246
DNA synthesis. *See* DNA replication
DNA synthesis, repair, 284
dnats mutations, 74–75, 81, 85, 96–97, **75, 86**
dnaX gene, 75, 88
dnaZ gene 75, 88
double helix, 7–8, **9**
Double Helix, The (James Watson), 2, 6, 205
Drosophila, heat shock response in, 274

E. coli, as model system, 9
E. coli, genetics of, 45–48, 52, 226, **45, 47, 48**
EDTA, 239–240, **240**
EF-G, 179, 184–185, **180, 184**
EF-Ts, 179, 184, **180, 184**
EF-Tu, 179, 185–186, **180, 185**
Eighth Day of Creation, The (Horace Judd), 2
electron microscopy, 96; and heteroduplex analysis, 250; and Holliday intermediates, 226, **226**; and λ integration, 243–244; in mapping transcription units, 215; and nucleoprotein structures, 98–100, 164, 243–244, **165, 243**; and phage structure, 49; provides evidence for splicing, 207–208, 209–210; and replication initiation, 98–100; and ribosomal structure, 196; to visualize DNA duplexes, 206

electrophoresis, gel, 189, 228; separation of DNA molecules by size, 149–150, **149, 150**; two-dimensional, 275, **275**
endolysin, 143, **143**
environmental signals, bacterial response to, 60, 102, 261, 273
enzymatic adaptation, 36, 39–40; renamed induced enzyme synthesis, 40
enzymes: in cell metabolism, 12; induction, 36; regulation of activity, 63–64
equilibrium dialysis, 121
erotic induction (zygotic induction), 45–46, 51, **47**
Escherichia coli. See E. coli
Escherichia coli and Salmonella: Cellular and Molecular Biology, 62
eukaryotes: cellular organization of, 203; gene cloning of, 344–347; gene expression in, 203, 344; gene regulation in, 162, 164–165
excision repair, error-free, 287
exons, 211–212, **212**
exonucleases: in DNA replication, 80, 82; RecBCD, 231; and study of λ cohesive sites, 323–324, **325**

f2 phage, 171–172
feedback inhibition, 63–64
F factor, 45, 47–48, **45, 48**
filter binding assay, 106, **107**
fingerprint, to characterize proteins, 19
F'*lac*, 55
flagellar genes, *Salmonella*, 245–246
Flp recombinase, 246
fmet tRNA_f, 173–175, 183, 185–186, **175**
focus assay, tumor virus, 299, **299**
foreign DNA, mechanisms limiting acquisition of by bacterial cells, 328, 340
formyl methionine, 173–175, **174**

G4 phage, 83, 86, 88, 92, **88**
gal operon (galactose operon), 107, 164, **107**
gal3 mutation, 249–250, **250**
genes: are carried on DNA, 2, 12–13, 102; definition of, 11; "one gene-one enzyme" hypothesis, 12
gene conversion, 223–224
gene duplication. *See* DNA replication
gene expression: assayed by gene fusion, 291, **292**; definition of, 102; in eukaryotes, 203, 344; regulation of (*see* gene regulation); temporal control of, 49, 102; universality of mechanisms, 10

gene fusions, 291, **292**
gene regulation: DNA looping, 163–164, **165**; eukaryotic, 162, 164–165; fundamental properties of, 102; glucose effect, 62; master regulators, 261–262; multi-operon, 62–63, 279; negative, 60–62, **61** (*see also* operon model; repressor); operon model for (*see* operon model); positive (*see* gene regulation, positive); primary and secondary switches, 268–269, 272; retroregulation, 271–272; salvational decisions, 273. *See also* regulatory mechanisms, diversity of
gene regulation, positive, 61; and alkaline phosphatase, 60–62: and *ara* operon, 62; and *lac* operon, 61, 63, 129–131, **131**; by cII protein, 263–266; genetic predictions for, 61–62, **61**; and glucose effect, 62–63, 129–132; and heat shock response, 275; in λ, 61, 62, 113–114, **114**, 115–117, 263–266; in *mal* operon, 63; Monod's opinion on, 60–62; and pleiotropic negative phenotypes, 61–62. *See also* transcriptional activators
genes, regulatory, definition of, 35
genetic code, 32–34, **32**; biochemical approach to solution, 21, 26–31, **28, 31**; "the coding problem", 13, 15–16; colinearity hypothesis and, 19–21, **19**; degeneracy of, 15, 32; information theory approach to solution, 13–21; and "one gene-one enzyme" hypothesis, 12; sequence hypothesis and, 15–17; solution depending on fusion of different experimental philosophies, 11, 13, 21; triplet nature of, 15, 17–19, **15, 18**
genetic diseases, transposons and, 260
genetic diversity, sources of, 220
genetic engineering, 319–321, **320**; amplification of genes by, 320, 327, 333, 340, 346–347; and cohesive sites, 322–327; engineered joining of DNA molecules, 333–338; and eukaryotic molecular biology, 344–347, **346**; industrial and medical ramifications of, 347–349; intellectual and experimental roots of, 321–333, 347; NIH guidelines for recombinant DNA, 342; polymerase chain reaction (PCR) and, 347, **347**; preparation of DNA fragments for, by shearing, 327, 338–339; safety concerns and regulatory policy, 340–344; site-specific recombinases and, 246–247; terminal transferase and,

lacY gene, 38–39, 52, 55–56. See also *lac* operon

lacZ gene, 11, 40, 52–53, 55–56, 291. See also *lac* operon

λ: chromosome of, 111, 233, **234**; cohesive (*cos*) sites, 322–324, **322**, **323**; discovery of, 42; helper phage and, 322, **323**, 327, 328, 339; as a model system, 41, 261–262; phage with higher DNA content killed by EDTA, 239; phage-specific general recombination, 236; as a temperate phage, 42, **43**; as a vector, 345. *See also under individual gene and protein names; see also* λ*gal* phage; λ integration and excision; λ lysis-lysogeny decision; λ lysogeny; λ lytic development; λ mutants; λ prophage induction; λ replication

Lambda II (Hendrix, *et al.*), 110

λ*dvgal* plasmid, 335–336

λ*gal* phage: as first example of molecular cloning, 321; in measuring gal mRNA , 106–107, **107**; source of λ early gene DNA, 114

λ integration and excision: attachment (*att*) sites, 233, 238, 242–243, **234**; Campbell model for, 111–112, 233–234, 252, 321–322, **112**; characterization of, 233–245, **234**, **237**, **238**; demonstration of, 234–237, **237**; proposed, 111–113, **112**; DNA-protein interactions in, 242–245; DNA supercoiling required for, 241; Int protein and, 233, 235–237, 238, 243–245, 271–272; *in vitro*, 239–242, **240**, **241**; nucleoprotein complexes in, 243–245, **243**; prophage excision, 237–238, 242–245, 321, **238**; protein requirements for, 233, 242, 271–272, **238**, **271**; regulation of, 271–272, **271**; Xis protein and, 238, 243–245, 271–272. *See also* recombination, site-specific

λ lysis-lysogeny decision, 43, 156, 262–272, **43**, **269**; transcription switch between cI and Cro, 269–271, **270**; ratio of Int to Xis, 271–272, **271**, **272**. *See also* λ lysogeny; λ lytic development

λ lysogeny, 42–45; regulation of, 45–46, 49–51, 53–55, 262, 265. *See also* λ lysis-lysogeny decision; lysogeny

λ lytic development, 36, 51, 113–117, **114**, **115**, **116**; regulation by antitermination, 134–135, 141–145. *See also* λ lysis-lysogeny decision

λ mutants: *b2* deletion, 235; *c* mutations, 44, 49–51, **50**; *c*IInd⁻, 123–124; *cy*⁻ mutant, 263; *hd* mutants, 109; isolated by Weigle, 43–44; plaque morphology mutants, 44; *red*⁻ mutants, 236; temperature-sensitive mutations, 109; virulent mutants, 55, 116. *See also N* gene mutations; *O* gene: mutations in; *P* gene gene: mutations in; and *Q* gene mutations

λ prophage induction, 36, 42–43; after mating, 45–46, **47**; cleavage of cI repressor, 290; by DNA damage, 286–287

λ replication, 73, 94, 115, **115**; initiation of, 96–99, 101, **100**; interaction of P and DnaB proteins during, 97, **96**, **97**, **100**; nucleoprotein structure and, 99, 101, **100**

λ repressor. *See* cI repressor

lambdology, as a field of study and as a community, 110–111, 113, 117–120

"Lambdology, The New", 119–120

lexA mutations, phenotype of, 288–290

LexA protein: binds DNA, 291; cleavage of, 290–291; is negative regulator of SOS response, 289–290

libraries, λ, 345–347, **346**

liquid holding recovery, 283

lysogeny, 40–42; in *Bacillus megaterium*, 42; in λ (*see* λ lysogeny)

M13 phage, replication of: primer requires RNA polymerase, 79–80, 86–87, **80**; SS to RF (SS to DS) replication, 79–80, 83, 86–87, 92, **80**, **87**

magnesium, and ribosome dissociation, 181

mal operon (maltose operon), 63

Meselson-Stahl experiment, 69–70, **70**

messenger RNA. *See* mRNA

metabolism, regulation of, 63

molecular biology: deceptive simplicity of, 26; fragility of many contemporary concepts in, 189; nature of, 68; principles of, 1, 13, 26, 58–60; progress in, 59–60, 76, 97; social context for basic research in, 318

mRNA: detection of, 108; determinants of ribosome binding, 192; identification of, 57–59; is the informational intermediate, 25–26, 57–59, 102–103; initiator sequences bound by ribosome, 191–194, **191**; instability, 25, 108, 199; properties of, 105–108; proposed, 25–26, 57; role in protein synthesis, 25–26. *See also* protein synthesis; transcription

mRNA, eukaryotic, 203, 204, 210–211, 345

Mu phage, 221, 291; characterization of, 251–254; MuA protein, 255–256, 258; MuB protein, 255–256; and retroviruses, 313–314; transposition of, 254–258, **255**, **259**

multi-drug resistance elements, 259–260

mutagenesis, 76; induced after DNA damage, 282, 286–289, 293–295, **294**; UmuCD pathway, 293–295, **294**; Weigle, 43, 288–289

mutagenic agents, and *E. coli*, 76

mutations: *amber*, 109, 133, 176–178, **133**, **177**; caused by IS elements, 249–250; caused by Mu phage, 252; *cis*-dominant, 55, **55**; complementation test for, 50, **50**; conditional-lethal, 74, 108–109; constitutive, 62, 132–133; pleiotropic negative phenotypes of, 61–62; screens for, 75–77; selections for, 74; suppression analysis of, 97, **96**, **97**; use of, to understand genes and processes, 12, 74. *See also* mutagenesis; nonsense mutations; polarity

negative regulation, 60–62, **61**. *See also* operon model; repressor

Neurospora crassa, 12

neutron scattering, 196

N gene mutations, 113–115, **114**, **115**. *See also* N protein

NIH peer-reviewed grant system, value of, 318

nitrophenyl galactoside (NPG), 38, **39**

nonsense codons, 32, 176–178, 186–187, **178**, **188**. *See also* nonsense mutations

nonsense mutations: create stop codons, 176–177, **177**; polar effect on downstream genes, 133, 141–142, **133**; show repressor is a protein, 59; suppression of, 109, 176–178, **177**

N protein: is an antiterminator, 128, 135, 142, **135**; does not activate transcription initiation, 134, **134**; and Int regulation, 271–272, **272**; multi-protein complex with, 143–144; positive regulator of early gene transcription, 115, **115**; purification of, 143, **143**. *See also* antitermination; λ lytic development; *N* gene mutants

NtrC activator, 164, **165**

nucleoprotein structures: advantages of, 101, 244; excitement of discovery, 244; implications of, 244; at replication origins, 96, 100–101, **95**, **100**; in site-specific recombination, 246; and transcriptional activation, 163–164

nucleotides, 6–7

nucleus, eukaryotic, 203

nus genes (*nusA, nusB, nusE*), 142–143

nus proteins, 143–145, **144**; NusA, 144–145; NusB, 144; NusE, 144; NusG, 144

nut site, 144–145

O gene, 97, 113, 115; mutations in, 113–115, **115**. *See also* O protein

oligonucleotides, separation of, 169

oncogene: analogy to λ*gal* transducing phage, 302–303, 310, 312; have homology to cellular genes, 310–312; mechanism of capture of, by retrovirus, 312; Temin as early supporter of, 308

oncogenesis, models for, 308–309, 312, **309**

one gene-one enzyme hypothesis, 12

operators: λ, 55, 116, 155–156, **116**, **155**; *lac*, 55, 146–148, **146**, **147**; properties of mutations in, 54–55

operon model: development of, 35–36, 53–56, 59–60, 116; influence of, 132–133, 135, 321; and molecular mechanisms, 56; universality of, 59

operons, **103**; as biochemical reality, 107–108; polarity in, 133, **133**; regulation by site-specific DNA binding proteins, 132. *See also ara* operon; *gal* operon; *his* operon; *lac* operon; operon model; *trp* operon.

O protein, 97–99, **100**

ori DNA. *See* origin of replication

origin of replication: chromosomal (*oriC*), 72–74, 95–98, 100–101; in genetic engineering, 320; of λ, 96–99, 101, **100**; nucleoprotein structure at, 96, 99–101, **95**, **100**

origin of transfer, Hfr, 46–48, **48**

Origins of Molecular Biology: A Tribute to Jacques Monod (Ullmann, A. and A. Lwoff, eds), 37

oriT (origin of transfer, F factor), 46–48, **48**

O-some, 99, **100**

ovalbumin RNA, splicing in, 211

P22 phage, 51, 333

PaJaMa experiments, 52–53, **52**

Pauling Principles, 4, 6

P gene, 97, 113, 115; mutations in, 97, 113–115, **115**. *See also* P protein

phage: developmental pathways of, 108–110; infection by, 49, **49**; lytic development of, 40, 49, 109–110, **110**; as model system, 8–10, 49, 108, 261–262; plaque assay for, 44, **44**; properties of, 8–9; restriction of growth, 328–329; structure of, 49, **49**; temperate, 41–42, 44, 49, 51, 235, **43**, **44**; virulent, 44, 57–58, **58**. *See also under names of individual phage; see also* phage, double-strand DNA; phage, RNA; phage, single-strand DNA

phage, double-strand DNA, 80

phage, RNA: isolation from sewers, 171–172; in protein synthesis studies, 174, 176, 191–194

phage, single-strand DNA: as model for genome replication, 83–84, 92–94; asymmetric replication of, 83, 93, **92**; RF to SS replication, 93–94, **80**, **92**, **93**, **94**; SS to RF (SS to DS) replication, 79–80, 83–84 86–88, 92–93, **83**, **85**, **86**, **87**, **88**, **89**. *See also* φX174 phage, M13 phage, G4 phage

Phage and the Origins of Molecular Biology (Cairns *et al.*), 42

phage group, founding of, 8

phenotype, 2

phenylalanine codon, 28

φ80*lac* phage, 106–107, 321

φX174 phage: A protein, 93–94, **93**, **94**; evidence for Holliday intermediates, 226, **226**; replication mechanism, 83; Rep protein, 93, **93**, **94**; RF to SS replication, 93, **92**, **93**, **94**; SS to RF replication, 83–84, 86, 88, **89**

photoreactivation, 283

plaque assay, for bacteriophage, 44, 297, **44**

plaque assay, for animal viruses, 297, 332, **297**

plasmids: and antibiotic resistance, 338, 339–340; as cloning vectors, 320, 339–340, **339**; pSC101, 339–340, **339**; R-factor, 335, 339; study of structure, 339

Pol I. *See* DNA polymerase I

Pol II. *See* DNA polymerase II

Pol III. *See* DNA polymerase III

polA gene, 76

polA mutant, 75–77, 81, 82

polarity, 133, 141–142, 249–250, **133**

polyA tail, 204, 207, 345, **205**, **208**

polyethylene glycol, 98

polymerase chain reaction (PCR), 347, **347**

polynucleotide chain, **7**

polynucleotide phosphorylase (PNPase), 29, 31, 67, 103

polypeptide chain, **3**. *See also* proteins

polyuridine, 28

positive regulation. *See* gene regulation, positive

positive regulators. *See* transcriptional activators

P protein, 97–99, **96**, **97**, **100**. *See also P* gene

promoters: features of, 156–158, 278, **157**, **278**; for heat shock genes, 278, **278**; identification of mutations in, 56; predictions for, 55

prophage: defined, 41; induction of, 41–42, 46, **47** (*see also* λ prophage induction); integration of, 235 (*see also* λ integration and excision)

protein-DNA interactions, 146–153, 155–162, **147**, **153**; helix-turn-helix motif 160–162, **161**. *See also* DNase I footprinting

protein-protein interactions, use of genetics to study, 97, **96**, **97**

protein purification: complementation assay, 85–86, **86**; fractionation, **85**

proteins: activity of, 3–4, 11, 12; folding of, 4–5, 281–282, **6**; specificity of, 3–4; structure of, 2, 4–5, 172, **3**, **6**

protein synthesis: acid-insolubility as assay for, 14; biochemical pathway for, 21–26, **25**; catalytic RNA in, 218; directionality of, 172–173; fmet tRNA$_f$ and, 173–175, 183, 185–186; initiation of, 33, 172–175, 181–182, 186, **33**, **182**, **187**; involvement in transcriptional regulation, 132–133; key questions regarding, 166, 171; mechanics of, 178–188; peptide bond formation, 185, 190, **183**; punctuation, 172, 178; reading frames, 33, **33**; RNA phage as experimental template, 171–172; role of GTP, 182–186, **184**, **185**, **187**; role of tRNA, 23–24, 167–168; selection of initiator AUG, 175, 192–194, **175**, **193**, **194**; steps in amino acid addition, 182–188, **183**; termination, 175–178, 186–187, **188**; in test tube, 14, 21–23, 27–28, 174, 176–177, 179–188; role of translation factors, 178–180, 183–188 (*see also:* translation elongation factors; translation initiation factors; translation release factors); translocation, 183, 185–186, **183**, **185**. *See also* ribosome proteolysis: by Hfl proteins, 268, 280; mediated by RecA, 228, 290–291, **291**; in regulation of cII, 266–268, **267**; and σ³², 279–280

provirus, 299–302; existence verified, 304–305; inactive copies in genome, 310, **311**; LTR sequences of, 305–306

p site. *See* promoters; *lac* operon

pulse label experiments, and DNA replication, 70–71, **71**

pulse-chase experiments: and discontinuous replication, **78**; and study of cII stability, 267–268, **267**

purines, 6

pyrimidine dimer, repair of, 283–285, **284, 286**

pyrimidines, 6

Q gene mutations, 113–115, **114, 115**

Q protein: as an antiterminator, 142–143, 145; is positive regulator of λ late gene RNA, 113–117, **114, 115**; purification of, 145. *See also* antitermination; λ lytic development

rII mutations, 17–19

rad51 mutants, 232

Rad51 protein, 232

radiation damage, 282–283

reading frame, 33, 175, **33, 175**

RecA-like proteins, 232–233

recA mutants, 226; phenotypes of, 227, 284–285, 288–289

RecA protein: activation of, 290, 293; and DNA repair, 227, 232, 284–285, **286**; and induced mutagenesis, 293–294, **294**; mediates cleavage of cI repressor, 228, 290–291; is positive regulator of SOS response, 286–290, **290**; and recombination, 226–231, **229**; stimulates self-cleavage of LexA and cI, 290–291, **291**. *See also* recombination; SOS response

RecBCD enzyme, 227, 230–231, **232**

RecBC pathway, 227, 230–231, **227**

recB gene, 227

recC gene, 227

recD gene, 227

recombinant DNA. *See* genetic engineering

recombination, 220–221, 232–233; eukaryotic, 232–233 ; extraordinary (illegitimate), 221; phage-specific, 236. *See also* recombination, general; recombination, site-specific

recombination, general, 220; branch migration in, 207, 224, 226, 230, **225, 229**; branch points in, 224, 225–226, **225**; Chi sites in, 231; evidence for breakage and rejoining during, 222–223, **223**; genes and enzymes involved in, 226–227, 230, **227**; heterozygosis and, 221–222, 224, **222**; Holliday model for, 221–226, **225, 226**; negative interference and, 222, 224, **222**; other pathways for, 231–233, 236; RecBC pathway, 227, 230–231, **227**; RecF pathway, 227, **227**; role of RecA, 226–231, **229**; role of Ruv proteins, 230.

recombination, site-specific, 220–221; Campbell model for λ integration, 111–112, 233–234, 252, 321–322, **112**; deletion and, 245–247, **246**; demonstration of, 234–237, **237**; DNA-protein interactions in λ integration, 242–245; DNA supercoiling required, 241; examples, 245–246; *in vitro*, 239–242, **240, 241**; Int protein, 233, 235–237, 238, 243–245, 271–272; inversion and, 245–247, **246**; λ attachment *(att)* sites, 233, 238, 242–243, **234**; λ integration, 111–112, 233–245, **112, 234, 237, 271**; λ prophage excision, 237–238, 242–245, 321, **238**; nucleoprotein complexes in, 243–245, **243**; practical applications, 246–247; protein requirements for λ integration/excision, 233, 242, 271–272, **238, 271**; regulation of, 271–272; Xis protein, 238, 243–245, 271–272

regulatory mechanisms, diversity of, 63–64, 162–165, **163, 165**. *See also* gene regulation; gene regulation, positive

regulatory proteins, DNA binding, 131–132, 146–148, 158–162, **147**; helix-turn-helix motif, 160–162, **161**; models and mechanisms, 162–165, **163, 165**

Rep protein, 93, **93, 94**

repressor: models for function, 53–55; Lac (*see* Lac repressor); λ (*see* cI repressor) nature of, 59

restriction and modification, 328–329, **328**

restriction enzymes, 153, 206, 305, 348; cleavage sites of, 330, 331–332, 335, 337–338, **330, 338**; *Eco*RI, 335, 337–338, **338**; as foundation of biotechnology, 347–348; genetic map of SV40 generated using, 331–333, **332**; *Hind*II, 331, 338; *Hind*III, 331; initial biochemical characterization, 329; nature of, 148–149; Type

I, 331; Type II, 331–332, 335; utility of, 148–149, 327, 330, 333, 349, **149**

restriction fragment length polymorphism (RFLP), 349

"retroregulation", 271–272, **272**

retrotransposons, 313

retroviruses: avian leukosis virus, 302; cancer causation in absence of oncogene, 312; HIV-1, 314–317; existence of provirus, 300, 301, 304–305; *int* coding sequence, 313; integration of, 303, 312–314, **314**; lentiviruses, 314; life cycle of, 302–308, **303**; long terminal repeats (LTRs), 303, 306, 308, **303, 307**; oncogenes, 308–312; package RNA, not DNA, into virus particles, 299–300; *pol* gene, 313, **302**; proteins encoded by, 302–303, **302**, 314–315; proviral structure, 303; quantitative assay for, 298–299; RNA genome of, 302–303, **302**; as transposons, 247, 312–314, **314**

reverse transcriptase, 345; demonstration of, 300–301, **301**; endonuclease activity associated with, 313; high error rate, 315; priming, 306, 308, **307**; RNaseH activity, 306, 308, **307**

Rho protein, 128, 145

ribosomal proteins, 189–190, 194–198

ribosomal RNA, 108; active role in protein synthesis, 185, 190, 197–198; of African toad, 340, 344; as informational intermediate, 24, 57, 105, 189; maturation of, 200–202, **202**; in mRNA binding, 192–194; operons for, 202; *Tetrahymena* self-splicing, 215–217, **215**

ribosome, 21–22, 102; 30S subunit of, 181–182, 186, 190, 192, 194–195, 196, **181, 182, 190, 197**; 50S subunit of, 181–182, 185, 186, 190, 195–197, **181, 182, 190, 197**; 70S, 181–182, **181, 182, 190, 197**; A site, 183–186, 196, **183**; changing concepts of, 189–190, 194–198; composition and structure of, 180–182, 189–190, 194–196, **190, 197**; dissociation, 180–182, **181, 182**; E site, 183, **183**; P site, 183, 185–186, **183**, 196; peptidyl transferase activity, 185, 187, 195–196, 198; stability of, 166; and translocation, 183, **183**, 185–186, **185**

ribozymes, 216; 218–219

rifampicin, 80, **80**

R-loop mapping: as evidence for splicing, 207, **208**; of intron in *Tetrahymena* rDNA, 215–216, **215**

RNA: base composition of, 21, 67; pools in cell, 108; role in protein synthesis, 21–26; secondary structure in, 138–141, 202–203, **139**; sequence determination, 146, 168–169

RNA, catalytic, 198, 200, 211, 215–219; RNaseP 200, 217–218; self-splicing 211, 215–217; 23S RNA 198

RNA polymerase, 30; activation by cII protein, 266; activity of, 104, **30**, **105**; and antitermination, 142–145; biochemistry of, 126–129; discovery of, 103–104; elongation by, 126–127, 142, 145; initiation of transcription, 126–127, 147, 156, **127**; long distance communication with regulatory proteins, 162–165; promoter recognition, 126–127, 156–158; role in DNA replication, 79–80, 86–87, **80**; sigma subunits, 126, 278–279, **127**, **278**; structure of, 126–127; transcript termination, 128. *See also* RNA synthesis; transcription

RNA polymerases, eukaryotic, 157–158

RNA processing, 199–202, 217–218

RNase III, 200–202, **202**

RNase P, *B. subtilis,* 218

RNase P, 200, **201**; 217–218

RNaseH, 306, **307**, 308

RNA slicing. *See* RNA processing

RNA synthesis, *in vitro,* 28–29. *See also* transcription

RNA tumor viruses, renamed retroviruses, 302

Rous Sarcoma Virus (RSV), 298–312, **301**, **302**, **311**; life cycle of, 302–308; mutants of, 302, 308–309; proteins encoded by, 302–303, **302**; quantitative assay for, 298–299; and reverse transcriptase, 300–301, **301**; RNA genome of, 302–303, **302**; *src* gene of, 302, **302**, 308–312, **311**

rpoH gene, 278. See also *htpR (hin)* gene

RuvA protein, 230

RuvB protein, 230

S-100 extract, 179, **180**

S-30 extract, 179, **180**

salivary gland chromosomes, *Drosophila*, 274

salvational responses, 261, 273, 289, 294. *See also* heat shock response; SOS response

science: advances in, 76; basic research and societal benefits, 319; communities in, 117–120; competition in, 118; different approaches to, 11, 13; development of models, 54, 59–60; discrepancies in data, 54; joys of, 74, 76–77; nature of discovery in, 244; social context for basic research, 318; and "the 60's," 117–118; universality of fundamental genetic mechanisms as a principle, 10. *See also* biochemistry; genetics; molecular biology

scientific conferences, 60–61, 117–120, 197, 342

selfish DNA, 258

sequence hypothesis, 15–17

serine, 167

sex, bacterial, 45–48, 52, 226, **45**, **47**, **48**

Shapiro model, for transposition, 254, 258, **255**

Shine-Dalgarno hypothesis, 192–194, **193**, **194**

sib site, 272, **272**

sigma factors, 126, 278–279, **127**, **278**; σ^32, 279–280

simple organisms as model systems, 8–10

single-strand binding protein. *See* Ssb

society, basic research and, 318–319

SOS response: and induced mutagenesis, 288–289, 292, 293–295, **294**; and prophage induction, 286–290, **287**, **288**, **289**, **290**; recombinational repair and, 284–285, **286**; regulation of, 286–293, **290**; role of LexA, 288–292, **290**; role of RecA, 284–294, **286**, **290**, **291**, **294**; signal for, 293; SOS-inducible genes, 291–292, **292**; a unitary response to UV, 286–290, **287**, **288**, **289**, **290**; Uvr pathway, **284**

spectrophotometer, 38

splicing, **205**; biochemistry of, 211–214, **212**; of eukaryotic tRNAs, 214; evidence for, 203–204, 206–211, **208**, **210**; involvement of catalytic RNA in spliceosome reaction, 214, 218; self-splicing reactions, 211, 215–217, **217**; snRNPs, 213–214; spliceosome, 211, 213–214, 218; universality of, 210–211

sRNA. *See* tRNA

Ssb, 89–90; DNA binding, 90, **90**; *E. coli,* 90; gene 32 protein, 89–90; and recombination, 230, 231, **229**; role in replication (general), 87–90, 93, 99–100, **90**, **91**, **95**; role in single-strand DNA phage replication, **87**, **88**, **89**, **93**, **94**

stable RNAs: processing of, 199–202; rRNA (*see* rRNA); tRNA (*see* tRNA)

Staphylococcus aureus DNA, 340

stop codons, 32, 176–178, 186–187, **178**, **188**. *See also* nonsense mutations

subunit equivalence, 4, 6

suicide approach, and *dnats* mutations, 74–75

sulA and *sulB* genes, 292

SV40 virus: DNA of, 331; and genetic engineering, 335–337; genetic map of, 331–333, **332**; replication of, 101, 333; as a research focus at Cold Spring Harbor, 205; splicing demonstrated in, 210; temporal program for growth defined, 206; transcript mapping, 333

T2 phage, 57, 104–105, 221

T4 phage: gene 32 protein, 89–90; genome of, 72; lytic development of, 109–110, **110**; and nonsense mutations, 176; replication, 90, 94; and study of mRNA, 58, 105

T7 phage, 94, 201–202

T-cells, 315, 316

telomere formation, 218

terminal transferase, 326–327, 336, **326**, **334**

Tetrahymena thermophila, 215–217, **215**

thymidine, 67

thymidine, radioactive: and *dnats* mutants, 75, **75**; in study of DNA replication, 67, 70–71, 77–79, **71**, **72**, **78**

thymine, 6, 67, **8**

tif mutant, 288–289

Tn3, 249

tobacco mosaic virus, 27–28

topoisomerases, 91, **91**

transcription, 102–104; activation at a distance, 162–163, **163**; coupling to translation, 132–133, 138–141; elongation, 142, 145; eukaryotic, 162, 164–164; initiation, 126–127, 156–157, **127**, 164; termination, 128, 132–145 (*see also* antitermination). *See also* RNA polymerase; RNA synthesis

transcriptional activators, 163–164, 278, **163**. *See also* cII protein; cyclicAMP receptor protein; gene regulation, positive; NtrC activator

transcription terminators, 128, 134–135, 142–145; intrinsic, 128, 138, 145; rho-dependent, 128, 135, 145. *See also* antitermination

transducing phage, 111, 321. *See also* λ*gal* phage; φ80*lac* phage

transfer RNA. *See* tRNA

transformation, bacterial, 13, 323, 338–339, 347, **323**

translation elongation factors, 179, 184–185, **180**, **184**, **185**. *See also under names of individual factors*

translation initiation factors, 179–180, 186, **181**, **187**

translation release factors, 180, 186–187, **188**

translation. *See* protein synthesis.

transposable element. *See* transposon

transposition, 221, 247–249; insertion (IS) elements, 249–251, **251**; *in vitro*, 256–258; mechanisms of, 248, 254–255, 257–258, **248**, **255**, **259**; of Mu phage, 251–258; and retroviruses, 247, 312–314, **314**; Shapiro model, 254, 258, **255**; and transposases, 247–248, 255, 258

transposons: and antibiotic resistance, 247, 249, 259–260, 338; definition of, 247, **248**; discovery in maize, 247; implications for organisms and species, 258–260, 313; and retroviruses, 312–314, **314**. *See also* Mu phage

trinucleotide (triplet) binding assay, 31, 173, 176, **31**

tRNA, 102, 108; activation of, 23, 171, **24**; cloverleaf structure of, 138, 170–171, **170**; codon-anticodon recognition, 167–168, 179, **168**; and histidine regulation, 132–133; identified as intermediate in protein synthesis, 23–24, **24**; maturation of, 199–200, **201**; primer for reverse transcriptase, 306; sequence and structure determination, 168–170

tRNAs, suppressor, 109, 177–178, 199, **177**

trp operon (tryptophan operon), regulation of, 135–141, **137**, **139**, **140**

tryptophan, role in *trp* operon regulation, 137, 139–141, **137**, **139**, **140**

tryptophan synthesis, and feedback inhibition, 63–64, **64**

umuC and *umuD* genes, 292, 293–294, **294**

UmuCD pathway, 293–295, **294**

uracil, 67

UV irradiation: and DNA damage, 283; and filamentation, 286–288, 292; and induction of λ prophage, 286–289; and mutagenesis, 288–289, 292, 293–294; multiple effects of, 286; and replication mutants, 76. *See also* DNA repair; SOS response

uvrA, *B*, and *C* genes, 284, 292

uvr mutants, 284–285, 287

UV sensitivity: *recA* mutants, 284–285; *uvr* and *recA* mutants have synergistic effect, 284–285, **285**; *uvr* mutants, 284

viruses, animal, 204–205; integration of, 312; plaque and focus assays for, 297–300; RNA tumor viruses renamed retroviruses, 302; tumor viruses, 299–302. *See also* adenovirus; retroviruses; SV40

Watson-Crick model, 6–8, 11; implications of, 13, 16, 65, 68–70, 90

women in science, difficulties of, 190–191, 192

X-ray crystallography: of Cro protein, 158–159; of Crp protein, 159–160; determination of protein structures, 4–5, 158–160, **159**; and DNA structure, 7

X-rays, DNA damage by, 282–283

Xis protein, 243–245

zygotic induction, 45–46, 51, **47**

Text:	10.5 /14 Bembo
Display:	Frutiger
Design:	Steve Renick
Figures:	Laura Southworth
Portraits:	Marc Nadel
Art editor:	Jon Tupy
Jacket art:	Marc Lenburg
Composition:	Penna Design and Production
Printing and binding:	Through Asia Pacific Offset